T0332719

Readings in Mathematics

Springer Science+Business Media, LLC

Reinhold Remmert

Theory of
Complex Functions

Translated by Robert B. Burckel

With 68 Illustrations

 Springer

Reinhold Remmert
Mathematisches Institut
  der Universität Münster
48149 Münster
Germany

Robert B. Burckel (*Translator*)
Department of Mathematics
Kansas State University
Manhattan, KS 66506
USA

Mathematics Subject Classification (1991): 30-01

Library of Congress Cataloging-in-Publication Data
Remmert, Reinhold.
    [Funktionentheorie. 1. English]
    Theory of complex functions / Reinhold Remmert ; translated by
Robert B. Burckel.
        p.   cm. — (Graduate texts in mathematics ; 122. Readings in
mathematics)
    Translation of: Funktionentheorie I. 2nd ed.
    ISBN 978-0-387-97195-7   ISBN 978-1-4612-0939-3 (eBook)
    DOI 10.1007/978-1-4612-0939-3
    1. Functions of complex variables. I. Title. II. Series:
Graduate texts in mathematics ; 122. III. Series: Graduate texts in
mathematics. Readings in mathematics.
    QA331.R4613   1990
    515'.9—dc20   90-9525

Printed on acid-free paper.

This book is a translation of the second edition of *Funktionentheorie I*, Grundwissen Mathematik 5, Springer-Verlag, 1989.

© 1991 Springer Science+Business Media New York
Originally published by Springer-Verlag New York Inc. in 1991

Camera-ready copy prepared using LaTeX.

9 8 7 6 5 4 (Fourth corrected printing, 1998)

ISBN 978-0-387-97195-7   SPIN 10689678

Preface to the English Edition

> Und so ist jeder Übersetzer anzusehen, dass er sich als Vermittler dieses allgemein-geistigen Handels bemüht und den Wechseltausch zu befördern sich zum Geschäft macht. Denn was man auch von der Unzulänglichkeit des Übersetzers sagen mag, so ist und bleibt es doch eines der wichtigsten und würdigsten Geschäfte in dem allgemeinem Weltverkehr. (And that is how we should see the translator, as one who strives to be a mediator in this universal, intellectual trade and makes it his business to promote exchange. For whatever one may say about the shortcomings of translations, they are and will remain most important and worthy undertakings in world communications.)
> J. W. von GOETHE, vol. VI of *Kunst und Alterthum*, 1828.

This book is a translation of the second edition of *Funktionentheorie I*, Grundwissen Mathematik 5, Springer-Verlag 1989. Professor R. B. BURCKEL did much more than just produce a translation; he discussed the text carefully with me and made several valuable suggestions for improvement. It is my great pleasure to express to him my sincere thanks.

Mrs. Ch. ABIKOFF prepared this TEX-version with great patience; Prof. W. ABIKOFF was helpful with comments for improvements. Last but not least I want to thank the staff of Springer-Verlag, New York. The late W. KAUFMANN-BÜHLER started the project in 1984; U. SCHMICKLER-HIRZEBRUCH brought it to a conclusion.

Lengerich (Westphalia), June 26, 1989

Reinhold Remmert

Preface to the Second German Edition

Not only have typographical and other errors been corrected and improvements carried out, but some new supplemental material has been inserted. Thus, e.g., HURWITZ's theorem is now derived as early at 8.5.5 by means of the minimum principle and Weierstrass's convergence theorem. Newly added are the long-neglected proof (without use of integrals) of Laurent's theorem by SCHEEFFER, via reduction to the Cauchy-Taylor theorem, and DIXON's elegant proof of the homology version of Cauchy's theorem. In response to an oft-expressed wish, each individual section has been enriched with practice exercises.

I have many readers to thank for critical remarks and valuable suggestions. I would like to mention specifically the following colleagues: M. BARNER (Freiburg), R. P. BOAS (Evanston, Illinois), R. B. BURCKEL (Kansas State University), K. DIEDERICH (Wuppertal), D. GAIER (Giessen), ST. HILDEBRANDT (Bonn), and W. PURKERT (Leipzig).

In the preparation of the 2nd edition, I was given outstanding help by Mr. K. SCHLÖTER and special thanks are due him. I thank Mr. W. HOMANN for his assistance in the selection of exercises. The publisher has been magnanimous in accommodating all my wishes for changes.

Lengerich (Westphalia), April 10, 1989

Reinhold Remmert

Preface to the First German Edition

> Wir möchten gern dem Kritikus gefallen: Nur nicht dem Kritikus vor allen. (We would gladly please the critic: Only not the critic above all.) G. E. LESSING.

The authors and editors of the textbook series "Grundwissen Mathematik"[1] have set themselves the goal of presenting mathematical theories in connection with their historical development. For function theory with its abundance of classical theorems such a program is especially attractive. This may, despite the voluminous literature on function theory, justify yet another textbook on it. For it is still true, as was written in 1900 in the prospectus for vol. 112 of the well-known series *Ostwald's Klassiker Der Exakten Wissenschaften*, where the German translation of Cauchy's classic "Mémoire sur les intégrales définies prises entre des limites imaginaires" appears: "Although modern methods are most effective in communicating the content of science, prominent and far-sighted people have repeatedly focused attention on a deficiency which all too often afflicts the scientific education of our younger generation. *It is this, the lack of a historical sense and of any knowledge of the great labors on which the edifice of science rests.*"

The present book contains many historical explanations and original quotations from the classics. These may entice the reader to at least page through some of the original works. "Notes about personalities" are sprinkled in "in order to lend some human and personal dimension to the science" (in the words of F. KLEIN on p. 274 of his *Vorlesungen über die Entwicklung der Mathematik im 19. Jahrhundert* — see [H8]). But the book is not a history of function theory; the historical remarks almost always reflect the contemporary viewpoint.

Mathematics remains the primary concern. What is treated is the material of a 4 hour/week, one-semester course of lectures, centering around

---

[1]The original German version of this book was volume 5 in that series (translator's note).

Cauchy's integral theorem. Besides the usual themes which no text on function theory can omit, the reader will find here

- RITT's theorem on asymptotic power series expansions, which provides a function-theoretic interpretation of the famous theorem of É. BOREL to the effect that any sequence of complex numbers is the sequence of derivatives at 0 of some infinitely differentiable function on the line.

- EISENSTEIN's striking approach to the circular functions via series of partial fractions.

- MORDELL's residue-theoretic calculations of certain Gauss sums.

In addition *cognoscenti* may here or there discover something new or long forgotten.

To many readers the present exposition may seem too detailed, to others perhaps too compressed. J. KEPLER agonized over this very point, writing in his *Astronomia Nova* in the year 1609: "Durissima est hodie conditio scribendi libros Mathematicos. Nisi enim servaveris genuinam subtilitatem propositionum, instructionum, demonstrationum, conclusionum; liber non erit Mathematicus: sin autem servaveris; lectio efficitur morosissima. (It is very difficult to write mathematics books nowadays. If one doesn't take pains with the fine points of theorems, explanations, proofs and corollaries, then it won't be a mathematics book; but if one does these things, then the reading of it will be extremely boring.)" And in another place it says: "Et habet ipsa etiam prolixitas phrasium suam obscuritatem, non minorem quam concisa brevitas (And detailed exposition can obfuscate no less than the overly terse)."

K. PETERS (Boston) encouraged me to write this book. An academic stipend from the Volkswagen Foundation during the Winter semesters 1980/81 and 1982/83 substantially furthered the project; for this support I'd like to offer special thanks. My thanks are also owed the Mathematical Research Institute at Oberwolfach for oft-extended hospitality. It isn't possible to mention here by name all those who gave me valuable advice during the writing of the book. But I would like to name Messrs. M. KOECHER and K. LAMOTKE, who checked the text critically and suggested improvements. From Mr. H. GERICKE I learned quite a bit of history. Still I must ask the reader's forebearance and enlightenment if my historical notes need any revision.

My colleagues, particularly Messrs. P. ULLRICH and M. STEINSIEK, have helped with indefatigable literature searches and have eliminated many deficiencies from the manuscript. Mr. ULLRICH prepared the symbol, name, and subject indexes; Mrs. E. KLEINHANS made a careful critical pass through the final version of the manuscript. I thank the publisher for being so obliging.

Lengerich (Westphalia), June 22, 1983   Reinhold Remmert

Notes for the Reader. Reading really ought to start with Chapter 1. Chapter 0 is just a short compendium of important concepts and theorems known to the reader by and large from calculus; only such things as are important for function theory get mentioned here.

A citation 3.4.2, e.g., means subsection 2 in section 4 of Chapter 3. Within a given chapter the chapter number is dispensed with and within a given section the section number is dispensed with, too. Material set in reduced type will not be used later. The subsections and sections prefaced with * can be skipped on the first reading. Historical material is as a rule organized into a special subsection in the same section were the relevant mathematics was presented.

Contents

Part A. Elements of Function Theory

Part B. The Cauchy Theory

Part C. Cauchy-Weierstrass-Riemann Function Theory

Historical Introduction

Wohl dem, der seiner Väter gern gedenkt (Blessings
on him who gladly remembers his forefathers)
– J. W. v. GOETHE

1. ... "Zuvörderst würde ich jemand, der eine neue Function in die Analyse
einführen will, um eine Erklärung bitten, ob er sie schlechterdings bloss auf
reelle Grössen (reelle Werthe des Arguments der Function) angewandt wissen will, und die imaginären Werthe des Arguments gleichsam nur als ein
Überbein ansieht – oder ob er meinem Grundsatz beitrete, dass man in dem
Reiche der Grössen die imaginären $a + b\sqrt{-1} = a + bi$ als gleiche Rechte
mit den reellen geniessend ansehen müsse. Es ist hier nicht von prakti-
schem Nutzen die Rede, sondern die Analyse ist mir eine selbständige Wissenschaft, die durch Zurücksetzung jener fingirten Grössen ausserordentlich
an Schönheit und Rundung verlieren und alle Augenblick Wahrheiten, die
sonst allgemein gelten, höchst lästige Beschränkungen beizufügen genöthigt
sein würde ... (At the very beginning I would ask anyone who wants to
introduce a new function into analysis to clarify whether he intends to
confine it to real magnitudes (real values of its argument) and regard the
imaginary values as just vestigial – or whether he subscribes to my fundamental proposition that in the realm of magnitudes the imaginary ones
$a + b\sqrt{-1} = a + bi$ have to be regarded as enjoying equal rights with the
real ones. We are not talking about practical utility here; rather analysis is, to my mind, a self-sufficient science. It would lose immeasurably
in beauty and symmetry from the rejection of any fictive magnitudes. At
each stage truths, which otherwise are quite generally valid, would have to
be encumbered with all sorts of qualifications...)."

C.F. GAUSS (1777–1855) wrote these memorable lines on December 18,
1811 to BESSEL; they mark the birth of function theory. This letter of
GAUSS' wasn't published until 1880 (*Werke* **8**, 90–92); it is probable that
GAUSS developed this point of view long before composing this letter. As

many details of his writing attest, GAUSS knew about the Cauchy integral theorem by 1811. However, GAUSS did not participate in the actual construction of function theory; in any case, he was familiar with the principles of the theory. Thus, e.g., he writes elsewhere (*Werke* **10**, 1, p. 405; no year is indicated, but sometime after 1831):

Reproduced with the kind permission of the *Niedersächsische Staats- und Universitätsbibliothek*, Göttingen.

"*Complete knowledge of the nature of an analytic function must also include insight into its behavior for imaginary values of the arguments. Often the latter is indispensable even for a proper appreciation of the behavior of the function for real arguments. It is therefore essential that the original determination of the function concept be broadened to a domain of magnitudes which includes both the real and the imaginary quantities, on an equal footing, under the single designation complex numbers.*"

2. The first stirrings of function theory are to be found in the 18th century with L. EULER (1707–1783). He had "eine für die meisten seiner Zeitgenossen unbegreifliche Vorliebe für die komplexen Größen, mit deren Hilfe es ihm gelungen war, den Zusammenhang zwischen den Kreisfunktionen und der Exponentialfunktion herzustellen. ... In der Theorie der elliptischen Integrale entdeckte er das Additionstheorem, machte er auf die Analogie dieser Integrale mit den Logarithmen und den zyklometrischen Funktionen aufmerksam. So hatte er alle Fäden in der Hand, daraus später

L. EULER 1707–1783

A.L. CAUCHY 1789–1857

B. RIEMANN 1826–1866

K. WEIERSTRASS 1815–1897

Line drawings by Martina Koecher

das wunderbare Gewebe der Funktionentheorie gewirkt wurde (... what for most of his contemporaries was an incomprehensible preference for the complex numbers, with the help of which he had succeeded in establishing a connection between the circular functions and the exponential function. ... In the theory of elliptic integrals he discovered the addition theorem and drew attention to the analogy between these integrals, logarithms and the cyclometric functions. Thus he had in hand all the threads out of which the wonderful fabric of function theory would later be woven)," G. FROBENIUS: *Rede auf L. Euler* on the occasion of Euler's 200th birthday in 1907; *Ges. Abhandl.* **3**, p.733).

Modern function theory was developed in the 19th century. The pioneers in the formative years were

A.L. CAUCHY (1789–1857), B. RIEMANN (1826–1866),
K. WEIERSTRASS (1815–1897).

Each gave the theory a very distinct flavor and we still speak of the CAUCHY, the RIEMANN, and the WEIERSTRASS points of view.

CAUCHY wrote his first works on function theory in the years 1814–1825. The function notion in use was that of his predecessors from the EULER era and was still quite inexact. To CAUCHY a holomorphic function was essentially a complex-differentiable function having a continuous derivative. CAUCHY's function theory is based on his famous integral theorem and on the residue concept. *Every* holomorphic function has a natural integral representation and is thereby accessible to the methods of analysis. The CAUCHY theory was completed by J. LIOUVILLE (1809–1882), [Liou]. The book [BB] of CH. BRIOT and J.-C. BOUQUET (1859) conveys a very good impression of the state of the theory at that time.

Riemann's epochal Göttingen inaugural dissertation *Grundlagen für eine allgemeine Theorie der Functionen einer veränderlichen complexen Größe* [R] appeared in 1851. To RIEMANN the geometric view was central: holomorphic functions are mappings between domains in the number plane ℂ, or more generally between Riemann surfaces, "entsprechenden kleinsten Theilen ähnlich sind (correspondingly small parts of each of which are similar)." RIEMANN drew his ideas from, among other sources, intuition and experience in mathematical physics: the existence of current flows was proof enough for him that holomorphic (= conformal) mappings exist. He sought – with a minimum of calculation – to understand his functions, not by formulas but by means of the "intrinsic characteristic" properties, from which the extrinsic representation formulas necessarily arise.

For WEIERSTRASS the point of departure was the power series; holomorphic functions are those which locally can be developed into convergent power series. Function theory is the theory of these series and is simply based in algebra. The beginnings of such a viewpoint go back to J.L. LAGRANGE. In his 1797 book *Théorie des fonctions analytiques* (2nd ed., Courcier, Paris 1813) he wanted to prove the proposition that every continuous function is developable into a power series. Since LAGRANGE we speak of *analytic* functions; at the same time it was supposed that these were precisely the functions which are useful in analysis. F. KLEIN writes "Die große Leistung von Weierstraß ist es, die im Formalen stecken gebliebene Idee von Lagrange ausgebaut und vergeistigt zu haben (The great achievement of Weierstrass is to have animated and realized the program implicit in Lagrange's formulas)" (cf. p.254 of the German original of [H8]). And CARATHÉODORY says in 1950 ([5], p.vii): WEIERSTRASS was able to "die Funktionentheorie arithmetisieren und ein System entwickeln, das an Strenge und Schönheit nicht übertroffen werden kann (arithmetize function theory and develop a system of unsurpassable beauty and rigor)."

3. The three methodologically quite different yet equivalent avenues to function theory give the subject special charm. Occasionally the impression arises that CAUCHY, RIEMANN and WEIERSTRASS were almost "ideological" proponents of their respective systems. But that was not the case. As early as 1831 CAUCHY was developing his holomorphic functions into power series and working with the latter. Any kind of rigid one-sidedness was alien to RIEMANN: he made use of whatever he found at hand; thus he too used power series in his function theory. And on the other hand WEIERSTRASS certainly didn't reject integrals on principle: as early as 1841 – two years before LAURENT – he developed holomorphic functions on annular regions into Laurent series via integral formulas [W1].

In 1898 in his article "L'oeuvre mathématique de Weierstrass", *Acta Math.* **22**, 1–18 (see pp. 6,7) H. POINCARÉ offered this evaluation: "La théorie de Cauchy contenait en germe à la fois la conception géometrique de Riemann et la conception arithmétique de Weierstrass, et il est aisé de comprendre comment elle pouvait, en se développant dans deux sens différents, donner naissance à l'une et à l'autre. ... La méthode de Riemann est avant tout une méthode de découverte, celle de Weierstrass est avant tout une méthode de démonstration. (Cauchy's theory contains at once a germ of Riemann's geometric conception and a germ of Weierstrass' arithmetic one, and it is easy to understand how its development in two different directions could give rise to the one or the other. ... The method of Riemann is above all a method of discovery, that of Weierstrass is above all a method of proof.)"

For a long time now the conceptual worlds of CAUCHY, RIEMANN and WEIERSTRASS have been inextricably interwoven; this has resulted not only in many simplifications in the exposition of the subject but has also made possible the discovery of significant new results.

During the last century function theory enjoyed very great triumphs in quite a short span of time. In just a few decades a scholarly edifice was erected which immediately won the highest esteem of the mathematical world. We might join R. DEDEKIND who wrote (cf. *Math. Werke* **1**,

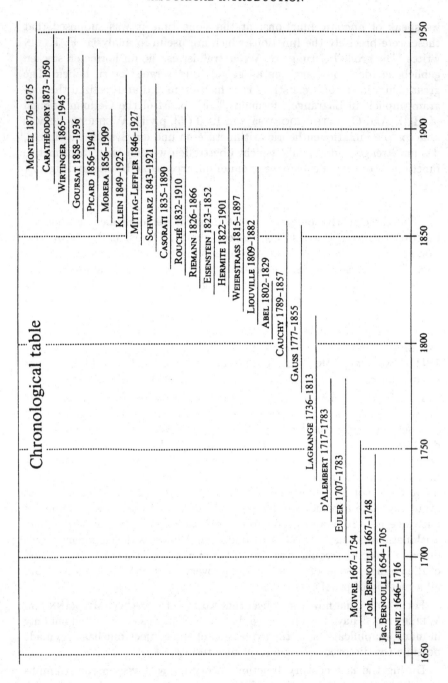

Chronological table

MONTEL 1876–1975
CARATHÉODORY 1873–1950
WIRTINGER 1865–1945
GOURSAT 1858–1936
PICARD 1856–1941
MORERA 1856–1909
KLEIN 1849–1925
MITTAG-LEFFLER 1846–1927
SCHWARZ 1843–1921
CASORATI 1835–1890
ROUCHÉ 1832–1910
RIEMANN 1826–1866
EISENSTEIN 1823–1852
HERMITE 1822–1901
WEIERSTRASS 1815–1897
LIOUVILLE 1809–1882
ABEL 1802–1829
CAUCHY 1789–1857
GAUSS 1777–1855
LAGRANGE 1736–1813
D'ALEMBERT 1717–1783
EULER 1707–1783
MOIVRE 1667–1754
JOH. BERNOULLI 1667–1748
JAC. BERNOULLI 1654–1705
LEIBNIZ 1646–1716

1950
1900
1850
1800
1750
1700
1650

pp. 105, 106): "Die erhabenen Schöpfungen dieser Theorie haben die Bewunderung der Mathematiker vor allem deshalb erregt, weil sie in fast beispielloser Weise die Wissenschaft mit einer außerordentlichen Fülle ganz neuer Gedanken befruchtet und vorher gänzlich unbekannte Felder zum ersten Male der Forschung erschlossen haben. Mit der Cauchyschen Integralformel, dem Riemannschen Abbildungssatz und dem Weierstraßschen Potenzreihenkalkül wird nicht bloß der Grund zu einem neuen Teile der Mathematik gelegt, sondern es wird zugleich auch das erste und bis jetzt noch immer fruchtbarste Beispiel des innigen Zusammenhangs zwischen Analysis und Algebra geliefert. Aber es ist nicht bloß der wunderbare Reichtum an neuen Ideen und großen Entdeckungen, welche die neue Theorie liefert; vollständig ebenbürtig stehen dem die Kühnheit und Tiefe der Methoden gegenüber, durch welche die größten Schwierigkeiten überwunden und die verborgensten Wahrheiten, die mysteria functiorum, in das hellste Licht gesetzt werden (The splendid creations of this theory have excited the admiration of mathematicians mainly because they have enriched our science in an almost unparalleled way with an abundance of new ideas and opened up heretofore wholly unknown fields to research. The Cauchy integral formula, the Riemann mapping theorem and the Weierstrass power series calculus not only laid the groundwork for a new branch of mathematics but at the same time they furnished the first and till now the most fruitful example of the intimate connections between analysis and algebra. But it isn't just the wealth of novel ideas and discoveries which the new theory furnishes; of equal importance on the other hand are the boldness and profundity of the methods by which the greatest of difficulties are overcome and the most recondite of truths, the *mysteria functiorum*, are exposed to the brightest light)."

Even from today's perspective nothing needs to be added to these exuberant statements. Function theory with its sheer inexhaustible abundance of beautiful and deep theorems is, as C.L. SIEGEL occasionally expressed it in his lectures, a one-of-a-kind gift to the mathematician.

Chapter 0

Complex Numbers and Continuous Functions

Nicht einer mystischen Verwendung von $\sqrt{-1}$ hat die Analysis ihre wirklich bedeutenden Erfolge des letzten Jahrhunderts zu verdanken, sondern dem ganz natürlichen Umstande, dass man unendlich viel freier in der mathematischen Bewegung ist, wenn man die Grössen in einer Ebene statt nur in einer Linie variiren läßt (Analysis does not owe its really significant successes of the last century to any mysterious use of $\sqrt{-1}$, but to the quite natural circumstance that one has infinitely more freedom of mathematical movement if he lets quantities vary in a plane instead of only on a line) – (Leopold KRONECKER, in [Kr].)

An exposition of function theory must necessarily begin with a description of the complex numbers. First we recall their most important properties; a detailed exposition can be found in the book *Numbers* [19], where the historical development is also extensively treated.

Function theory is the theory of complex-differentiable functions. Such functions are, in particular, continuous. Therefore we also discuss the general concept of continuity. Furthermore, we introduce concepts from topology which will see repeated use. "Die Grundbegriffe und die einfachsten Tatsachen aus der mengentheoretischen Topologie braucht man in sehr verschiedenen Gebieten der Mathematik; die Begriffe des topologischen und des metrischen Raumes, der Kompaktheit, die Eigenschaften stetiger Abbildungen u. dgl. sind oft unentbehrlich... (The basic ideas and simplest facts of set-theoretic topology are needed in the most diverse areas of mathematics; the concepts of topological and metric spaces, of compactness, the

9

properties of continuous functions and the like are often indispensable...).”
P. ALEXANDROFF and H. HOPF wrote this sentence in 1935 in their treatise
Topologie I (Julius Springer, Berlin, p.23). It is valid for many mathemat-
ical disciplines, but especially so for function theory.

§1   The field $\mathbb{C}$ of complex numbers

The field of real numbers will always be denoted by $\mathbb{R}$ and its theory is
supposed to be known by the reader.

1. The field $\mathbb{C}$. In the 2-dimensional $\mathbb{R}$-vector space $\mathbb{R}^2$ of ordered pairs
$z := (x, y)$ of real numbers a multiplication, denoted as usual by juxtapo-
sition, is introduced by the decree

$$(x_1, y_1)(x_2, y_2) := (x_1 x_2 - y_1 y_2, x_1 y_2 + x_2 y_1).$$

$\mathbb{R}^2$ thereby becomes a (commutative) *field* with $(1, 0)$ as unit element, the
additive structure being coordinate-wise, and the multiplicative inverse of
$z = (x, y) \neq 0$ being the pair $\left(\frac{x}{x^2+y^2}, \frac{-y}{x^2+y^2}\right)$, denoted as usual by z^{-1}.
This field is called *the field $\mathbb{C}$ of complex numbers.*

The mapping $x \mapsto (x, 0)$ of $\mathbb{R} \to \mathbb{C}$ is a *field embedding* (because, e.g.,
$(x_1, 0)(x_2, 0) = (x_1 x_2, 0)$). We identify the real number x with the complex
number $(x, 0)$. Via this identification $\mathbb{C}$ becomes a *field extension* of $\mathbb{R}$ with
the unit element $1 := (1, 0) \in \mathbb{C}$. We further define

$$i := (0, 1) \in \mathbb{C};$$

this notation was introduced in 1777 by EULER: “... formulam $\sqrt{-1}$ littera
i in posterum designabo” (*Opera Omnia* (1) **19**, p.130). Evidently we have
$i^2 = -1$. The number i is often called the *imaginary unit* of $\mathbb{C}$. Every
number $z = (x, y) \in \mathbb{C}$ admits a *unique* representation

$$(x, y) = (x, 0) + (0, 1)(y, 0), \text{ that is, } z = x + iy \text{ with } x, y \in \mathbb{R};$$

this is the usual way to write complex numbers. One sets

$$\Re z := x, \ \Im z := y$$

and calls x and y the *real part* and the *imaginary part*, respectively, of
z. The number z is called *real*, respectively, *pure(ly) imaginary* if $\Im z = 0$,
respectively, $\Re z = 0$; the latter meaning that $z = iy$.

Ever since GAUSS people have visualized complex numbers geometrically
as points in the *Gauss(ian) plane* with rectangular coordinates, the addition
being then vector addition (cf. the figure on the left).

The multiplication of complex numbers, namely

$$(x_1 + iy_1)(x_2 + iy_2) = (x_1 x_2 - y_1 y_2) + i(x_1 y_2 + y_1 x_2),$$

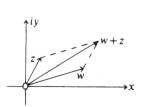
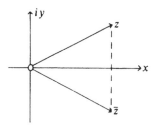

is just what one would expect from the distributive law and the fact that $i^2 = -1$. As to the geometric significance of this multiplication in terms of polar coordinates, cf. 5.3.1 below and 3.6.2 of the book *Numbers*.

$\mathbb{C}$ is identified with $\mathbb{R}^2$ since $z = x + iy$ is the row vector (x, y); but it is sometimes more convenient to make the identification of z to the column vector $\binom{x}{y}$. The plane $\mathbb{C} \setminus \{0\}$ punctured at 0 is denoted by $\mathbb{C}^\times$. With respect to the multiplication in $\mathbb{C}$, $\mathbb{C}^\times$ is a group (*the multiplicative group of the field $\mathbb{C}$*).

For each number $z = x + iy \in \mathbb{C}$ the number $\overline{z} := x - iy \in \mathbb{C}$ is called the (*complex*) *conjugate* of z. The mapping $z \mapsto \overline{z}$ is called the *reflection* in the real axis (see the right-hand figure above). The following elemental rules of calculation prevail:

$$\overline{z + w} = \overline{z} + \overline{w}, \quad \overline{zw} = \overline{z}\,\overline{w}, \quad \overline{\overline{z}} = z, \quad \Re z = \frac{1}{2}(z + \overline{z}),$$

$$\Im z = \frac{1}{2i}(z - \overline{z}), \quad z \in \mathbb{R} \Leftrightarrow z = \overline{z}, \quad z \in i\mathbb{R} \Leftrightarrow z = -\overline{z}.$$

The conjugation operation is a field automorphism of $\mathbb{C}$ which leaves $\mathbb{R}$ element-wise fixed.

2. $\mathbb{R}$-linear and $\mathbb{C}$-linear mappings of $\mathbb{C}$ into $\mathbb{C}$. Because $\mathbb{C}$ is an $\mathbb{R}$-vector space as well as a $\mathbb{C}$-vector space, we have to distinguish between $\mathbb{R}$-linear and $\mathbb{C}$-linear mappings of $\mathbb{C}$ into $\mathbb{C}$. Every $\mathbb{C}$-linear mapping has the form $z \mapsto \lambda z$ with $\lambda \in \mathbb{C}$ and is $\mathbb{R}$-linear. Conjugation $z \mapsto \overline{z}$ is $\mathbb{R}$-linear but not $\mathbb{C}$-linear. Generally:

A mapping $T : \mathbb{C} \to \mathbb{C}$ is $\mathbb{R}$-linear if and only if it satisfies

$$T(z) = T(1)x + T(i)y = \lambda z + \mu \overline{z} \text{ , for all } z = x + iy \in \mathbb{C}$$

with

$$\lambda := \frac{1}{2}(T(1) - iT(i)), \ \mu := \frac{1}{2}(T(1) + iT(i)).$$

An $\mathbb{R}$-linear mapping $T : \mathbb{C} \to \mathbb{C}$ is then $\mathbb{C}$-linear when $T(i) = iT(1)$; in this case it has the form $T(z) = T(1)z$.

Proof. $\mathbb{R}$-linearity means that for $z = x + iy$, $x, y \in \mathbb{R}$, $T(z) = xT(1) + yT(i)$. Upon writing $\frac{1}{2}(z + \bar{z})$ for x and $\frac{1}{2i}(z - \bar{z})$ for y, the first assertion follows; the second assertion is immediate from the first.

If $\mathbb{C}$ is identified with $\mathbb{R}^2$ via $z = x+iy = \begin{pmatrix} x \\ y \end{pmatrix}$, then every *real 2×2 matrix*

$A = \begin{pmatrix} a & b \\ c & d \end{pmatrix}$ induces an $\mathbb{R}$-linear *right-multiplication* mapping $T : \mathbb{C} \to \mathbb{C}$
defined by

$$\begin{pmatrix} x \\ y \end{pmatrix} \longmapsto \begin{pmatrix} a & b \\ c & d \end{pmatrix} \begin{pmatrix} x \\ y \end{pmatrix} = \begin{pmatrix} ax + by \\ cx + dy \end{pmatrix}.$$

It satisfies

(∗) $T(1) = a + ic, \qquad T(i) = b + id.$

Theorems of linear algebra ensure that every $\mathbb{R}$-linear map is realized this way: The mapping T and the matrix A determine each other via (∗). We claim

Theorem. *The following statements about a real matrix*

$$A = \begin{pmatrix} a & b \\ c & d \end{pmatrix}$$

are equivalent:

  i) *The mapping $T : \mathbb{C} \to \mathbb{C}$ induced by A is $\mathbb{C}$-linear.*

 ii) *The entries $c = -b$ and $d = a$, that is, $A = \begin{pmatrix} a & -c \\ c & a \end{pmatrix}$ and $T(z) =$*
     $(a + ic)z$.

Proof. The decisive equation $b + id = T(i) = iT(1) = i(a + ic)$ obtains exactly when $c = -b$ and $d = a$.   □

It is apparent from the preceding discussion that an $\mathbb{R}$-linear mapping $T : \mathbb{C} \to \mathbb{C}$ can be described in three ways: by means of a real 2×2 matrix, in the form $T(z) = T(1)x + T(i)y$, or in the form $T(z) = \lambda z + \mu \bar{z}$. These three possibilities will find expression later in the theory of differentiable functions $f = u+iv$, where, besides the real partial derivatives u_x, u_y, v_x, v_y (which correspond to the matrix elements a, b, c, d), the complex partial derivatives f_x, f_y (which correspond to the numbers $T(1), T(i)$) and $f_z, f_{\bar{z}}$ (which correspond to λ, μ) will be considered. The conditions $a = d, b = -c$ of the theorem are then a manifestation of the Cauchy-Riemann differential equations $u_x = v_y, u_y = -v_x$; cf. Theorem 1.2.1.

3. Scalar product and absolute value. For $w = u + iv, z = x + iy \in \mathbb{C}$ the equations

$$\langle w, z \rangle := \Re(w\bar{z}) = ux + vy = \Re(\bar{w}z) = \langle z, w \rangle$$

codify the *euclidean scalar product* in the real vector space $\mathbb{C} = \mathbb{R}^2$ with respect to the basis $1, i$. The non-negative real number

$$|z| := = \sqrt{\langle z, z \rangle} = +\sqrt{z\bar{z}} = +\sqrt{x^2 + y^2}$$

measures the *euclidean length* of z and is called the *absolute value*, and sometimes the *modulus* of z. It and the scalar product satisfy

$$|\bar{z}| = |z|, \ |\Re z| \le |z|, \ |\Im z| \le |z| \ \text{and} \ z^{-1} = \frac{1}{|z|^2}\bar{z} \quad \text{if } z \ne 0,$$

$$\langle aw, az \rangle = |a|^2 \langle w, z \rangle \quad , \quad \langle \bar{w}, \bar{z} \rangle = \langle w, z \rangle \qquad \text{for all } w, z \in \mathbb{C}.$$

Routine calculations immediately reveal the identity

$$\langle w, z \rangle^2 + \langle iw, z \rangle^2 = |w|^2 |z|^2, \qquad \text{for all } w, z \in \mathbb{C},$$

which contains as a special case the

Cauchy-Schwarz Inequality:

$$|\langle w, z \rangle| \le |w||z|, \qquad \text{for all } w, z \in \mathbb{C}.$$

Likewise direct calculation yields the

Law of Cosines:

$$|w + z|^2 = |w|^2 + |z|^2 + 2\langle w, z \rangle \qquad \text{for all } w, z \in \mathbb{C}.$$

Two vectors w, z are called *orthogonal* or *perpendicular* if $\langle w, z \rangle = 0$. Because $\langle z, cz \rangle = \Re(\bar{z}cz) = |z|^2\Re c$, z and $cz \in \mathbb{C}^\times$ are orthogonal exactly when c is purely imaginary. The following *rules* are fundamental for calculating with the absolute value:

1) $|z| \ge 0$ and $|z| = 0 \Leftrightarrow z = 0$

2) $|wz| = |w| \cdot |z|$ (*product rule*)

3) $|w + z| \le |w| + |z|$ (*triangle inequality*).

Here 1) and 2) are direct and 3) is gotten by means of the Law of Cosines and the Cauchy-Schwarz inequality (cf. also 3.4.2 in *Numbers* [19]) as follows:

$$|w+z|^2 = |w|^2 + |z|^2 + 2\langle w, z \rangle \le |w|^2 + |z|^2 + 2|w||z| = (|w| + |z|)^2. \qquad \square$$

The product rule implies the *division rule*:

$$|w/z| = |w|/|z| \text{ for all } w, z \in \mathbb{C}, z \neq 0.$$

The following variations of the triangle inequality are often useful:

$$|w| \geq |z| - |w - z| \,, \quad |w + z| \geq ||w| - |z|| \,, \quad ||w| - |z|| \leq |w - z|.$$

Rules 1)–3) are called *evaluation rules*. A map $| \cdot | : K \rightarrow \mathbb{R}$ of a (commutative) field K into $\mathbb{R}$ which satisfies these rules is called a *valuation* on K; a field together with a valuation in called a *valued field*. Thus $\mathbb{R}$ and $\mathbb{C}$ are valued fields.

From the Cauchy-Schwarz inequality it follows that

$$-1 \leq \frac{\langle w, z \rangle}{|w||z|} \leq 1 \text{ for all } w, z \in \mathbb{C}^{\times}.$$

According to (non-trivial) results of calculus, for each $w, z \in \mathbb{C}^{\times}$ therefore a unique real number φ, with $0 \leq \varphi \leq \pi$, exists satisfying

$$\cos \varphi = \frac{\langle w, z \rangle}{|w||z|};$$

φ is called *the angle* between w and z, symbolically $\angle(w, z) = \varphi$.

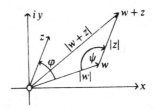

Because $\langle w, z \rangle = |w||z| \cos \varphi$ and $\cos \varphi = -\cos \psi$ (due to $\psi + \varphi = \pi$ — see the accompanying figure), the Law of Cosines can be written in the form

$$|w + z|^2 = |w|^2 + |z|^2 - 2|w||z| \cos \psi,$$

familiar from elementary geometry.

With the help of the absolute value of complex numbers and the fact that every non-negative real number r has a non-negative square-root $\sqrt{r}$, square-roots of *any* complex number can be exhibited. Direct verification confirms that

for $a, b \in \mathbb{R}$ and $c := a + ib$ the number

$$\xi := \sqrt{\frac{1}{2}(|c| + a)} + i\eta \sqrt{\frac{1}{2}(|c| - a)} \,,$$

with $\eta := \pm 1$ so chosen that $b = \eta|b|$, satisfies $\xi^2 = c$.

Zeros of arbitrary quadratic polynomials $z^2 + cz + d \in \mathbb{C}[z]$ are now determined by transforming into a "pure" polynomial $(z + \frac{1}{2}c)^2 + d - \frac{1}{4}c^2$ (that is, by completing the square). Not until 9.1.1 will we show that *every* non-constant complex polynomial has zeros in $\mathbb{C}$ (the Fundamental Theorem of Algebra); for more on the problem of solvability of complex equations, compare also Chapter 3.3.5 and Chapter 4 of *Numbers* [19].

4. Angle-preserving mappings. In the function theory of RIEMANN, angle-preserving mappings play an important role. In preparation for the considerations of Chapter 2.1, we look at $\mathbb{R}$-linear injective (consequently also bijective) mappings $T : \mathbb{C} \to \mathbb{C}$. We write simply Tz instead of $T(z)$. We call T *angle-preserving* if

$$|w||z|\langle Tw, Tz\rangle = |Tw||Tz|\langle w, z\rangle \quad \text{for all } w, z \in \mathbb{C}.$$

The terminology is justified by rephrasing this equality in the previously introduced language of the angle between two vectors. So translated, it says that $\angle(Tw, Tz) = \angle(w, z)$ for all $w, z \in \mathbb{C}^\times$. Angle-preserving mappings admit a simple characterization.

Lemma. *The following statements about an $\mathbb{R}$-linear map $T : \mathbb{C} \to \mathbb{C}$ are equivalent:*

i) *T is angle-preserving.*

ii) *There exists an $a \in \mathbb{C}^\times$ such that either $Tz = za$ for all $z \in \mathbb{C}$ or $Tz = a\overline{z}$ for all $z \in \mathbb{C}$.*

iii) *There exists a number $s > 0$ such that $\langle Tw, Tz\rangle = s\langle w, z\rangle$ for all w, $z \in \mathbb{C}$.*

Proof. i) $\Rightarrow$ ii)   Because T is injective, $a := T1 \in \mathbb{C}^\times$. For $b := a^{-1}Ti \in \mathbb{C}$ it then follows that

$$0 = \langle i, 1\rangle = \langle Ti, T1\rangle = \langle ab, a\rangle = |a|^2 \Re b,$$

that is, b is *purely imaginary*: $b = ir$, $r \in \mathbb{R}$. We see that $Tz = T1 \cdot x + Ti \cdot y = a(x + iry)$ and so $\langle T1, Tz\rangle = \langle a, a(x + iry)\rangle = |a|^2 x$. Therefore, on account of the angle-preserving character of T (take $w := 1$ in the defining equation), it follows that for all $z \in \mathbb{C}$

$$|x + iy||a|^2 x = |1||z|\langle T1, Tz\rangle = |T1||Tz|\langle 1, z\rangle = |a||a(x + iry)|x,$$

that is, $|x + iry| = |x + iy|$ for all z with $x \neq 0$. This implies that $r = \pm 1$ and we get $Tz = a(x \pm iy)$, that is, $Tz = az$ for all z or $Tz = a\overline{z}$ for all z.

ii) $\Rightarrow$ iii)   Because $\langle aw, az \rangle = |a|^2 \langle w, z \rangle$ and $\langle \overline{w}, \overline{z} \rangle = \langle w, z \rangle$, in either case $\langle Tw, Tz \rangle = s \langle w, z \rangle$ holds with $s := |a|^2 > 0$.

iii) $\Rightarrow$ i)   Because $|Tz| = \sqrt{s}|z|$ for all z, T is injective; furthermore this equality and that in iii) give

$$|w||z|\langle Tw, Tz \rangle = |w||z|s\langle w, z \rangle = |Tw||Tz|\langle w, z \rangle. \qquad \square$$

The lemma just proved will be applied in 2.1.1 to the $\mathbb{R}$-linear differential of a real-differentiable mapping.

In the theory of the euclidean vector spaces, a linear self-mapping $T : V \to V$ of a vector space V with euclidean scalar product $\langle \ , \ \rangle$ is called a *similarity* if there is a real number $r > 0$ such that $|Tv| = r|v|$ holds for all $v \in V$; the number r is called the *similarity constant* or the *dilation factor* of T. (In case $r = 1$, T is called *length-preserving = isometric*, or an *orthogonal transformation*.) Because of the Law of Cosines, a similarity then also satisfies

$$\langle Tv, Tv' \rangle = r^2 \langle v, v' \rangle \quad \text{for all } v, v' \in V.$$

Every similarity is *angle-preserving*, that is, $\angle(Tv, Tv') = \angle(v, v')$, if one again defines $\angle(v, v')$ as the value in $[0, \pi]$ of the arccosine of $|v|^{-1}|v'|^{-1}\langle v, v' \rangle$ (and the latter one can do because the Cauchy-Schwarz inequality is valid in every euclidean space).

Above we showed that conversely in the special case $V = \mathbb{C}$ every angle-preserving (linear) mapping is a similarity. Actually this converse prevails in every finite-dimensional euclidean space, a fact usually proved in linear algebra courses.

Exercises

Exercise 1. Let $T(z) := \lambda z + \mu \overline{z}$, $\lambda, \mu \in \mathbb{C}$. Show that

a) T is bijective exactly when $\lambda \overline{\lambda} \neq \mu \overline{\mu}$. *Hint*: You don't necessarily have to show that T has determinant $\lambda \overline{\lambda} - \mu \overline{\mu}$.

b) T is isometric, i.e., $|T(z)| = |z|$ for all $z \in \mathbb{C}$, precisely when $\lambda \mu = 0$ and $|\lambda + \mu| = 1$.

Exercise 2. Let $a_1, \ldots, a_n$, $b_1, \ldots, b_n \in \mathbb{C}$ and satisfy $\sum_{\nu=1}^{n} a_\nu^j = \sum_{\nu=1}^{n} b_\nu^j$ for all $j \in \mathbb{N}$. Show that there is a permutation π of $\{1, 2, \ldots, n\}$ such that $a_\nu = b_{\pi(\nu)}$ for all $\nu \in \{1, 2, \ldots, n\}$.

Exercise 3. For $n > 1$ consider real numbers $c_0 > c_1 > \cdots > c_n > 0$. Prove that the polynomial $p(z) := c_0 + c_1 z + \cdots + c_n z^n$ in $\mathbb{C}$ has no zero whose modulus does not exceed 1. *Hint*: Consider $(1 - z)p(z)$ and note (i.e., prove) that for $w, z \in \mathbb{C}$ with $w \neq 0$ the equality $|w - z| = ||w| - |z||$ holds exactly when $z = \lambda w$ for some $\lambda \geq 0$.

Exercise 4. a) Show that from $(1 + |v|^2)u = (1 + |u|^2)v$, $u, v \in \mathbb{C}$, it follows that either $u = v$ or $\overline{u}v = 1$.

b) Show that for $u, v \in \mathbb{C}$ with $|u| < 1$, $|v| < 1$ and $\bar{u}v \neq u\bar{v}$, we always have

$$|(1 + |u|^2)v - (1 + |v|^2)u| > |u\bar{v} - \bar{u}v|.$$

c) Show that for $a, b, c, d \in \mathbb{C}$ with $|a| = |b| = |c|$ the complex number

$$(a - b)(c - d)(\bar{a} - \bar{d})(\bar{c} - \bar{b}) + i(c\bar{c} - d\bar{d})\Im(c\bar{b} - c\bar{a} - a\bar{b})$$

is real.

§2 Fundamental topological concepts

Here we collect the topological language and properties which are indispensable for function theory (e.g., "open", "closed", "compact"). *Too much topology at the beginning is harmful, but our program would fail without any topology at all.* There is a quotation from R. DEDEKIND's book *Was sind und was sollen die Zahlen* (Vieweg, Braunschweig, 1887; English trans. by W. W. BEMAN, *Essays in the Theory of Numbers*, Dover, New York, 1963) which is equally applicable to set-theoretic topology, even though the latter had not yet appeared on the scene in Dedekind's time: "Die größten und fruchtbarsten Fortschritte in der Mathematik und anderen Wissenschaften sind vorzugsweise durch die Schöpfung und Einführung neuer Begriffe gemacht, nachdem die häufige Wiederkehr zusammengesetzter Erscheinungen, welche von den alten Begriffen nur mühselig beherrscht werden, dazu gedrängt hat (The greatest and most fruitful progress in mathematics and other sciences is made through the creation and introduction of new concepts; those to which we are impelled by the frequent recurrence of compound phenomena which are only understood with great difficulty in the older view)." Since only metric spaces ever occur in function theory, we limit ourselves to them.

1. Metric spaces. The expression

$$|w - z| = \sqrt{(u - x)^2 + (v - y)^2}$$

measures the *euclidean* distance between the points $w = u + iv$ and $z = x + iy$ in the plane $\mathbb{C}$ (figure below).
  The function

$$\mathbb{C} \times \mathbb{C} \to \mathbb{R}, \quad (w, z) \mapsto |w - z|$$

has, by virtue of the evaluation rules of 1.3, the properties

$$|w - z| \geq 0, \qquad |w - z| = 0 \Leftrightarrow w = z, \qquad |w - z| = |z - w| \text{ (symmetry)}$$

$$|w - z| \leq |w - w'| + |w' - z| \quad \text{(triangle inequality)} .$$

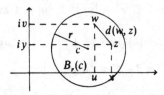

If X is any set, a function

$$d : X \times X \to \mathbb{R}, \quad (x, y) \mapsto d(x, y)$$

is called a *metric on* X if it has the three preceding properties; that is, if for all $x, y, z \in X$ it satisfies

$$d(x, y) \geq 0, \qquad d(x, y) = 0 \Leftrightarrow x = y,$$

$$d(x, y) = d(y, x), \qquad d(x, z) \leq d(x, y) + d(y, z).$$

X together with a metric is called a *metric space*. In $X = \mathbb{C}, d(w, z) :=$ $|w - z|$ is called the *euclidean metric* of $\mathbb{C}$.

In a metric space X with metric d the set

$$B_r(c) := \{x \in X : d(x, c) < r\}$$

is called the *open ball of radius* $r > 0$ *with center* $c \in X$; in the case of the euclidean metric in $\mathbb{C}$ the balls

$$B_r(c) = \{z \in \mathbb{C} : |z - c| < r\}, \qquad r > 0$$

are called *open discs about* c, traditionally but less precisely, *circles about* c.

The *unit disc* $B_1(0)$ plays a distinguished role in function theory. Recalling that the German word for "unit disc" is *Einheitskreisscheibe*, we will use the notation

$$\mathbb{E} := B_1(0) = \{z \in \mathbb{C} : |z| < 1\}.$$

Besides the euclidean metric the set $\mathbb{C} = \mathbb{R}^2$ carries a second natural metric. By means of the usual metric $|x - x'|$, $x, x' \in \mathbb{R}$ on $\mathbb{R}$ we define the *maximum metric* on $\mathbb{C}$ as

$$\hat{d}(w, z) := \max\{|\Re w - \Re z|, |\Im w - \Im z|\}, \qquad w, z \in \mathbb{C}.$$

It takes only a minute to show that this really is a metric in $\mathbb{C}$. The "open balls" in this metric are the open squares [*Quadrate* in German] $Q_r(c)$ of center c and side-length $2r$.

In function theory we work primarily with the euclidean metric, whereas in the study of functions of two real variables it is often more advantageous to use the maximum metric. Analogs of both of these metrics can be introduced into any n-dimensional real vector space $\mathbb{R}^n, 1 \leq n < \infty$.

2. Open and closed sets. A subset U of a metric space X is called *open* (*in X*), if for every $x \in U$ there is an $r > 0$ such that $B_r(x) \subset U$. The empty set and X itself are open. The *union of arbitrarily many* and the *intersection of finitely many* open sets are each open (proof!). The "open balls" $B_r(c)$ of X are in fact open sets.

Different metrics can determine the same system of open sets; this happens, for example, with the euclidean metric and the maximum metric in $\mathbb{C} = \mathbb{R}^2$ (more generally in $\mathbb{R}^n$). The reason is that every open disc contains an open square of the same center and vice-versa. □

A set $C \subset X$ is called *closed* (*in X*) if its complement $X \backslash C$ is open. The sets

$$\overline{B}_r(c) := \{x \in X : d(x, c) \le r\}$$

are closed and consequently we call them *closed balls* and in the case $X = \mathbb{C}$, *closed discs*.

Dualizing the statements for open sets, we have that the *union of finitely many* and the *intersection of arbitrarily many* closed sets are each closed. In particular, for every set $A \subset X$ the intersection $\overline{A}$ of all the closed subsets of X which contain A is itself closed and is therefore the smallest closed subset of X which contains A; it is called the *closed hull* of A or the *closure* of A in X. Notice that $\overline{\overline{A}} = \overline{A}$.

A set $W \subset X$ is called a *neighborhood of the set $M \subset X$*, if there is an open set V with $M \subset V \subset W$. The reader should note that *according to this definition a neighborhood is not necessarily open*. But an open set is a neighborhood of each of its points and this property characterizes "openness".

Two different points $c, c', \in X$ always have a pair of disjoint neighborhoods:

$$B_\epsilon(c) \cap B_\epsilon(c') = \emptyset \quad \text{for } \epsilon := \tfrac{1}{2} d(c, c') > 0.$$

This is the Hausdorff "separation property" (named for the German mathematician and writer Felix HAUSDORFF; born in 1868 in Breslau; from 1902 professor in Leipzig, Bonn, Greifswald, and then Bonn; his 1914 treatise *Grundzüge der Mengenlehre* (Veit & Comp., Leipzig) contains the foundations of set-theoretic topology; died by his own hand in Bonn in 1942 as a result of racial persecution; as a writer he published in his youth under the pseudonym Paul MONGRÉ, among other things poems and aphorisms).

3. Convergent sequences. Cluster points. Following Bourbaki we define $\mathbb{N} := \{0, 1, 2, 3, \ldots\}$. Let $k \in \mathbb{N}$. A mapping $\{k, k+1, k+2, \ldots\} \to X, n \mapsto c_n$ is called a *sequence* in X; it is briefly denoted (c_n) and generally $k = 0$. A *subsequence* of (c_n) is a mapping $\ell \mapsto c_{n_\ell}$ in which $n_1 \le n_2 \le \cdots$ is an infinite subset of $\mathbb{N}$. A sequence (c_n) is called *convergent in X*, if there is a point $c \in X$ such that *every neighborhood of c contains almost all* (that is, *all but finitely many*) terms c_n of the sequence; such a point c

is called *a limit* of the sequence, in symbols

$$c = \lim_{n \to \infty} c_n \text{ or, more succinctly, } c = \lim c_n.$$

Non-convergent sequences are called *divergent*.

The separation property ensures that every convergent sequence has *exactly one limit*, so that the implication $c = \lim c_n$ and $c' = \lim c_n \Rightarrow c = c'$, to which our notation already commits us, does in fact obtain. Also

Every subsequence (c_{n_ℓ}) of a convergent sequence (c_n) is convergent and $\lim_{\ell \to \infty} c_{n_\ell} = \lim_{n \to \infty} c_n.$

If d is a metric on X then $c = \lim c_n$ if and only if to every $\epsilon > 0$ there corresponds an $n_\epsilon \in \mathbb{N}$ such that $d(c_n, c) < \epsilon$ for all $n \geq n_\epsilon$; for $X = \mathbb{C}$ with the euclidean metric this is written in the form

$$|c_n - c| < \epsilon, \text{ i.e., } c_n \in B_\epsilon(c), \text{ for all } n \geq n_\epsilon.$$

A set $M \subset X$ is *closed in X* exactly when M contains the limit of each convergent sequence (c_n) of $c_n \in M$.

A point $p \in X$ is called a *cluster point* or *point of accumulation of the set* $M \subset X$ if $U \cap (M \setminus \{p\}) \neq \emptyset$ for every neighborhood U of p. Every neighborhood of a cluster point p of M contains infinitely many points of M and there is always a sequence (c_n) in $M \setminus \{p\}$ with $\lim c_n = p$.

A subset A of a metric space X is called *dense in X* if every non-empty open subset of X contains points of A; this occurs exactly when $\bar{A} = X$. A subset A of X is certainly dense in X if every point of X is a cluster point of A and in this case every point $x \in X$ is the limit of a sequence in A (proof!).

In $\mathbb{C}$ the set $\mathbb{Q} + i\mathbb{Q}$ of all "rational" complex numbers is *dense and countable*. [Recall that a set is called *countable* if it is the image of $\mathbb{N}$ under some map.]

4. Historical remarks on the convergence concept. Great difficulties attended the precise codification of this concept in the 19th century. The limit concept has its origin in the *method of exhaustion* of antiquity. LEIBNIZ, NEWTON, EULER and many others worked with infinite series and sequences without having a precise definition of "limit". For example, it didn't trouble EULER to write (motivated by $\sum_0^\infty x^\nu = (1 - x)^{-1}$)

$$1 - 1 + 1 - 1 + - \ldots = \frac{1}{2}.$$

Even in his *Cours d'analyse* [C] CAUCHY, in defining limits, still used such expressions as "successive values", or "gets indefinitely close" or "as small as one wants". These admittedly suggestive and convenient locutions were first rendered precise by WEIERSTRASS, beginning about 1860 in his Berlin lectures, where the $\epsilon - \delta$ inequalities, still in use today, were formulated. With these the "arithmetization of analysis" began, in this age of rigor.

The ideas of Weierstrass at first reached the mathematical public only through transcriptions and re-copyings of his lectures by his auditors. Only gradually did textbooks adopt the ideas, one of the first being *Vorlesungen über Allgemeine Arithmetik. Nach den neueren Ansichten*, worked out by O. Stolz in Innsbruck, Teubner-Verlag, Leipzig 1885.

5. Compact sets. As in calculus, compact sets also play a central role in function theory. We will introduce the idea of a compact (metric) space, beginning with the classical

Equivalence Theorem. *The following statements concerning a metric space X are equivalent:*

i) *Every open covering $\mathcal{U} = \{U_i\}_{i \in I}$ of X contains a finite sub-covering (Heine-Borel property).*

ii) *Every sequence (x_n) in X contains a convergent subsequence (Weierstrass-Bolzano property).*

We will consider the proof already known to the reader from his prior study of calculus. By way of clarification let us just remind him that an open covering $\mathcal{U}$ of X means any family $\{U_i\}_{i \in I}$ of open sets U_i such that $X = \bigcup_{i \in I} U_i$. In arbitrary topological spaces (which won't come up at all in this book) statements i) and ii) remain meaningful but they are not always equivalent.

X is called *compact* if conditions i) and ii) are fulfilled. A *subset K* of X is called *compact* or a *compactum (in X)* if K is a compact metric space when the metric of X is restricted to K. The reader should satisfy himself that

Every compactum in X is closed in X and in a compact space X every closed subset is compact.

We also highlight the easily verified

Exhaustion property of open sets in $\mathbb{C}$: every open set D in $\mathbb{C}$ is the union of a countably infinite family of compact subsets of D.

Exercises

Exercise 1. Let X be the set of all bounded sequences in $\mathbb{C}$. Show that

a) $d_1((a_n), (b_n)) := \sup\{|a_k - b_k| : k \in \mathbb{N}\}$ and $d_2((a_n), (b_n)) :=$ $\sum_{k=0}^{\infty} 2^{-k} |a_k - b_k|$ define two metrics on X.

b) Do the open sets defined by these two metrics coincide?

Exercise 2. Let $X := \mathbb{C}^{\mathbb{N}}$ be the set of all sequences in $\mathbb{C}$. Show that

a) $d((a_n), (b_n)) := \sum_{k=0}^{\infty} 2^{-k} \frac{|a_k - b_k|}{1 + |a_k - b_k|}$ defines a metric on X;

b) a sequence $x_k = (a_n^{(k)})$ in X converges in this metric to $x = (a_n)$ if and only if for each $n \in \mathbb{N}$, $a_n = \lim_k a_n^{(k)}$.

§3 Convergent sequences of complex numbers

In the subsections of this section we examine the special metric space $X = \mathbb{C}$. Complex sequences can be added, multiplied, divided and conjugated. The limit laws which hold for reals carry over verbatim to complexes, because the absolute value function $|\ |$ has the same properties on $\mathbb{C}$ as it does on $\mathbb{R}$. The field $\mathbb{C}$ inherits from the field $\mathbb{R}$ the (metric) completeness which Cauchy's convergence criterion expresses.

If there is no possibility of misunderstanding, we will designate a sequence (c_n) briefly as c_n. If we have to indicate that the sequence starts with the index k, then we write $(c_n)_{n \geq k}$. A convergent sequence with limit 0 is called a *null sequence*.

1. Rules of calculation. If the sequence c_n converges to $c \in \mathbb{C}$ then almost all terms c_n of the sequence are inside each disc $B_\epsilon(c)$ around c. For every $z \in \mathbb{C}$ with $|z| < 1$ the sequence z^n of *powers* converges: $\lim z^n = 0$; for all z with $|z| > 1$ the sequence z^n diverges.

A sequence c_n is called *bounded* if there is a real number $M > 0$ (called a "bound") such that $|c_n| \leq M$ for all n. Just as for real sequences it follows that

Every convergent sequence of complex numbers is bounded.   □

For convergent sequences c_n, d_n the expected *limit laws* prevail:

L.1 *For all $a, b \in \mathbb{C}$ the sequence $ac_n + bd_n$ converges:*

$$\lim(ac_n + bd_n) = a \lim c_n + b \lim d_n \quad (\mathbb{C}\text{-linearity}).$$

L.2 *The "product sequence" $c_n d_n$ converges:*

$$\lim(c_n d_n) = (\lim c_n)(\lim d_n).$$

L.3 *If $\lim d_n \neq 0$, then there is a $k \in \mathbb{N}$ such that $d_n \neq 0$ for all $n \geq k$, and the quotient sequence $(c_n/d_n)_{n \geq k}$ converges to $(\lim c_n)/(\lim d_n)$.*

Remark. Rules L.1 and L.2 admit a very elegant formulation in the language of algebra. For *arbitrary* sequences c_n, d_n of complex numbers we can define the *sum sequence* and the *product sequence* by setting

$$a(c_n) + b(d_n) := (ac_n + bd_n) \quad \text{for all } a, b \in \mathbb{C}; \qquad (c_n)(d_n) := (c_n d_n).$$

The limit laws L.1 and L.2 can then be reformulated as:

The collection of all convergent sequences forms a (commutative) $\mathbb{C}$-algebra $\mathcal{A}$ (more precisely, a $\mathbb{C}$-subalgebra of the $\mathbb{C}$-algebra of all sequences) with zero element $(0)_n$ and unit element $(1)_n$. The mapping $\lim : \mathcal{A} \to \mathbb{C}$, $(c_n) \mapsto \lim c_n$ is a $\mathbb{C}$-algebra homomorphism.

[Here perhaps we should recall for the reader's convenience: a *$\mathbb{C}$-algebra* $\mathcal{A}$ is a $\mathbb{C}$-vector space between whose elements a *multiplication* $\mathcal{A} \times \mathcal{A} \to \mathcal{A} : (a, a') \mapsto aa'$ is defined which satisfies the two *distributive laws* $(\lambda a + \mu b)a' = \lambda(aa') + \mu(ba'), a'(\lambda a + \mu b) = \lambda(a'a) + \mu(a'b)$. A $\mathbb{C}$-vector-space homomorphism $f : \mathcal{A} \to \mathcal{B}$ between $\mathbb{C}$-algebras $\mathcal{A}$ and $\mathcal{B}$ is called a $\mathbb{C}$-*algebra homomorphism* if it is *multiplicative*: $f(aa') = f(a)f(a')$ for all $a, a' \in \mathcal{A}$.]

The limit laws L.1 – L.3 are supplemented with the following:

L.4 *The sequence $|c_n|$ of absolute values of a convergent sequence is convergent and $\lim |c_n| = |\lim c_n|$.*

L.5 *The sequence $\bar{c}_n$ conjugate to a convergent sequence is convergent and $\lim \bar{c}_n = \overline{(\lim c_n)}$.*

Proofs are immediate from the inequality $||c_n| - |c|| \leq |c_n - c|$ and the equality $|\bar{c}_n - \bar{c}| = |c_n - c|$, respectively, where $c := \lim c_n$.

Every sequence c_n determines its *sequence of real parts* $\Re c_n$ and its *sequence of imaginary parts* $\Im c_n$. The question of convergence in $\mathbb{C}$ can always be reduced to two convergence questions in $\mathbb{R}$:

Theorem. *For a sequence c_n of complex numbers the following are equivalent:*

  i) c_n *is convergent.*

  ii) *The real sequences* $\Re c_n$ *and* $\Im c_n$ *are each convergent.*

In case of convergence, $\lim c_n = \lim(\Re c_n) + i \lim(\Im c_n)$.

Proof. i) $\Rightarrow$ ii)   This is clear from L.1 and L.5 together with the equations

$$\Re c_n = \frac{1}{2}(c_n + \bar{c}_n), \qquad \Im c_n = \frac{1}{2i}(c_n - \bar{c}_n).$$

  ii) $\Rightarrow$ i)   L.1 yields this as well as the equality relating the three limits.
$\square$

2. Cauchy's convergence criterion. Characterization of compact sets in $\mathbb{C}$. A sequence c_n is called a *Cauchy sequence* if to every $\epsilon > 0$ there corresponds a $k_\epsilon \in \mathbb{N}$ such that $|c_n - c_m| < \epsilon$ for all $n, m \geq k_\epsilon$. As in $\mathbb{R}$ we have in $\mathbb{C}$ the fundamental

Convergence criterion of CAUCHY. *For any complex sequence* (c_n) *the following are equivalent:*

  i) (c_n) *is convergent.*

  ii) (c_n) *is a Cauchy sequence.*

Proof. i) $\Rightarrow$ ii)   Given $\epsilon > 0$, choose $k \in \mathbb{N}$ such that $|c_n - c| < \frac{1}{2}\epsilon$ for all $n \geq k$, where $c := \lim c_n$. Then

$$|c_m - c_n| \leq |c_m - c| + |c - c_n| < \epsilon \text{ for all } m, n \geq k.$$

  ii) $\Rightarrow$ i)   The inequalities

$$|\Re c_m - \Re c_n| \leq |c_m - c_n|\,, \; |\Im c_m - \Im c_n| \leq |c_m - c_n|,$$

valid for all m and n, show that along with (c_n) the real sequences $(\Re c_n)$ and $(\Im c_n)$ are each Cauchy sequences. Because of the completeness of $\mathbb{R}$, they converge to numbers a and b in $\mathbb{R}$. Then by L.1 the sequence (c_n) in $\mathbb{C}$ converges to $a + bi$.
$\square$

  The notion of a Cauchy sequence can be defined in every metric space X: A sequence (c_n), $c_n \in X$, is called a *Cauchy sequence in* X if for every $\epsilon > 0$ there is a $k_\epsilon \in \mathbb{N}$ such that $d(c_m, c_n) < \epsilon$ for all $m, n \geq k_\epsilon$. *Convergent sequences are always Cauchy sequences* (as the above proof in $\mathbb{C}$ essentially demonstrates). Those metric spaces in which the converse holds are called *complete*. Thus $\mathbb{C}$, as well as $\mathbb{R}$, is a complete valued field.

  Compacta in $\mathbb{C}$ admit a simple characterization.

Theorem. *The following assertions about a set* $K \subset \mathbb{C}$ *are equivalent:*

i) K *is compact.*

ii) K *is bounded and closed in* $\mathbb{C}$.

We will consider this equivalence to be known to the reader from calculus: he has, of course, to consider $\mathbb{C}$ as $\mathbb{R}^2$. By the same token this equivalence, which rests ultimately on the completeness of $\mathbb{R}$, is valid in every $\mathbb{R}^n$, $1 \leq n < \infty$. It is not, however, valid in *every* complete metric space.

A special case of the foregoing theorem is

WEIERSTRASS-BOLZANO Theorem. *Every bounded sequence of complex numbers has a convergent subsequence.*

Exercises

Exercise 1. For which $z \in \mathbb{C}$ do the following limits exist?

a) $\lim_n \frac{z^n}{n!}$

b) $\lim_n (\frac{z}{n})^n$

c) $\lim_n z^n$.

Exercise 2. Let (c_n) be a *bounded* sequence of complex numbers. Show that this sequence converges to $c \in \mathbb{C}$ if and only if each of its convergent subsequences has limit c.

Exercise 3. a) Let $(a_n)_{n \geq 0}$ and $(b_n)_{n \geq 0}$ be convergent sequences of complex numbers. Show that

$$\lim_n \frac{a_0 b_n + a_1 b_{n-1} + \cdots + a_n b_0}{n+1} = (\lim_k a_k)(\lim_m b_m).$$

b) If the sequence $(a_n)_{n \geq 0}$ converges to $a \in \mathbb{C}$, then $\lim_n \frac{1}{n} \sum_{k=0}^{n-1} a_k = a$. Does the converse of this hold?

Exercise 4. Show that the metric spaces in Exercises 1 and 2 of §2 are complete.

§4 Convergent and absolutely convergent series

Convergent series $\sum a_\nu$ in $\mathbb{C}$ are defined, just as in $\mathbb{R}$, via their sequences of partial sums. Among the various forms of limit processes the easiest to deal with are convergent series: together with the approximants $s_n :=$ $a_0 + \ldots + a_n$ a "correction term" which leads from s_n to the next approximant $s_{n+1} = s_n + a_{n+1}$ is always given. This makes working with series more convenient than working with sequences. In the 19th century people worked principally with series and hardly at all with sequences: the insight that convergent sequences are really the fundamental cells which generate all the limit processes of analysis only took hold at the beginning of the 20th century.

The *absolutely convergent* series $\sum a_\nu$, those for which $\sum |a_\nu| < \infty$, are especially important in the theory of series. The most important convergence criterion for such series is the *majorant criterion* or *comparison test* (subsection 2). As with real numbers there is for absolutely convergent complex series a rearrangement theorem (subsection 3) and a product theorem (subsection 6).

1. Convergent series of complex numbers. If $(a_\nu)_{\nu \geq k}$ is a sequence of complex numbers, then the sequence $(s_n)_{n \geq k}$, $s_n := \sum_{\nu=k}^{n} a_\nu$ of *partial sums* is called an (*infinite*) *series* with *terms* a_ν. One writes $\sum_{\nu=k}^{\infty} a_\nu$ or $\sum_{k}^{\infty} a_\nu$ or $\sum_{\nu \geq k} a_\nu$ or simply $\sum a_\nu$. Generally k is either 0 or 1.

A series $\sum a_\nu$ is said to be *convergent* if the sequence (s_n) of partial sums converges; otherwise the series is said to be *divergent*. In the case of convergence it is customary to write, suggestively

$$\sum a_\nu := \lim s_n.$$

As in the case of $\mathbb{R}$, the symbol $\sum a_\nu$ thus does double duty: it denotes both the *sequence* of partial sums and the *limit* of this sequence (when it exists).

The standard example of an infinite series, which is used again and again to provide majorizations, is the *geometric series* $\sum_{\nu \geq 0} z^\nu$. Its partial sums (finite geometric series) are

$$\sum_{0}^{n} z^\nu = \frac{1 - z^{n+1}}{1 - z} \qquad \text{for every } z \neq 1.$$

Because (z^{n-1}) is a null sequence for all $z \in \mathbb{C}$ with $|z| < 1$, it follows that

$$\sum_{0}^{\infty} z^\nu = \frac{1}{1 - z} \qquad \text{for all } z \in \mathbb{C} \text{ with } |z| < 1. \qquad \square$$

Because $a_n = s_n - s_{n-1}$, $\lim a_n = 0$ holds for every convergent series $\sum a_\nu$.

The limit laws L.1 and L.5 carry over immediately to series:

$$\sum_{\nu \geq k} (aa_\nu + bb_\nu) = a \sum_{\nu \geq k} a_\nu + b \sum_{\nu \geq k} b_\nu , \qquad \overline{\sum_{\nu \geq k} a_\nu} = \sum_{\nu \geq k} \overline{a_\nu},$$

a special case of which is:

The complex series $\sum_{\nu \geq k} a_\nu$ converges if and only if each of the real series $\sum_{\nu \geq k} \Re a_\nu$ and $\sum_{\nu \geq k} \Im a_\nu$ converges, and then

$$\sum_{\nu \geq k} a_\nu = \sum_{\nu \geq k} \Re a_\nu + i \sum_{\nu \geq k} \Im a_\nu.$$

Moreover, we trivially have for convergent series

$$\sum_{k}^{\infty} a_\nu = \sum_{k}^{\ell} a_\nu + \sum_{\ell+1}^{\infty} a_\nu \qquad \text{for all } \ell \in \mathbb{N} \text{ with } \ell \geq k.$$

For infinite series there is also a

CAUCHY convergence criterion. *A series $\sum a_\nu$ converges precisely when to every $\varepsilon > 0$ there corresponds an $n_\epsilon \in \mathbb{N}$ such that*

$$\left| \sum_{m+1}^{n} a_\nu \right| < \varepsilon \qquad \text{for all } m, n \text{ with } n > m \geq n_\epsilon.$$

This is clear because the equality $\sum_{m+1}^{n} a_\nu = s_n - s_m$ means that the condition in this criterion is exactly that (s_n) be a Cauchy sequence.

2. Absolutely convergent series. The majorant criterion. The limit of a convergent series can be changed by a rearrangement which alters the positions of infinitely many terms. Manipulations of this kind can be routinely carried out only on series which are absolutely convergent.

A series $\sum a_\nu$ is called absolutely convergent if the series $\sum |a_\nu|$ of nonnegative real numbers is convergent.

The completeness of $\mathbb{C}$ makes it possible, as with $\mathbb{R}$, to infer the convergence of the series $\sum a_\nu$ from that of the series $\sum |a_\nu|$. Since $|\sum_{m+1}^{n} a_\nu| \leq \sum_{m+1}^{n} |a_\nu|$, it follows immediately from Cauchy's convergence criterion for series that

Every absolutely convergent series $\sum a_\nu$ is convergent; and $|\sum a_\nu| \leq \sum |a_\nu|$.

Furthermore it is clear that

Every subseries $\sum_{\ell=0}^\infty a_{\nu_\ell}$ of an absolutely convergent series $\sum_{\nu=0}^\infty a_\nu$ is absolutely convergent. [In fact, it can even be shown that a series converges absolutely if and only if every one of its subseries converges.]

Quite fundamental is the simple

Majorant criterion or **Comparison test.** *Let $\sum_{\nu \geq k} t_\nu$ be a convergent series with non-negative real terms t_ν; let $(a_\nu)_{\nu \geq k}$ be a sequence of complex numbers which for almost all $\nu \geq k$ satisfy $|a_\nu| \leq t_\nu$. Then $\sum_{\nu \geq k} a_\nu$ is absolutely convergent.*

Proof. There is an $n_1 \geq k$ such that for all $n > m \geq n_1$

$$\sum_{m+1}^n |a_\nu| \leq \sum_{m+1}^n t_\nu.$$

Because $\sum t_\nu$ converges, the claim follows from the Cauchy criterion. □

The series $\sum t_\nu$ is called a *majorant* of $\sum a_\nu$. The most frequently occurring majorants are the geometric series $\sum cq^\nu, 0 < q < 1, 0 \leq c \in \mathbb{R}$.
□

Calculating with absolutely convergent series is significantly simpler than calculating with series that are merely convergent, because series with positive terms are easier to handle. Because $\max(|\Re a|, |\Im a|) \leq |a| \leq |\Re a| + |\Im a|$, it further follows (from the majorant criterion) that

The complex series $\sum a_\nu$ is absolutely convergent precisely when each of the real series $\sum \Re a_\nu$ and $\sum \Im a_\nu$ is absolutely convergent.

3. The rearrangement theorem. *If $\sum_{\nu \geq 0} a_\nu$ is absolutely convergent, then every "rearrangement" of this series also converges and to the same limit:*

$$\sum_{\nu \geq 0} a_{\tau(\nu)} = \sum_{\nu \geq 0} a_\nu \qquad \text{for every bijection } \tau \text{ of } \mathbb{N}.$$

Proof. The proof that most readers have doubtlessly seen for $\mathbb{R}$ works as well in $\mathbb{C}$. It runs as follows: Let $s := \sum_{\nu \geq 0} a_\nu$. For each $\varepsilon > 0$ let $\nu_\varepsilon \in \mathbb{N}$ be such that $\sum_{\nu > \nu_\varepsilon} |a_\nu| < \varepsilon$. Let $F_\varepsilon := \tau^{-1}\{0, 1, \ldots, \nu_\varepsilon\}$ and

$n_\epsilon := \max F_\epsilon$. Then $\{0, 1, \ldots, k\} \supset F_\epsilon$, so $\tau(\{0, 1, \ldots, k\}) \supset \tau(F_\epsilon) = \{0, 1, \ldots, \nu_\epsilon\}$ whenever $n \geq n_\epsilon$. For such n

$$
\begin{aligned}
\left| \sum_{\nu=0}^{n} a_{\tau(\nu)} - s \right| &= \left| \sum_{\nu \in F_\epsilon} a_{\tau(\nu)} - s + \sum_{\nu \in \{0,1,\ldots,n\}\setminus F_\epsilon} a_{\tau(\nu)} \right| \\
&= \left| \sum_{\mu \in \tau(F_\epsilon)} a_\mu - s + \sum_{\mu \in \tau\{0,1,\ldots,n\}\setminus \tau(F_\epsilon)} a_\mu \right| \\
&\leq \left| \sum_{\mu=0}^{\nu_\epsilon} a_\mu - s \right| + \sum_{\mu \in \tau\{0,1,\ldots,n\}\setminus\{0,1,\ldots,\nu_\epsilon\}} |a_\mu| \\
&\leq 2 \sum_{\nu > \nu_\epsilon} |a_\nu| < 2\epsilon. \qquad \square
\end{aligned}
$$

In the literature the rearrangement theorem is sometimes also called the *commutative law for infinite series*. Generalizations of this commutative law will be found in the classical book *Theory and Applications of Infinite Series* by KNOPP [15].

4. Historical remarks on absolute convergence. In 1833 CAUCHY observed [*Œuvres* (2) **10**, 68-70] that a convergent real series whose terms are not all positive could have a divergent subseries. In a famous 1837 work on number theory, "Beweis des Satzes, daß jede unbegrenzte arithmetische Progression, deren erstes Glied und Differenz ganze Zahlen ohne gemeinschaftlichen Teiler sind, unendlich viele Primzahlen enthält (Proof that every unbounded arithmetic progression whose first term and common difference are integers without common divisors contains infinitely many primes)" (*Werke* **1**, p.319) DIRICHLET presented the (conditionally) convergent series

$$1 - \frac{1}{2} + \frac{1}{3} - \frac{1}{4} + - \ldots \quad \text{and} \quad 1 + \frac{1}{3} - \frac{1}{2} + \frac{1}{5} + \frac{1}{7} - \frac{1}{4} + - \ldots,$$

which are rearrangements of one another and which have different sums, namely $\log 2$ and $\frac{3}{2} \log 2$, respectively. In this same work (p.318) DIRICHLET proved the rearrangement theorem for series with real terms. In his 1854 Habilitationsschrift *Über die Darstellbarkeit einer Function durch eine trigonometrische Reihe* (*Concerning the representation of a function by a trigonometric series*) [*Werke*, p.235], where among other things he introduced his integral, RIEMANN wrote that by 1829 DIRICHLET knew "daß die unendlichen Reihen in zwei wesentlich verschiedene Klassen zerfallen, je nachdem sie, wenn man sämtliche Glieder positiv macht, convergent bleiben oder nicht. In den ersteren können die Glieder beliebig versetzt werden, der Werth der letzteren dagegen ist von der Ordnung der Glieder abhängig (that infinite series fall into two essentially different classes, according to whether they remain convergent or not after all terms have been made positive. In series of the first type the terms may be arbitrarily permuted; by contrast, the value of a series of the second type depends on the order of its terms)". RIEMANN then proves his rearrangement theorem: *A convergent series (of real terms) which is not absolutely convergent* "kann durch geeignete Anordnung der Glieder einen beliebig gegebenen (reellen) Werth C erhalten (*can converge to an arbitrary given real value C after appropriate re-ordering of its terms*)". The discovery of this apparent paradox contributed essentially to a re-examination and rigorous founding (focussing on the sequence of partial sums) of the theory of infinite series. On November 15, 1855 RIEMANN [*Werke*, Nachträge p. 111] made note of the fact that: "Die Erkenntnis des Umstandes, daß die unendlichen Reihen in zwei Klassen zerfallen (je nachdem der Grenzwert unabhängig von der Anordnung ist oder nicht), bildet einen Wendepunkt in der Auffassung des Unendlichen in der Mathematik (The recognition of the fact that infinite series fall into two classes (according to whether the limit is independent of the ordering of the terms or not) constitutes a turning-point in the conceptualization of the infinite in mathematics)".

5. Remarks on Riemann's rearrangement theorem. This theorem does not carry over from reals to complexes without modification. For example, if $\sum a_\nu$ is convergent but not absolutely convergent then of course at least one of the real series $\sum \Re a_\nu$, $\sum \Im a_\nu$ is not absolutely convergent and by Riemann's theorem, given $r \in \mathbb{R}$, there is a bijection τ of $\mathbb{N}$ such that one of the series $\sum \Re a_{\tau(\nu)}$, $\sum \Im a_{\tau(\nu)}$ converges to r; but *prima facie* nothing is known about the convergence of the other one.

Let us understand by the phrase *rearrangement-induced sums* of an infinite series $\sum a_\nu$ ($a_\nu \in \mathbb{C}$) the set L of all $c \in \mathbb{C}$ to which correspond some bijection τ of $\mathbb{N}$ such that $\sum a_{\tau(\nu)} = c$. It can be shown that exactly one of the following alternatives always prevails:

1) *L is empty (so-called "proper" divergence).*

2) *L is a single point ($\Leftrightarrow \sum a_\nu$ is absolutely convergent).*

3) *L is a straight line in $\mathbb{C}$.*

4) *L coincides with $\mathbb{C}$.*

Each of the four cases can indeed be realized: E.g., $L = \mathbb{R} + i$ for the series $\sum_1^\infty \left[\frac{(-1)^\nu}{\nu} + \frac{i}{\nu(\nu+1)} \right]$ and $L = \mathbb{C}$ occurs for all series $\sum a_\nu$ in which $a_{2\nu} \in \mathbb{R}$ and $a_{2\nu+1} \in i\mathbb{R}$ for all ν and each of $\sum a_{2\nu}$, $\sum a_{2\nu-1}$ is convergent but not absolutely so.

A generalization of Riemann's rearrangement theorem was formulated in 1905 by P. LÉVY ("Sur les séries semi-convergentes", *Nouv. Annales* (4), **5**, p.506) and in 1913/14 E. STEINITZ, the founder of abstract field theory, gave a logically satisfactory treatment in his paper "Bedingt konvergente Reihen und konvexe Systeme", *Jour. für Reine und Angew. Math.* **143**, p.128ff and **144**, p.1ff. [The Steinitz replacement theorem, which is often used in linear algebra to prove the invariance of the cardinality of bases and which is a much-dreaded examination question, is to be found on p.133 of the first part of this work.] What STEINITZ proved is:

If $v_\nu \in \mathbb{R}^m (1 \le m < \infty)$ for all ν, then the set of rearrangement-induced sums of the series $\sum v_\nu$ is either empty or an affine subspace of $\mathbb{R}^m$.

A very accessible modern account of this will be found in P. ROSENTHAL "The remarkable theorem of Lévy and Steinitz," *Amer. Math. Monthly* **94**(1987), pp. 342-351.

In this connection the 1917 paper "Bedingt konvergente Reihen" published by W. GROSS in the *Monatshefte für Mathematik* **28**, pp. 221-237 is also worth reading.   □

In analysis a series $\sum a_\nu$ such that every rearrangement $\sum a_{\tau(\nu)}$ converges and to the same limit is very often called *unconditionally convergent*. According to 2), in $\mathbb{C}$ the unconditionally convergent series coincide with the absolutely convergent ones. For arbitrary Banach spaces this is no longer the case; in fact we have the following rather surprising characterization:

Theorem. *The following statements about a Banach space V are equivalent:*

i) *The class of unconditionally convergent series $\sum v_\nu$, $v_\nu \in V$, coincides with the class of absolutely or normally convergent ones, i.e., those for which $\sum \|v_\nu\| < \infty$.*

ii) *The space V is finite-dimensional.*

This was proved in 1950 by A. DVORETZKY and C.A. ROGERS (*Proc. Nat. Acad. Sci. USA* **36**, pp. 192-197). The problem of determining all vector spaces for which the two classes of series coincide had been brought up by S. BANACH on p. 240 of his classical book *Théorie des Opérations Linéaires* (Monografie Matematyczne 1, Warsaw 1932; English translation by F. JELLETT, North-Holland Publ. Co., Amsterdam, 1987). A. PIETSCH gives a simple proof on p. 68 of his book *Nuclear Locally Compact Spaces* (Springer-Verlag, New York & Berlin 1972); he deduces the result from the fact that nuclear mappings are necessarily pre-compact (see p. 52).

6. A theorem on products of series. If $\sum_0^\infty a_\mu$, $\sum_0^\infty b_\nu$ are two series, then any series $\sum_0^\infty c_\lambda$ in which the terms c_λ run through all possible products $a_\mu b_\nu$ exactly once, is called *a product series* of $\sum a_\mu$ and $\sum b_\nu$. The most important product series in the *Cauchy product* $\sum p_\lambda$ in which $p_\lambda := \sum_{\mu+\nu=\lambda} a_\mu b_\nu$. Such sums are suggested by formally multiplying out the power series product $\left(\sum_{\mu \ge 0} a_\mu X^\mu\right) \left(\sum_{\nu \ge 0} b_\nu X^\nu\right)$ and collecting together the coefficients of like powers of X.

A product theorem. *Let $\sum_0^\infty a_\mu$, $\sum_0^\infty b_\nu$ be absolutely convergent series.*

Then every product series $\sum_0^\infty c_\lambda$ of the two is absolutely convergent and

$$\left(\sum_0^\infty a_\mu\right)\left(\sum_0^\infty b_\nu\right) = \sum_0^\infty c_\lambda.$$

Proof. For each $\ell \in \mathbb{N}$ there is an $m \in \mathbb{N}$ such that $\{c_0, ..., c_\ell\} \subset \{a_\mu b_\nu : 0 \leq \mu, \nu \leq m\}$. It follows that

$$\sum_0^\ell |c_\lambda| \leq \left(\sum_0^m |a_\mu|\right)\left(\sum_0^m |b_\nu|\right) \leq \left(\sum_0^\infty |a_\mu|\right)\left(\sum_0^\infty |b_\nu|\right) < \infty.$$

Consequently, the series $\sum_0^\infty c_\lambda$ is absolutely convergent and to evaluate $c := \sum_0^\infty c_\lambda$ we can use any ordering of the terms $a_\mu b_\nu$ which results from multiplying out the products $(a_0 + a_1 + ... + a_n)(b_0 + b_1 + ... + b_n)$ one after another $(n = 0, 1, 2, ...)$. This observation means that

$$c = \lim_{n \to \infty} \left(\sum_0^n a_\mu\right)\left(\sum_0^n b_\nu\right) = \left(\sum_0^\infty a_\mu\right)\left(\sum_0^\infty b_\nu\right). \qquad \square$$

Absolute convergence is essential to the validity of this product theorem: The Cauchy product of the convergent (but not absolutely convergent) series $\sum \frac{(-1)^\nu}{\sqrt{\nu+1}}$ with itself is divergent. Indeed, because $0 \leq (x - y)^2 = x^2 - 2xy + y^2$ for real x and y, we have $2\sqrt{\mu+1}\sqrt{\nu+1} \leq \mu + 1 + \nu + 1$ and so if $\mu + \nu = \lambda$, then $(-1)^\lambda a_\mu a_\nu = \frac{1}{\sqrt{\mu+1}\sqrt{\nu+1}} \geq \frac{2}{\lambda+2}$ and $(-1)^\lambda p_\lambda = \sum_{\mu+\nu=\lambda}(-1)^\lambda a_\mu a_\nu \geq \sum_{\mu+\nu=\lambda} \frac{2}{\lambda+2} = (\lambda + 1) \cdot \frac{2}{\lambda+2} \geq 1$ for all $\lambda \in \mathbb{N}$.

The above product theorem for complex series occurs on p.237 of Cauchy's 1821 *Cours d'analyse* [C].

If both $\sum a_\mu, \sum b_\nu$ are convergent and at least one of them is absolutely convergent, then their Cauchy product converges and has sum equal to $\left(\sum_0^\infty a_\mu\right) \times \left(\sum_0^\infty b_\nu\right)$ – theorem of F. MERTENS, 1875 (see KNOPP [15], §45). In 7.4.4 we will become acquainted with the product theorem for convergent power series and from it deduce an 1826 product theorem of ABEL, the hypotheses of which are quite different from those of the Cauchy product theorem above.

Exercises

Exercise 1. Investigate the convergence and absolute convergence of the following series:

a) $\sum_{n \geq 1} \frac{i^n}{n}$

b) $\sum_{n \geq 1} \frac{(2+i)^n}{(1+i)^{2n}}$

c) $\sum_{n\geq 1}((z - n - \frac{1}{2})^2 - \frac{1}{4})^{-1}$, $z \in \mathbb{C} \setminus \mathbb{N}$.

Find the limit of the series in c).

Exercise 2. Let (a_n) be a sequence of complex numbers with $\Re a_n \geq 0$ for almost all n. Prove that if both the series $\sum a_n$ and $\sum a_n^2$ converge, then the second one must actually converge absolutely. Does the converse hold?

Exercise 3. Let (a_n) be a sequence of complex numbers, non-zero for all n beyond some n_0. Show that if there is a real number $A < -1$ such that the sequence $n^2(|\frac{a_{n+1}}{a_n}| - 1 - \frac{A}{n})$, $n > n_0$, is bounded, then the series $\sum a_n$ is absolutely convergent. *Hint.* Set $c := 1 - A$, $d := 1 + \frac{c}{2}$, $b_n := n^{-d}$ and show that $\lim n(|\frac{a_{n+1}}{a_n}| - 1) < \lim n(|\frac{b_{n+1}}{b_n}| - 1)$ [why do these limits exist?] and infer that for an appropriate finite constant C, $|a_n| \leq Cb_n$ for all n.

Exercise 4. Let $(a_n)_{n\geq 0}$ and $(b_n)_{n\geq 0}$ be sequences of complex numbers and suppose that

i) the sequence of partial sums $S_m := \sum_{n=0}^{m} a_n$ is bounded;

ii) $\lim b_n = 0$;

iii) the sum $\sum_{n=1}^{\infty} |b_n - b_{n-1}|$ is finite.

Show that then the series $\sum a_n b_n$ is convergent. *Hint.* Use "Abel summation": $\sum_{k=n}^{m} a_k b_k = \sum_{k=n}^{m} (S_k - S_{k-1})b_k$ for $n > 0$.

Exercise 5. To each $(m, n) \in \mathbb{N}^2$ associate a complex number $a_{m,n}$. Suppose the numbers $a_{m,n}$ are "somehow" organized into a sequence c_k. Show that the following statements are equivalent:

i) The series $\sum c_k$ is absolutely convergent.

ii) For each $n \in \mathbb{N}$ the series $\sum_m a_{m,n}$ converges absolutely and the series $\sum_n(\sum_m |a_{m,n}|)$ converges.

If i) and ii) are fulfilled then the series $\sum_n a_{m,n}$ is absolutely convergent for every $m \in \mathbb{N}$ and the series $\sum_m(\sum_n a_{m,n})$, $\sum_n(\sum_m a_{m,n})$ are convergent with

$$\sum_m(\sum_n a_{m,n}) = \sum_n(\sum_m a_{m,n}) = \sum_k c_k.$$

§5   Continuous functions

The main business of analysis is the study of functions. The concept of function will be taken for granted here and the words *function* and *mapping* will be used synonymously. Functions with *domain* X and *range* in Y are indicated by

$$f : X \to Y, x \mapsto f(x) \quad \text{or} \quad f : X \to Y \quad \text{or} \quad f(x) \quad \text{or just } f.$$

In what follows X, Y, Z are always metric spaces and d_X, d_Y, d_Z are their metrics.

1.   The continuity concept. A mapping $f : X \to Y$ is said to be *continuous at the point* $a \in X$ if the f-*pre-image* (also called the f-*inverse-image*) $f^{-1}(V) := \{x \in X : f(x) \in V\}$ of every neighborhood V of $f(a)$ in Y is a neighborhood of a in X. In terms of the metrics, we have the

(ε, δ)-Criterion. $f : X \to Y$ *is continuous at the point* a *if and only if for every* $\varepsilon > 0$ *there exists a* $\delta = \delta_\varepsilon > 0$ *such that*

$$d_Y(f(x), f(a)) < \varepsilon \text{ whenever } x \in X \text{ satisfies } d_X(x, a) < \delta.$$

As in calculus, it is convenient to use the following terminology and notation: the function $f : X \to Y$ *converges to (approaches)* b *under approach to* a, or in symbols:

$$\lim_{x \to a} f(x) = b \quad \text{or} \quad f(x) \to b \quad \text{as } x \to a$$

if, corresponding to each neighborhood V of b in Y, there is a neighborhood U of a in X such that $f(U \setminus \{a\}) \subset V$. *It should be noted that it is the punctured neighborhood* $U \setminus \{a\}$ *which is involved in* X. We now obviously have

f is continuous at a if and only if the limit $\lim_{x \to a} f(x) \in Y$ *exists and coincides with the function value $f(a)$.*

Also useful in practice is the

Sequence criterion. $f : X \to Y$ *is continuous at a if and only if* $\lim f(x_n) = f(a)$ *for every sequence* (x_n) *of points* $x_n \in X$ *which converges to* a.

Two mappings $f : X \to Y$ and $g : Y \to Z$ may be put together to form a third $g \circ f : X \to Z$ according to the rule $(g \circ f)(x) := g(f(x))$ for all x. This *composition* of mappings inherits continuity from its component functions.

If $f : X \to Y$ is continuous at $a \in X$ and $g : Y \to Z$ is continuous at $f(a) \in Y$, then $g \circ f : X \to Z$ is continuous at a.

A function $f : X \to Y$ is simply called *continuous* if it is continuous at each point of X. For example, the *identity mapping* id $: X \to X$ is continuous, for any metric space X.

Well known (and essentially trivial) is the

Continuity criterion. *The following statements are equivalent:*

i) *f is continuous.*

ii) *The inverse-image $f^{-1}(V)$ of every open set V in Y is open in X.*

iii) *The inverse-image $f^{-1}(C)$ of every closed set C in Y is closed in X.*

In particular every fiber $f^{-1}(f(x))$, $x \in X$, associated with a continuous mapping $f : X \to Y$ is closed in X. Continuity and compactness relate to each other quite well:

Theorem. *Let $f : X \to Y$ be continuous and K be a compact subset of X. Then $f(K)$ is a compact subset of Y.*

Proof. Let (y_n) be any sequence in $f(K)$. Then for every n there exists an $x_n \in K$ with $f(x_n) = y_n$. Because K is compact, 2.5 ensures that some subsequence (x'_n) of (x_n) converges to some $a \in K$. Because f is continuous, $y'_n := f(x'_n)$ satisfies

$$\lim y'_n = \lim f(x'_n) = f(a) \in f(K).$$

Consequently, (y'_n) is a subsequence of (y_n) which converges to a limit in $f(K)$.
□

Contained in this theorem is the fact that continuous real-valued functions $f : X \to \mathbb{R}$ attain maxima and minima on each compactum in X. It was WEIERSTRASS (in his Berlin lectures from 1860 on) who first put in evidence the fundamental role of this fact (in the case where $X = \mathbb{R}$).

2. The $\mathbb{C}$-algebra $\mathcal{C}(X)$. In this section we take $Y := \mathbb{C}$. Complex-valued functions $f : X \to \mathbb{C}$, $g : X \to \mathbb{C}$ can be *added* and *multiplied*:

$$(f + g)(x) := f(x) + g(x), \quad (f \cdot g)(x) := f(x)g(x), \quad x \in X.$$

Every complex number c determines the corresponding *constant function* $X \to \mathbb{C}$, $x \mapsto c$. This function is again denoted by c. The function $\bar{f}$ *conjugate to f* is defined by

$$\bar{f}(x) := \overline{f(x)}, \quad x \in X.$$

The rules of computation for the conjugation mapping $\mathbb{C} \to \mathbb{C}$, $z \mapsto \bar{z}$ (cf. 1.1) apply without change to $\mathbb{C}$-valued functions. Thus

$$\overline{f+g} = \bar{f} + \bar{g}, \quad \overline{fg} = \bar{f}\bar{g}, \quad \bar{\bar{f}} = f.$$

The *real part* and the *imaginary part* of f are defined by

$$(\Re f)(x) := \Re f(x), \quad (\Im f)(x) := \Im f(x), \quad x \in X.$$

These are real-valued functions and throughout we will write

$$u := \Re f, \quad v := \Im f.$$

Then we have

$$f = u + iv, \quad u = \frac{1}{2}(f + \bar{f}), \quad v = \frac{1}{2i}(f - \bar{f}), \quad f\bar{f} = u^2 + v^2.$$

The limit laws from 3.1 together with the sequence criterion immediately imply:

If $f, g : X \to \mathbb{C}$ are both continuous at $a \in X$, then so are the sum $f + g$, the product fg, and the conjugated function $\bar{f}$.

Contained in this is the fact that:

A function f is continuous at a if and only if its real part u and its imaginary part v are both continuous at a.

We will designate the set of all continuous $\mathbb{C}$-valued functions on X by $C(X)$. Since constant functions are certainly continuous, we have a natural inclusion $\mathbb{C} \subset C(X)$. Recalling the concept of a $\mathbb{C}$-algebra from the discussion in 3.1, it is clear from all the foregoing that:

$C(X)$ is a commutative $\mathbb{C}$-algebra with unit. There is an $\mathbb{R}$-linear, involutory (that is, equal to its own inverse) automorphism $C(X) \to C(X)$, $f \mapsto \bar{f}$.
The function f is in $C(X)$ if and only if each of $\Re f$ and $\Im f$ is in $C(X)$.

If g is *zero-free on X*, meaning that $0 \notin g(X)$, then the function $X \to \mathbb{C}$ defined by

$$x \mapsto f(x)/g(x)$$

is called the *quotient function* of f by g and it is designated simply f/g. The limit laws from 3.1 imply that:

$f/g \in C(X)$ for every $f \in C(X)$ and every zero-free $g \in C(X)$.

The zero-free functions in $C(X)$ are (in the sense of that word in algebra) just the *units* of the ring $C(X)$, that is, exactly those elements $e \in C(X)$ for which there exists a (necessarily unique) $\hat{e} \in C(X)$ satisfying $e\hat{e} = 1$.

3. Historical remarks on the concept of function. During the Leibniz and Euler period it was predominantly *real-valued functions of real variables* which were studied. From them mathematicians slowly groped toward the idea of *complex-valued functions of complex variables*. Thus in 1748 EULER intended his famous formula $e^{iz} = \cos z + i \sin z$ ([E],§138) only for real values of z. GAUSS was the first to see clearly - as his letter to BESSEL shows - that many properties of the classical functions are only fully understood when complex arguments are allowed. (Cf. Section 1 of the Historical Introduction.)

The word "function" occurs in 1692 with LEIBNIZ as a designation for certain magnitudes (like abscissæ, radii of curvature, etc.) which depend on the points of a curve, these points thought of as changing or varying. As early as 1698 in a letter to LEIBNIZ, Joh. BERNOULLI spoke of "*beliebigen Funktionen der Ordinaten* (arbitrary functions of the ordinates)" and in 1718 he designated as function any "*aus einer Veränderlichen und irgendwelchen Konstanten zusammengesetzte Größe* (magnitude which is built up from a variable and any constants whatsoever)". In his *Introductio* [E] EULER called any *analytic expression*, involving a variable and constants, a function.

Extension of the function concept was made necessary by the investigations of D'ALEMBERT, EULER, Daniel BERNOULLI and LAGRANGE on the problem of the vibrating string; thus EULER was led to abandon the idea of an *a priori* analytic expression and to introduce so-called *arbitrary functions*. Nevertheless it was only through the efforts of DIRICHLET that the presently accepted definition of function as unambiguous assignment of values or *mapping* became established. In 1829 in his paper, *Sur la convergence des séries trigonométriques qui servent a représenter une fonction arbitraire...* (English translation by R. FUJISAWA in *Memoirs on Infinite Series*, Tokio Mathematical and Physical Society (1891), Tokyo) he presented a function $\varphi(x)$ which is "égale à une constante déterminée c lorsque la variable x obtient une valeur rationelle, et égale à une autre constante d, lorsque cette variable est irrationelle (equal to a certain constant c whenever the variable x takes on a rational value and equal to another constant d whenever this variable is irrational)" - see his *Werke* **1**, p.132. And in his paper of 1837 he wrote, concerning the extent of the function concept *Über die Darstellung ganz willkürlicher Funktionen durch Sinus- und Cosinusreihen:* "Es ist gar nicht nöthig, daß $f(x)$ im ganzen Intervalle nach demselben Gesetze von x abhängig sei, ja man braucht nicht einmal an eine durch mathematische Operationen ausdrückbare Abhängigkeit zu denken. (It is certainly not necessary that the law of dependence of $f(x)$ on x be the same throughout the interval; in fact one need not even think of the dependence as given by explicit mathematical operations.)" - see his *Werke* 1, p.135. On p.227 ff. of his 1854 Habilitationsschrift cited in 4.4 RIEMANN gave a detailed discussion of the historical development of the function concept up to that time. An interesting survey that updates this is Dieter RÜTHING's "Some definitions of the concept of function from Joh. Bernoulli to N. Bourbaki," *Math. Intelligencer* **6**, no. 4 (1984), 72-77.

4. Historical remarks on the concept of continuity. LEIBNIZ and EULER used (intuitively) a very strong notion of continuity: for them *continuous* amounted almost to *analytic* or *generated by analytic functions*. (On this point see, for example, C. TRUESDELL, "The rational mechanics of flexible or elastic bodies", in the Comments on Euler's Mechanics, Euler's *Opera Omnia* (2) **11**, part 2, especially pp.243-249.) The presently accepted definition and its very precise arithmetic formulation had to wait until the work of BOLZANO, CAUCHY and WEIERSTRASS in the 19th century. Even in 1837 DIRICHLET gave a definition (*Werke* **1**, p.135) which says that "sich $f(x)$ mit x ebenfalls allmählich verändert ($f(x)$ changes gradually when x does so)." In the 20th century, starting already with HAUSDORFF on p.359 of his 1914 book *Grundzüge der Mengenlehre*, the idea of continuous mappings between topological spaces has become a matter of course.

LEIBNIZ believed that a continuity principle underlay all the laws of nature. The law of continuity "Natura non facit saltus" runs like a red thread through all his work in philosophy, physics and mathematics. In the *Initia rerum Mathematicarum metaphysica* (*Math. Schriften* **VII**, 17-29) it says: "... Kontinuität aber kommt der Zeit wie der Ausdehnung, den Qualitäten wie den Bewegungen, überhaupt aber jedem Übergange in der Natur zu, da ein solcher niemals sprungweise vor sich geht (... Continuity however is attributable to time as much as to spatial extension, to qualities just as to motion, actually to every transition in nature, since these never proceed by leaps)." LEIBNIZ applied his continuity principle also, for example, to *biology* and in this seems to have anticipated DARWIN somewhat; in a letter to VARIGNON he writes: "Die zwingende Kraft des Kontinuitätsprinzip steht für mich so fest, daß ich nicht im geringsten über die Entdeckung von *Mittelwesen* erstaunt wäre, die in manchen Eigenthümlichkeiten, etwa in ihrer Ernährung und Fortpflanzung, mit ebenso großem Rechte als Pflanzen wie als Tiere gelten können ... . (The continuity principle carries such conviction for me that I wouldn't in the least be astonished at the discovery of intermediate life-forms many of whose characteristics, like their methods of feeding and reproduction, would give them equal claim to being plants or animals ...)." The continuity postulate later became known as Leibniz' dogma.

Exercises

Exercise 1. Let X, Y be metric spaces, $f : X \to Y$ a mapping.

    a) Show that f is continuous exactly when $f(\overline{A}) \subset \overline{f(A)}$ for every subset A of X.

b) f is called a *homeomorphism* if it is bijective and both f and f^{-1} are continuous. Show that a bijection f is a homeomorphism if and only if $f(\overline{A}) = \overline{f(A)}$ for every subset A of X.

Exercise 2. Let X be a metric space and $f : X \to \mathbb{R}$ be a mapping. Show that f is continuous exactly when, for every $b \in \mathbb{R}$, the pre-images $f^{-1}((-\infty, b))$ and $f^{-1}((b, \infty))$ are open in X. Is there any similar criterion for mappings from X into $\mathbb{C}$?

Exercise 3. Let X, Y be metric spaces with metrics d_X, d_Y and let $f : X \to Y$ be continuous. Show that if X is *compact*, then f is *uniformly continuous*, that is, for each $\varepsilon > 0$ there exists a $\delta(\varepsilon) > 0$ such that $d_Y(f(u), f(v)) < \varepsilon$ whenever $u, v \in X$ satisfy $d_X(u, v) < \delta(\varepsilon)$.

§6   Connected spaces. Regions in $\mathbb{C}$

In 1851 RIEMANN introduced the concept of connectedness in his dissertation ([R],p.9) as follows:

"Wir betrachten zwei Flächentheile als zusammenhängend oder Einem Stücke angehörig, wenn sich von einem Punkt des einen durch das Innere der Fläche eine Linie nach einem Punkte des andern ziehen lässt. (We consider two parts of a surface as being connected or as belonging to a single piece, if from a point of the one a curve can be drawn in the interior of the surface to a point of the other.)"

In contemporary language this is the concept of path-connectedness. Since the evolution of set-theoretic topology at the beginning of the 20th century a more general notion of connectedness, which contains Riemann's as a special case, has emerged. Both concepts can be used to advantage in function theory and will be discussed in this section.

X and Y will always denote metric spaces. For $a, b \in \mathbb{R}$ with $a \leq b$, $[a, b]$ denotes the compact interval $\{x \in \mathbb{R} : a \leq x \leq b\}$ in $\mathbb{R}$.

1. Locally constant functions. Connectedness concept. A function $f : X \to \mathbb{C}$ is called *locally constant in* X if every point $x \in X$ lies in some neighborhood $U \subset X$ such that $f|U$ is constant. Generally locally constant functions are not constant. For example, if B_0, B_1 are two disjoint open balls in some metric space, $X := B_0 \cup B_1$ and $f : X \to \mathbb{C}$ is the function which throughout B_j has the value j $(j = 0, 1)$, then f is locally constant in X but not constant. And this example is representative of a general

Theorem. *For any metric space X the following are equivalent:*

i) *Every locally constant function $f : X \to \mathbb{C}$ is constant.*

ii) *The only non-empty subset of X which is both open and closed is X itself.*

Proof. i) $\Rightarrow$ ii)   Suppose A is a non-empty subset of X which is both open and closed. (The neologism "clopen" is a popular abbreviation for "both closed and open".) Then its "characteristic" or "indicator" function $1_A : X \rightarrow \mathbb{C}$ defined by

$$1_A(x) := 1 \text{ for } x \in A \, , \, 1_A(x) := 0 \text{ for } x \in X \setminus A$$

is locally constant, since both A and $X \setminus A$ are open. Therefore 1_A is constant. Since $A \neq \emptyset$, the constant value must be 1, whence $X \setminus A = \emptyset$, $A = X$.

ii) $\Rightarrow$ i)   Fix $c \in X$. The fiber $A := f^{-1}(f(c))$ is non-empty and is open in X because f is locally constant. Because locally constant functions are trivially continuous, A is also closed in X. It follows that $A = X$, that is, $f(x) = f(c)$ for all $x \in X$. $\qquad\qquad\qquad\qquad\qquad\qquad\qquad\qquad\square$

Mathematical experience has shown that the equivalent properties i) and ii) of the preceding theorem optimally capture the intuitively clear yet vague conception of a space being connected. So we make these properties into the following definition: a metric space X is defined to be *connected* if it has properties i) and ii). A theorem which is immediate from this definition is that a continuous mapping $f : X \rightarrow Y$ from a connected space X has a connected image set $f(X)$. From calculus we borrow the important

Theorem. *Each closed interval and each open interval in the real number line $\mathbb{R}$ is connected.*

2. Paths and path connectedness. Any continuous mapping $\gamma : [a, b] \rightarrow X$ of a closed interval in $\mathbb{R}$ into a metric space X is called a *path in X* from the *initial point* $\gamma(a)$ to the *terminal point* $\gamma(b)$; we say that γ *joins* the points $\gamma(a)$ and $\gamma(b)$ *in X*. Paths are also called *curves*. A path is called *closed* if its initial and terminal points coincide. The image set $|\gamma| := \gamma([a, b]) \subset X$ is called the *trajectory* or the *trace* or the *impression* of the curve. Because γ is continuous, $|\gamma|$ is a compact set. A path is more than just its trajectory: the latter is traversed according to the law $\gamma(t)$ [t thought of as a time parameter]. Nevertheless it is convenient to allow the abuse of language whereby we sometimes write γ for $|\gamma|$.

If $\gamma_j : [a_j, b_j] \rightarrow X$, $j = 1, 2$, are paths in X and if the terminal point $\gamma_1(b_1)$ of γ_1 coincides with the initial point $\gamma_2(a_2)$ of γ_2, then the *path-sum* $\gamma_1 + \gamma_2$ *of γ_1 and γ_2 in X* is defined as the continuous mapping

$$\gamma : [a_1, b_2 - a_2 + b_1] \rightarrow X, \quad t \mapsto \begin{cases} \gamma_1(t) & \text{for } t \in [a_1, b_1] \\ \gamma_2(t + a_2 - b_1) & \text{for } t \in [b_1, b_2 - a_2 + b_1]. \end{cases}$$

The path-sum $\gamma_1 + \gamma_2 + \cdots + \gamma_n$ of finitely many paths $\gamma_1, \gamma_2, \ldots, \gamma_n$ (whose initial and terminal points are appropriately related) is defined correspondingly. One verifies immediately that *path-addition is associative* so that, in fact, no parentheses need be used. Path-addition is naturally *not commutative*.

A space X is called *path-connected* if, for every pair of points $p, q \in X$, there is a path γ in X with initial point p and terminal point q. This would be an infelicitous choice of language were it not for the fact that

Every path-connected space X is connected.

Proof. Let $U \neq \emptyset$ be a subset of X which is both open and closed. Fix $p \in U$ and consider any $q \in X$. Let the path $\gamma : [a, b] \to X$ join p to q. Since γ is continuous, $\gamma^{-1}(U)$ is an open and closed subset of the real interval $[a, b]$ and it is non-empty (because it contains $a \in \gamma^{-1}(p)$). Because $[a, b]$ is connected, it follows that $\gamma^{-1}(U) = [a, b]$ and so $q = \gamma(b) \in U$. This shows that $X \subset U$, so $U = X$. $\square$

The converse of the fact just proved is not true: As an example, consider the space $X := \{iy : -1 \leq y \leq 1\} \cup \{z = x + iy : 0 < x \leq \frac{1}{4}, y = \sin(x^{-1})\}$ with the metric induced from $\mathbb{C}$. It is *connected* but not *path-connected*, as the reader is encouraged to convince himself via a sketch.

3. Regions in $\mathbb{C}$. The path in $\mathbb{C}$, $\gamma : [0, 1] \to \mathbb{C}$, defined by $\gamma(t) := (1 - t)z_0 + tz_1$ is called the *(line) segment from z_0 to z_1* and is designated by $[z_0, z_1]$. Intervals $[a, b]$ in $\mathbb{R}$ are segments via $t \mapsto (1 - t)a + tb$. A *polygon* or *polygonal path from $p \in \mathbb{C}$ to $q \in \mathbb{C}$* is a finite sum $P = [z_0, z_1] + [z_1, z_2] + \cdots + [z_n, z_{n+1}]$ of segments with $z_0 = p$ and $z_{n+1} = q$. P is called *axis-parallel* if each segment is parallel to one of the two coordinate axes, that is, if for every ν either $\Re z_\nu = \Re z_{\nu+1}$ or $\Im z_\nu = \Im z_{\nu+1}$. Of course every polygon *is* a path.

Non-empty open sets in $\mathbb{C}$ are called *domains* and will be denoted by D throughout.

Theorem. *The following statements concerning domains $D \subset \mathbb{C}$ are equivalent:*

i) *D is connected.*

ii) *For every pair of points $p, q \in D$ there is an axis-parallel polygon $P \subset D$ from p to q.*

iii) *D is path-connected.*

Proof. i) $\Rightarrow$ ii)   Fix some $p \in D$. We define a function $f : D \to \mathbb{C}$ as follows: $f(w) := 1$ if there exists an axis-parallel polygon in D from p to w and otherwise $f(w) := 0$. Now consider any open disc $B \subset D$. For every pair of points z, $w \in B$ there is evidently an axis-parallel polygon P_{zw} in B from z to w. If $f(z) = 1$ for at least one point $z \in B$, then $f(w) = 1$ for all $w \in B$. From this it is clear that either $f|B = 1$ or $f|B = 0$. This shows that f is locally-constant and so, by connectedness of D, f is constant. Since $f(p) = 1$, this constant value is $1 : f(w) = 1$ for all $w \in D$, that is, for every point $w \in D$ there is an axis-parallel polygon in D from p to w. As p is arbitrary in D, this establishes the path-connectedness of D.

ii) $\Rightarrow$ iii)   This is trivial because axis-parallel polygons are paths.

iii) $\Rightarrow$ i)   Clear from 2.   □

Remark. The openness of D in $\mathbb{C}$ was only used in the (non-constructive!) proof that i) $\Rightarrow$ ii) and there only to ensure that every point of D have a path-connected neighborhood (namely, a disc, in which in fact any two points can even be joined by an axis-parallel polygon). Spaces with this property are called *locally path-connected*. Evidently then our proof of i) $\Rightarrow$ ii) establishes the more general fact that:

Every connected and locally path-connected space, e.g., every connected domain in $\mathbb{R}^n, 1 \leq n < \infty$, is path-connected.

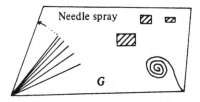

Connected domains in $\mathbb{C}$ are called *regions* and are traditionally denoted with G (after the corresponding German word *Gebiet*). Thus in a region $G \subset \mathbb{C}$ every pair of points can be joined by an axis-parallel polygon in G. However, regions may appear very complicated, containing for example many "needle spray" and "spiral" excisions, as in the figure above. Of course *all discs $B_r(c)$, as well as $\mathbb{C}$ and $\mathbb{C}^\times$, are regions.*

Regions play a much more important role in function theory than in real analysis. It will later be shown, after we have available the Identity Theorem, that the *topological property of connectedness of a domain $D \subset \mathbb{C}$* is equivalent to the *algebraic* property that the ring $\mathcal{O}(D)$ of holomorphic functions on D have no zero-divisors.

4. Connected components of domains. Two points $p, q \in D$ are called "path-equivalent" if there is a path in D from p to q. In this way an equivalence relation is defined in D. The associated equivalence classes are called *connected components* (or simply *components*) of D. The terminology is justified by the fact that

Every component G of D is a region in $\mathbb{C}$ and D has at most countably many components.

Proof. a) Consider $c \in G$. Since D is open, there is an $r > 0$ such that $B_r(c) \subset D$. But every point of $B_r(c)$ is equivalent to c (!), from which $B_r(c) \subset G$ follows. Thus G is open. By definition G is path-connected, hence a region.

b) Every domain D contains a countable dense subset; e.g., the rational complex numbers in D constitute such a set. The set of all open discs centered at these points and having rational radii is a countable cover $U_0, U_1, \cdots$ of D by path-connected sets U_j. Thus each U_j lies in a unique component G_j and every component G of D contains some point of the dense set (because G is open) and therefore also contains some U_k. The map $j \mapsto G_j$ therefore sends $\mathbb{N}$ *onto* (though not necessarily injectively) the set of all components of D. □

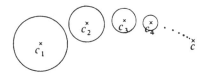

The figure above illustrates a *bounded* domain with *infinitely many* components: The centers c_n converge to c and the respective radii r_n are so small that the various discs $U(c_j) := B_{r_j}(c_j)$ are disjoint. The components of $D := \bigcup_{j=1}^{\infty} U(c_j)$ are exactly the discs $U(c_j)$. □

Everything just said about components is equally valid for domains in $\mathbb{R}^n$ and even more generally for any *locally connected space* which has a countable dense subset.

5. Boundaries and distance to the boundary. If D is a domain in $\mathbb{C}$, the set

$$\partial D := \overline{D} \setminus D$$

is called *the boundary of D* and the points of ∂D are called *boundary points of D*. The boundary ∂D is always *closed* in $\mathbb{C}$. For discs we have $\partial B_r(c) = \{z \in \mathbb{C} : |z - c| = r\}$. A point $a \in \mathbb{C} \setminus D$ is a boundary point of D precisely if $a = \lim z_n$ for some sequence of $z_n \in D$. Notice that unless $D = \emptyset$ or $D = \mathbb{C}$, the set ∂D is always non-empty: otherwise we would have $D = \overline{D}$, so D would be both open and closed in the connected space $\mathbb{C}$. If D is a non-void, proper subset of $\mathbb{C}$, then for every point $c \in D$ we may define

$$d_c(D) := \inf\{|c - z| : z \in \partial D\}.$$

This number is *positive* and is called the *boundary-distance of c in D*. For $D = \mathbb{C}$ we set $d_c(\mathbb{C}) := \infty$ for all c. The (extended) number $d := d_c(D)$

is the maximum radius r such that the disc $B_r(c)$ lies wholly in D. When $d < \infty$ (i.e., $D \neq \mathbb{C}$) there is at least one boundary point p of D on $\partial B_d(c)$ (cf. the figure).

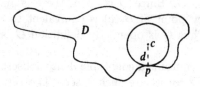

The boundary-distance will play an important role in the development of holomorphic functions into power series (Chapter 7.3.2).

Exercises

Exercise 1 (distance between sets). For non-empty subsets M, N of $\mathbb{C}$ define

$$d(M, N) := \inf\{|z - w| : z \in M, \quad w \in N\}.$$

Show that if A is closed, K compact, then there exists $z \in A$, $w \in K$ such that $|z - w| = d(A, K)$. In particular, if in addition $A \cap K = \emptyset$, then $d(A, K) > 0$.

Exercise 2. Let $f : D \to D'$ be a mapping between domains in $\mathbb{C}$. Show that:

a) If W is a component of D, then $f(W)$ is contained in a unique component of D'.

b) If additionally, f is a homeomorphism, then $f(W)$ is a component of D'.

Exercise 3. A subset M of $\mathbb{C}$ is called *convex* if for any points $w, z \in M$ the whole segment $[w, z]$ joining them lies in M. Convex sets are thus path-connected.

a) Show that any intersection of convex subsets of $\mathbb{C}$ is a convex set.

b) Let G be a region in $\mathbb{C}$. Show that $\overline{G}$ is convex if G is. What about the converse?

c) Let G be a convex region in $\mathbb{C}$ and $c \in \partial G$. Show that there is a (real) line L in $\mathbb{C}$ with $c \in L$ and $L \cap G = \emptyset$.

Exercise 4. Let D be a domain in $\mathbb{C}$ and V an open, convex n-gon ($n \geq 3$) which, together with its boundary, lies in D. Show that there exists an open convex n-gon V' in $\mathbb{C}$ such that $\overline{V} \subset V' \subset D$.

Chapter 1

Complex-Differential Calculus

A cornerstone of our thinking is that in the infinitely small every function becomes linear (from an unknown mathematical physicist, 1915).

1. The adage leading off this chapter is the kernel of all differential calculi. Notwithstanding that this cornerstone was pulverized by Riemann's and Weierstrass' discovery of (real-valued) everywhere continuous nowhere differentiable functions on $\mathbb{R}$, it is still a valuable principle for creative mathematicians and physicists.

The concept of complex-differentiability will be introduced exactly as was that of differentiability on $\mathbb{R}$. Complex-valued functions which are complex-differentiable throughout a domain in $\mathbb{C}$ are called *holomorphic* and *function theory* is usually understood to be the study of such functions. None of the early works on this subject studied holomorphic functions *per se* but rather the theory was initially fuelled by the rich legacy of special functions bequeathed by the Euler era. The first works to treat function theory as an independent mathematical discipline originated with CAUCHY, although even he had no plan for founding a general theory of complex-differentiable functions. The main concern of his first great treatise [C₁] *Mémoire sur les intégrales définies* from the year 1814 as well as that of his second, considerably shorter paper [C₂] *Mémoire sur les intégrales définies, prises entre des limites imaginaires* from the year 1825, is, as these titles indicate, the integral calculus in $\mathbb{C}$; scarcely anything is said in them, consciously anyway, about complex differential calculus—it is just used, uncritically. With Eulerian prescience CAUCHY differentiates functions of a complex variable according to the rules known for functions on $\mathbb{R}$; he implicitly makes use of the existence and continuity of the first derivative. (We will have more to say on this point in 7.1.3–7.1.5.)

2. CAUCHY's *Cours d'analyse* of 1821 (see [C]) prepared the way for function theory. Here we recognize very clearly a striving to free the function concept from the restrictions of "effective representation." Today we can scarcely understand the conceptual difficulties which had to be overcome at that time. Even 30 years later in 1851 we find RIEMANN emphasizing repeatedly in his dissertation that he is concerned with complex-differentiable "Funktionen einer veränderlichen complexen Grösse unabhängig von einem Ausdruck für dieselben (functions of a complex variable independently of any particular expression for them)"; he gives the following definition of holomorphic function ([R],p.5): "Eine veränderliche complexe Grösse w heißt eine Function einer anderen veränderlichen complexen Grösse z, wenn sie mit ihr sich so ändert, dass der Werth des Differentialquotienten $\frac{dw}{dz}$ unabhängig von dem Werthe des Differentials dz ist. (A variable complex quantity w is called a function of another variable complex quantity z if it changes with z in such a manner that the value of the differential quotient $\frac{dw}{dz}$ is independent of the value of the differential dz.)"

For a long time now it has been customary in real and complex analysis not to treat differential and integral calculus together and simultaneously but rather one after the other and (for reasons of economy or pedagogy (?)) to begin with differential calculus. We will also proceed in this way and first discuss in some detail the fundamental concept of complex-differentiability. At the center of our considerations are the famous CAUCHY–RIEMANN differential equations

$$\frac{\partial u}{\partial x} = \frac{\partial v}{\partial y} \quad \text{and} \quad \frac{\partial u}{\partial y} = -\frac{\partial v}{\partial x}$$

for complex-differentiable functions $f = u + iv$, together with the interpretation of these equations as affirming the $\mathbb{C}$-linearity of the differential of f. These seem to reduce the theory of complex-differentiable functions to being a part of the theory of real partial differential equations. Nevertheless, many complex analysts feel that methods of real analysis should be proscribed and in the present volume we will more or less conform to this principle of methodological purity; all the more readily since the path through the reals is often the more arduous one. (For example, it is very tedious to derive the differentiation rules by reduction to the real case.)

3. In Sections 1, 2 and 3 of this chapter the usual material of complex differential calculus will be treated. In doing so emphasis will be laid on the fact that the *complex-differentiable functions* are precisely the *real-differentiable functions which have a $\mathbb{C}$-linear* (and not just an $\mathbb{R}$-linear) *differential*; the Cauchy-Riemann differential equations describe nothing more or less than this complex linearity (Theorem 2.1). In Section 4 we define and discuss the partial derivatives with respect to z and $\bar{z}$. This makes it possible to organize the *two* Cauchy-Riemann equations into the single equation

$$\frac{\partial f}{\partial \bar{z}} = 0;$$

in 4.4 we will go a little deeper into the technique of differentiation with respect to z and $\bar{z}$.

§1 Complex-differentiable functions

Just as in the real case, complex-differentiability will be reduced to a continuity issue by a *linearization condition*. But we also want to see that complex-differentiability is far more than a mere analog of real-differentiability. A simple discussion of difference quotients will lead immediately to the Cauchy-Riemann equations

$$u_x = v_y \text{ and } u_y = -v_x \quad \text{for } f = u + iv$$

from which will follow, in particular, that the real and imaginary parts of a complex-differentiable function satisfy Laplace's potential-equation $\Delta u = 0$ and are consequently *harmonic* functions.

1. Complex-differentiability. A function $f : D \to \mathbb{C}$ is called *complex-differentiable* at $c \in D$ if there exists a funtion $f_1 : D \to \mathbb{C}$ which is continuous at c and satisfies

$$f(z) = f(c) + (z - c)f_1(z) \quad \text{for all } z \in D \quad (\mathbb{C}\text{-}linearization).$$

Such a function f_1, if it exists at all, is *uniquely determined* by f:

$$f_1(z) = \frac{f(z) - f(c)}{z - c} \quad \text{for } z \in D \setminus \{c\} \quad (\textit{difference quotient})$$

and then, upon setting $h := z - c$, the continuity of f_1 at c entails that

$$\lim_{h \to 0} \frac{f(c + h) - f(c)}{h} = f_1(c) \quad (\textit{differential quotient}).$$

The number $f_1(c) \in \mathbb{C}$ is called the *derivative* (*with respect to* z) *of* f *at* c; we write

$$\frac{df}{dz}(c) := f'(c) := f_1(c).$$

Complex-differentiability of f *at* c *implies continuity of* f *at* c, since f is a sum of products of the functions $f(c)$, $z - c$ and f_1 which are all continuous at c. The fact that formally the definition of differentiability is the same for real and complex variables will, in 3.1, immediately produce the expected differentiation rules.

One proves directly that

> If f is complex-differentiable at c, then for every $\varepsilon > 0$ there is a $\delta > 0$ such that $|f(c+h) - f(c) - f'(c)h| \leq \varepsilon|h|$ for all $h \in \mathbb{C}$ satisfying $|h| \leq \delta$.

Examples. 1) Each power function $z^n, n \in \mathbb{N}$, is complex-differentiable everywhere in $\mathbb{C}$:

$$z^n = c^n + (z - c)f_1(z) \quad \text{with } f_1(z) := z^{n-1} + cz^{n-2} + \cdots + c^{n-2}z + c^{n-1}$$

shows that $(z^n)' = nz^{n-1}$ for every $z \in \mathbb{C}$. More generally, all polynomials $p(z) \in \mathbb{C}[z]$ are everywhere complex-differentiable and rational functions $g(z) \in \mathbb{C}(z)$ are complex-differentiable at every point where they are defined, i.e., at every point not a zero of the denominator (cf. 3.2).

2) The conjugation function $f(z) := \bar{z}$, $z \in \mathbb{C}$, is not complex-differentiable at any point, because the difference quotient

$$\frac{f(c+h) - f(c)}{h} = \frac{\bar{h}}{h}, \quad h \neq 0$$

at $c \in \mathbb{C}$ has the value 1 for $h \in \mathbb{R}$ and the value -1 for $h \in i\mathbb{R}$, and consequently has no limit.

3) The functions $\Re z, \Im z, |z|$ are complex-differentiable *nowhere* in $\mathbb{C}$. This is shown by first considering real and then purely imaginary h, as in 2).

2. The Cauchy-Riemann differential equations. We write $c = a + ib = (a, b)$, $z = x + iy = (x, y)$. If $f(z) = u(x, y) + iv(x, y)$ is complex-differentiable, then

$$f'(c) = \lim_{h \to 0} \frac{f(c+h) - f(c)}{h} = \lim_{h \to 0} \frac{f(c+ih) - f(c)}{ih}.$$

Choosing h *real*, it follows that

$$
\begin{aligned}
f'(c) &= \lim_{h \to 0} \frac{u(a+h, b) - u(a, b)}{h} + i \lim_{h \to 0} \frac{v(a+h, b) - v(a, b)}{h} \\
&= \lim_{h \to 0} \frac{u(a, b+h) - u(a, b)}{ih} + i \lim_{h \to 0} \frac{v(a, b+h) - v(a, b)}{ih}.
\end{aligned}
$$

Thus the *partial* derivatives with respect to x and y of the real-valued functions u and v all exist at the point c; and, using the usual notations $u_x(c), \cdots, v_y(c)$ for these derivatives, the equations

$$f'(c) = u_x(c) + iv_x(c) = \frac{1}{i}(u_y(c) + iv_y(c))$$

obtain.

With this we have proven:

A necessary condition for the complex-differentiability of $f = u + iv$ at c is that the real part u and the imaginary part v of f each be differentiable with respect to x and with respect to y and that the "Cauchy-Riemann differential equations"

$$(*) \qquad u_x(c) = v_y(c), \qquad u_y(c) = -v_x(c)$$

obtain. When this happens, $f'(c) = u_x(c) + iv_x(c) = v_y(c) - iu_y(c)$.

The equations $(*)$ are just the analytic manifestation of the geometric insight that the difference quotient of f has to have the same limiting value for approach to c parallel to the real axis and for approach parallel to the imaginary axis. Deeper aspects of this *naive* but *mnemonically useful* approach to the Cauchy-Riemann equations will be examined in 2.1.

3. Historical remarks on the Cauchy-Riemann differential equations. CAUCHY [C₁] obtained the equations in 1814 while discussing the interchange of the order of integration in a real double integral (on the top of p.338 in a more general form, on the bottom of p.339 in the well-known form). He emphasizes (p.338) that his differential equations contain the whole theory of the passage from reals to complexes: "Ces deux équations renferment toute la théorie du passage du réel à l'imaginaire, et il ne nous reste plus qu'à indiquer la manière de s'en servir." Nevertheless, CAUCHY did not make these equations the foundation of his function theory.

RIEMANN put these differential equations at the beginning of his function theory and consistently built on them. He recognized "in der partiellen Differentialgleichung die wesentliche Definition einer [komplex differenzierbaren] Function von einer complexen Veränderlichen ... Wahrscheinlich sind diese, für seine ganze spätere Laufbahn maassgebenden Ideen zuerst in den Herbstferien 1847 [als 21-jähriger] gründlich von ihm verarbeitet (in the partial differential equation the essential definition of a [complex-differentiable] function of a complex variable ... Probably these ideas, which were decisive for the course of his whole later life, were first worked out by him during the autumn holidays of 1847 [as a 21-year-old])" – quoted from R. DEDEKIND: *Bernhard Riemann's Lebenslauf* (p.544 of Riemann's *Werke*). However, neither CAUCHY nor RIEMANN was the first to discover these equations; they occur previously in 1752 in D'ALEMBERT's theory of fluid flow *Essai d'une nouvelle théorie de la résistance des fluides* (David, Paris); and in the work of EULER and LAGRANGE.

In 1851 RIEMANN argues succinctly ([R], pp.6,7) thus: "Bringt man den Differentialquotienten

$$(1) \qquad \frac{du + dvi}{dx + dyi}$$

in die Form

$$(2) \qquad \frac{\left(\frac{\partial u}{\partial x} + \frac{\partial v}{\partial x}i\right) dx + \left(\frac{\partial v}{\partial y} - \frac{\partial u}{\partial y}i\right) dyi}{dx + dyi},$$

so erhellt, dass er und zwar nur dann für je zwei Werthe von dx und dy denselben Werth haben wird, wenn

$$(3) \qquad \frac{\partial u}{\partial x} = \frac{\partial v}{\partial y} \quad \text{and} \quad \frac{\partial v}{\partial x} = -\frac{\partial u}{\partial y}$$

ist. Diese Bedingungen sind also hinreichend und nothwendig, damit $w = u + vi$ eine Function von $z = x + yi$ sei. Für die einzelnen Glieder dieser Function fliessen aus ihnen die folgenden:

$$(4) \qquad \frac{\partial^2 u}{\partial x^2} + \frac{\partial^2 u}{\partial y^2} = 0 \;, \quad \frac{\partial^2 v}{\partial x^2} + \frac{\partial^2 v}{\partial y^2} = 0,$$

welche für die Untersuchung der Eigenschaften, die Einem Gliede einer solchen Function einzeln betrachtet zukommen, die Grundlage bilden. (If one brings the differential quotient (1) into the form (2), it becomes evident that it will have the same value for every two values of dx and dy, if and indeed only if the equations (3) hold. These latter conditions are thus necessary and sufficient in order that $w = u + vi$ be a function of $x + yi$. Out of them flow equations (4) for the respective terms of the function w. These equations form the basis for investigating the properties possessed by any one term of such a function when that term is considered individually.)"

§2 Complex and real differentiability

The customary graphic significance of the real derivative of a real function as the "slope of the tangent" is not feasible in the realm of the complexes, because the graph of a complex function $w = f(z)$ is a "surface" in the 4 real-dimensional complex (w, z)-space $\mathbb{C}^2$. There is nevertheless a geometric interpretation of the complex differential quotient $f'(c)$. To explicate it we need the fundamental concept of the real vector differential calculus:

A mapping $f : D \to \mathbb{R}^n$ of a domain $D \subset \mathbb{R}^m$ is called real-differentiable at the point $c \in D$, if for some $\mathbb{R}$-linear mapping $T : \mathbb{R}^m \to \mathbb{R}^n$

(1)
$$\lim_{h \to 0} \frac{|f(c + h) - f(c) - T(h)|}{|h|} = 0.$$

(Here $|\quad|$ represent fixed norms in $\mathbb{R}^m$ and $\mathbb{R}^n$.)

As is well known and elementary, T is then *uniquely determined* and is called *the differential $Tf(c)$*, sometimes also *the tangent mapping* of f at c. It is clear from (1) that real-differentiability at c entails continuity at c.

If bases are given in $\mathbb{R}^m$ and $\mathbb{R}^n$ and if the component functions of f in these bases are $f_\nu(x_1, \cdots, x_m)$, $1 \le \nu \le n$, then the real-differentiability of f at c implies that all the partial derivatives $\frac{\partial f_\nu}{\partial x_\mu}(c)$ exist, $1 \le \mu \le m$, $1 \le \nu \le n$ and that the differential $Tf(c)$ is implemented by writing elements of $\mathbb{R}^m$ as column vectors and multiplying them on the left by the $n \times m$ Jacobian matrix

$$\left(\frac{\partial f_\nu}{\partial x_\mu}(c) \right)_{\substack{\mu=1,\ldots,m \\ \nu=1,\ldots,n}}$$

1. Characterization of complex-differentiable functions.

We apply the theory from the real realm just sketched to complex-valued functions (thus $m = n = 2$ and $\mathbb{R}^2 = \mathbb{C}$). If $f : D \to \mathbb{C}$ is complex-differentiable at c, then (cf. 1.1)

$$\lim_{h \to 0} \frac{f(c + h) - f(c) - f'(c)h}{h} = 0.$$

From this and (1) it follows immediately that complex-differentiable mappings are real-differentiable and have $\mathbb{C}$-*linear* differentials. This $\mathbb{C}$-linearity of the differential is significant for complex-differentiability and is the deeper reason why the Cauchy-Riemann differential equations hold; we have namely, if we again set $z = x + iy$, $f = u + iv$:

Theorem. *The following statements about a function $f : D \to \mathbb{C}$ are equivalent:*

  i) *f is complex-differentiable at $c \in D$.*

 ii) *f is real-differentiable at c and the differential $Tf(c) : \mathbb{C} \to \mathbb{C}$ is complex-linear.*

iii) *f is real-differentiable at c and the Cauchy-Riemann equations $u_x(c) = v_y(c)$, $u_y(c) = -v_x(c)$ hold.*

If i) – iii) *prevail, then* $f'(c) = u_x(c) + iv_x(c) = v_y(c) - iu_y(c)$.

Proof. i) $\Leftrightarrow$ ii)   This is clear on the basis of the relevant definitions.
  ii) $\Leftrightarrow$ iii)   The differential $Tf(c)$ is given by the 2×2 matrix

$$\begin{pmatrix} u_x(c) & u_y(c) \\ v_x(c) & y_y(c) \end{pmatrix}$$

According to Theorem 0.1.2 the $\mathbb{R}$-linear mapping $\mathbb{C} \to \mathbb{C}$ determined by this matrix is $\mathbb{C}$-linear if and only if $u_x(c) = v_y(c)$ and $u_y(c) = -v_x(c)$.
The equation for $f'(c)$ was already proven in 1.2. $\square$

In order to be able to apply this theorem, we need a criterion for the real-differentiability of $f = u + iv$ at c. We will occupy ourselves with this question below.

2. A sufficiency criterion for complex-differentiability. Together with $f : D \to \mathbb{C}$ and $g : D \to \mathbb{C}$, all the mappings $af + bg : D \to \mathbb{C}$ $(a, b \in \mathbb{C})$ will be real-differentiable at c and from equation (1) in the introductory material of this section we infer that

$$(T(af + bg))(c) = a(Tf)(c) + b(Tg)(c).$$

Also, together with f, the conjugated function $\bar{f}$ will be real-differentiable at c and, if $Tf(c)(h)$ is given by $\lambda h + \mu \bar{h}$, then $T\bar{f}(c)(h)$ is given by $\bar{\mu} h + \bar{\lambda} \bar{h}$. It follows that

The function $f = u + iv : D \to \mathbb{C}$ *is real-differentiable at* $c \in D$ *if and only if each of the functions* $u : D \to \mathbb{R}$, $v : D \to \mathbb{R}$ *is real-differentiable at* c.

To prove this we just direct our attention to the equations $u = \frac{1}{2}(f + \bar{f})$, $v = \frac{1}{2i}(f - \bar{f})$ and to the fact that a real-valued function $D \to \mathbb{R}$ is real-differentiable at c if and only if the corresponding complex-valued function $D \to \mathbb{R} \hookrightarrow \mathbb{C}$ is. $\square$

A function $u : D \to \mathbb{R}$ is called *continuously (real-) differentiable* if the partial derivatives u_x, u_y exist throughout D and are continuous functions there. In the real-differential calculus it is shown with the help of the Mean Value Theorem that

Every continuously differentiable function $u : D \to \mathbb{R}$ *is real-differentiable at each point of* D.

The continuity requirement on u_x and u_y is essential here, as the well-known example $u(z) := xy|z|^{-2}$ for $z = (x, y) \neq (0, 0)$ and $u(0, 0) := 0$

shows. Here u_x and u_y exist everywhere with $u_x(0,0) = u_y(0,0) = 0$, yet u is not even continuous at $(0,0)$.

By means of Theorem 1 we now deduce a criterion which is quite handy in applications:

Sufficiency criterion for complex-differentiability. *If u, v are continuously differentiable real-valued functions in D, then the complex-valued function $f := u + iv$ is real-differentiable at every point of D.*

If furthermore $u_x = v_y$ and $u_y = -v_x$ throughout D, then f is complex-differentiable at every point of D.

This criterion is almost always called on when one wants to describe complex-differentiable functions via statements about their real and imaginary parts.

3. Examples involving the Cauchy-Riemann equations.

1) The function $f(z) := x^3y^2 + ix^2y^3$ is, according to 2, real-differentiable throughout $\mathbb{C}$. The Cauchy-Riemann equations hold at the point $c = (a, b)$ exactly when $3a^2b^2 = 3a^2b^2$ and $2a^3b = -2ab^3$, i.e., when $ab(a^2 + b^2) = 0$; which, since a, b are real, amounts to $ab = 0$. In summary, the points at which f is complex-differentiable are the points on the two coordinate axes.

2) We will assume the reader is acquainted with the *real exponential function* e^t and the *real trigonometric functions*, $\cos t, \sin t, t \in \mathbb{R}$. The function

$$\tilde{e}(z) := e^x \cos y + ie^x \sin y$$

is real-differentiable at every $z = x + iy$ in $\mathbb{C}$, by 2, and the Cauchy-Riemann equations clearly hold at every point. Thus $\tilde{e}(z)$ is complex-differentiable in $\mathbb{C}$ and $\tilde{e}'(z) = u_x(z) + iv_x(z) = \tilde{e}(z)$. In 5.1.1 we will see that $\tilde{e}(z)$ is the *complex exponential function* $\exp z = \sum_0^\infty \frac{z^\nu}{\nu!}$.

3) This example is for readers acquainted with the *real logarithm function* $\log t, t > 0$, and the *real arctangent function* $\arctan t, t \in \mathbb{R}$; the notation refers to the *principal branch*, that is, the values of arctangent lying between $-\pi/2$ and $\pi/2$. From 2 and the properties of these functions, namely the identities $\log'(t) = t^{-1}$ and $\arctan'(t) = (1 + t^2)^{-1}$, we see that

$$\tilde{\ell}(z) := \frac{1}{2}\log(x^2 + y^2) + i\arctan\frac{y}{x}$$

is real-differentiable throughout $\mathbb{C} \setminus \{z \in \mathbb{C} : \Re z = 0\}$ and satisfies the Cauchy-Riemann equations there as well. Thus $\tilde{\ell}(z)$ is complex-differentiable everywhere to the left and to the right of the imaginary axis. A direct calculation shows that

$$\tilde{\ell}'(z) = u_x(z) + iv_x(z) = \frac{1}{z}, \qquad z \in \mathbb{C} \text{ with } \Re z \neq 0.$$

In 5.4.4 we will see that $\tilde{\ell}(z)$ coincides in the right half-plane with the principal branch of the *complex logarithm function* and that

$$\tilde{\ell}(z) = \log z = \sum_{1}^{\infty} \frac{(-1)^{\nu-1}}{\nu}(z-1)^{\nu} \quad \text{for } z \in B_1(1).$$

In the last two examples complex-differentiable functions were fashioned out of transcendental real functions with the help of the Cauchy-Riemann equations. However, in this book — as in classical function theory generally — this mode of constructing complex-differentiable functions will not be pursued any further.

4) If $f = u + iv$ is *complex-differentiable in D, then throughout D*

$$|f'|^2 = \det\begin{pmatrix} u_x & u_y \\ v_x & v_y \end{pmatrix} = u_x^2 + v_x^2 = u_y^2 + v_y^2,$$

a fact which follows from $|f'|^2 = f'\overline{f'} = u_x^2 + v_x^2$ on account of $u_x = v_y$, $u_y = -v_x$. $|f'(z)|^2$ is thus the value of the *Jacobian functional determinant* of the mapping $(x,y) \mapsto (u(x,y), v(x,y))$; this determinant is *never negative* and in fact is positive at every point $z \in D$ where $f'(z) \neq 0$. In example 2) we see, e.g., that

$$|\tilde{e}'(z)|^2 = e^{2x}\cos^2 y + e^{2x}\sin^2 y = e^{2\Re z}.$$

4*. Harmonic functions. Not all real-valued real-differentiable functions $u(x,y)$ occur as real parts of complex-differentiable functions. The Cauchy-Riemann equations lead at once to a quite restrictive necessary condition on u for this to happen. To formulate it, recall that the twice continuous (real-) differentiability of u in D means that the partial derivatives u_x and u_y are differentiable and the four second-order partial derivatives $u_{xx}, u_{xy}, u_{yx}, u_{yy}$ are continuous in D. As a consequence of this continuity $u_{xy} = u_{yx}$ in D, another well-known fact from the real differential calculus, often proved via the Mean Value Theorem. Now the aforementioned necessary condition reads

Theorem. *If $f = u + iv$ is complex-differentiable in D and if u and v are twice continuously real-differentiable in D, then*

$$u_{xx} + u_{yy} = 0 , \; v_{xx} + v_{yy} = 0 \quad \text{in } D.$$

Proof. Because f is complex-differentiable throughout D, $u_x = v_y$ and $u_y = -v_x$ in D. More partial differentiation yields $u_{xx} = v_{yx}, u_{xy} = v_{yy}$, $u_{yy} = -v_{xy}, u_{yx} = -v_{xx}$. It follows that $u_{xx} + u_{yy} = v_{yx} - v_{xy}$ and $v_{xx} + v_{yy} = -u_{yx} + u_{xy}$ in D. Since all the second-order partial derivatives of u and v are continuous in D, we have, as noted above, that $u_{xy} = u_{yx}$ and $v_{xy} = v_{yx}$; and so the claimed equalities follow. $\square$

The supplemental assumption of the twice continuous differentiability of u and v in this theorem is actually superfluous, because it turns out that every complex-differentiable function is infinitely often complex-differentiable (cf. 7.4.1).

In the literature the differential polynomial

$$\Delta := \frac{\partial^2}{\partial x^2} + \frac{\partial^2}{\partial y^2}$$

is known as the *Laplace operator*. For every twice real-differentiable function $u : D \to \mathbb{R}$ the function $\Delta u = u_{xx} + u_{yy}$ is defined in D. u is called a *potential-function in D* if u satisfies the *potential-equation* $\Delta u = 0$ in D. (The language is motivated by considerations from physics, especially electrostatics, because functions with $\Delta u = 0$ arise as potentials in physics.) Potential-functions are also known as *harmonic functions*.

The essence of the theorem is that the real and imaginary parts of complex-differentiable functions are potential-functions. Thus simple examples of potential-functions can be obtained from the examples in the preceding section; e.g., $\Im z^2 = 2xy$, $\Re z^3 = x^3 - 3xy^2$ are harmonic in $\mathbb{C}$. Furthermore, the functions

$$\begin{array}{llll} \Re\tilde{e}(z) & = & e^x \cos y \,, & \Im\tilde{e}(z) & = & e^x \sin y \\ \Re\tilde{\ell}(z) & = & \log|z| \,, & \Im\tilde{\ell}(z) & = & \arctan \frac{y}{x} \end{array}$$

are harmonic in their domains of definition. The function $x^2 + y^2 = |z|^2$ is *not harmonic* and so not the real part of any complex-differentiable function. (Take a look at $x^2 - y^2 = \Re z^2$; alternatively, look at $\Delta(x^2 + y^2)$.) □

For every harmonic polynomial $u(x, y) \in \mathbb{R}[x, y]$ one can directly write down a complex polynomial $p(z) \in \mathbb{C}[z]$ whose real part is $u(x, y)$; namely, $p(z) := 2u(\frac{1}{2}z, \frac{1}{2i}z) - u(0, 0)$. The reader should clarify this for himself with a few examples; he might even give a proof of the general assertion.

Harmonic functions of two variables played a big role in classical mathematics and gave essential impulses to it. In this connection let us only recall here the famous

DIRICHLET Boundary-Value Problem. *A real-valued continuous function g on the boundary $\partial \mathbb{E} = \{z \in \mathbb{C} : |z| = 1\}$ of the unit disc is given. A continuous function u on $\mathbb{E} \cup \partial \mathbb{E}$ is sought having the properties that $u|\partial \mathbb{E} = g$ and $u|\mathbb{E}$ is a potential-function in $\mathbb{E}$.*

It can be shown that there is always exactly one such function u. □

The theory of holomorphic (see next section for the definition) functions has gotten valuable stimulus from the theory of harmonic functions. Some properties of harmonic functions (integral formulas, maximum principle, convergence theorems, etc.) are shared by holomorphic functions. But nowadays it is customary to develop the theory of holomorphic functions completely and then to derive from it the fundamental properties of harmonic functions of two variables.

Exercises

Exercise 1. Where are the following functions complex-differentiable?

a) $f(x + iy) = x^4y^5 + ixy^3$

b) $f(x + iy) = y^2 \sin x + iy$

c) $f(x + iy) = \sin^2(x + y) + i \cos^2(x + y)$

d) $f(x + iy) = -6(\cos x + i \sin x) + (2 - 2i)y^3 + 15(y^2 + 2y)$.

Exercise 2. Let G be a region in $\mathbb{C}$ and $f = u + iv$ be complex-differentiable in G. Show that a function $\hat{v} : G \to \mathbb{R}$ satisfies $u + i\hat{v}$ complex-differentiable in G, if and only if, $v - \hat{v}$ is constant.

Exercise 3. For each of the given functions $u : \mathbb{C} \to \mathbb{R}$ find all functions $v : \mathbb{C} \to \mathbb{R}$ such that $u + iv$ is complex-differentiable:

a) $u(x + iy) = 2x^3 - 6xy^2 + x^2 - y^2 - y$

b) $u(x + iy) = x^2 - y^2 + e^{-y} \sin x - e^y \cos x$.

Exercise 4. Show that for integer $n \geq 1$ the function $u : \mathbb{C}^\times \to \mathbb{R}$, $z \mapsto \log |z^n|$ is harmonic but is not the real part of any function which is complex-differentiable in $\mathbb{C}^\times$.

Exercise 5. Show that every harmonic function $u : \mathbb{C} \to \mathbb{R}$ is the real part of some complex-differentiable function on $\mathbb{C}$.

§3 Holomorphic functions

Now we introduce the fundamental idea of all of function theory. A function $f : D \to \mathbb{C}$ is called *holomorphic in the domain* D if f is complex-differentiable at every point of D; we say f is *holomorphic at* $c \in D$ if there is an open neighborhood U of c lying in D such that the restriction $f|U$ of f to U is holomorphic in U.

The set of all points at which a function is holomorphic, is always *open* in $\mathbb{C}$. A function which is holomorphic at c is complex-differentiable at c

but a function which is complex-differentiable at c need not be holomorphic at c. For example, the function

$$f(z) := x^3 y^2 + i x^2 y^3 \text{ , where } z = x + iy \text{ ; } x, y \in \mathbb{R},$$

is, according to 2.3, complex-differentiable at the points of the coordinate axes but nowhere else. So this function is not holomorphic at any point of $\mathbb{C}$.

The set of all holomorphic functions in the domain D is always denoted by $\mathcal{O}(D)$. We naturally have the inclusions

$$\mathbb{C} \subset \mathcal{O}(D) \subset \mathcal{C}(D);$$

the first because constant functions are (complex-) differentiable everywhere in $\mathbb{C}$ and the second because complex-differentiability implies continuity.

1. Differentiation rules are proved as in the case of real-differentiation; doing so provides some evidence that the definition of complex-differentiability in use today offers considerable advantages over Riemann's definition via his differential equations.

Sum- and Product-rule. *Let $f : D \to \mathbb{C}$ and $g : D \to \mathbb{C}$ be holomorphic in D. Then for all $a, b \in \mathbb{C}$ the functions $af + bg$ and $f \cdot g$ are holomorphic in D, with*

$$(af + bg)' = af' + bg' \qquad (sum\text{-}rule),$$

$$(f \cdot g)' = f'g + fg' \qquad (product\text{-}rule).$$

We will be content to recall how the proof of the product-rule goes. By hypothesis, for each $c \in D$ there are functions $f_c, g_c : D \to \mathbb{C}$ which are continuous at c and satisfy

$$f(z) = f(c) + (z - c)f_c(z) \text{ , } g(z) = g(c) + (z - c)g_c(z) \text{ , } z \in D.$$

Multiplication yields, for all $z \in D$,

$$(f \cdot g)(z) = (f \cdot g)(c) + (z - c)[f_c(z)g(c) + f(c)g_c(z) + (z - c)(f_c \cdot g_c)(z)].$$

Since the square-bracketed expression is a function of $z \in D$ which is evidently continuous at c, the complex-differentiability at c of the product function $f \cdot g$ is confirmed, with moreover $(f \cdot g)'(c)$ being the value of that function at c, viz.,

$$(f \cdot g)'(c) = f_c(c)g(c) + f(c)g_c(c) = f'(c)g(c) + f(c)g'(c).$$

From the sum- and product-rules follows, as with real-differentiability:

Every complex polynomial $p(z) = a_0 + a_1 z + \cdots + a_n z^n \in \mathbb{C}[z]$ is holomorphic in $\mathbb{C}$ and satisfies $p'(z) = a_1 + 2a_2 z + \cdots + n a_n z^{n-1} \in \mathbb{C}[z]$.

As in the reals, we also have a

Quotient rule. *Let f, g be holomorphic and g zero-free in D. Then the quotient function $\frac{f}{g} : D \to \mathbb{C}$ is holomorphic in D and*

$$\left(\frac{f}{g}\right)' = \frac{f'g - fg'}{g^2} \qquad (Quotient\text{-}rule).$$

Differentiation of the composite function $h \circ g$ is codified in the

Chain-rule. *Let $g \in \mathcal{O}(D), h \in \mathcal{O}(D')$ be holomorphic functions with $g(D) \subset D'$. Then the composite function $h \circ g : D \to \mathbb{C}$ is holomorphic in D and*

$$(h \circ g)'(z) = h'(g(z)) \cdot g'(z), \quad z \in D \qquad (Chain\text{-}rule).$$

The quotient- and chain-rules are proved just as for real-differentiability.

On the basis of Theorem 2.1 a function $f = u + iv$ is holomorphic in the domain $D \subset \mathbb{C}$ exactly when f is real-differentiable and satisfies the Cauchy-Riemann equations $u_x = v_y, u_y = -v_x$ throughout D. But these differentiability hypotheses may be dramatically weakened. For example, we have

A continuous function $f : D \to \mathbb{C}$ is already holomorphic in D if through each point $c \in D$ there are two distinct straight lines L, L' along which the limits

$$\lim_{z \in L, \, z \to c} \frac{f(z) - f(c)}{z - c} \quad , \quad \lim_{z \in L', \, z \to c} \frac{f(z) - f(c)}{z - c}$$

exist and are equal.

This theorem is due to D. MENCHOFF: "Sur la généralisation des conditions de Cauchy-Riemann," *Fund. Math.* **25**(1935), 59-97. As a special case, that in which every L is parallel to the x-axis and every L' is parallel to the y-axis, we have the so-called LOOMAN-MENCHOFF theorem:

A continuous function $f : D \to \mathbb{C}$ is already holomorphic in D if the partial derivatives u_x, u_y, v_x, v_y of the real-valued functions $u := \Re f, v := \Im f$ exist and satisfy the Cauchy-Riemann equations $u_x = v_y, u_y = -v_x$ throughout D.

The hypothesis about the continuity of f, or some weaker surrogate, is needed, as the following example shows:

$$f(z) := \exp(-z^{-4}) \qquad \text{for } z \in \mathbb{C}^\times , \ f(0) := 0.$$

(At the trouble-point $z = 0$, the two partial derivatives of f exist and are 0 by an elementary use of the Mean Value Theorem of real analysis, since $\frac{\partial f}{\partial x}(x) = 4x^{-5}f(x)$ $[x \neq 0]$ and $\frac{\partial f}{\partial y}(iy) = 4y^{-3}f(y)$ $[y \neq 0]$ imply that $\lim_{x \to 0} \frac{\partial f}{\partial x}(x) = \lim_{y \to 0} \frac{\partial f}{\partial y}(iy) = 0$.) On the other hand, if we ask that f be continuous throughout D but only ask that the Cauchy-Riemann equations hold at one point, then complex-differentiability at that point cannot be inferred, as the example

$$f(z) := |z|^{-4}z^5 \quad \text{for } z \in \mathbb{C}^\times , \, f(0) := 0$$

shows.

Actually the result which is usually designated as the Looman-Menchoff theorem contains even weaker differentiability hypotheses than those stated above: the partial derivatives need only exist on a set whose complement in D is countable and the Cauchy-Riemann equations need only hold on a set whose complement in D has area 0. A very accessible proof of this, together with a full history and bibliography of other possible weakenings of the differentiability hypotheses will be found in J. D. GRAY and S. A. MORRIS, "When is a function that satisfies the Cauchy-Riemann equations analytic?", *Amer. Math. Monthly* **85**(1978), 246-256. Another elementary account, which deals with Menchoff's first theorem as well, is K. MEIER, "Zum Satz von Looman-Menchoff," *Comm. Math. Helv.* **25**(1951), 181-195; some simplifications of this paper will be found in M. G. ARSOVE, "On the definition of an analytic function," *Amer. Math. Monthly* **62**(1955), 22-25.

2. The $\mathbb{C}$-algebra $\mathcal{O}(D)$. The differentiation rules yield directly that:

For every domain D in $\mathbb{C}$ *the set* $\mathcal{O}(D)$ *of all functions which are holomorphic in D is a $\mathbb{C}$-subalgebra of the $\mathbb{C}$-algebra* $\mathcal{C}(D)$. *The units of* $\mathcal{O}(D)$ *are exactly the zero-free functions.*

For the exponential function $\tilde{e}(z)$ and the logarithm function $\tilde{\ell}(z)$ of examples 2) and 3), respectively, in 2.3 we have $\tilde{e}(z) \in \mathcal{O}(\mathbb{C})$ and $\tilde{\ell}(z) \in \mathcal{O}(\mathbb{C} \setminus i\mathbb{R})$.

In contrast to $\mathcal{C}(D)$, the $\mathbb{C}$-algebra $\mathcal{O}(D)$ does not contain the conjugate $\bar{f}$ of each of its functions f; we saw, e.g., that $z \in \mathcal{O}(D)$ but $\bar{z} \notin \mathcal{O}(D)$. Also in general, if $f \in \mathcal{O}(D)$ then none of $\Re f$, $\Im f$ of $|f|$ belongs to $\mathcal{O}(D)$; for example, each of $\Re z$, $\Im z$ and $|z|$ is not complex-differentiable at *any* point of $\mathbb{C}$.

Every *polynomial* in z is *holomorphic* in $\mathbb{C}$; every rational function (meaning *quotient of polynomials*) is holomorphic in the *complement of the zero-set of its denominator*. Further examples of holomorphic functions can only be secured via *limiting processes* and so are no longer considered elementary functions. In 4.3.2 we will see that power series inside their circles of convergence furnish an inexhaustible reservoir of holomorphic functions.

If f is a holomorphic function in D, then

$$f' : D \to \mathbb{C} , \, z \mapsto f'(z)$$

is another function defined on D. It is called the (first) derivative of f in D. If one thinks about differentiable functions on $\mathbb{R}$ like $x|x|$, then there is no reason to expect f' to be holomorphic in D. But a fundamental theorem of function theory, which we will get from the Cauchy integral formula (but not until 7.4.1), says exactly this, that f' is holomorphic in D whenever f is. As a consequence, every holomorphic function in D turns out to be *infinitely often (complex-) differentiable in D*, that is, all the derivatives $f', \cdots, f^{(m)}, \cdots$ exist. Here, as in the reals, we understand by the mth *derivative* $f^{(m)}$ of f (in case it exists) the first derivative of $f^{(m-1)}$, $m = 1, 2, \cdots$; thus $f^{(0)} := f$ and $f^{(m)} := (f^{(m-1)})'$. The same proof (induction) used for functions on $\mathbb{R}$ will also establish Leibniz' product-rule for higher derivatives of holomorphic functions

$$(f \cdot g)^{(m)} = \sum_{k+\ell=m} \frac{m!}{k!\ell!} f^{(k)} g^{(\ell)}.$$

3. Characterization of locally constant functions. *The following statements about a function $f : D \to \mathbb{C}$ are equivalent:*

  i) f *is locally constant in D.*

  ii) f *is holomorphic in D and $f'(z) = 0$ for all $z \in D$.*

First Proof. Only ii) $\Rightarrow$ i) needs to be verified. Let $u := \Re f$, $v := \Im f$. Since $f' = u_x + iv_x$ and $u_x = v_y$, $v_x = -u_y$, our hypothesis ii) means that $u_x(z) = u_y(z) = v_x(z) = v_y(z) = 0$ for all $z \in D$. From a well-known theorem of real analysis it then follows that each of u and v, and therewith also $f = u + iv$, is locally constant in D.

The theorem from real analysis used above is proved via the Mean Value Theorem, but its use can easily be circumvented by another elementary (compactness) argument:

Second proof. Consider any $B = B_r(b) \subset D$ and any $z \in B$. Let L denote the line segment from b to z and let $\varepsilon > 0$ be given. For each $c \in L$ there is a disc $B_\delta(c) \subset D$, $\delta = \delta(c) > 0$, such that (cf. 1.1 and remember that $f' \equiv 0$):

$$|f(w) - f(c)| \le \varepsilon|w - c| \qquad \text{for all } w \in B_\delta(c).$$

Because finitely many of the discs $B_\delta(c)$ suffice to cover the compactum L, there is a *succession* of points $z_0 = b, z_1, \cdots, z_n = z$ on L such that

$$|f(z_\nu) - f(z_{\nu-1})| \le \varepsilon|z_\nu - z_{\nu-1}|, \qquad 1 \le \nu \le n.$$

It follows that

$$|f(z) - f(b)| = \left| \sum_1^n [f(z_\nu) - f(z_{\nu-1})] \right| \le \sum_1^n |f(z_\nu) - f(z_{\nu-1})|$$

$$\le \varepsilon \sum_1^n |z_\nu - z_{\nu-1}| = \varepsilon|z - b|.$$

Here $\varepsilon > 0$ is arbitrary, so $f(z) = f(b)$ follows. This is true for each $z \in B$, that is, $f|B$ is constant.

After studying complex integral calculus (cf. 6.3.2), we will give a third proof of this theorem, using primitives. On the basis of 0.6 the theorem can also be expressed thus:

For a region G in $\mathbb{C}$, a function $f : G \to \mathbb{C}$ is constant in G if and only if it is holomorphic in G and f' vanishes everywhere in G.

We will illustrate the result of this paragraph by two examples.

1) *Every function f which is holomorphic in D and assumes only real values, respectively, only purely imaginary values, is locally constant in D.*

Proof. In case $u := \Re f = f$, we have $v := \Im f = 0$ and so the Cauchy-Riemann equations give $u_x = v_y = 0 = v_x$ and so $f' = u_x + iv_x = 0$ in D. By the theorem f is locally constant in D. If, on the other hand, we have $f = i\Im f$ throughout D, then we apply what we just learned to if in the role of f to conclude the local constancy of f. □

2) *Every holomorphic function which has constant modulus in D is locally constant in D.*

Proof. Suppose $f = u + iv$ is holomorphic in D and $u^2 + v^2 = c$ is constant in D. Then differentiation of this equation with respect to y gives $uu_y + vv_y = 0$ and so, since $u_y = -v_x$, $uv_x = vv_y$. Since also $uu_x + vv_x = 0$ and $u_x = v_y$, we get

$$
\begin{aligned}
0 &= u \cdot (uu_x + vv_x) = u^2 u_x + v \cdot (uv_x) = u^2 u_x + v \cdot (vv_y) \\
&= (u^2 + v^2)u_x = cu_x.
\end{aligned}
$$

Similarly, $cv_x = 0$. If $c = 0$ then, of course, f is constant (equal 0) in D. If $c \neq 0$, we now have $f' = u_x + iv_x = 0$ in D, so that f is locally constant in D by the theorem. □

In 8.5.1 we will prove the Open Mapping Theorem for holomorphic functions; this theorem contains both of the above examples as trivial cases.

4. Historical remarks on notation. The word "holomorphic" was introduced in 1875 by BRIOT and BOUQUET, [BB], 2nd ed., p.14. In their

1st edition (cf. pp. 3,7 and 11) they used instead of "holomorphic" the designation "synectic", which goes back to CAUCHY. Other synonyms in the older literature are "monogenic", "monodromic", "analytic" and "regular". These and other terms originally described various properties, like having vanishing integral over every closed curve (cf. Chapter 6) or having local power series expansions or satisfying the Cauchy-Riemann equations, etc., and so were not at first recognized as synonyms. When the theory of functions reached maturity, these properties were all seen to be equivalent (and the reader will see this presently); so it is appropriate that most of these terms have now faded into oblivion. "Analytic" is still sometimes used as a synonym for "holomorphic", but usually it has a more technical meaning having to do with the Weierstrass continuation process.

Actually, as late as 1851 CAUCHY still had no exact definition of the class of functions for which his theory was valid $\cdots$ "La théorie des fonctions de variables imaginaires présente des questions délicates qu'il importait de résoudre $\cdots$ (The theory of functions of an imaginary variable presents delicate questions which it was important to resolve $\cdots$)"; thus begins a *Comptes Rendus* note on February 10, 1851 bearing the title "Sur les fonctions de variables imaginaires" (*Œuvres* (1) **11**, pp. 301-304). See pp. 169, 170 of BOTTAZZINI [H₄].

The notation $\mathcal{O}(D)$ is used – since about 1952 – by the French school around Henri CARTAN, especially in the function theory of several variables. It is sometimes said that $\mathcal{O}$ was chosen to honor the great Japanese mathematician OKA, and it is sometimes even maintained that the $\mathcal{O}$ reflects the French pronunciation of the word *holomorphic*. Nevertheless, the choice of the symbol $\mathcal{O}$ appears to have been purely accidental. In a letter of March 22, 1982 to the author of this book, H. CARTAN wrote: "Je m'étais simplement inspiré d'une notation utilisée par van der Waerden dans son classique traité 'Moderne Algebra' (cf. par exemple §16 de la 2e édition allemande, p. 52)". [I was simply inspired by a notation used by van der Waerden in his classic treatise 'Moderne Algebra' (cf. for example, §16 Vol. I of the English translation).]

Exercises

Exercise 1. Let G be a region in $\mathbb{C}$. Determine all holomorphic functions f on G for which $(\Re f)^2 + i(\Im f)^2$ is also holomorphic on G.

Exercise 2. Suppose that $f = u + iv$ is holomorphic in the region $G \subset \mathbb{C}$ and that for some pair of non-zero complex numbers a and b, $au + bv$ is constant in G. Show that then f itself is constant in G.

Exercise 3. Let $f = u + iv$ be holomorphic in the region G and satisfy $u = h \circ v$ for some differentiable function $h : \mathbb{R} \to \mathbb{R}$. Show that f is constant.

Exercise 4. Let D, D' be domains in $\mathbb{C}$, $g : D \to \mathbb{C}$ continuous with $g(D) \subset D'$, and $h \in \mathcal{O}(D')$. Show that if h' is zero-free on $g(D)$ and $h \circ g$ is holomorphic in D, then g is holomorphic in D. *Hint:* For fixed $c \in D$ consider the $\mathbb{C}$-linearization of $h \circ g$ at c and that of h at $g(c)$.

§4 Partial differentiation with respect to x, y, z and $\bar{z}$

If $f : D \to \mathbb{C}$ is real-differentiable at c and $T = Tf(c)$, then the limit relation

$$\lim_{h \to 0} \frac{|f(c + h) - f(c) - T(h)|}{|h|} = 0$$

is valid without the absolute value signs. (But in general, for functions into $\mathbb{R}^m$, stripping away the absolute values results in a meaningless division by the vector h.) Upon setting $z = c + h$, this observation becomes

Differentiability criterion. $f : D \to \mathbb{C}$ *is real-differentiable at c precisely when there exist a (uniquely determined) $\mathbb{R}$-linear map $T : \mathbb{C} \to \mathbb{C}$ and a function $\hat{f} : D \to \mathbb{C}$, which is continuous at c with $\hat{f}(c) = 0$, such that*

$$f(z) = f(c) + T(h) + h\hat{f}(z).$$

If we write the $\mathbb{R}$-linear differential

$$T(h) = \begin{pmatrix} u_x(c) & u_y(c) \\ v_x(c) & v_y(c) \end{pmatrix} \begin{pmatrix} \Re h \\ \Im h \end{pmatrix}$$

of $f = u + iv$ at c in the form

(1) $$T(h) = T(1)\Re h + T(i)\Im h$$

or in the form

(2) $$T(h) = \lambda h + \mu \bar{h},$$

then we are led almost automatically to introduce, besides the partial derivatives of u and v with respect to x and y, also the partial derivatives of f itself with respect to x and y and even with respect to z and $\bar{z}$ (subsection 1 below). Because, thanks to 0.1.2, there are among the quantities $u_x(c), \cdots, v_y(c), T(1), T(i), \lambda, \mu$ the relations

$$(*) \quad \begin{aligned} T(1) &= u_x(c) + iv_x(c), & T(i) &= u_y(c) + iv_y(c) \\ \lambda &= \tfrac{1}{2}(T(1) - iT(i)), & \mu &= \tfrac{1}{2}(T(1) + iT(i)), \end{aligned}$$

certain identities between these *formally introduced* derivatives and the familiar derivatives of u and v are immediately obtained.

It should be emphasized that this section consists largely of introducing some new terminology and re-interpreting preceding results in this language.

1. The partial derivatives $f_x, f_y, f_z, f_{\bar{z}}$. If f is real-differentiable at c and $T = Tf(c)$ is the differential of f at c, then the coefficients defined in (1) and (2) are denoted

$$f_x(c) := \tfrac{\partial f}{\partial x}(c) := T(1) \qquad f_y(c) := \tfrac{\partial f}{\partial y}(c) := T(i);$$

$$f_z(c) := \tfrac{\partial f}{\partial z}(c) := \lambda \qquad f_{\bar{z}}(c) := \tfrac{\partial f}{\partial \bar{z}}(c) := \mu$$

and called the *partial derivatives of f at c with respect to x, y, z and $\bar{z}$, respectively.* Thus we have

$$Tf(c)(h) = f_x(c)\Re h + f_y(c)\Im h = f_z(c)h + f_{\bar{z}}(c)\bar{h} = \begin{pmatrix} u_x(c) & u_y(c) \\ v_x(c) & u_y(c) \end{pmatrix} \begin{pmatrix} \Re h \\ \Im h \end{pmatrix}.$$

There is good motivation for having chosen the symbols $f_x, f_y, f_z, f_{\bar{z}}$:

Theorem. *The following statements about $f : D \to \mathbb{C}$ are equivalent:*

　i) *f is real-differentiable at $c = a + ib$.*

　ii) *There are functions $\hat{f}_1, \hat{f}_2 : D \to \mathbb{C}$, each continuous at c, such that*

$$f(z) = f(c) + (z - c)\hat{f}_1(z) + (\bar{z} - \bar{c})\hat{f}_2(z) \qquad \text{for all } z \in D.$$

　iii) *There are functions $f_1, f_2 : D \to \mathbb{C}$, each continuous at c, such that*

$$f(z) = f(c) + (x - a)f_1(z) + (y - b)f_2(z) \qquad \text{for all } z \in D.$$

When these conditions are fulfilled,

$$f_z(c) = \hat{f}_1(c), \quad f_{\bar{z}}(c) = \hat{f}_2(c), \quad f_x(c) = f_1(c), \quad f_y(c) = f_2(c).$$

Proof. i) $\Rightarrow$ ii) The equation $f(z) = f(c) + T(z - c) + (z - c)\hat{f}(z)$ which features in the differentiability criterion above proves the claim, once we write T in the form $Th = \lambda h + \mu \bar{h}$ and define $\hat{f}_1(z) := \lambda + \hat{f}(z), \hat{f}_2(z) := \mu$.

ii) $\Rightarrow$ iii) Set $f_1 := \hat{f}_1 + \hat{f}_2, f_2 := i(\hat{f}_1 - \hat{f}_2)$ and recall that $z - c = x - a + i(y - b)$.

iii) $\Rightarrow$ i) The mapping $T(h) := f_1(c)\Re h + f_2(c)\Im h$ is $\mathbb{R}$-linear. We define $\hat{f} : D \to \mathbb{C}$ by $\hat{f}(c) := 0$ and

$$\hat{f}(z) := \frac{(x - a)(f_1(z) - f_1(c)) + (y - b)(f_2(z) - f_2(c))}{z - c} \quad \text{for } z \neq c.$$

Since $|x - a| \leq |z - c|$ and $|y - b| \leq |z - c|$, it follows that

$$|\hat{f}(z)| \leq |f_1(z) - f_1(c)| + |f_2(z) - f_2(c)| \quad \text{for } z \in D \setminus \{c\}$$

and consequently $\hat{f}$ is continuous at c. The identity $f(z) = f(c) + T(z - c) + (z - c)\hat{f}(z)$ is immediate from the definitions of $\hat{f}$ and T.

2. Relations among the derivatives $u_x, u_y, v_x, v_y, f_x, f_y, f_z, f_{\bar{z}}$. Here we will consider functions $f : D \to \mathbb{C}$ which are real-differentiable in D. For such an $f = u + iv$ the eight partial derivatives $u_x, u_y, v_x, v_y, f_x, f_y, f_z, f_{\bar{z}}$ are all well-defined functions in D. The following four identities are immediate from the equations $(*)$ in the introductory remarks to this section:

$$(3) \quad f_x = u_x + iv_x \,, \ f_y = u_y + iv_y \,, \ f_z = \frac{1}{2}(f_x - if_y) \,, \ f_{\bar{z}} = \frac{1}{2}(f_x + if_y).$$

The equations here for f_x and f_y are scarcely surprising, on account of $f = u + iv$. The equations for f_z and $f_{\bar{z}}$, at first so strangely charming, are better understood via the following mnemonic device: since $x = \frac{1}{2}(z + \bar{z})$ and $y = -\frac{i}{2}(z - \bar{z})$, think of $f = f(x, y)$ as a function of z and $\bar{z}$ and *regard $z, \bar{z}$ as though they were independent variables.* The differentiation rules give, formally

$$\frac{\partial x}{\partial z} = \frac{\partial x}{\partial \bar{z}} = \frac{1}{2} \,, \quad \frac{\partial y}{\partial z} = \frac{-i}{2} \,, \quad \frac{\partial y}{\partial \bar{z}} = \frac{i}{2}$$

and then the chain rule implies that

$$f_z = \frac{\partial f}{\partial x}\frac{\partial x}{\partial z} + \frac{\partial f}{\partial y}\frac{\partial y}{\partial z} = \frac{1}{2}f_x - \frac{i}{2}f_y;$$

$$f_{\bar{z}} = \frac{\partial f}{\partial x}\frac{\partial x}{\partial \bar{z}} + \frac{\partial f}{\partial y}\frac{\partial y}{\partial \bar{z}} = \frac{1}{2}f_x + \frac{i}{2}f_y. \qquad \square$$

From equations (3) we also obtain "inversion formulas":

$$(4) \quad \begin{aligned} u_x &= \tfrac{1}{2}(f_x + \bar{f}_x) & u_y &= \tfrac{1}{2}(f_y + \bar{f}_y) \\ v_x &= \tfrac{1}{2i}(f_x - \bar{f}_x) & v_y &= \tfrac{1}{2i}(f_y - \bar{f}_y) \\ f_x &= f_z + f_{\bar{z}} & f_y &= i(f_z - f_{\bar{z}}). \end{aligned}$$

For more particulars on the differential calculus with respect to z and $\bar{z}$ compare subsection 4.

3. The Cauchy-Riemann differential equation $\partial f/\partial\bar{z}=0$.

In 1.2 we got the Cauchy-Riemann equations $u_x = v_y$, $u_y = -v_x$ for the holomorphic function $f = u + iv$ from the identity $f' = u_x + iv_x = i^{-1} \cdot (u_y + iv_y)$. The latter can now also be written as

$$f' = f_x = i^{-1}f_y \quad \text{in case } f \in \mathcal{O}(D);$$

the condition for holomorphy can be compressed into the single equation

$$if_x = f_y.$$

(Already in 1857 in his work "Theorie der ABELschen Functionen" RIE-MANN himself combined the two differential equations $u_x = v_y$ and $u_y = -v_x$ into the single equation $i\frac{\partial w}{\partial x} = \frac{\partial w}{\partial y}$, where $w = u + iv$ [cf. *Werke*, p. 88].) If now we utilize the derivatives $f_z, f_{\bar{z}}$, we see

Theorem. *A real-differentiable function $f : D \to \mathbb{C}$ is holomorphic in D if and only if*

$$\frac{\partial f}{\partial \bar{z}}(c) = 0$$

for every $c \in D$. In this case $\frac{\partial f}{\partial z}$ coincides in D with the derivative f' of f.

This is nothing but the equivalence i) $\Leftrightarrow$ iii) of theorem 2.1. Of course the claims here follow as well directly from the preceding theorem.   □

The Cauchy-Riemann equations for the function $\bar{f} = u - iv$ conjugate to f are $u_x = -v_y$ and $u_y = v_x$, and these may be written (proof!) as the *single* equation $f_z = 0$. It follows that $\bar{f}'(c) = \overline{f_{\bar{z}}(c)}$ for $c \in D$ and so under the same hypotheses and with the same proof, *mutatis mutandis*, as the preceding theorem, we get:

$\bar{f} : D \to \mathbb{C}$ *is holomorphic in D if and only if $f_z \equiv 0$ in D; when this occurs, $f_{\bar{z}}(c)$ coincides with the derivative of $\bar{f}$ at $c \in D$.*

This fact also follows easily from Theorem 1.

4. Calculus of the differential operators ∂ and $\bar{\partial}$.

The theory developed above becomes especially elegant if we systematically and consistently utilize partial *differentiation with respect to z and with respect to $\bar{z}$*. The differential calculus of these operations, though largely irrelevant for classical function theory, is unusually fascinating; it goes back to H. POINCARÉ and was developed principally by W. WIRTINGER; it is even often called, especially in the German literature, the *Wirtinger calculus*. It is quite indispensable in the function theory of several variables.

Besides the customary "real" differential operators $\frac{\partial}{\partial x}$ and $\frac{\partial}{\partial y}$, one is motivated by the formulas

$$\frac{\partial f}{\partial z} = \frac{1}{2}\left(\frac{\partial f}{\partial x} - i\frac{\partial f}{\partial y}\right), \qquad \frac{\partial f}{\partial \bar{z}} = \frac{1}{2}\left(\frac{\partial f}{\partial x} + i\frac{\partial f}{\partial y}\right),$$

to introduce the "complex" differential operators

$$\partial := \frac{\partial}{\partial z} := \frac{1}{2}\left(\frac{\partial}{\partial x} - i\frac{\partial}{\partial y}\right), \qquad \bar{\partial} := \frac{\partial}{\partial \bar{z}} := \frac{1}{2}\left(\frac{\partial}{\partial x} + i\frac{\partial}{\partial y}\right).$$

Then the equations

$$\frac{\partial}{\partial x} = \partial + \bar{\partial}, \qquad \frac{\partial}{\partial y} = i(\partial - \bar{\partial})$$

hold.

The differential calculus of $\partial, \bar{\partial}$ rests on what at first appears to be a rather absurd

Thesis. *In differentiating with respect to the conjugate complex variables z and $\bar{z}$ we can treat them as though they were independent variables.*

The Cauchy-Riemann equation $\bar{\partial}f = 0$ is, in this view, interpreted as saying that

Holomorphic functions are independent of $\bar{z}$ and depend only on z.

As soon as one is convinced of the correctness and the power of this calculus and has mastered it, he is apt to be reminded of what Jacobi had to say about the significance of algorithms (see A. KNESER, "Euler und die Variationsrechnung", *Festschrift zur Feier des 200. Geburtstages Leonhard Euler*, Teubner Verlag, 1907, p. 24): "da es nämlich in der Mathematik darauf ankommt, Schlüsse auf Schlüsse zu häufen, so wird es gut sein, so viele Schlüsse als möglich in ein Zeichen zusammenzuhäufen. Denn hat man dann ein für alle Mal den Sinn der Operation ergründet, so wird der sinnliche Anblick des Zeichens das ganze Räsonnement ersetzen, das man früher bei jeder Gelegenheit wieder von vorn anfangen mußte (because in mathematics we pile inferences upon inferences, it is a good thing whenever we can subsume as many of them as possible under one symbol. For once we have understood the true significance of an operation, just the sensible apprehension of its symbol will suffice to obviate the whole reasoning process that earlier we had to engage anew each time the operation was encountered)."

Formally, differentiation with respect to z and $\bar{z}$ proceeds according the the same rules as ordinary partial differentiation. Designating by f and g real-differentiable functions from D into $\mathbb{C}$, we maintain that

1) ∂ and $\bar{\partial}$ are $\mathbb{C}$-linear mappings (Sum-rule) for which the Product- and Quotient-rule hold.

2) $\bar{\partial} f = \overline{\partial \bar{f}}$, $\bar{\partial}\,\bar{f} = \overline{\partial f}$.

3) $f \in \mathcal{O}(D) \Leftrightarrow \bar{\partial} f = 0$ and $\partial f = f'$; $\bar{f} \in \mathcal{O}(D) \Leftrightarrow \partial f = 0$ and $\bar{f}' = \overline{\bar{\partial} f}$.

Proof. ad 1) We confine ourselves to a few words about the product-rules: $\partial(fg) = \partial f \cdot g + f \cdot \partial g$ and $\bar{\partial}(fg) = \bar{\partial} f \cdot g + f \cdot \bar{\partial} g$. Consider $c \in D$. Theorem 1 furnishes functions f_1, f_2, g_1, g_2 in D which are continuous at c and satisfy

$$
\begin{aligned}
f(z) &= f(c) + (z - c)f_1(z) + (\bar{z} - \bar{c})f_2(z) \\
g(z) &= g(c) + (z - c)g_1(z) + (\bar{z} - \bar{c})g_2(z).
\end{aligned}
$$

Abbreviating $f_1(z)$ to f_1, etc., it follows that

$$
\begin{aligned}
f(z)g(z) = f(c)g(c) \;\; &+ \;\; (z - c)[f_1 g(c) + f(c)g_1 + (z - c)f_1 g_1 + (\bar{z} - \bar{c})f_1 g_2] \\
&+ \;\; (\bar{z} - \bar{c})[f_2 g(c) + f(c)g_2 + (\bar{z} - \bar{c})f_2 g_2 + (z - c)f_2 g_1].
\end{aligned}
$$

Since all the functions occurring on the right side are continuous at c, the product-rules follow at once from the relevant definitions.

ad 2) From $f = f(c) + (z-c)f_1 + (\bar{z}-\bar{c})f_2$ follows $\bar{f} = \bar{f}(c) + (z-c)\bar{f}_2 + (\bar{z}-\bar{c})\bar{f}_1$. Because the continuity at c of f_1, f_2 entails that of $\bar{f}_2, \bar{f}_1$, the claim follows.

ad 3) The first statement follows from Theorem 4.3, the second then follows from 2). □

Remark. Naturally the product- and quotient-rules can be deduced from the corresponding rules for partial differentiation with respect to x and y by using the transformation equations from 2 to express $f_z, f_{\bar{z}}$ and f_x, f_y in terms of each other. However the calculations would be unpleasant; moreover, such a procedure would not contribute to understanding why ∂ and $\bar{\partial}$ behave like partial derivatives. □

The chain rules read as follows:

4) *If $g : D \to \mathbb{C}$, $h : D' \to \mathbb{C}$ are real-differentiable in D and D', respectively and $g(D) \subset D'$, then $h \circ g : D \to \mathbb{C}$ is also real-differentiable; writing w for the variable in D', we then have, for all $c \in D$*

$$
\frac{\partial(h \circ g)}{\partial z}(c) = \frac{\partial h}{\partial w}(g(c)) \cdot \frac{\partial g}{\partial z}(c) + \frac{\partial h}{\partial \bar{w}}(g(c)) \cdot \frac{\partial \bar{g}}{\partial z}(c),
$$

$$
\frac{\partial(h \circ g)}{\partial \bar{z}}(c) = \frac{\partial h}{\partial w}(g(c)) \cdot \frac{\partial g}{\partial \bar{z}}(c) + \frac{\partial h}{\partial \bar{w}}(g(c)) \cdot \frac{\partial \bar{g}}{\partial \bar{z}}(c).
$$

Here too the most convenient proof is an imitation of the proof in the real case, using Theorem 1; we will however forego the details. □

Naturally we could also consider *mixed higher order partial derivatives* like

$$
f_{xx}, f_{xy}, \cdots, f_{xz}, f_{yz}, f_{zz} := \partial^2 f := \partial(\partial f), f_{z\bar{z}} := \bar{\partial}\partial f := \bar{\partial}(\partial f).
$$

For them we have

5) *If $f : D \to \mathbb{C}$ is twice continuously differentiable with respect to x and y, then*

$$\partial\bar{\partial}f = \bar{\partial}\partial f = \frac{1}{4}(f_{xx} + f_{yy}).$$

Proof. It suffices to deal with the case of a real-valued function $f = u$. From $2u_z = u_x - iu_y$ and the known identities like $u_{xy} = u_{yx}$ follows

$$4\bar{\partial}\partial u = 2u_{x\bar{z}} - 2iu_{y\bar{z}} = u_{xx} + iu_{xy} - i(u_{yx} + iu_{yy}) = u_{xx} + u_{yy}$$

and $4\partial\bar{\partial}u = u_{xx} + u_{yy}$ follows analogously.   □

We will conclude this brief glimpse at the Wirtinger calculus with an amusing application to function theory. As a preliminary to that we show

If f, g are twice complex-differentiable in D, then

$$\bar{\partial}\partial(f \cdot \bar{g}) = f' \cdot \overline{g'} \qquad in\ D.$$

Proof. Because $f, g \in \mathcal{O}(D)$,

$$\partial(f\bar{g}) = \partial f \cdot \bar{g} + f \cdot \partial\bar{g} = f'\bar{g} + f\overline{\bar{\partial}g} = f'\bar{g}.$$

Furthermore, $f' \in \mathcal{O}(D)$ then entails that

$$\bar{\partial}\partial(f\bar{g}) = \bar{\partial}(f'\bar{g}) = \bar{\partial}f' \cdot \bar{g} + f' \cdot \bar{\partial}\bar{g} = f'\overline{\partial g} = f'\overline{g'}.$$ □

The following not so obvious result can now be derived rather expeditiously:

If $f_1, f_2, \ldots, f_n$ are twice complex-differentiable in D and if the function $|f_1|^2 + |f_2|^2 + \cdots + |f_n|^2$ is locally constant in D, then each of the functions $f_1, f_2, \ldots, f_n$ must be locally constant in D.

Proof. On account of the local constancy, $0 = \bar{\partial}\partial(\sum_1^n f_\nu \bar{f}_\nu) = \sum_1^n f'_\nu\overline{f'_\nu}$. Since $f'_\nu\overline{f'_\nu} \geq 0$, it follows that $f'_\nu = 0$ in D. According to 3.3 each function f_ν is then locally constant in D.   □

Exercises

Exercise 1. Let D be a domain in $\mathbb{C}$ and $f : D \to \mathbb{C}$ a real-differentiable function. Suppose that for some $c \in D$ the limit

$$\lim_{h \to 0} \left| \frac{f(c+h) - f(c)}{h} \right|$$

exists. Prove that either f or $\bar{f}$ is complex-differentiable at c.

Exercise 2. Determine all the points in $\mathbb{C}$ at which the following functions are complex-differentiable:

a) $f(z) = |z|^2(|z|^2 - 2)$

b) $f(z) = \sin(|z|^2)$

c) $f(z) = z(z + \bar{z}^2)$.

Exercise 3. Let f be real-differentiable in the domain $D \subset \mathbb{C}$. Show that its Jacobian functional determinant satisfies

$$\det \begin{pmatrix} u_x & u_y \\ v_x & v_y \end{pmatrix} = \det \begin{pmatrix} \partial f & \bar{\partial} f \\ \partial \bar{f} & \bar{\partial} \bar{f} \end{pmatrix} = |f_z|^2 - |f_{\bar{z}}|^2.$$

Chapter 2

Holomorphy and Conformality. Biholomorphic Mappings

Der Umstand, dass das Verständnis mehrerer Arbeiten Riemanns anfänglich nur einem kleinen Leserkreis zugänglich war, findet wohl darin seine Erklärung, dass RIEMANN es unterlassen hat, bei der Veröffentlichung seiner allgemeinen Untersuchungen das Eigenthümliche seiner Betrachtungsweise an der vollständigen Durchführung specieller Beispiele ausführlich zu erläutern. (That several of Riemann's works were at first comprehensible to only a small readership is explained by the fact that in the publication of his general investigations RIEMANN failed to illustrate his novel ideas thoroughly enough by carrying through the complete analysis of special examples.) – Hermann Amandus SCHWARZ, 1869.

1. The investigation of length-preserving, respectively, angle-preserving mappings between surfaces in $\mathbb{R}^3$ is one of the interesting problems addressed in classical differential geometry. This problem is important for cartography: every page of an atlas is a mapping of a part of the (spherical) surface of the earth into a plane. We know that there cannot be any length-preserving atlases; but by contrast there are indeed angle-preserving atlases (e.g., those based on stereographic projection). The first goal of this chapter is to show that for domains in the plane $\mathbb{R}^2 = \mathbb{C}$ angle-preserving mappings and holomorphic functions are essentially the same thing (Sec-

tion 1). The interpretation of holomorphic functions as angle-preserving (= conformal) mappings was advocated especially by RIEMANN (cf. 1.5). It provides the best way to "intuitively comprehend" such functions. One examines in detail how paths behave under such mappings. The invariance, under the mapping, of the angles in which curves intersect each other frequently makes possible a good description of the function. "The conformal mapping associated with an analytic function affords an excellent visualization of the properties of the latter; it can well be compared to the visualization of a real function by its graph" (AHLFORS [1], p. 89).

2. In Riemann's function theory a central role is played by the biholomorphic mappings. Such mappings are angle-preserving in both domain and range. The question of whether two domains D, D' in $\mathbb{C}$ are *biholomorphically equivalent*, that is, whether a biholomorphic mapping $f : D \xrightarrow{\sim} D'$ exists, even though it has been solved in only a few cases, has proved to be extremely fruitful. In section 2 we will present some significant examples of biholomorphic mappings. Perhaps surprisingly it will turn out that hidden among the examples we already have at hand are some extremely interesting biholomorphic mappings. Thus we will show, among other things, that as simple a function as $\frac{z-i}{z+i}$ maps the unbounded upper half-plane biholomorphically onto the bounded unit disc.

The biholomorphic mappings of a domain D onto itself constitute a group, the so-called *automorphism group* Aut D of D. The precise determination of this (generally non-commutative) group is an important and fascinating challenge for Riemann's function theory; but only in exceptional cases is it possible. In section 3 it will be shown that among the fractional linear functions $\frac{az+b}{cz+d}$ are to be found automorphisms of both the upper half-plane and of the unit disc. These automorphisms are so numerous that any two points in the region in question can be carried into one another by one of them. This so-called homogeneity will be used later in 9.2.2 together with SCHWARZ's lemma to show that every automorphism of the upper half-plane or of the unit disc is fractional linear.

§1   Holomorphic functions and angle-preserving mappings

In 0.1.4 we introduced for $\mathbb{R}$-linear mappings $T : \mathbb{C} \to \mathbb{C}$ the concept of angle preservation. Now a real-differentiable mapping $f : D \to \mathbb{C}$ will be called *angle-preserving at the point* $c \in D$ if the differential $Tf(c) : \mathbb{C} \to \mathbb{C}$ is an angle-preserving $\mathbb{R}$-linear mapping. f will be called simply *angle-preserving* in D if it is so at every point of D. In subsection 3 we will go more deeply into the geometric interpretation of the concept of angle-preserving, but first we will show that angle-preserving and holomorphic are "almost" the same thing.

1. Angle-preservation, holomorphy and antiholomorphy. Since all mappings of the form $h \mapsto \lambda h$, $\lambda \in \mathbb{C}^\times$, and $h \mapsto \mu\bar{h}$, $\mu \in \mathbb{C}^\times$, are angle-preserving, according to Lemma 0.1.4, it follows immediately that

If $f : D \to \mathbb{C}$, respectively, $\bar{f} : D \to \mathbb{C}$ is holomorphic in D and if for each $c \in D$, $f'(c) \neq 0$, respectively, $\bar{f}\,'(c) \neq 0$, then f is angle-preserving in D.

Proof. Because of 1.4.3 and the present hypotheses, $Tf(c) : \mathbb{C} \to \mathbb{C}$, which generally is given by $h \mapsto f_z(c)h + f_{\bar{z}}(c)\bar{h}$, here has either the form $h \mapsto f'(c)h$ or the form $h \mapsto \overline{f'(c)}\bar{h}$. □

A function $f : D \to \mathbb{C}$ is called *anti-holomorphic* (or *conjugate-holomorphic*) in D if $\bar{f} : D \to \mathbb{C}$ is holomorphic in D; this occurs exactly if $f_z(c) = 0$ for all $c \in D$. Thus the above implies that *holomorphic and anti-holomorphic functions having zero-free derivatives are angle-preserving.*

In order to prove the converse we have to hypothesize the continuity of f_z and $f_{\bar{z}}$ in D — a state of affairs that occurs exactly when both f_x and f_y, or equivalently all four of u_x, u_y, v_x, and v_y exist and are continuous throughout D, that is, when the real and imaginary parts of f are each continuously real-differentiable in D. Such an f is called *continuously real-differentiable* and is, in particular, real-differentiable in D (recall 1.2.2). In order to make the proof of the converse as simple as possible, we confine our attention to regions.

Theorem. *For a region G in $\mathbb{C}$ the following assertions about a continuously real-differentiable function $f : G \to \mathbb{C}$ are equivalent:*

i) *Either f is holomorphic throughout G and f' is zero-free in G, or f is anti-holomorphic throughout G and $\bar{f}\,'$ is zero-free in G.*

ii) *f is angle-preserving in G.*

Proof. Only ii) $\Rightarrow$ i) remains to be proved. According to Lemma 0.1.4 the differential $Tf(c) : \mathbb{C} \to \mathbb{C}$,

$$h \mapsto f_z(c)h + f_{\bar{z}}(c)\bar{h}, \qquad c \in G,$$

is angle-preserving if

$$either \quad f_{\bar{z}}(c) = 0 \quad and \quad f_z(c) \neq 0, \quad or \quad f_z(c) = 0 \quad and \quad f_{\bar{z}}(c) \neq 0.$$

The function

$$\frac{f_z(c) - f_{\bar{z}}(c)}{f_z(c) + f_{\bar{z}}(c)}, \qquad c \in G,$$

is consequently well defined, and takes only the values 1 and -1. Since this function is continuous by hypothesis, it must map the connected set G onto a connected image in $\{-1, 1\}$, i.e., it must be constant in G. This means that either $f_{\bar{z}}$ vanishes everywhere and f_z nowhere in G, or that this situation is reversed.   □

It is clear that holomorphic (or anti-holomorphic) mappings *cannot* be angle-preserving at any zero of their derivatives f_z (or $f_{\bar{z}}$); thus under the mapping $z \mapsto z^n (n > 1)$ angles at the origin are increased n-fold.

2. Angle- and orientation-preservation, holomorphy. In function theory anti-holomorphic functions are rather unwelcome and in order to effectively legislate them out of statement i) in theorem 1 we introduce the concept of orientation-preserving mappings. A real-differentiable function $f = u + iv$ is called *orientation-preserving* at $c \in D$ if the Jacobian determinant

$$\det \begin{pmatrix} u_x & u_y \\ v_x & v_y \end{pmatrix}$$

is positive at c. (Cf. also M. KOECHER: *Lineare Algebra und analytische Geometrie*, Grundwissen Mathematik, Bd. 2, Springer-Verlag (1985), Berlin.) As example 4) in 1.2.3 shows, a holomorphic function f is orientation-preserving at every point c where $f'(c) \neq 0$. The Jacobian determinant of an anti-holomorphic function is never positive (proof!); accordingly such functions are nowhere orientation-preserving. In the light of this it is clear from theorem 1 that

Theorem. *The following assertions about a real-differentiable function $f : D \to \mathbb{C}$ are equivalent:*

  i) *f is holomorphic and f' is zero-free in D.*

  ii) *f is both angle-preserving and orientation-preserving in D.*

Remarks on terminology. More often in English (and in French) when discussing holomorphic functions one sees the word *"conformal"* ("conforme") instead of "angle-preserving". However, the term "conformal" is sometimes also used as a synonym for "biholomorphic" (defined in §2 below). As we shall see later, a holomorphic mapping is angle-preserving if and only if it is locally biholomorphic, and so the difference between the two usages of "conformal" comes down to local versus global injectivity. When "conformal" is used in the latter global sense, the term "locally conformal" is then naturally expropriated for "angle-preserving". Angle-preserving anti-holomorphic mappings are sometimes called *indirectly conformal* or *anti-conformal*. As there seems to be no unanimity on these usages, the reader has to proceed with caution in the literature.

3. Geometric significance of angle-preservation. First let's recall the geometric significance of the tangent mapping $Tf(c)$ at $c \in D$ of a real-differentiable mapping $f : D \to \mathbb{C}$. We consider paths $\gamma : [a, b] \to D$, $t \mapsto \gamma(t) = x(t) + iy(t)$ which pass through c, say, $\gamma(\xi) = c$ for some ξ satisfying $a < \xi < b$. Say that γ is *differentiable at* ξ if the derivatives $x'(\xi)$ and $y'(\xi)$ exist and in that case set $\gamma'(\xi) := x'(\xi) + iy'(\xi)$. [Paths with these and other differentiability properties will play a central role in our later development of the integral calculus of functions in domains in $\mathbb{C}$.] In case $\gamma'(\xi) \neq 0$, the path has a *tangent* (*line*) at c, given by the mapping

$$\mathbb{R} \to \mathbb{C}, \ t \mapsto c + \gamma'(\xi)t, \qquad t \in \mathbb{R}.$$

The mapping

$$f \circ \gamma : [a, b] \to \mathbb{C}, \ t \mapsto f(\gamma(t)) = u(x(t), y(t)) + iv(x(t), y(t))$$

is called the *image path* (*of* γ *under* $f = u + iv$). Along with γ, $f \circ \gamma$ is also differentiable at ξ and indeed (the chain rule once again!)

$$\begin{aligned}(f \circ \gamma)'(\xi) &= u_x(c)x'(\xi) + u_y(c)y'(\xi) + i(v_x(c)x'(\xi) + v_y(c)y'(\xi)) \\ &= Tf(c)(\gamma'(\xi)).\end{aligned}$$

In case $(f \circ \gamma)'(\xi) \neq 0$, the image path has a tangent at $f(c)$; this "image tangent" is then given by

$$\mathbb{R} \to \mathbb{C}, \ t \mapsto f(c) + Tf(c)(\gamma'(\xi))t.$$

In somewhat simplified language, if we call $\gamma'(\xi)$ the tangent direction (of the path γ at c), then (see the figures on the left on the next page)

The differential $Tf(c)$ maps tangent directions of differentiable paths onto tangent directions of the image paths.

In particular this makes the denomination "tangent mapping" for the differential $Tf(c)$ understandable.

After these preparations it is easy to explain, using a somewhat naive interpretation of the angle between curves, the significance of the term "angle-preserving": If γ_1, γ_2 are two differentiable paths through c with tangent directions $\gamma_1'(c), \gamma_2'(c)$ at c, then $\angle(\gamma_1'(c), \gamma_2'(c))$ measures the *angle*

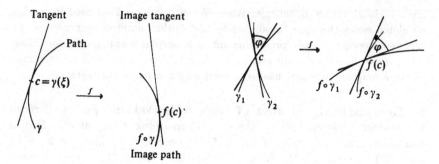

of intersection φ between these paths at c. The angle-preserving character of f at c therefore means that : *If two paths γ_1, γ_2 intersect at the point c in the angle φ, then the image curves $f \circ \gamma_1, f \circ \gamma_2$ intersect at the image point $f(c)$ in the same angle φ.*

Evidently two cases now have to be distinguished: the angle φ "together with its sense, or direction of rotation" is conserved (as in the right-hand figures above) or "the sense of φ is reversed", as evidently occurs under the conjugation mapping $z \mapsto \bar{z}$. This "reversal of orientation" in fact is manifested by all anti-holomorphic mappings and by contrast "conservation of orientation" occurs with all holomorphic mappings. In fact (cf. 5):

Angle- and orientation-preservation together amount to "angle-preserva-tion with conservation of direction"; this is what Riemann called "Aehn-lichkeit in den kleinsten Theilen (similarity in the smallest parts)".

4. Two examples. Under a holomorphic mapping, paths which intersect orthogonally have orthogonally intersecting image paths. In particular, an "orthogonal net" is mapped onto another such net. We offer here two simple but very instructive examples of this highly graphic state of affairs.

First example. The mapping $f : \mathbb{C}^\times \to \mathbb{C}^\times$, $z \mapsto z^2$ is holomorphic and $f'(c) = 2c \neq 0$ at every point $c \in \mathbb{C}^\times$. Consequently f is angle-preserving. We have

$$u = \Re f = x^2 - y^2 , \qquad v = \Im f = 2xy.$$

The lines $x = a$, parallel to the y-axis, and $y = b$, parallel to the x-axis, are thus mapped onto the parabolas $v^2 = 4a^2(a^2 - u)$ and $v^2 = 4b^2(b^2 + u)$, respectively, which all have their foci at the origin. The parabolas of the first family open to the left and those of the second family to the right; and parabolas from the two families intersect at right angles. The "level-lines" $u = a$ and $v = b$ are, on the other hand, hyperbolas in the (x, y)-plane which have the diagonals, respectively, the coordinate axes as asymptotes and intersect each other orthogonally.

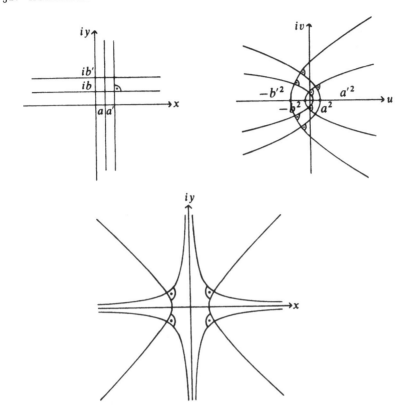

Second example. The mapping $q : \mathbb{C}^\times \to \mathbb{C}$, $z \mapsto \frac{1}{2}(z + z^{-1})$ is holomorphic. Since $q'(z) = \frac{1}{2}(1 - z^2)$, q is angle-preserving throughout $\mathbb{C}^\times \setminus \{-1, 1\}$. If we set $r := |z|$, $\xi := x/r$ and $\eta := y/r$, then

$$ u = \Re q = \frac{1}{2}(r + r^{-1})\xi , \qquad v = \Im q = \frac{1}{2}(r - r^{-1})\eta. $$

From which we infer (because $\xi^2 + \eta^2 = 1$) that

$$ \frac{u^2}{[\frac{1}{2}(r + r^{-1})]^2} + \frac{v^2}{[\frac{1}{2}(r - r^{-1})]^2} = 1 \quad \text{and} \quad \frac{u^2}{\xi^2} - \frac{v^2}{\eta^2} = 1. $$

These equations show that the q-image of every circle $|z| = r < 1$ is an *ellipse* in the (u, v)-plane having *major diameter* $r + r^{-1}$ and *minor diameter* $r^{-1} - r$; and the q-image of every radial segment $z = ct$, $0 < t < 1$, c fixed and $|c| = 1$, is *half of a branch of a hyperbola*.

All these ellipses and hyperbolas are confocal (the common foci being -1 and 1). Because every circle $|z| = r$ cuts every radial segment $z = ct$ orthogonally, every ellipse cuts every hyperbola orthogonally. The mapping q will be put to use in 12.1.6 to effect an "integration-free" proof of LAURENT's theorem.

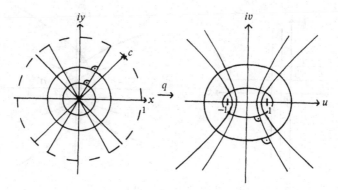

5. Historical remarks on conformality. In the classical literature angle-preserving mappings were designated by the locution "in den kleinsten Theilen ähnlich (similar in the smallest parts)". The first work to deal with such mappings was written in 1825 by GAUSS (*Werke* 4, 189-216): *Allgemeine Auflösung der Aufgabe : Die Theile einer gegebenen Fläche auf einer andern gegebenen Fläche so abzubilden, dass die Abbildung dem Abgebildeten in den kleinsten Theilen ähnlich wird (Als Beantwortung der von der königlichen Societät der Wissenschaften in Copenhagen für 1822 aufgegebenen Preisfrage*; partial English translation by H. P. EVANS in Vol. II of D. E. SMITH, *A Source Book in Mathematics*, Dover Publications Inc., New York (1958); full English translation in *Philosophical Magazine* 4 (1828), 104-113 and 206-215.) GAUSS recognized, among other things, that angle-preserving mappings between domains in the plane $\mathbb{R}^2 = \mathbb{C}$ were just those that could be described by holomorphic or anti-holomorphic functions (although of course, he did not use the language of function theory).

With RIEMANN the geometric significance of holomorphic functions as angle-preserving mappings is strongly in the foreground: he represents the numbers $z = x + iy$ and $w = u + iv$ as points in two planes A and B, respectively, and writes in 1851 ([R], p.5): "Entspricht jedem Werthe von z ein bestimmter mit z sich stetig ändernder Werth von w, mit andern Worten, sind u und v stetige Functionen von x, y, so wird jedem Punkt der Ebene A ein Punkt der Ebene B, jeder Linie, allgemein zu reden, eine Linie, jedem zusammenhängenden Flächenstücke ein zusammenhängendes Flächenstück entsprechen. Man wird sich also diese Abhängigkeit der Größe w von z vorstellen können als eine Abbildung der Ebene A auf der Ebene B. (If to every value of z there corresponds a definite value of w which changes continuously with z, in other words, if u and v are continuous functions of x and y, then to every point in the plane A will correspond a point in the plane B, to every line (generally speaking) will correspond a line and to every connected piece of area will correspond a connected piece of area. Thus one can imagine this dependence of the magnitude w on z as a mapping of the plane A on the plane B.)" Thereupon, in half a page, he confirms that in the case of holomorphy "zwischen den kleinsten Theilen der Ebene A und ihres Bildes auf der Ebene B Aehnlichkeit statt [findet] (similarity obtains between the smallest parts of the plane A and their images on the plane B)".

Angle-preserving mappings play no role at all in the work of CAUCHY and WEIERSTRASS.

Exercises

In the first two exercises f denotes the holomorphic mapping $z \mapsto z^{-1}$ of $\mathbb{C}^\times$ onto $\mathbb{C}^\times$. This map is its own inverse and preserves angles and orientation.

Exercise 1. Let L be a circle in $\mathbb{C}$ with center $M \in \mathbb{C}$ and radius $R > 0$. Show that:

a) If $M = 0$, then $f(L)$ is the circle around 0 of radius R^{-1}.

b) If $M \neq 0$ and $R \neq |M|$, then $f(L)$ is the circle around $\overline{M}/(|M|^2 - R^2)$ of radius $R/||M|^2 - R^2|$.

c) If $M \neq 0$ and $R = |M|$, then $f(L \setminus \{0\})$ is the straight line through $(2M)^{-1}$ which is perpendicular to the segment joining 0 and $(2M)^{-1}$, that is, $f(L \setminus \{0\}) = \{z \in \mathbb{C} : \langle z - (2M)^{-1}, (2M)^{-1} \rangle = 0\} = \{x + iy , x, y \in \mathbb{R} : x \Re(2M)^{-1} + y \Im(2M)^{-1} = |2M|^{-2}\}$.

d) Let $a \in \mathbb{C}^\times$, H be the (real) line through 0 and a, H' that through 0 and a^{-1}. Show that $f(H \setminus \{0\}) = H' \setminus \{0\}$.

Since f is self-inverse, the image under f of every straight line and every circle in $\mathbb{C}$ can be determined with the aid of Exercise 1.

Exercise 2. Determine the image $f(G)$ of the following regions G:

a) $G := \mathbb{E} \cap \mathbb{H}$, with $\mathbb{H}$ the open upper half-plane.

b) $G := \mathbb{E} \cap B_1(1)$.

c) G the open triangle with vertices $0, 1$ and i.

d) G the open square with vertices $0, 1, 1 + i$ and i.

Hint. Use Exercise 1 above and Exercise 2 from Chapter 0, §6.

Exercise 3. Let $q : \mathbb{C}^\times \to \mathbb{C}$ be given by $q(z) := \frac{1}{2}(z + z^{-1})$. Show that:

a) q is surjective.

b) For $c \in \mathbb{C}^\times$, $q(c)$ is real if and only if either $c \in \mathbb{R} \setminus \{0\}$ or $|c| = 1$.

c) $q(\partial \mathbb{E}) = [-1, 1]$.

d) For each point $w \in \mathbb{C} \setminus \{-1, 1\}$ the q-preimage $q^{-1}(\{w\}) = \{z \in \mathbb{C}^\times : q(z) = w\}$ consists of exactly two points. If $w \in \mathbb{C} \setminus [-1, 1]$, then one of these points lies in $\mathbb{E}^\times := \mathbb{E} \setminus \{0\}$, the other in $\mathbb{C} \setminus \mathbb{E}$.

e) q maps $\mathbb{E}^\times$ bijectively onto $\mathbb{C} \setminus [-1, 1]$.

f) q maps the upper half-plane $\mathbb{H} := \{z \in \mathbb{C} : \Im z > 0\}$ bijectively onto $\mathbb{C} \setminus \{x \in \mathbb{R} : |x| \geq 1\}$.

Exercise 4. Let $q : \mathbb{C}^\times \to \mathbb{C}$ be as in Exercise 3 and s, R, σ real numbers satisfying the following relations:

$$s > 1 + \sqrt{2}, \qquad R = \frac{1}{2}(s - s^{-1}), \qquad \sigma = R + \sqrt{R^2 - 1}.$$

Confirm the following inclusions:

$$\{z \in \mathbb{C} : \sigma^{-1} < |z| < \sigma\} \subset q^{-1}(B_R(0)) \subset \{z \in \mathbb{C} : s^{-1} < |z| < s\}.$$

§2   Biholomorphic mappings

A holomorphic function $f \in \mathcal{O}(D)$ is called a *biholomorphic mapping* of D onto D' if $D' := f(D)$ is a domain and the mapping $f : D \to D'$ has an inverse mapping $f^{-1} : D' \to D$ which is holomorphic in D'. In such cases we write suggestively

$$f : D \xrightarrow{\sim} D'.$$

The inverse mapping is itself then biholomorphic. Biholomorphic mappings are injective. In 9.4.1 we will see that for every holomorphic injection $f : D \to \mathbb{C}$ the image $f(D)$ is *automatically open* in $\mathbb{C}$ and the (set-theoretic) inverse mapping $f^{-1} : f(D) \to D$ is *automatically holomorphic in $f(D)$*. Trivial but useful is the remark that

$f \in \mathcal{O}(D)$ is a biholomorphic mapping of D onto D' precisely when there exists a $g \in \mathcal{O}(D')$ such that:

$$f(D) \subset D' , \ g(D') \subset D , \ f \circ g = \mathrm{id}_{D'} \quad \text{and} \quad g \circ f = \mathrm{id}_D.$$

Proof. Because $f \circ g$ is the identity mapping (on D'), we have $f(D) = D'$ and because $g \circ f$ is the identity mapping (on D), we have $g(D') = D$; but then g is the inverse mapping $f^{-1} : D' \to D$ of $f : D \to D'$. $\square$

The reader won't have any trouble proving the

Theorem on compositions. *If $f : D \overset{\sim}{\to} D'$ and $g : D' \overset{\sim}{\to} D''$ are biholomorphic mappings, then the compositie mapping $g \circ f : D \to D''$ is biholomorphic.*

1. Complex 2×2 matrices and biholomorphic mappings. To every complex matrix $A = \begin{pmatrix} a & b \\ c & d \end{pmatrix}$ with $(c, d) \neq (0, 0)$ is associated the *fractional linear rational* function

$$h_A(z) := \frac{az + b}{cz + d} \in \mathbb{C}(z).$$

It satisfies $h'_A(z) = \frac{\det A}{(cz+d)^2}$, where $\det A := ad - bc$; so in case $\det A = 0$, h_A is constant.

In what follows only functions h_A whose matrices A have $\det A \neq 0$, i.e., are invertible matrices, will be considered. The set of matrices having non-zero determinants forms a group under matrix multiplication, the *general linear group*, and is denoted by $GL(2, \mathbb{C})$; the neutral element of this group is the identity matrix $E := \begin{pmatrix} 1 & 0 \\ 0 & 1 \end{pmatrix}$.

We take note of two fundamental rules of calculation:

1) $h_A = \text{id} \Leftrightarrow A = aE$ for some $a \in \mathbb{C}^\times$.

For all $A, B \in GL(2, \mathbb{C})$ a "substitution rule" holds:

2) $h_{AB} = h_A \circ h_B$, i.e., $h_{AB}(z) = h_A(h_B(z))$.

The proofs are just simple calculations. □

In case $A = \begin{pmatrix} a & b \\ 0 & d \end{pmatrix}$, $h_A \in \mathcal{O}(\mathbb{C})$ and the mapping $h_A : \mathbb{C} \to \mathbb{C}$ is biholomorphic. More interesting however is the case $c \neq 0$. A direct verification shows that:

In case $A = \begin{pmatrix} a & b \\ c & d \end{pmatrix} \in GL(2, \mathbb{C})$ and $c \neq 0$, $h_A \in \mathcal{O}(\mathbb{C} \setminus \{-c^{-1}d\})$; the mapping $h_A : \mathbb{C} \setminus \{-c^{-1}d\} \overset{\sim}{\to} \mathbb{C} \setminus \{ac^{-1}\}$ is biholomorphic and $h_{A^{-1}}$ is its inverse mapping.

2. The biholomorphic Cayley mapping $\mathbb{H} \overset{\sim}{\to} \mathbb{E}$, $z \mapsto \dfrac{z - i}{z + i}$. The *upper half-plane*

$$\mathbb{H} := \{z \in \mathbb{C} : \Im z > 0\}$$

is an *unbounded* region in $\mathbb{C}$. We will show that nevertheless $\mathbb{H}$ can be *biholomorphically mapped onto the (bounded) unit disc* $\mathbb{E}$. To this end we introduce the matrices

$$C := \begin{pmatrix} 1 & -i \\ 1 & i \end{pmatrix} , C' := \begin{pmatrix} i & i \\ -1 & 1 \end{pmatrix} \in GL(2, \mathbb{C});$$

they satisfy

(*) $$CC' = C'C = 2iE.$$

To C and C' are associated the rational functions

$$h_C(z) = \frac{z-i}{z+i} \in \mathcal{O}(\mathbb{C} \setminus \{-i\}) , \quad h_{C'}(z) = i\frac{1+z}{1-z} \in \mathcal{O}(\mathbb{C} \setminus \{1\}).$$

By virtue of (*) and the rules of calculations from subsection 1, we have: $h_C \circ h_{C'} = h_{C'} \circ h_C = \text{id}$. And direct computation confirms that

$$1 - |h_C(z)|^2 = \frac{4\Im z}{|z+i|^2} \qquad \text{for } z \neq -i,$$

$$\Im h_{C'}(z) = \frac{1-|z|^2}{|1-z|^2} \qquad \text{for } z \neq 1.$$

These equations imply that

$$1 - |h_C(z)|^2 > 0 \qquad \text{for } \Im z > 0 ; \qquad \Im h_{C'}(z) > 0 \qquad \text{for } |z| < 1$$

and consequently $h_C(\mathbb{H}) \subset \mathbb{E}$ and $h_{C'}(\mathbb{E}) \subset \mathbb{H}$. Together with the above descriptions of the composites of h_C, $h_{C'}$ with one another, this says that h_C maps $\mathbb{H}$ biholomorphically onto $\mathbb{E}$ with $h_C^{-1} = h_{C'}$. This discussion has proved the

Theorem. *The mapping* $h_C : \mathbb{H} \xrightarrow{\sim} \mathbb{E}, z \mapsto \dfrac{z-i}{z+i}$ *is biholomorphic with inverse mapping* $h_{C'} : \mathbb{E} \xrightarrow{\sim} \mathbb{H}, z \mapsto i\dfrac{1+z}{1-z}$.

For historical reasons the mappings $h_C, h_{C'}$ are called the *Cayley mappings* of $\mathbb{H}$ onto $\mathbb{E}$ and $\mathbb{E}$ onto $\mathbb{H}$, respectively.

3. Remarks on the Cayley mapping. A critical reader might ask: "How would the function $h_C(z) = \frac{z-i}{z+i}$ ever occur to one as a candidate for a biholomorphic mapping $\mathbb{H} \xrightarrow{\sim} \mathbb{E}$?"

The general question: "Are $\mathbb{H}$ and $\mathbb{E}$ biholomorphically equivalent?" offers no hint of the role of this function. But once it is known, the rest is routine verifications; the real mathematical contribution consists just in writing down this function.

With hindsight we can discern certain simple heuristic considerations that could have led to the function h_C. First suppose one already recognizes the fractional linear transformations as an interesting class of functions (which itself requires some mathematical experience) and is prescient enough to look for a biholomorphic mapping $\mathbb{H} \xrightarrow{\sim} \mathbb{E}$ among them. Then it isn't wild to speculate that the sought-for function h will map the boundary of $\mathbb{H}$, i.e., the real axis, into the boundary of $\mathbb{E}$, i.e., into the unit circle. (Or *is* that plausible?: the circle is compact but the line is not!) So we try to map some pre-assigned points of $\mathbb{R}$ to certain points of the circle, say (here we have considerable freedom),

$$h(0) := -1 \; ; \qquad h(1) := -i \; ; \qquad h(\infty) := 1 \text{ , that is, } \lim_{|x| \to \infty} h(x) = 1.$$

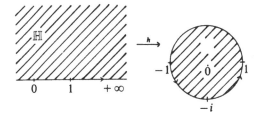

For a function of the form $h(z) = \frac{az+b}{cz+d}$ these requirements translate into the equations

$$-1 = h(0) = \frac{b}{d} \; ; \qquad -i = h(1) = \frac{a+b}{c+d} \; ; \qquad 1 = h(\infty) = \frac{a+b/\infty}{c+d/\infty} = \frac{a}{c}.$$

From which one sees that $b = -d$, $a = c$ and so (via the second equation) $a - d = -ia - id$, that is, $d = ia$. With these specifications the function h turns out to be the Cayley mapping.

4*. Bijective holomorphic mappings of $\mathbb{H}$ and $\mathbb{E}$ onto the slit plane. It is quite surprising what can be done with just the Cayley mapping h_C and the squaring function z^2. Let $\mathbb{C}^-$ denote the *complex plane slit along the negative real axis*, i.e.,

$$\mathbb{C}^- := \mathbb{C} \setminus \{z \in \mathbb{C} : \Re z \leq 0 \, , \, \Im z = 0\}.$$

First we claim that

The mapping $q : \mathbb{H} \to \mathbb{C}^-$, $z \mapsto -z^2$ is holomorphic and bijective.

(Were we to have slit $\mathbb{C}$ along the positive real axis, then we would consider z^2 here instead of $-z^2$. The reason for slitting along the negative real axis has to do with the complex logarithm function to be introduced later (in 5.4.4): this can only be done in a slit plane and we are reluctant to discard the positive real axis where the classical real logarithm has lived all along.)

Proof. There is no $c \in \mathbb{H}$ such that $t := q(c)$ is real and non-positive, since from $q(c) = -c^2 = t \in \mathbb{R}$ follows $c^2 = -t \geq 0$ and therewith $c \in \mathbb{R}$, $c \notin \mathbb{H}$. It follows that $q(\mathbb{H}) \subset \mathbb{C}^-$. For $c, c' \in \mathbb{H}$, $q(c) = q(c')$ occurs if and only if $c' = \pm c$ and since not both c and $-c$ can lie in $\mathbb{H}$, it must be true that $c' = c$; thus $q : \mathbb{H} \to \mathbb{C}^-$ is injective.

Every point $w \in \mathbb{C}^-$ has a q-pre-image in $\mathbb{H}$, because the quadratic equation $z^2 = -w$ has two distinct roots, one of which lies in $\mathbb{H}$ — cf. the explicit square-root formula at the end of 0.1.3.   □

As a simple consequence we note that

The mapping $p : \mathbb{E} \to \mathbb{C}^-$, $z \mapsto \left(\frac{z+1}{z-1}\right)^2$ is holomorphic and bijective.

Proof. Via $q \circ h_{C'}$ the region $\mathbb{E}$ is mapped holomorphically and bijectively onto $\mathbb{C}^-$, and $q(h_{C'}(z)) = -\left(i\frac{1+z}{1-z}\right)^2 = \left(\frac{z+1}{z-1}\right)^2$, that is, $q \circ h_{C'} = p$.   □

Scarcely anyone would trust that such a seemingly simple function as p could really map the *bounded unit disc* bijectively and conformally onto the *whole plane minus its negative real axis* (cf. the figure).

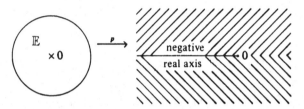

It is appropriate to mention here that the mappings $q : \mathbb{H} \to \mathbb{C}^-$ and $p : \mathbb{E} \to \mathbb{C}^-$ are even biholomorphic. This is because the inverse mapping $q^{-1} : \mathbb{C}^- \to \mathbb{H}$ is automatically holomorphic, as will be proved in 9.4.1.

Remark. There is no biholomorphic mapping of the unit disc $\mathbb{E}$ onto the whole plane $\mathbb{C}$, a fact which follows from LIOUVILLE's theorem (cf. 8.3.3). However, the famous theorem announced by RIEMANN in 1851 ([R], p. 40) says that *every simply connected region other than $\mathbb{C}$ can be biholomorphically mapped onto $\mathbb{E}$.* But we won't reach this remarkable theorem until the second volume.

Exercises

In what follows Q_1 denotes the first quadrant $\{z \in \mathbb{H} : \Re z > 0\}$ of the complex plane.

Exercise 1. Show that the Cayley transformation $h_C : \mathbb{H} \to \mathbb{E}$, $z \mapsto \frac{z-i}{z+i}$, maps Q_1 biholomorphically onto $\{w \in \mathbb{E} : \Im w < 0\}$.

Exercise 2. Supply holomorphic, bijective and angle-preserving mappings of Q_1 onto $\mathbb{E} \setminus (-1, 0]$ and of Q_1 onto $\mathbb{E}$.

Exercise 3. Let $f(z) := \frac{az+b}{cz+d}$ with $c \neq 0$ and $ad - bc \neq 0$. Further, let L be a circle or a (real) line in $\mathbb{C}$. With the aid of Exercise 1 of §1 determine the images $f(L)$, respectively, $f(L \setminus \{-d/c\})$. (Case distinctions have to be made!) *Hint.* First prove that f can be written as

$$f(z) = \frac{bc - ad}{c^2}(z + d/c)^{-1} + \frac{a}{c}.$$

Exercise 4. For $M > R > 0$ form the punctured upper half-plane $\mathbb{H} \setminus \overline{B_R(iM)}$. With the aid of a fractional linear transformation map it biholomorphically onto an annulus of the form $\{w \in \mathbb{C} : p < |w| < 1\}$. *Hint.* First look for a $c < 0$ such that $z \mapsto (z - ic)^{-1}$ maps the boundary of $B_R(iM)$ and the real axis onto concentric circles. Use Exercise 1 of §1.

Exercise 5. Find a map of $\mathbb{E}$ onto $\{w \in \mathbb{C} : \Im w > (\Re w)^2\}$ which is holomorphic, bijective and angle-preserving.

§3 Automorphisms of the upper half-plane and the unit disc

A biholomorphic mapping $h : D \overset{\sim}{\to} D$ of a domain D onto itself is called an *automorphism* of D. The set of all automorphisms of D will be denoted by Aut D. Pursuant to the remarks in the introduction of §2, it is clear that

Aut D *is a group with respect to the composition of mappings and the identity mapping* id *is its neutral element.*

The group Aut $\mathbb{C}$ contains, for example, all "affine linear" mappings $z \mapsto az + b$, $a \in \mathbb{C}^\times$, $b \in \mathbb{C}$ and because of this it is not commutative. Aut $\mathbb{C}$ is actually rather simple, for these affine linear maps exhaust it; but we will only see that via the theorem of CASORATI and WEIERSTRASS in 10.2.2.

In this section we will study the groups Aut $\mathbb{H}$ and Aut $\mathbb{E}$ exclusively. First we consider the upper half-plane $\mathbb{H}$ (in subsection 1) and then (in subsection 2) transfer the results about $\mathbb{H}$ to $\mathbb{E}$ via Cayley mappings. In subsection 3 we give a somewhat different representation of the automorphisms of $\mathbb{E}$; finally (subsection 4) we show that $\mathbb{H}$ and $\mathbb{E}$ are homogeneous with respect to their automorphisms.

1. Automorphisms of $\mathbb{H}$. The sets $GL^+(2, \mathbb{R})$ and $SL(2, \mathbb{R})$ of all *real* 2 × 2 matrices with *positive* determinant, respectively, with determinant 1, are each groups with respect to matrix multiplication (proof!). We write

$A = \begin{pmatrix} \alpha & \beta \\ \gamma & \delta \end{pmatrix}$ for the typical such matrix and designate with $h_A(z) = \frac{\alpha z + \beta}{\gamma z + \delta}$, as we did in 2.1, the fractional linear transformation associated with A. Then

$$(1) \quad \Im h_A(z) = \frac{\det A}{|\gamma z + \delta|^2} \Im z \quad \text{for every } A = \begin{pmatrix} \alpha & \beta \\ \gamma & \delta \end{pmatrix} \in GL^+(2, \mathbb{R}).$$

Proof. Because A is real

$$
\begin{aligned}
2i \Im h_A(z) &= h_A(z) - \overline{h_A(z)} \\
&= \frac{\alpha z + \beta}{\gamma z + \delta} - \frac{\alpha \bar{z} + \beta}{\gamma \bar{z} + \delta} = \frac{\alpha \delta - \beta \gamma}{|\gamma z + \delta|^2}(z - \bar{z}) \\
&= 2i \frac{\det A}{|\gamma z + \delta|^2} \Im z.
\end{aligned}
$$

From (1) follows immediately

Theorem. *For every matrix $A \in GL^+(2, \mathbb{R})$ the mapping $h_A : \mathbb{H} \to \mathbb{H}$ is an automorphism of the upper half-plane and $h_{A^{-1}}$ is its inverse mapping.*

Proof. Since $A, A^{-1} \in GL^+(2, \mathbb{R})$, h_A and $h_{A^{-1}}$ are holomorphic in $\mathbb{H}$. On account of (1), $h_A(\mathbb{H}) \subset \mathbb{H}$ and $h_{A^{-1}}(\mathbb{H}) \subset \mathbb{H}$. Finally, $h_A \circ h_{A^{-1}} = h_{A^{-1}} \circ h_A = \text{id}$ puts h_A in $\text{Aut } \mathbb{H}$.   □

Furthermore (with the help of the rules in 1.2) we now get:

The mapping $GL^+(2, \mathbb{R}) \to \text{Aut } \mathbb{H}$ given by $A \to h_A$ is a group homomorphism whose kernel consists of the matrices λE, $\lambda \in \mathbb{R} \setminus \{0\}$. The restriction of this homomorphism to the subgroup $SL(2, \mathbb{R})$ has the same image group and its kernel consists of just the two matrices $\pm E$.

2. Automorphisms of $\mathbb{E}$. If $f : D \xrightarrow{\sim} D'$ is biholomorphic, then the mapping $h \mapsto f \circ h \circ f^{-1}$ effects a *group isomorphism* of $\text{Aut } D$ onto $\text{Aut } D'$ (proof!). Thus, knowing f, f^{-1} and automorphisms of D, we can construct automorphisms of D'. Applying this process to the Cayley mapping $h_C : \mathbb{H} \to \mathbb{E}$ together with its inverse $h_{C'}$ shows, in view of theorem 1, that all the functions

$$h_C \circ h_A \circ h_{C'} = h_{CAC'} \quad \text{with } A \in SL(2, \mathbb{R})$$

are automorphisms of $\mathbb{E}$. This observation leads to the following

Theorem. *The set* $M := \{B := \begin{pmatrix} a & b \\ \bar{b} & \bar{a} \end{pmatrix} : a, b \in \mathbb{C}, \det B = 1\}$ *is a subgroup of* $SL(2, \mathbb{C})$ *and the mapping from* $SL(2, \mathbb{R})$ *into* M *defined by* $A \mapsto \frac{1}{2i}CAC'$ *is a group isomorphism.*

The mapping $B \mapsto h_B(z) = \dfrac{az + b}{\bar{b}z + \bar{a}}$ *of* M *into* $\operatorname{Aut} \mathbb{E}$ *is a group homomorphism whose kernel consists of the two matrices* $\pm E$.

Proof. Since $\det C = \det C' = 2i$, the multiplicativity of the determinant gives $\det(\frac{1}{2i}CAC') = 1$. Therefore the mapping

$$\varphi : SL(2, \mathbb{R}) \to SL(2, \mathbb{C}), \quad A \mapsto \frac{1}{2i}CAC'$$

is well defined and, thanks to the fact $CC' = C'C = 2iE$, a group monomorphism. The first claim in the theorem therefore follows as soon as we have shown that the image of φ is M. To this end note that for $A = \begin{pmatrix} \alpha & \beta \\ \gamma & \delta \end{pmatrix}$

$$
\begin{aligned}
CAC' &= \begin{pmatrix} 1 & -i \\ 1 & i \end{pmatrix}\begin{pmatrix} \alpha & \beta \\ \gamma & \delta \end{pmatrix}\begin{pmatrix} i & i \\ -1 & 1 \end{pmatrix} \\
(2) \qquad &= i\begin{pmatrix} \alpha + \delta + i(\beta - \gamma) & \alpha - \delta - i(\beta + \gamma) \\ \alpha - \delta + i(\beta + \gamma) & \alpha + \delta - i(\beta - \gamma) \end{pmatrix}.
\end{aligned}
$$

Upon setting $a := \frac{1}{2}[(\alpha + \delta) + i(\beta - \gamma)]$ and $b := \frac{1}{2}[(\alpha - \delta) - i(\beta + \gamma)]$, it follows that

$$B := \varphi(A) = \begin{pmatrix} a & b \\ \bar{b} & \bar{a} \end{pmatrix} \quad, \text{ and so image of } \varphi \subset M.$$

The other inclusion $M \subset$ image of φ requires that for every $B = \begin{pmatrix} a & b \\ \bar{b} & \bar{a} \end{pmatrix} \in M$ there be an $A = \begin{pmatrix} \alpha & \beta \\ \gamma & \delta \end{pmatrix} \in SL(2, \mathbb{R})$ with $2iB = CAC'$. To realize such an A it suffices to set

$$\alpha := \Re(a + b), \quad \beta := \Im(a - b), \quad \gamma := -\Im(a + b), \quad \delta := \Re(a - b).$$

The matrix A so defined is then real and satisfies $CAC' = 2iB$, thanks to (2), which holds for *all real* 2×2 matrices A. Of course, $\det B = 1$ means that $\det A = 1$, so $A \in SL(2, \mathbb{R})$, as desired.

To verify the second claim, recall that by 2.1(1) $h_B = h_C \circ h_A \circ h_{C'}$ whenever $B = \frac{1}{2i}CAC'$. Therefore (1) says that for all $B \in M$ the functions h_B are automorphisms of $\mathbb{E}$. The homomorphic property of the mapping $M \to \operatorname{Aut} \mathbb{E}$ and the assertion about its kernel now follow from 2.1(2) and 2.1(1). $\qquad \square$

The subgroup M of $SL(2,\mathbb{C})$, which according to our theorem is isomorphic to $SL(2,\mathbb{R})$, is often designated by $SU(1,1)$.

3. The encryption $\eta\frac{z-w}{\overline{w}z-1}$ for automorphisms of $\mathbb{E}$.

The automorphisms of $\mathbb{E}$ furnished by theorem 2 can be written in another way; to do it we need a matrix-theoretic

Lemma. *To every matrix* $W = \begin{pmatrix} \eta & -\eta w \\ \overline{w} & -1 \end{pmatrix}, \eta \in \partial\mathbb{E}, w \in \mathbb{E}$, *corresponds a matrix* $B \in M$ *such that* $W = sB$ *for some* $s \in \mathbb{C}^\times$.

Proof. Since $w \in \mathbb{E}$, $1 - |w|^2 > 0$ and $\frac{-\eta}{1-|w|^2} \in \mathbb{C}^\times$. So by 0.1.3 there exists an $a \in \mathbb{C}^\times$ with $a^2 = \frac{-\eta}{1-|w|^2}$. For $b := -wa$, then $|a|^2 - |b|^2 = |\eta| = 1$, and so $B := \begin{pmatrix} a & b \\ \overline{b} & \overline{a} \end{pmatrix} \in M$. If we set $s := \eta a^{-1} = -a(1 - |w|^2)$, then $s\overline{a} = -1$ (so $s \in \mathbb{C}^\times$!) and $sB = W$. $\square$

Bearing in mind that the functions h_W and h_B coincide when W, B are related as in the lemma, it follows directly from theorem 2 that

Theorem. *Every function* $z \mapsto \eta\frac{z-w}{\overline{w}z-1}$ *with* $\eta \in \partial\mathbb{E}$ *and* $w \in \mathbb{E}$*, defines an automorphism of* $\mathbb{E}$.

In case $w = 0$ the automorphism is $z \mapsto \eta z$, a rotation about the origin. A special role is played by the automorphisms

$$(1) \qquad\qquad g : \mathbb{E} \xrightarrow{\sim} \mathbb{E}, \qquad z \mapsto \frac{z-w}{\overline{w}z-1}, \ w \in \mathbb{E}.$$

For them $g(0) = w$, $g(w) = 0$ and $g \circ g = \text{id}$. The latter equality following from 2.1(1) and the calculation

$$\begin{pmatrix} 1 & -w \\ \overline{w} & -1 \end{pmatrix} \begin{pmatrix} 1 & -w \\ \overline{w} & -1 \end{pmatrix} = (1 - |w|^2)E.$$

Because of the property $g \circ g = \text{id}$, the automorphisms g are called *involutions* of $\mathbb{E}$.

4. Homogeneity of $\mathbb{E}$ and $\mathbb{H}$.

A domain D in $\mathbb{C}$ is called *homogeneous with respect to a subgroup* L *of* Aut D, if for every two points $z, \hat{z} \in D$ there is an automorphism $h \in L$ with $h(z) = \hat{z}$. It is also said in such cases that the group L *acts transitively* on D.

Lemma. *If there is a point* $c \in D$ *whose orbit* $\{g(c) : g \in L\}$ *fills* D, *then* D *is homogeneous with respect to* L.

Proof. Given $z, \hat{z} \in D$ there exist $g, \hat{g} \in L$ such that $g(c) = z$ and $\hat{g}(c) = \hat{z}$. Then $h := \hat{g} \circ g^{-1} \in L$ satisfies $h(z) = \hat{z}$. □

Theorem. *The unit disc* $\mathbb{E}$ *is homogeneous with respect to the group* Aut $\mathbb{E}$.

Proof. The point 0 has full orbit: for each $w \in \mathbb{E}$ the function $g(z) := \frac{z-w}{\bar{w}z-1}$ lies in Aut $\mathbb{E}$ (theorem 3) and satisfies $g(0) = w$. The theorem therefore follows from the lemma. □

If D is homogeneous with respect to Aut D and if D' is biholomorphically equivalent to D, then D' is homogeneous with respect to Aut D' (the verification of which is entrusted to the reader). From this and the fact that the Cayley mapping $h_{C'}$ effects a biholomorphic equivalence of $\mathbb{E}$ with $\mathbb{H}$, it is clear that

The upper half-plane $\mathbb{H}$ *is homogeneous with respect to the group* Aut $\mathbb{H}$.

This also follows directly from the lemma, with $c = i$. Every $w \in \mathbb{H}$ has the form $w = \rho + i\sigma^2$ with $\rho, \sigma \in \mathbb{R}$ and $\sigma \neq 0$. For $A := \begin{pmatrix} \sigma & \rho\sigma^{-1} \\ 0 & \sigma^{-1} \end{pmatrix} \in SL(2, \mathbb{R})$, the associated function h_A satisfies $h_A(i) = w$. □

A region G in $\mathbb{C}$ is generally not homogeneous; in fact Aut G most of the time is just $\{\mathrm{id}\}$. But we won't become acquainted with any examples of this until 10.2.4.

Exercises

Exercise 1. Show that the unbounded regions $\mathbb{C}$ and $\mathbb{C}^\times$ are homogeneous with respect to their automorphism groups.

Exercise 2. Let L be a circle in $\mathbb{C}$, a, b two points in $\mathbb{C} \setminus L$. Show that there is a fractional linear transformation f whose domain includes $L \cup \{a\}$ and which satisfies $f(a) = b$, $f(L) = L$.

Exercise 3 (cf. also 9.2.3). Let $g(z) := \eta \frac{z-w}{\bar{w}z-1}$, $\eta \in \partial\mathbb{E}$, $w \in \mathbb{E}$ be an automorphism of $\mathbb{E}$. Show that if g is not the identity map, then g fixes at most one point in $\mathbb{E}$.

Chapter 3

Modes of Convergence in Function Theory

> Die Annäherung an eine Grenze durch Operationen,
> die nach bestimmten Gesetzen *ohne Ende* fortgesetzt
> werden — dies ist der eigentliche Boden, auf welchem
> die transscendenten Functionen erzeugt werden. (The
> approach to a limit via operations which proceed ac-
> cording to definite laws but *without termination* —
> this is the real ground on which the transcendental
> functions are generated.) — GAUSS 1812.

1. Outside of the *polynomials* and *rational functions*, which arise from applying the four basic species of calculation finitely often, there really aren't
any other interesting holomorphic functions available at first. Further functions have to be generated by (possibly multiple) limit processes; thus, for
example, the exponential function $\exp z$ is the limit of its *Taylor polynomials* $\sum_0^n z^\nu/\nu!$ or, as well, the limit of the Euler sequence $(1 + z/n)^n$.
The technique of getting *new* functions via limit processes was described
by GAUSS as follows (*Werke* **3**, p.198): "Die transscendenten Functionen
haben ihre wahre Quelle allemal, offen liegend oder versteckt, im Unendlichen. Die Operationen des Integrirens, der Summationen unendlicher
Reihen $\cdots$ oder überhaupt die Annäherung an eine Grenze durch Operationen, die nach bestimmten Gesetzen *ohne Ende* fortgesetzt werden —
dies ist der eigentliche Boden, auf welchem die transscendenten Functionen erzeugt werden $\cdots$" (The transcendental functions all have their true
source, overtly or covertly, in the infinite. The operation of integration,
the summation of infinite series $\cdots$ or generally the approach to a limit via
operations which proceed according to definite laws but *without termina-
tion* — this is the real ground on which the transcendental functions are
generated $\cdots$)

The point of departure for all limit processes on functions is the concept of *pointwise* convergence, which is as old as the infinitesimal calculus itself. If X is any non-empty set and f_n a sequence of complex-valued functions $f_n : X \to \mathbb{C}$, then this sequence is said to be *convergent at the point* $a \in X$ if the sequence $f_n(a)$ of complex numbers converges in $\mathbb{C}$. The sequence f_n is called *pointwise convergent in a subset* $A \subset X$ if it converges at every point of A: then via

$$f(x) := \lim f_n(x) \ , \ x \in A \ , \text{ the } \textit{limit function } f : A \to \mathbb{C}$$

of the sequence in A is defined; we write, somewhat sloppily, $f := \lim f_n$. This concept of (pointwise) convergence is the most naive one in analysis, and is sometimes also referred to as *simple* convergence.

2. Among real-valued functions simple examples show how *pointwise convergent sequences can have bad properties*: the continuous functions x^n on the interval $[0, 1]$ converge pointwise there to a limit function which is *discontinuous* at the point 1. Such pathologies are eliminated by the introduction of the idea of *locally uniform convergence*. But it is well known that locally uniformly convergent sequences of real-valued functions have quirks too when it comes to differentiation: Limits of differentiable functions are not generally differentiable themselves. Thus, e.g., according to the Weierstrass approximation theorem *every* continuous $f : I \to \mathbb{R}$ from any compact interval $I \subset \mathbb{R}$ is uniformly approximable on I by polynomials, in particular by differentiable functions. A further example of misbehavior is furnished by the functions $n^{-1} \sin(n!x), x \in \mathbb{R}$; they converge uniformly on $\mathbb{R}$ to 0 but their derivatives $(n-1)! \cos(n!x)$ don't converge at any point of $\mathbb{R}$.

For function theory the concept of pointwise convergence is likewise unsuitable. Here however such compelling examples as those above cannot be adduced: *We don't know any simple sequence of holomorphic functions in the unit disc* $\mathbb{E}$ *which is pointwise convergent to a limit function that is not holomorphic.* Such sequences can be constructed with the help of Runge's approximation theorem, but we won't encounter that until the second volume. At that point we will also see, however, why explicit examples are difficult to come by: pointwise convergence of holomorphic functions is necessarily "almost everywhere" locally uniform. In spite of this somewhat pedagogically unsatisfactory situation, one is well advised to emphasize locally uniform convergence from the very beginning. For example, this will allow us later to extend rather effortlessly the various useful theorems from the real domain on the interchange of limit operations and orders of integration. It is nevertheless surprising how readily mathematicians accept this received view; perhaps it's because in our study of the infinitesimal calculus we became fixated so early on the concept of local uniform convergence that we react almost like Pavlov's dogs.

As soon as one knows the Weierstrass convergence theorem, which among other things ensures the unrestricted validity of the relation $\lim f_n' = (\lim f_n)'$ for local uniform convergence, any residual doubt dissipates: no undesired limit functions intrude; *local uniform convergence is the optimal convergence mode for sequences of holomorphic functions.*

3. Besides sequences we also have to consider series of holomorphic functions. But in calculating with even locally uniformly convergent series caution has to be exercised: in general such series need not converge absolutely and so cannot, without further justification, be rearranged at will. WEIERSTRASS confronted these difficulties with his majorant criterion. Later a virtue was made of necessity and the series satisfying the majorant criterion were formally recognized with the name *normally convergent* (cf. 3.3). Normally convergent series are in particular locally uniformly and absolutely convergent; every rearrangement of a normally convergent series is itself normally convergent. In 4.1.2 we will see that because of the classical Abel convergence criterion, power series always converge normally inside their discs of convergence. *Normal convergence is the optimal convergence mode for series of holomorphic functions.*

In this chapter we plan to discuss in some detail the concepts of *locally uniform, compact* and *normal* convergence. X will always designate a metric space.

§1   Uniform, locally uniform, and compact convergence

1. Uniform convergence. A sequence of functions $f_n : X \to \mathbb{C}$ is said to be *uniformly convergent in $A \subset X$ to $f : A \to \mathbb{C}$* if every $\varepsilon > 0$ has an $n_0 = n_0(\varepsilon) \in \mathbb{N}$ such that

$$|f_n(x) - f(x)| < \varepsilon \qquad \text{for all } n \geq n_0 \text{ and all } x \in A;$$

when this occurs the *limit function f* is uniquely determined.

A series $\sum f_\nu$ of functions *converges uniformly in A* if the sequence $s_n = \sum^n f_\nu$ of partial sums converges uniformly in A; as with numerical series, the symbol $\sum f_\nu$ is also used to denote the limit function.

Uniform convergence in A implies ordinary convergence. In uniform convergence there is associated with every $\varepsilon > 0$ an index $n_0(\varepsilon)$ which is independent of the location of the point x in A, while in mere pointwise convergence in A this index generally also depends (perhaps quite strongly) on the individual x.

The theory of uniform convergence becomes especially transparent upon introduction of the *supremum semi-norm*

$$|f|_A := \sup\{|f(x)| : x \in A\}$$

for subsets $A \subset X$ and functions $f : X \to \mathbb{C}$. The set $V := \{f : X \to \mathbb{C} : |f|_A < \infty\}$ of all complex-valued functions on X which are *bounded on A* is a $\mathbb{C}$-vector space; the mapping $f \mapsto |f|_A$ is a *"semi-norm"* on V; more precisely,

$$|f|_A = 0 \Leftrightarrow f|A = 0 \,, \qquad |cf|_A = |c||f|_A$$
$$|f + g|_A \leq |f|_A + |g|_A \qquad f, g \in V \,, c \in \mathbb{C}.$$

A sequence f_n converges uniformly in A to f exactly when

$$\lim |f_n - f|_A = 0. \qquad\qquad \square$$

Without any effort at all we prove two important

Limit rules. *Let f_n, g_n be sequences of functions in X which converge uniformly in A. Then*

L1 *For all $a, b \in \mathbb{C}$ the sequence $af_n + bg_n$ converges uniformly in A and*

$$\lim(af_n + bg_n) = a \lim f_n + b \lim g_n \qquad (\mathbb{C}\text{-linearity}).$$

L2 *If the functions $\lim f_n$ and $\lim g_n$ are both bounded in A, then the product sequence $f_n g_n$ also converges uniformly in A and*

$$\lim(f_n g_n) = (\lim f_n)(\lim g_n).$$

Of course there is a corresponding version of L1 for series $\sum f_\nu, \sum g_\nu$. To see that the supplemental hypothesis in L2 is necessary, look at $X := \{x \in \mathbb{R} : 0 < x < 1\}, f_n(x) = g_n(x) := \frac{1}{x} + \frac{1}{n}, f(x) = g(x) := \frac{1}{x}$. Then $|f_n - f|_X = |g_n - g|_X = \frac{1}{n}$, yet $|f_n g_n - fg|_X \geq f_n g_n(\frac{1}{n^2}) - fg(\frac{1}{n^2}) = 2n + \frac{1}{n^2}$.

2. Locally uniform convergence. The sequence of powers z^n converges in every disc $B_r(0), r < 1$, uniformly to the function 0, because $|z^n|_{B_r(0)} = r^n$. Nevertheless the convergence is *not uniform in the unit disc* $\mathbb{E}$: for every $0 < \varepsilon < 1$ and every $n \geq 1$ there is a point $c \in \mathbb{E}$, for example, $c := \sqrt[n]{\varepsilon}$, with $|c^n| \geq \varepsilon$. This kind of convergence behavior is symptomatic of many sequences and series of functions. It is one of WEIERSTRASS's significant contributions to have clearly recognized and high-lighted this convergence phenomenon: uniform convergence on the whole space X is usually not an issue; what's important is only that uniform convergence prevail "in the small".

A sequence of functions $f_n : X \to \mathbb{C}$ is called *locally uniformly convergent in X* if every point $x \in X$ lies in a neighborhood U_x in which the sequence f_n converges uniformly.

A series $\sum f_\nu$ is called *locally uniformly convergent in X* when its associated sequence of partial sums is locally uniformly convergent in X.

Uniform convergence naturally implies locally uniform convergence. The limit rules stated above carry over at once to locally uniformly convergent sequences and series.

Limit functions of (simply) convergent sequences of continuous functions are not in general continuous. From his study of calculus the reader knows that under locally uniform convergence the limit function does inherit continuity. More generally we have

Continuity theorem. *If the sequence $f_n \in C(X)$ converges locally uniformly in X, then the limit function $f = \lim f_n$ is likewise continuous on X, i.e., $f \in C(X)$.*

Proof (as for functions on $\mathbb{R}$). Let $a \in X$ be given. For all $x \in X$ and for all indices n

$$|f(x) - f(a)| \leq |f(x) - f_n(x)| + |f_n(x) - f_n(a)| + |f_n(a) - f(a)|.$$

By hypothesis a lies in a neighborhood U of uniform convergence, so given $\varepsilon > 0$ there exists an n such that $|f - f_n|_U < \varepsilon$ and the above inequality implies that for this n and all $x \in U$, $|f(x) - f(a)| \leq 2|f - f_n|_U + |f_n(x) - f_n(a)| < 2\varepsilon + |f_n(x) - f_n(a)|$. Finally, the continuity of f_n at a means that there is a $\delta > 0$ such that $|f_n(x) - f_n(a)| < \varepsilon$ for all $x \in B_\delta(a)$. Therefore for all x in the neighborhood $B_\delta(a) \cap U$ of a we have $|f(x) - f(a)| < 3\varepsilon$.   □

For series, the continuity theorem asserts that the sum of a locally uniformly convergent series of continuous functions on X is itself a continuous function on X.

3. Compact convergence. Clearly if the sequence $f_n : X \to \mathbb{C}$ converges uniformly in each of finitely many subsets $A_1, \ldots, A_b$ of X, then it also converges uniformly in the union $A_1 \cup A_2 \cup \ldots \cup A_b$ of these sets. An immediate consequence is

If the sequence f_n converges locally uniformly in X, then it converges uniformly on each compact subset K of X.

Proof. Every point $x \in K$ has an open neighborhood U_x in which f_n is uniformly convergent. The open cover $\{U_x : x \in K\}$ of the compact set K admits a finite subcover, say $U_{x_1}, \ldots, U_{x_b}$. Then f_n converges uniformly in $U_{x_1} \cup \ldots \cup U_{x_b}$, and all the more so in the subset K of this union.   □

We will say that a sequence or a series *converges compactly* in X if it converges uniformly on every compact subset of X. Thus we have just seen that:

Local uniform convergence implies compact convergence.

This important theorem, which leads from uniform convergence "in the small" to uniform convergence "in the large" was originally proved by WEIERSTRASS for intervals in $\mathbb{R}$. WEIERSTRASS gave a direct proof for this "local-to-global principle" (you have to bear in mind that at that time neither the general concept of compactness nor the HEINE–BOREL theorem was available). $\square$

In important cases the converse of the above statement is also true. Call X *locally compact* if each of its points has at least one compact neighborhood. Then it is trivial to see that

If X is locally compact, then every compactly convergent sequence or series in X is locally uniformly convergent in X.

In locally compact spaces the concepts of local uniform convergence and compact convergence therefore coincide. Since domains in $\mathbb{C}$ are locally compact, it is not necessary in any function-theoretic discussion to distinguish between these notions.

We have some preference for the expression "compactly convergent" because it's shorter than "locally uniformly convergent." In the literature the misleading locution "uniformly convergent in the interior of X" is also to be found.

4. On the history of uniform convergence. The history of the concept of uniform convergence is a paradigm in the history of ideas in modern mathematics. In 1821 CAUCHY maintained in his *Cours d'analyse* ([C], p.120) that convergent series of continuous functions f_n always have continuous limit functions f. CAUCHY considers the equation

$$f(x) - \sum_0^n f_\nu(x) = \sum_{n+1}^\infty f_\nu(x),$$

where the *finite* series on the left is certainly continuous. The error in the proof lay in the implicit assumption that, independently of x, n may be chosen so large that the *infinite* series on the right (for all x) becomes sufficiently small ("deviendra insensible, si l'on attribue à n une valeur très considérable").

It was ABEL who in 1826 in his paper on the binomial series was the first to criticize Cauchy's theorem; he writes ([A], p.316): "Es scheint mir aber, daß dieser Lehrsatz Ausnahmen leidet. So ist z. B. die Reihe

$$(*) \qquad \sin\varphi - \frac{1}{2}\sin 2\varphi + \frac{1}{3}\sin 3\varphi - \dots \text{usw.}$$

unstetig für jeden Werth $(2m + 1)\pi$ von x, wo m eine ganze Zahl ist. Bekanntlich giebt es eine Menge von Reihen mit ähnlichen Eigenschaften (It appears to me, however, that this theorem suffers exceptions. Thus, e.g., the series $(*)$ is discontinuous for every value $(2m + 1)\pi$ of x, where m is a whole number. As is well known, there is an abundance of series with similar properties.)" ABEL is considering here a *Fourier series* which converges throughout $\mathbb{R}$ to a limit function f which is discontinuous at the points $(2m+1)\pi$. [In fact, $f(x) = \frac{1}{2}x$ for $-\pi < x < \pi$ and $f(-\pi) = f(\pi) = 0$.] ABEL discusses this Fourier series in detail on p.337 of his paper and also in a letter from Berlin on January 16, 1826 to his teacher and friend HOLMBOE (*Œuvres* **2**, pp. 258 ff). We will return to this letter when we discuss the problem of term-by-term differentiation of infinite series in 4.3.3.

The first mathematician to have used the idea of uniform convergence seems to have been Christoph GUDERMANN, WEIERSTRASS' teacher in Münster. (He was born in 1798 in Winneburg near Hildesheim, was a secondary school teacher in Cleve and Münster, published works on elliptic functions and integrals in Crelle's Journal between 1838 and 1843, and died in Münster in 1852.) In 1838 in the midst of his investigations of modular functions he wrote on pp.251-252, Volume 18 of *Jour. für Reine und Angew. Math.*: "Es ist ein bemerkenswerther Umstand, daß... die so eben gefundenen Reihen *einen im Ganzen gleichen Grad der Convergenz* haben. (It is a fact worth noting that ... the series just found have *all the same convergence rate*.)"

In 1839/40 WEIERSTRASS was the only auditor as GUDERMANN lectured on modular functions. Here he may have encountered the new type of convergence for the first time. The designation "uniform convergence" originates with WEIERSTRASS, who by 1841 was working routinely with uniformly convergent series in the paper "Zur Theorie der Potenzreihen" [W_2] which was written that year in Münster but not published until 1894. There one finds the statement (pp.68-69): "Da die betrachtete Potenzreihe ... *gleichmässig convergirt*, so lässt sich aus ihr nach Annahme einer beliebigen positiven Grösse δ eine endliche Anzahl von Gliedern so herausheben, dass die Summe aller übrigen Glieder für jedes der angegebenen Werthsysteme ... ihrem absoluten Betrage nach $< \delta$ ist. (Because the power series under consideration ... *converges uniformly*, given an arbitrary positive quantity δ, a finite number of terms of the series can be discarded so that the sum of all the remaining terms is, for every value in the specified domain ... in absolute value $< \delta$.)"

The realization of the central role of the concept of uniform convergence in analysis came about slowly in the last century. The mathematical world gradually, through WEIERSTRASS' lectures *Introduction to Analysis* at Berlin in the winter 1859/60 and in the summer 1860, became aware of the incisiveness and indispensability of this concept. Still WEIERSTRASS wrote on March 6, 1881 to H. A. SCHWARZ: "Bei den Franzosen hat namentlich meine letzte Abhandlung mehr Aufsehen gemacht, als sie eigentlich verdient; man scheint endlich einzusehen, welche Bedeutung der Begriff der gleichmässigen Convergenz hat. (My latest treatise created more of a sensation among the French than it really deserves; people finally seem to realize the significance of the concept of uniform convergence.)"

In 1847, independently of WEIERSTRASS and of each other, PH. L. SEIDEL (*München Akad. Wiss. Abh.* **7**(1848), 379-394) and Sir G. G. STOKES (*Trans. Camb. Phil. Soc.* **8**(1847), 533-583) introduced concepts which correspond to that of the uniform convergence of a series. But their contributions exercised no further influence on the development of the idea. In an article which is well worth reading, entitled "Sir George Stokes and the concept of uniform convergence" (*Proc. Camb. Phil. Soc.* **19**(1916-1919), 148-156), the renowned British analyst G. H. HARDY compares the definitions of these three mathematicians; he says: "Weierstrass's discovery was the earliest, and he alone fully realized its far-reaching importance as one of the fundamental ideas of analysis."

5*. Compact and continuous convergence. A sequence of functions $f_n : X \to \mathbb{C}$ is said to be *continuously convergent* in X, if for *every* convergent sequence $\{x_n\} \subset X$, the limit $\lim_{n \to \infty} f_n(x_n)$ exists in $\mathbb{C}$. In particular (use constant sequences of x's), the sequence f_n converges pointwise on X to a limit function $f : X \to \mathbb{C}$. If the two sequences $\{x'_n\}$, $\{x''_n\}$ converge to the same limit x in X, then the sequences $\{f_n(x'_n)\}$, $\{f_n(x''_n)\}$ have the same limit in $\mathbb{C}$. This can be seen by interlacing $\{x'_{2n}\}$ and $\{x''_{2n+1}\}$ into a single convergent sequence $\{x_n\}$ and using the hypothesized convergence of $\{f_n(x_n)\}$. Of course the common value is $f(x)$. Notice too that $\lim_k f_{n_k}(x_k) = f(x)$ for any subsequence $\{f_{n_k}\}$ of $\{f_n\}$. To see this, define $x'_m := x_1$ for $1 \leq m \leq n_1$ and $x'_m := x_k$ for $n_{k-1} < m \leq n_k$ and all $k > 1$. Then $\lim_m x'_m = x$, so $\lim_m f_m(x'_m) = f(x)$; hence $\lim_k f_{n_k}(x'_{n_k}) = \lim_k f_{n_k}(x_k) = f(x)$. It is almost immediate that

If the sequence $\{f_n\}$ converges continuously on X to f, then f is continuous on X (even if the f_n are not themselves continuous).

Proof. Consider any $x \in X$, any sequence $\{x_n\} \subset X$ convergent to x and any $\varepsilon > 0$. There is a strictly increasing sequence $n_k \in \mathbb{N}$ such that $|f_{n_k}(x_k) - f(x_k)| < \varepsilon/2$. Since, as noted above, $\lim_k f_{n_k}(x_k) = f(x)$, there exists a k_ε such that $|f_{n_k}(x_k) - f(x)| < \varepsilon/2$ for all $k \geq k_\varepsilon$. The continuity of f at x follows:

$$|f(x_k) - f(x)| \leq |f(x_k) - f_{n_k}(x_k)| + |f_{n_k}(x_k) - f(x)| < \varepsilon \quad \text{for all } k \geq k_\varepsilon.$$

We now show that in important cases "continuous" and "compact" convergence are identical:

Theorem. *The following assertions concerning a sequence $f_n : X \to \mathbb{C}$*

are equivalent:

i) *The sequence f_n converges compactly in X to a function $f \in C(X)$.*

ii) *The sequence f_n is continuously convergent in X.*

Proof. i) $\Rightarrow$ ii) Let x_n be a convergent sequence in X and let $x = \lim x_n$. The set $L := \{x, x_1, x_2, \ldots\}$ is a compact subset of X. Consequently, $\lim |f - f_n|_L = 0$ and so in particular $f(x_n) - f_n(x_n)$ is a null sequence in $\mathbb{C}$. But $f(x) - f(x_n)$ is also a null sequence, since f is continuous at x. The sum of these two null sequences, viz., $f(x) - f_n(x_n)$ is then also a null sequence.

ii) $\Rightarrow$ i) Let f be the limit function. We have seen above that $f \in C(X)$. Suppose there is a compact $K \subset X$ such that $|f - f_n|_K$ is not a null sequence. This means that there is an $\varepsilon > 0$ and a subsequence n' of indices such that $|f - f_{n'}|_K > \varepsilon$ for all n'. In turn the latter means that there are points $x_{n'} \in K$ such that

$$(*) \qquad\qquad |f(x_{n'}) - f_{n'}(x_{n'})| > \varepsilon \qquad \text{for all } n'.$$

Because K is compact, we may assume, by passing to a further subsequence if necessary, that the sequence $x_{n'}$ converges, say to x. But then $\lim f(x_{n'}) = f(x)$ because of the continuity of f at x and $\lim f_{n'}(x_{n'}) = f(x)$ by hypothesis ii) and the remarks preceding this theorem. Subtraction gives that $\lim[f(x_{n'}) - f_{n'}(x_{n'})] = 0$, contradicting $(*)$. $\qquad\square$

As an application of this theorem we get an easy proof of the

Composition Theorem. *Let D, D' be domains in $\mathbb{C}$, $f_n \in C(D)$, $g_n \in C(D')$ sequences of continuous functions which converge compactly in their respective domains to $f \in C(D)$, $g \in C(D')$. Suppose that $f_n(D) \subset D'$ for all n and $f(D) \subset D'$. Then the composite sequence $g_n \circ f_n$ is a well-defined sequence in $C(D)$ and it converges compactly in D to $g \circ f \in C(D)$.*

Proof. For every sequence $x_n \in D$ with $\lim x_n = x \in D$, we have $\lim f_n(x_n) = f(x)$ and then $\lim g_n(f_n(x_n)) = g(f(x))$, by the preceding theorem. But then one more application of the theorem assures us that $g_n \circ f_n$ converges compactly to $g \circ f$. $\qquad\square$

Historical note. In 1929 C. CARATHÉODORY made a plea for continuous convergence; in his paper "Stetige Konvergenz und normale Familien von Funktionen", *Math. Annalen* **101**(1929), 515–533 [= *Gesammelte Math. Schriften* **4**, 96–118], on pp. 96–97 he writes: "Mein Vorschlag geht dahin, jedesmal, wo es vorteilhaft ist — und es ist, wie ich glaube, mit ganz wenigen Ausnahmen immer vorteilhaft —, den Begriff der [lokal] gleichmäßigen Konvergenz in der Funktionentheorie durch den Begriff der "stetigen Konvergenz" zu ersetzen ..., dessen Handhabung unvergleichlich einfacher ist. Im allgemeinen ist allerdings die stetige Konvergenz enger als die gleichmäßige; für den in der Funktionentheorie allein in Betracht kommenden Fall, in dem die Funktionen der Folge stetig sind, decken sich die beiden Begriffe vollkommen. (My suggestion is this: every time it is advantageous, and I believe that with few exceptions it is always so, one should replace the concept of [local] uniform convergence in function theory with that of "continuous convergence" ... whose manipulation is incomparably simpler. Granted that in the framework of general topology continuous convergence is more narrow than uniform convergence; but the only case which comes up in function theory is that of sequences of continuous functions and for them the two concepts in fact coincide.)" Nevertheless it has not become customary to check compact convergence by confirming that continuous convergence obtains. The concept of continuous convergence was introduced in 1921 by H. HAHN in his book *Theorie der Reellen Funktionen*, Julius Springer, Berlin; cf. pp. 238 ff. Actually the idea and the terminology showed up earlier in R. COURANT, "Über eine Eigenschaft der Abbildungsfunktionen bei konformer Abbildung", *Nachrichten Königl. Gesell. Wissen. Göttingen*, Math.-phys. Kl. (1914), 101-109, esp. p. 106.

Exercises

Exercise 1. a) For $n \in \mathbb{N}$ let $f_n : \mathbb{C} \setminus \partial \mathbb{E} \to \mathbb{C}$ be defined by $f_n(z) := \frac{1}{1+z^n}$. Show that for each $0 < r < 1$ the sequence $\{f_n\}$ converges uniformly on $\overline{B_r}(0) \cup (\mathbb{C} \setminus B_{1/r}(0))$ but not uniformly on $\mathbb{C} \setminus \partial \mathbb{E}$.

b) Where does the sequence of functions $f_n : \mathbb{C} \setminus \partial \mathbb{E} \to \mathbb{C}$ defined by $f_n(z) := \frac{z^n}{1+z^{2n}}$ converge uniformly?

Exercise 2. A sequence of polynomials

$$p_n(z) := a_{n,0} + a_{n,1}z + a_{n,2}z^2 + \cdots + a_{n,d}z^d, \quad a_{n,j} \in \mathbb{C},$$

all of degree not exceeding some $d \in \mathbb{N}$, is given. Show that the following statements are equivalent.

   i) The sequence $\{p_n\}$ converges compactly in $\mathbb{C}$.

   ii) There exist $d + 1$ *distinct* points $c_0, \ldots, c_d \in \mathbb{C}$ such that $\{p_n\}$ converges on $\{c_0, \ldots, c_d\}$.

   iii) For each $0 \leq j \leq d$ the sequence of coefficients $\{a_{n,j}\}_{n \in \mathbb{N}}$ converges.

In case of convergence the limit function is the polynomial $a_0 + a_1 z + \cdots + a_d z^d$, where $a_j := \lim_{n \to \infty} a_{n,j}$, $0 \leq j \leq d$.

Exercise 3. A sequence of automorphisms

$$f_k(z) := \eta_k \frac{z - w_k}{\overline{w}_k z - 1}, \quad w_k \in \mathbb{E}, \quad \eta_k \in \partial\mathbb{E}$$

of the unit disc is given. Prove the equivalence of the following statements:

i) The sequence $\{f_k\}$ converges compactly in $\mathbb{E}$ to a non-constant function.

ii) There exist $c_0, c_1 \in \mathbb{E}$ such that $\lim f_n(c_0)$ and $\lim f_n(c_1)$ exist and are different.

iii) Each of the sequences $\{\eta_k\}$ and $\{w_k\}$ converges and $\lim w_k \in \mathbb{E}$.

Hint for iii) $\Rightarrow$ i). Show that $\{f_k\}$ converges compactly in $\mathbb{E}$ to $f(z) := \eta \frac{z-w}{\overline{w}z-1}$, where $w := \lim w_k$, $\eta := \lim \eta_k$. To this end you can use theorem 5*.

Exercise 4. Let D be a domain in $\mathbb{C}$ and for each $n \geq 1$ set $K_n := \{z \in D : |z| \leq n , d_z(\partial D) \geq n^{-1}\}$. Each K_n is compact, $K_n \subset K_m$ if $n \leq m$ and $D = \cup_{n\geq 1} K_n$. Show that:

a) For all $f, g \in \mathcal{C}(D)$ the series $d(f, g) := \sum_{n\geq 1} 2^{-n} \frac{|f-g|_{K_n}}{1+|f-g|_{K_n}}$ converges.

b) A metric is defined in $\mathcal{C}(D)$ by d.

c) A sequence $\{f_k\} \subset \mathcal{C}(D)$ converges compactly on D to $f \in \mathcal{C}(D)$, if and only if it converges to f with respect to the metric d, that is, if and only if $\lim_k d(f_k, f) = 0$.

Exercise 5. Let K be a compact metric space, d_K its metric, $\{f_n\}$ a uniformly convergent sequence of *continuous* complex-valued functions on K, with limit function f. Show that the sequence is (uniformly) *equicontinuous*, meaning that for every $\varepsilon > 0$ there exists $\delta > 0$ such that $|f_n(x) - f_n(y)| < \varepsilon$ for all $n \in \mathbb{N}$ whenever $x, y \in K$ satisfy $d_K(x, y) < \delta$. *Hint. f* and each f_n is uniformly continuous on K (cf. Exercise 3 of Chapter 0, §5).

§2 Convergence Criteria

In analogy with the situation for numerical sequences a sequence of functions $f_n : X \to \mathbb{C}$ is said to be a *Cauchy sequence* (with respect to the supremum semi-norm) on $A \subset X$, if for each $\varepsilon > 0$ there exists an $n_\varepsilon \in \mathbb{N}$ such that

$$|f_m - f_n|_A < \varepsilon \qquad \text{for all } m, n \geq n_\varepsilon.$$

In subsection 1 below we carry over the Cauchy convergence criterion to sequences and series of functions; and in subsection 2 we do the same for the majorant criterion 0.4.2.

1. Cauchy's convergence criterion. *The following assertions about a sequence $f_n : X \to \mathbb{C}$ and a non-empty subset A of X are equivalent:*

i) *f_n is uniformly convergent in A.*

ii) *f_n is a Cauchy sequence in A.*

Proof. We will check only the non-trivial implication ii) $\Rightarrow$ i). Since $|f_m(x) - f_n(x)| \leq |f_m - f_n|_A$ for every $x \in A$, each of the numerical sequences $f_n(x)$, $x \in A$, is a Cauchy sequence. Therefore the sequence f_n is pointwise convergent in A; let $f := \lim f_n$. For all n, m and all $x \in A$ we have

$$|f_n(x) - f(x)| \leq |f_n(x) - f_m(x)| + |f_m(x) - f(x)| \leq |f_n - f_m|_A + |f_m(x) - f(x)|.$$

If now $\varepsilon > 0$ is given, then there is an n_ε such that $|f_n - f_m|_A < \varepsilon$ for all $n, m \geq n_\varepsilon$. For each $x \in A$ there further exists an $m = m(x) \geq n_\varepsilon$ such that $|f_m(x) - f(x)| < \varepsilon$. It follows that $|f_n(x) - f(x)| < 2\varepsilon$ for all $n \geq n_\varepsilon$ and this holds at each $x \in A$. That is, $|f_n - f|_A \leq 2\varepsilon$ for all $n \geq n_\varepsilon$. $\square$

The re-formulation for series is obvious:

Cauchy's convergence criterion for series. *The following assertions concerning an infinite series $\sum f_\nu$ of functions $f_\nu : X \to \mathbb{C}$ and a non-empty subset A of X are equivalent:*

i) *$\sum f_\nu$ is uniformly convergent in A.*

ii) *For every $\varepsilon > 0$ there exists an $n_\varepsilon \in \mathbb{N}$ such that $|f_{m+1}(x) + \cdots + f_n(x)| < \varepsilon$ for all $n > m \geq n_\varepsilon$ and all $x \in A$.*

CAUCHY introduced this criterion in his 1853 work "Note sur les séries convergentes dont les divers termes sont des fonctions continues ..." (*Œuvres* (1) **11**, 30-36; Théorème II, p.34). Here for the first time he worked with the notion of uniform convergence, without, however, using the word "uniforme". Here too (p. 31) he acknowledges that his continuity theorem is incorrect but dismisses the matter with "il est facile de voir comment on doit modifier l'énoncé du théorème (it is easy to see how one should modify the statement of the theorem)."

For continuous convergence of a sequence of functions there is also a Cauchy convergence criterion. The reader may check for himself that: *A sequence of continuous functions $f_n : X \to \mathbb{C}$ converges continuously in X if and only if $f_n(x_n)$ is a numerical Cauchy sequence for every convergent sequence x_n in X.*

2. Weierstrass' majorant criterion. Cauchy's criterion for the uniform convergence of a series is seldom used in practice. But particularly easy to handle in many applications is the

Majorant criterion (or M-test) of Weierstrass. *Let $f_\nu : X \to \mathbb{C}$ be a sequence of functions, A a non-void subset of X and suppose that there is a sequence of real numbers $M_\nu \geq 0$ such that*

$$|f_\nu|_A \leq M_\nu \quad \text{for all } \nu \in \mathbb{N} \text{ and } \quad \sum M_\nu < \infty.$$

Then the series $\sum f_\nu$ converges uniformly in A.

Proof. For all $n > m$ and all $x \in A$

$$\left| \sum_{m+1}^n f_\nu(x) \right| \leq \sum_{m+1}^n |f_\nu(x)| \leq \sum_{m+1}^n M_\nu.$$

Since $\sum M_\nu < \infty$, for each $\varepsilon > 0$ there is an $n_\varepsilon \in \mathbb{N}$ such that $\sum_{m+1}^n M_\nu < \varepsilon$ for all $n > m \geq n_\varepsilon$. This means that $|f_{m+1}(x) + \cdots + f_n(x)| < \varepsilon$ for all $n > m \geq n_\varepsilon$ and all $x \in A$. Therefore by Cauchy's criterion $\sum f_\nu$ converges uniformly in A.   □

WEIERSTRASS consigned his criterion to a footnote (on p. 202) in his 1880 treatise *Zur Functionenlehre* [W4].

Exercises

Exercise 1. Let p_n be a sequence of complex polynomials, $p_n(z) \in \mathbb{C}[z]$. Show that $\{p_n\}$ converges uniformly in $\mathbb{C}$, if and only if for some $N \in \mathbb{N}$ and some convergent sequence $\{c_n\}$ of complex numbers we have $p_n = p_N + c_n$ for all $n > N$. In case of convergence the limit function is then the polynomial $p_N + \lim c_n$.

Exercise 2 (cf. also exercise 4 to Chapter 0, §4). Let X be any non-void set, $f_\nu : X \to \mathbb{C}$ ($\nu \geq 0$) a sequence of functions with the following properties:

(i) $\{f_\nu\}$ converges uniformly in X to the 0 function.

(ii) The series $\sum_{\nu \geq 1} |f_\nu - f_{\nu-1}|$ converges uniformly on X.

a) Show that if $a_n \in \mathbb{C}$ and the sequence $s_m := \sum_{\nu=0}^{m} a_\nu$ $(m \in \mathbb{N})$ is bounded, then the series $\sum a_\nu f_\nu$ converges uniformly in X.

b) Under what conditions on a second sequence of functions $g_\nu : X \to \mathbb{C}$ does the series $\sum g_\nu f_\nu$ converge uniformly on X?

Exercise 3. Show that the series $\sum_{n \geq 1} \frac{(-1)^n}{z+n}$ converges compactly in $\mathbb{C} \setminus \{-1, -2, -3, \ldots\}$.

Exercise 4. Show that the series $\sum_{n \geq 1} \frac{(-1)^n}{z^2+n}$ converges uniformly in the strip $S := \{z \in \mathbb{C} : |\Im z| \leq 1/2\}$ but is not absolutely convergent at any point of S.

Exercise 5. Show that the metric space $\mathcal{C}(D)$ in Exercise 4 of §1 is complete.

§3 Normal convergence of series

Although the series $\sum_{1}^{\infty} \frac{(-1)^{\nu-1}}{x^2+\nu}$ is locally uniformly convergent in $\mathbb{R}$ it is possible to create divergent series from it by rearrangement. In order to calculate comfortably and without qualms we need (in analogy with series of complex numbers) a convergence notion for series $\sum f_\nu$ of functions which precludes such phenomena and guarantees that *every re-arrangement of the series* and *every subseries will converge locally uniformly*. These desiderata are secured by the concept of normal convergence, which we now discuss.

1. Normal convergence. A series $\sum f_\nu$ of functions $f_\nu : X \to \mathbb{C}$ is called *normally convergent* in X if each point of x has a neighborhood U which satisfies $\sum |f_\nu|_U < \infty$. "Normal" here refers to the presence of (semi-) norms and has none of the common parlance significance ("expected, average, representative, according to the rule") of that word. We should emphasize that normal convergence is only defined for series and not for sequences generally. On the basis of Weierstrass' majorant criterion we see that

Every series which is normally convergent in X is locally uniformly convergent in X.

From the continuity theorem 1.2 it follows in particular that

If $f_\nu \in C(X)$ and $f = \sum f_\nu$ converges normally in X, then f is continuous on X.

Quite trivial is the observation that

Every subseries of a series which converges normally in X will itself converge normally in X.

But we also have the indispensable

Rearrangement theorem. *If $\sum_0^\infty f_\nu$ converges normally in X to f, then for every bijection $\tau : \mathbb{N} \to \mathbb{N}$ the rearranged series $\sum_0^\infty f_{\tau(\nu)}$ also converges normally in X to f.*

Proof. Each point $x \in X$ has a neighborhood U for which $\sum |f_\nu|_U < \infty$. Consequently for every bijection τ of $\mathbb{N}$, the rearrangement theorem for series of complex numbers (0.4.3) ensures that $\sum |f_{\tau(\nu)}|_U < \infty$. And by applying that result to the numerical series $\sum f_\nu(x)$ we are further assured that $\sum f_{\tau(\nu)}(x) = f(x)$. This is true of every $x \in X$, and so the normal convergence in X of $\sum f_{\tau(\nu)}$ to f is proved. □

The rearrangement theorem can be sharpened as follows:

Let $\mathbb{N} = \cup_{k \geq 0} \mathbb{N}_k$ be a decomposition of the natural numbers into mutually disjoint non-void subsets and suppose that $\sum_0^\infty f_\nu$ converges normally in X to f. Then for each k the series $\sum_{\nu \in \mathbb{N}_k} f_\nu$ converges normally in X to a function $g_k : X \to \mathbb{C}$ and the series $\sum_{k \geq 0} g_k$ converges normally in X to f.

The reader should carry out the proof of this; see Exercise 2 below.

Along with $\sum f_\nu$ and $\sum g_\nu$, every series of the form $\sum (af_\nu + bg_\nu)$ will also be normally convergent in X $(a, b \in \mathbb{C})$. From the product theorem 0.4.6 there also follows immediately

Product theorem for normally convergent series. *If $f = \sum f_\mu$ and $g = \sum g_\nu$ converge normally in X, then every series $\sum h_\kappa$ in which $h_0, h_1, \ldots$ run through every product $f_\mu g_\nu$ exactly once, converges normally in X to fg.*

We write $fg = \sum f_\mu g_\nu$ and in particular, $fg = \sum p_\lambda$ with $p_\lambda := \sum_{\mu + \nu = \lambda} f_\mu g_\nu$ (the Cauchy product).

2. Discussion of normal convergence. Normal convergence is by definition a *local* property. Nevertheless:

If $\sum f_\nu$ is normally convergent in X, then $\sum |f_\nu|_K < \infty$ for every compactum K in X.

And there is a converse in the following sense

If X is locally compact and $\sum |f_\nu|_K < \infty$ for every compactum K in X, then $\sum f_\nu$ is normally convergent in X.

The proofs of both these statements are trivial.   □

For discs $B_s(c)$ in $\mathbb{C}$, every compactum $K \subset B_s(c)$ lies in $B_r(c)$ for some $r < s$, so we have as a particular case of the above

If $f_\nu : B_s(c) \to \mathbb{C}$ is a sequence which satisfies $\sum |f_\nu|_{B_r(c)} < \infty$ for each $0 < r < s$, then $\sum f_\nu$ converges normally in $B_s(c)$.   □

Normal convergence is more than local uniform convergence, as the series $\sum_1^\infty \frac{(-1)^\nu}{z+\nu}, z \in \mathbb{C} \setminus \{-1, -2, -3, \cdots\}$ shows. The reader should corroborate this.

In spite of the last example above, it is always possible to force a locally uniformly convergent series to be a normally convergent one by judicious insertion of parentheses:

Suppose that $f = \sum f_\nu$ is locally uniformly convergent in X. Then every point of X has a neighborhood U with this property: there exists a sequence $0 = n_0 < n_1 < n_2 < \cdots$ for which the "re-grouped" series $\sum F_\nu$, where $F_\nu := f_{n_\nu} + f_{n_\nu+1} + \cdots + f_{n_{\nu+1}-1}$, converges to f and satisfies $\sum |F_\nu|_U < \infty$.

Proof. Let $\varepsilon_1 > \varepsilon_2 > \varepsilon_3 > \cdots$ be a sequence of positive real numbers with $\sum \varepsilon_\nu < \infty$. By definition of local uniform convergence, there is a neighborhood U of the given point and indices $0 < n_1 < n_2 < \cdots$ such that

$$|t_\nu - f|_U \le \varepsilon_\nu \qquad \text{for} \quad t_\nu := \sum_0^{n_\nu - 1} f_\mu \, , \nu = 1, 2, \cdots$$

Setting $F_0 := t_1, F_\nu := t_{\nu+1} - t_\nu$, it is clear that $f = \sum F_\nu$. And because $F_\nu = (t_{\nu+1} - f) - (t_\nu - f)$ for all $\nu \ge 1$, it follows that $|F_\nu|_U \le \varepsilon_{\nu+1} + \varepsilon_\nu < 2\varepsilon_\nu$ for $\nu \ge 1$ and so $\sum |F_\nu|_U \le |F_0|_U + 2\sum_{\nu \ge 1} \varepsilon_\nu < \infty$.   □

It should be noted that the sequence $n_0, n_1, \cdots$ gotten here depends on the particular neighborhood U.

3. Historical remarks on normal convergence. The French mathematician René BAIRE (1874-1932), known for contributions to measure theory [Baire sets] and topology [Baire category theorem], introduced this convergence concept in 1908 in the second volume (pp. 29 ff.) of his work *Leçons sur les théories générales de l'analyse* (Gauthier-Villars, Paris 1908). He was guided by Weierstrass' majorant criterion 2.2 and in the introduction to this volume (p. VII) he says: "Bien qu'à mon avis l'introduction de termes nouveaux ne doive se faire qu'avec une extrême prudence, il m'a paru indispensable de caractériser par une locution brève le cas le plus simple et de beaucoup le plus courant des séries uniformément convergentes, celui des séries dont les termes sont moindres en module que des nombres positifs formant série convergente (ce qu'on appelle quelquefois *critère de Weierstrass*). J'appelle ces séries *normalement* convergentes, et j'espère qu'on voudra bien excuser cette innovation. Un grand nombre de démonstrations, soit dans la théorie des séries, soit plus loin dans la théorie des produits infinis, sont considérablement simplifiées quand on met en avant cette notion, beaucoup plus maniable que la propriété de convergence uniforme. (Although in my opinion the introduction of new terms must only be made with extreme prudence, it appeared indispensable to me to characterize by a brief phrase the simplest and by far the most prevalent case of uniformly convergent series, that of series whose terms are smaller in modulus than the positive numbers forming a convergent series (what one sometimes calls the *Weierstrass criterion*). I call these series *normally* convergent, and I hope that people will be willing to excuse this innovation. A great number of demonstrations, be they in the theory of series or somewhat further along in the theory of infinite products, are considerably simplified when one advances this notion, which is much more manageable than that of uniform convergence.)"

Thus BAIRE practically apologized for introducing a new concept into mathematics! This convergence is most often encountered and studied in the more general context of normed linear spaces — *functional analysis* — a framework and direction of analysis which was being founded about the time Baire's book appeared.

Exercises

Exercise 1. Show that the series $\sum_{n \geq 1} \frac{z^{2n}}{1 - z^n}$ is normally convergent in $\mathbb{E}$.

Exercise 2. a) Formulate and prove for normally convergent series an assertion corresponding to that in Exercise 5 of Chapter 0, §4 for absolutely convergent series of complex numbers.

b) From a) deduce the sharpened form of the rearrangement theorem stated in subsection 1.

Chapter 4

Power Series

> Die Potenzreihen sind deshalb besonders bequem, weil man mit ihnen *fast* wie mit Polynomen rechnen kann (Power series are therefore especially convenient because one can compute with them *almost* as with polynomials).—C. CARATHÉODORY

The series of functions which are the most important and fruitful in function theory are the power series, series which as early as 1797 had been considered by LAGRANGE in his *Théorie des fonctions analytiques*. In this chapter the elementary theory of convergent power series will be discussed. This theory used to be known also as *algebraic analysis* (from the subtitle *Analyse algébrique* of Cauchy's *Cours d'analyse* [C]). Also of interest in this connection is the article of the same title by G. FABER and A. PRINGSHEIM in the *Encyklopädie der Mathematischen Wissenschaften* II, 3.1, pp. 1-46 (1908).

In section 1 we first show that every power series has a well-defined "radius of convergence" R and that inside its "circle of convergence" $B_R(c)$ it converges normally. The calculation of R can generally be accomplished by means of either the CAUCHY-HADAMARD formula or the quotient rule. In section 2 we determine the radii of convergence of various important power series like the *exponential series* $\exp z$, the *logarithmic series* $\lambda(z)$ and the *binomial series* $b_\sigma(z)$. In section 3 we show that a convergent power series represents a holomorphic function in its disc of convergence. (The all-important converse of this cannot however be proved until 7.3.1.) With this the "preliminaries to Weierstrass' function theory" have been attended to and there are no further obstacles to our constructing many interesting holomorphic functions. In particular, the functions $\exp z$, $\lambda(z)$ and $b_\sigma(z)$ are holomorphic in their discs of convergence; in the unit disc

$\mathbb{E}$ the relation $b_\sigma(z) = \exp(\sigma\lambda(z))$ holds among them. We will later bring this into the suggestive form $(1 + z)^\sigma = e^{\sigma \log(1+z)}$, but the usual proof of this for real σ and z does not work when they are complex because $\log z$ is no longer the inverse of the function $\exp z$.

In section 4 we make an excursion into algebra and study the ring $\mathcal{A}$ of *all convergent power series*. This ring proves to be a "*discrete valuation ring*" and consequently is arithmetically simpler than the ring $\mathbb{Z}$ of integers or the polynomial ring $\mathbb{C}[z]$. $\mathcal{A}$ *is a (unique) factorization domain and has just one prime element*. So the adage of CARATHÉODORY at the head of the chapter is even an understatement: in many respects calculation with power series is actually easier than with polynomials.

§1 Convergence criteria

Fixing $c \in \mathbb{C}$, any function series of the form

$$\sum_0^\infty a_\nu (z - c)^\nu$$

with $a_\nu \in \mathbb{C}$ is called a *(formal) power series* with *center* c and *coefficients* a_ν.

The power series form a $\mathbb{C}$-*algebra*: the number $a \in \mathbb{C}$ is identified with $a + \sum_1^\infty 0(z - c)^\nu$ and for $f = \sum_0^\infty a_\nu (z - c)^\nu$, $g = \sum_0^\infty b_\nu (z - c)^\nu$ the *sum* and the *product* are defined by

$$f + g := \sum_0^\infty (a_\nu + b_\nu)(z - c)^\nu$$
$$f \cdot g := \sum_0^\infty p_\lambda (z - c)^\lambda, \text{ where } \quad p_\lambda := \sum_{\mu+\nu=\lambda} a_\mu b_\nu.$$

This multiplication is just the Cauchy multiplication (cf. 0.4.6).

To simplify our statements, we frequently assume $c = 0$ if that involves no loss of generality. We abbreviate $B_r(0)$ to B_r and — following our earlier convention — write $\sum a_\nu z^\nu$ throughout instead of $\sum_0^\infty a_\nu z^\nu$.

1. Abel's convergence lemma. Every power series trivially converges at its center. So a power series is only called *convergent* if there is some other point $z_1 \neq c$ at which it converges. We will show

Convergence lemma (ABEL). *Suppose that for the power series* $\sum a_\nu z^\nu$ *there are positive real numbers* s *and* M *such that* $|a_\nu| s^\nu \leq M$ *for all* ν. *Then this power series is normally convergent in the open disc* B_s.

Proof. Consider an arbitrary r with $0 < r < s$ and set $q := rs^{-1}$. Then $|a_\nu z^\nu|_{B_r} = |a_\nu|r^\nu \le Mq^\nu$ for all ν. Since $\sum q^\nu < \infty$, on account of $0 < q < 1$, it follows that $\sum |a_\nu z^\nu|_{B_r} \le M \sum q^\nu < \infty$. As this is the case for every $r < s$, normal convergence in B_s follows (cf. 3.3.2).   □

Corollary. *If the series $\sum a_\nu z^\nu$ converges at $z_0 \ne 0$, then it converges normally in the open disc $B_{|z_0|}$.*

For the sequence $|a_\nu||z_0|^\nu$ is a null sequence and consequently is bounded.

2. Radius of convergence. The geometric series $\sum z^\nu$ converges in the unit disc $\mathbb{E}$ and diverges at every point of $\mathbb{C} \setminus \mathbb{E}$. This convergence behavior is representative of what happens in general.

Convergence theorem for power series. *Let $\sum a_\nu z^\nu$ be a power series and denote by R the supremum of all real numbers $t \ge 0$ for which $|a_\nu|t^\nu$ is a bounded sequence. Then*

1) *The series converges normally in the open disc B_R.*
2) *The series diverges at every point $w \in \mathbb{C} \setminus \overline{B_R}$.*

Proof. We have $0 \le R \le \infty$ and there is nothing to prove if $R = 0$. So suppose $R > 0$. The sequence $|a_\nu|s^\nu$ is bounded, for each s with $0 < s < R$. By the convergence lemma $\sum a_\nu z^\nu$ is consequently normally convergent in B_s for each such s. Since B_R is the union of these open subdiscs, normal convergence holds true in B_R.

For each w with $|w| > R$ the sequence $|a_\nu||w|^\nu$ is unbounded, so the series $\sum a_\nu w^\nu$ is necessarily divergent.   □

Power series are the simplest normally convergent series of continuous functions. The limit function to which $\sum a_\nu z^\nu$ converges is continuous in B_R (cf. 3.3.1); we will designate this function (as well as the power series itself) by f.

The quantity $R \in [0, \infty]$ determined by the convergence theorem is called the *radius of convergence*, and the set B_R is called the *disc of convergence* (sometimes less precisely, the *circle of convergence*) of the power series. In the subsections immediately following we will find some criteria for determining the radius of convergence.

3. The CAUCHY-HADAMARD formula. *The radius of convergence of the power series $\sum a_\nu(z - c)^\nu$ is*

$$R = \frac{1}{\overline{\lim} \sqrt[\nu]{|a_\nu|}}.$$

Here we have to recall that for every sequence of real numbers r_ν its *limit superior* is defined by $\overline{\lim} r_\nu := \lim_{\nu \to \infty} [\sup\{r_\nu, r_{\nu+1}, \ldots\}]$. We use the conventions $1/0 := \infty, 1/\infty := 0$.

For the proof of the Cauchy-Hadamard formula, set $L := (\overline{\lim} \sqrt[\nu]{|a_\nu|})^{-1}$. The desired inequalities $L \leq R$ and $R \leq L$ follow if we can show that: *for every r with $0 < r < L$, $r \leq R$ holds*, and *for every s with $L < s < \infty$, $s \geq R$ holds*.

First consider $0 < r < L$, so that $r^{-1} > \overline{\lim} \sqrt[\nu]{|a_\nu|}$. From the definition of $\overline{\lim}$ there is then a $\nu_0 \in \mathbb{N}$ such that $\sqrt[\nu]{|a_\nu|} < r^{-1}$ for all $\nu \geq \nu_0$. The sequence $|a_\nu| r^\nu$ is therefore bounded, that is, $r \leq R$.

Now consider $L < s < \infty$, so that $s^{-1} < \overline{\lim} \sqrt[\nu]{|a_\nu|}$. From the definition of $\overline{\lim}$ there is then an *infinite* subset M of $\mathbb{N}$ such that for all $m \in M$, $s^{-1} < \sqrt[m]{|a_m|}$, that is, $|a_m| s^m > 1$. The sequence $|a_\nu| s^\nu$ is thus certainly not a null sequence and so we must have $s \geq R$.   □

By means of the limit superior formula we at once find for the series

$$\sum \nu^\nu z^\nu, \quad \sum z^\nu \quad \text{and} \quad \sum z^\nu / \nu^\nu$$

the respective radii of convergence $R = 0$, $R = 1$ and $R = \infty$, with respective discs of convergence $B_R = \emptyset$, $B_R = \mathbb{E}$ and $B_R = \mathbb{C}$. Nevertheless, the Cauchy-Hadamard formula is not always optimally suited to determining the radius of convergence (a case in point being the exponential series $\sum z^\nu / \nu!$). Frequently very helpful in such cases is the

4. Ratio criterion. *Let $\sum a_\nu (z - c)^\nu$ be a power series with radius of convergence R and $a_\nu \neq 0$ for all but finitely many values of ν. Then*

$$\underline{\lim} \frac{|a_\nu|}{|a_{\nu+1}|} \leq R \leq \overline{\lim} \frac{|a_\nu|}{|a_{\nu+1}|};$$

In particular $R = \lim \frac{|a_\nu|}{|a_{\nu+1}|}$ whenever this limit exists.
[Recall the definition

$$\underline{\lim} \, r_\nu := \lim_{\nu \to \infty} [\inf\{r_\nu, r_{\nu+1}, \ldots\}]$$

of the *limit inferior* of a sequence r_ν of real numbers, which is analogous to that of the limit superior. Always $\underline{\lim} r_\nu \leq \overline{\lim} r_\nu$ and $\lim r_\nu$ exists precisely when these are equal, in which case it coincides with their common value.]

Proof. Set $S := \underline{\lim} \frac{|a_\nu|}{|a_{\nu+1}|}$, $T := \overline{\lim} \frac{|a_\nu|}{|a_{\nu+1}|}$. Then what we must show is that $s \leq R$ for every s with $0 < s < S$ and $t \leq R$ for every t with $T < t < \infty$.

First consider $0 < s < S$. From the definition of $\underline{\lim}$ there must be an $\ell \in \mathbb{N}$ such that

$$|a_\nu a_{\nu+1}^{-1}| > s, \text{ that is, } |a_{\nu+1}|s > |a_\nu| \qquad \text{for all } \nu \geq \ell.$$

Setting $A := |a_\ell|s^\ell$, it follows by induction that $|a_{\ell+m}|s^{\ell+m} \leq A$ for all $m \geq 0$. The sequence $|a_\nu|s^\nu$ is consequently bounded; that is, $s \leq R$.

Now consider $T < t < \infty$. According to the definition of $\overline{\lim}$ there is then an $\ell \in \mathbb{N}$ such that

$$a_\nu \neq 0 \text{ and } |a_\nu a_{\nu+1}^{-1}| < t, \text{ that is, } |a_{\nu+1}|t > |a_\nu| \qquad \text{for all } \nu \geq \ell.$$

Setting $B := |a_\ell|t^\ell$, it follows by induction that $|a_{\ell+m}|t^{\ell+m} \geq B$ for all $m \geq 0$. Since $B > 0$, this means that $|a_\nu|t^\nu$ is not a null sequence, that is, $t \geq R$.   □

This ratio criterion for power series contains the well-known ratio test for convergence of numerical series $\sum a_\nu$, $a_\nu \in \mathbb{C}^\times$: because from $|a_{\nu+1}a_\nu^{-1}| \leq q < 1$ for almost all ν it follows that $\underline{\lim} |a_\nu a_{\nu+1}^{-1}| \geq q^{-1}$, so that the series $\sum a_\nu z^\nu$ has radius of convergence $R \geq q^{-1} > 1$ and consequently converges absolutely at $z = 1$.

Warning. It is possible that infinitely many coefficients in a power series vanish. For such "lacunary series"

$$\sum_{\lambda=0}^{\infty} a_{n_\lambda} z^{n_\lambda}, \qquad a_{n_\lambda} \in \mathbb{C}^\times, \; n_0 < n_1 < n_2 < \cdots$$

in which $n_{\lambda+1} > n_\lambda + 1$ infinitely often, consideration of the sequence $|a_{n_\lambda} \cdot a_{n_{\lambda+1}}^{-1}|$ does not generally lead to determination of the radius of convergence. For example, in the geometric series $\sum 2^{2\nu} z^{2\nu}$, $a_{2\nu} = 2^{2\nu}$, $a_{2\nu+1} = 0$ and $a_{2\nu} \cdot a_{2\nu+2}^{-1} = 1/4$, for all ν; yet according to CAUCHY-HADAMARD the radius of convergence of this series is $1/2$.

5.   On the history of convergent power series.   EULER calculated quite routinely with them; e.g., in [E], §335 ff. he was already implicitly using the ratio criterion (cf. also in this connection 7.3.3). LAGRANGE wanted to base all of analysis on power series. In 1821 CAUCHY proved the first general proposition about them; thus in [C], pp. 239/40 he showed that every power series, real or complex, converges in a well-determined circular disc $B \subset \mathbb{C}$ and diverges everywhere in $\mathbb{C} \setminus \overline{B}$. He also proved the formulas

$$R = \frac{1}{\overline{\lim} \sqrt[\nu]{|a_\nu|}} \qquad \text{and} \qquad R = \lim \frac{|a_\nu|}{|a_{\nu+1}|},$$

the latter under the explicit assumption that the limit exists ("Scolie" on p. 240). The limit superior representation was re-discovered in 1892 by

J. S. HADAMARD (French mathematician, co-prover of the prime number theorem, 1865-1963), who was apparently unaware of Cauchy's formula; his paper is in *Jour. Math. Pures et Appl.* (4) **8**, p. 108.

ABEL published his basic convergence lemma in 1826 in his landmark work [A] concerning binomial series; his formulation was as follows:

Lehrsatz IV. *Wenn die Reihe* $f(\alpha) = \nu_0 + \nu_1\alpha + \nu_2\alpha^2 + \cdots + \nu_m\alpha^m + \cdots$ *für einen gewissen Werth* δ *von* α *convergirt, so wird sie auch für jeden kleineren Werth von* α *convergiren,* ...

This is the essence of our corollary to the convergence lemma. It is further interesting to read what ABEL had to say about mathematical rigor in connection with convergence questions, Cauchy's *Cours d'analyse* notwithstanding; thus ABEL begins his exposition with the note-worthy words: "Untersucht man das Raisonnement, dessen man sich gewöhnlich bedient, wo es sich um unendliche Reihen handelt, genauer, so wird man finden, daß es im ganzen wenig befriedigend, und daß also die Zahl derjenigen Sätze von unendlichen Reihen, die als streng begründet angesehen werden können, nur sehr geringe ist. (If one examines more closely the reasoning which is usually employed in the treatment of infinite series, he will find that by and large it is unsatisfactory and that the number of propositions about infinite series which can be regarded as rigorously confirmed is small indeed.)"

Power series were just ancillary with CAUCHY and RIEMANN; they were first given primacy by WEIERSTRASS. They were already on center-stage in his 1840 work *Über die Entwicklung der Modular-Functionen.* [This was a written homework assignment in connection with the examination for prospective high-school teachers; it is dated "Westernkotten in Westfalen, im Sommer 1840." GUDERMANN, WEIERSTRASS's teacher, in his evaluation of it wrote: "Der Kandidat tritt hierdurch ebenbürtig in die Reihe ruhmgekrönter Erfinder. (With this work the candidate enters the ranks of famous inventors as a co-equal.)" GUDERMANN urged publication of the exam project as soon as possible and that would have happened had the philosophy faculty of the royal academy at Münster/Westphalia at that time had the authority to grant degrees. "Dann würden wir die Freude haben, Weierstrass zu unsern Doktoren zu zählen (Then we would have the pleasure of counting Weierstrass among our doctoral graduates)", so we read in the 1897 rector's address of Weierstrass' former pupil W. KILLING (whose name was later immortalized in Lie theory). Not until 1894, fifty-four years after it was written, did WEIERSTRASS publish his exam work; his *Mathematische Werke* begin with this work.] For WEIERSTRASS function theory was synonymous with the theory of functions represented by power series.

Exercises

Exercise 1. Determine the radius of convergence of each of the following:

a) $\displaystyle\sum_{n=1}^{\infty} \left(\frac{7n^4 + 2n^3}{5n^4 + 23n^3} \right)^n z^n$

b) $\displaystyle\sum_{n=1}^{\infty} \frac{z^{2n}}{c^n}, \quad c \in \mathbb{C}^{\times}$

c) $\displaystyle\sum_{n=1}^{\infty} \frac{n!}{2^n (2n)!} (z-1)^n$

d) $\displaystyle\sum_{n=0}^{\infty} (n^2 + a^n) z^n, \quad a \in \mathbb{C}$

e) $\displaystyle\sum_{n=0}^{\infty} \frac{(n!)^k}{(kn)!} z^n, \quad k \in \mathbb{N}, k \geq 1$

f) $\displaystyle\sum_{n=0}^{\infty} (\sin n) z^n$.

Exercise 2. Let $R > 0$ be the radius of convergence of the power series $\sum a_n z^n$. Determine the radius of convergence of the following series:

$$\sum a_n z^{2n}, \quad \sum a_n^2 z^n, \quad \sum a_n^2 z^{2n}, \quad \sum \frac{a_n}{n!} z^n.$$

Exercise 3. Let $R_1, R_2 > 0$ be the radii of convergence of the series $\sum a_n z^n$ and $\sum b_n z^n$, respectively. Show that

a) The radius of convergence R of the series $\sum (a_n + b_n) z^n$ satisfies $R \geq \min\{R_1, R_2\}$ and that equality holds if $R_1 \neq R_2$.

b) The radius of convergence R of the series $\sum a_n b_n z^n$ satisfies $R \geq R_1 R_2$.

c) If $a_n \neq 0$ for all n, then the radius of convergence R of the series $\sum (a_n)^{-1} z^n$ satisfies $R \leq R_1^{-1}$. Give an example where this inequality is strict.

Exercise 4. The power series $\sum_{n=1}^{\infty} \frac{z^n}{n^2}$ has radius of convergence 1. Show that the function it represents is injective in $B_{2/3}(0)$. *Hint.* For $z, w \in \mathbb{C}$ and integer $n \geq 2$

$$z^n - w^n = (z - w)(z^{n-1} + z^{n-2} w + \cdots + z w^{n-2} + w^{n-1}).$$

(With more refined estimates it can be shown that the function is injective in an even larger disc.)

§2 Examples of convergent power series

With the aid of the ratio criterion we will determine the radii of convergence of some important power series. We also briefly allude to convergence behavior on the boundary and discuss the famous Abel limit theorem, though in fact it is not particularly relevant to function theory proper.

1. The exponential and trigonometric series. Euler's formula. The most important power series after the geometric one is the *exponential series*

$$\exp z := \sum \frac{z^\nu}{\nu!} = 1 + z + \frac{z^2}{2!} + \frac{z^3}{3!} + \cdots$$

Its radius of convergence is determined from the ratio criterion (in which $a_\nu := \frac{1}{\nu!}$) to be $R = \lim \frac{|a_\nu|}{|a_{\nu+1}|} = \lim(\nu+1) = \infty$; that is, the series converges *throughout* $\mathbb{C}$. On account of the estimate

$$\left| \sum_{\nu \geq n} \frac{z^\nu}{\nu!} \right| \leq \frac{|z|^n}{n!} \left(1 + \frac{|z|}{n+1} + \frac{|z|^2}{(n+1)(n+2)} + \cdots \right)$$

we have

$$\left| \exp z - \sum_{0}^{n-1} \frac{z^\nu}{\nu!} \right| \leq \frac{2}{n!}|z|^n \quad \text{for } n \geq 1 \quad \text{and} \quad |z| \leq 1 + \frac{1}{2}(n-1).$$

Remark. The CAUCHY-HADAMARD formula is not particularly suitable for determining R because it involves the not obviously accessible $\overline{\lim}(\sqrt[\nu]{\nu!})^{-1}$. However, now that we know $R = \infty$, that formula tells us that $\overline{\lim}(\sqrt[\nu]{\nu!})^{-1} = 0$.

The *cosine series* and the *sine series*

$$\cos z := \sum_{0}^{\infty} \frac{(-1)^\nu}{(2\nu)!} z^{2\nu} = 1 - \frac{z^2}{2!} + \frac{z^4}{4!} - \frac{z^6}{6!} + - \cdots \,,$$

$$\sin z := \sum_{0}^{\infty} \frac{(-1)^\nu}{(2\nu+1)!} z^{2\nu+1} = z - \frac{z^3}{3!} + \frac{z^5}{5!} - \frac{z^7}{7!} + - \cdots$$

likewise converge *everywhere* in $\mathbb{C}$, because $\sum \frac{|z|^{2\nu}}{(2\nu)!}$ and $\sum \frac{|z|^{2\nu+1}}{(2\nu+1)!}$ are subseries of the convergent series $\sum \frac{|z|^\nu}{\nu!}$, for each $z \in \mathbb{C}$.

We have thus defined in $\mathbb{C}$ three complex-valued functions $\exp z$, $\cos z$ and $\sin z$ which coincide on $\mathbb{R}$ with the functions bearing the same names from the infinitesimal calculus. We speak of the *complex exponential function* and the *complex cosine* and *sine functions*. These functions will be extensively discussed in Chapter 5. Here we only want to note the famous

Euler formula: $\boxed{\exp iz = \cos z + i \sin z \quad \textit{for all } z \in \mathbb{C}}$

which EULER announced in 1748 for real arguments ([E], §138). The formula follows from the identity

$$\sum_{\nu=0}^{2m+1} \frac{(iz)^\nu}{\nu!} = \sum_{\mu=0}^{m}(-1)^\mu \frac{z^{2\mu}}{(2\mu)!} + i\sum_{\mu=0}^{m}(-1)^\mu \frac{z^{2\mu+1}}{(2\mu+1)!}$$

(which is itself immediate because $i^2 = -1$) by passage to the limit as $m \to \infty$. □

Because cosine is an *even* and sine an *odd* function:

$$\cos(-z) = \cos z\ , \quad \sin(-z) = -\sin z\ , \quad z \in \mathbb{C},$$

we have $\exp(-iz) = \cos z - i\sin z$ and from this by addition and subtraction with the original formula, we get the Euler representations

$$\cos z = \frac{1}{2}[\exp iz + \exp(-iz)]\ , \quad \sin z = \frac{1}{2i}[\exp iz - \exp(-iz)].$$

2. The logarithmic and arctangent series. The power series

$$\lambda(z) := \sum_{1}^{\infty} \frac{(-1)^{\nu-1}}{\nu} z^\nu = z - \frac{z^2}{2} + \frac{z^3}{3} - + \cdots$$

is called the *logarithmic series*; it has radius of convergence $R = 1$, since $\frac{|a_\nu|}{|a_{\nu+1}|} = \frac{\nu+1}{\nu}$. (And once again CAUCHY-HADAMARD leads to a non-trivial corollary: $1 \le \varliminf \sqrt[\nu]{\nu} \le \varlimsup \sqrt[\nu]{\nu} = 1$ and so $\lim \sqrt[\nu]{\nu}$ exists and equals 1.)

In 5.4.2 we will see that the function defined in the unit disc $\mathbb{E}$ by this series is the principal branch $\log(1 + z)$ of the logarithm. Nicolaus MERCATOR (real name, KAUFMANN; born 1620 in Holstein, lived in London; one of the first members of the Royal Society; went to France in 1683 and designed the fountains at Versailles; died 1687 in Paris; not to be confused with the inventor of the mercator projection 100 years earlier) found the logarithmic series in 1668 in the course of his quadrature of the hyperbola, thus

$$\log(1+x) = \int_0^x \frac{dt}{1+t} = \int_0^x (1-t+t^2-t^3+-\cdots)dt = x - \frac{x^2}{2} + \frac{x^3}{3} - \frac{x^4}{4} + \cdots$$

The power series

$$a(z) := \sum_{1}^{\infty} \frac{(-1)^{\nu-1}}{2\nu-1} z^{2\nu-1} = z - \frac{z^3}{3} + \frac{z^5}{5} - + \cdots$$

is called the *arctangent series*; it has radius of convergence $R = 1$ (why?) and in the unit disc it represents the inverse function of the tangent function (cf. 5.2.5). The arctangent series was discovered in 1671 by J. GREGORY (1638-1675, Scottish mathematician) but did not come to public attention until 1712.

3. The binomial series. In the year 1669 the 26-year-old Isaac NEWTON (1643-1727; 1689 MP for Cambridge University; 1699 superintendent of the royal mint; 1703 president of the Royal Society; 1705 knighted) in a work entitled "De analysi per aequationes numero terminorum infinitas" (Vol. II, pp. 206-247 of *The Mathematical Papers of Isaac Newton*) wrote that for *every* real number s the *binomial series*

$$\sum_0^\infty \binom{s}{\nu} x^\nu = 1 + sx + \frac{s(s-1)}{2}x^2 + \cdots + \frac{s(s-1)\cdots(s-n+1)}{n!}x^n + \cdots$$

represents the *binomial* $(1+x)^s$ for all real numbers x satisfying $-1 < x < 1$. ABEL [A] considered this series for arbitrary complex exponents $\sigma \in \mathbb{C}$ and for complex arguments z; he shows that for all $\sigma \in \mathbb{C} \setminus \mathbb{N}$ the series has radius of convergence 1 and again represents the binomial $(1+z)^\sigma$ in the unit disc, provided this power function is "properly" defined.

For each $\sigma \in \mathbb{C}$ we define, just as for real numbers, the *binomial coefficients*, as

$$\binom{\sigma}{0} := 1 \, , \quad \binom{\sigma}{n} := \frac{\sigma(\sigma-1)\cdots(\sigma-n+1)}{n!} \, , \, n = 1, 2, \ldots$$

They clearly satisfy

$$(*) \qquad \binom{\sigma}{n+1} = \frac{\sigma-n}{n+1}\binom{\sigma}{n} \qquad \text{for all } \sigma \in \mathbb{C}, \text{ all } n \in \mathbb{N}.$$

The *binomial series* for $\sigma \in \mathbb{C}$ is given by

$$b_\sigma(z) := \sum_0^\infty \binom{\sigma}{\nu} z^\nu = 1 + \sigma z + \binom{\sigma}{2}z^2 + \cdots + \binom{\sigma}{n}z^n + \cdots.$$

Of course if σ is a non-negative integer, then $\binom{\sigma}{\nu} = 0$ for all $\nu > \sigma$ and then we get (in any field of characteristic 0) the *binomial formula*

$$(1+z)^\sigma = \sum \binom{\sigma}{\nu} z^\nu \qquad \text{for all } z \in \mathbb{C} \, , \sigma \in \mathbb{N}$$

which is just a polynomial of degree σ. For all other σ however, $\binom{\sigma}{\nu} \neq 0$ for all ν. In these cases the binomial series is an infinite power series. For example, since $\binom{-1}{\nu} = (-1)^\nu$, we have for $\sigma = -1$ the *alternating geometric series*

$$b_{-1}(z) = \sum \binom{-1}{\nu} z^\nu = \sum(-z)^\nu = \frac{1}{1+z} \qquad \text{for all } z \in \mathbb{E}.$$

Generally

The binomial series has radius of convergence 1 for every $\sigma \in \mathbb{C} \setminus \mathbb{N}$.

Proof. The coefficient $a_\nu := \binom{\sigma}{\nu}$ is non-zero for all ν and by $(*)$

$$\frac{a_\nu}{a_{\nu+1}} = \frac{\nu+1}{\sigma-\nu} = -\frac{1+1/\nu}{1-\sigma/\nu} \qquad \text{for all } \nu \geq 1,$$

which shows that $R = \lim \frac{|a_\nu|}{|a_{\nu+1}|} = 1$. □

And we should take note of a *multiplication formula*

$$(1+z)b_{\sigma-1}(z) = b_\sigma(z),$$

valid for all $z \in \mathbb{E}$ and all $\sigma \in \mathbb{C}$; a formula which should come as no surprise since, as noted, $b_\sigma(z)$ will turn out to be the power $(1+z)^\sigma$.

Proof. The arithmetically confirmed identity

$$\binom{\sigma-1}{\nu} + \binom{\sigma-1}{\nu-1} = \binom{\sigma}{\nu} , \ \nu \geq 1$$

gives

$$\sum_0^\infty \binom{\sigma-1}{\nu} z^\nu + \sum_1^\infty \binom{\sigma-1}{\nu-1} z^\nu = \sum_0^\infty \binom{\sigma}{\nu} z^\nu.$$

Now shift the summation index in the second sum. □

Since conjugation is a continuous (isometric!) map of $\mathbb{C}$, for every power series $f(z) = \sum a_\nu z^\nu$ we have $\overline{f(z)} = \sum \bar{a}_\nu \bar{z}^\nu$, and thus $\overline{f(z)} = f(\bar{z})$ in case all the coefficients a_ν are *real*. In particular, we have

$$\overline{\exp z} = \exp \bar{z} , \quad \overline{\cos z} = \cos \bar{z} , \quad \overline{\sin z} = \sin \bar{z} \quad \text{for all } z \in \mathbb{C},$$
$$\overline{\lambda(z)} = \lambda(\bar{z}) \quad \text{for all } z \in \mathbb{E} , \quad \overline{b_\sigma(z)} = b_\sigma(\bar{z}) \quad \text{for all } z \in \mathbb{E},$$

the last provided that $\sigma \in \mathbb{R}$.

4*. Convergence behavior on the boundary. The convergence behavior of a power series on the periphery of the open disc of convergence can be quite different from case to case: *convergence (absolute) can occur everywhere*, as for example with the series $\sum_1^\infty \frac{z^\nu}{\nu^2}$, whose $R = 1$; *convergence can occur nowhere*, as for example with the series $\sum z^\nu$, whose $R = 1$; and *points of convergence can co-exist with points of divergence*, as occurs for example with the "logarithmic" series $\sum_1^\infty \frac{(-1)^{\nu-1}}{\nu} z^\nu$, whose $R = 1$ — at $z = 1$ it is the alternating harmonic series and so converges, but at $z = -1$ it is the ordinary harmonic series and so diverges. Actually it can be shown that this series converges at every point of the boundary circle except $z = -1$.

There is an extensive literature dealing with convergence behavior on the boundary. For example, in 1911 the Russian mathematician N. LUSIN constructed a power series $\sum c_\nu z^\nu$ with radius of convergence 1 which satisfies $c_\nu \to 0$ and diverges at every $z \in \mathbb{C}$ of modulus 1. And the Polish mathematician W. SIERPIŃSKI produced in 1912 a power series with radius of convergence 1 which converges at the point $z = 1$ but diverges at every other point on the unit circle. The reader interested in such matters will find these last two examples, and more, presented in detail in the beautiful booklet [Lan].

5*. Abel's continuity theorem. In 1827 ABEL formulated the following problem (*Jour. für die Reine und Angew. Math.*, Vol. 2, p. 286; also *Œuvres* Vol. 1, p. 618): "En supposant la série

$$(*) \qquad\qquad fx = \alpha_0 + \alpha_1 x + \alpha_2 x^2 + \cdots$$

convergente pour toute valeur positive moindre que la quantité positive α, on propose de trouver la limite vers laquelle converge la valeur de la fonction fx, en faisant converger x vers la limite α. (Supposing the series $(*)$ to be convergent for every positive quantity x less than a certain positive quantity α, I propose to find the limit to which the values of the function fx converge when x is made to converge to the limit α.)" This problem is essentially that of determining the behavior under radial approach to the boundary of the convergence disc of the limit function of a power series. For the case where the power series itself converges at the boundary point being considered, ABEL had already solved his problem in his 1826 paper [A]; namely, the complete formulation of his theorem which was alluded to in 1.5 runs as follows:

Lehrsatz IV. *If the series $f(\alpha) = \nu_0 + \nu_1\alpha + \nu_2\alpha^2 + \cdots + \nu_m\alpha^m + \cdots$ converges for a certain value δ of α, then it also converges for every smaller value of α and is of such a nature that $f(\alpha - \beta)$ approaches the limit $f(\alpha)$ as β decreases to 0, provided that α is less than or equal to δ.*

Since $\alpha = \delta$ is explicitly allowed, it is being said here that the function $f(\alpha)$ is continuous in the *closed* interval $[0, \delta]$, that is,

$$\lim_{\alpha \to \delta - 0} f(\alpha) = \nu_0 + \nu_1\delta + \nu_2\delta^2 + \cdots$$

holds. This proposition, which is certainly not trivial and was derived by ABEL using an elegant trick, is called the *Abel continuity theorem* or *limit theorem*. [The theorem had already been stated and used by GAUSS ("Disquisitiones generales circa seriem ...," 1812; *Werke* **III**, p. 143) but his proof contained a gap involving the uncritical interchange of two limit processes.] The applications of this theorem in the calculus are well enough known. E.g., once one has that the series

$$x - \frac{x^2}{2} + \frac{x^3}{3} - \frac{x^4}{4} + - \cdots \text{ and } x - \frac{x^3}{3} + \frac{x^5}{5} - \frac{x^7}{7} + - \cdots$$

converge in the interval $(-1, 1)$ and represent the functions $\log(1+x)$ and $\arctan x$, respectively, then from their convergence at $x = 1$ and the continuity theorem (and the fact that $\log(1 + x)$ and $\arctan x$ are each continuous at $x = 1$) he gets the equation $\log 2 = 1 - \frac{1}{2} + \frac{1}{3} - \frac{1}{4} + - \cdots$ and the famous

Leibniz formula:

$$\frac{\pi}{4} = \arctan 1 = 1 - \frac{1}{3} + \frac{1}{5} - \frac{1}{7} + - \cdots$$

The limit theorem can also be phrased this way: for a boundary point at which the series converges, the function defined by the series has a limit under *radial* approach to this point and it is the sum of the series at this point. In 1875 O. STOLZ proved the following generalization (*Zeitschrift Math. Phys.* **20**, 369-376):

Let $\sum a_\nu (z - c)^\nu$ be a power series with radius of convergence R; suppose that it converges at the point $b \in \partial B_R(c)$. Then the series converges uniformly in every closed triangle Δ which has one vertex at b and its other two vertices in $B_R(c)$ (cf. the figure). In particular, the function

$$f : \Delta \to \mathbb{C} \, , \, \zeta \mapsto \sum a_\nu(\zeta - c)^\nu$$

is continuous throughout Δ and, consequently, satisfies

$$\lim_{\substack{z \to b \\ z \in \Delta}} f(z) = \sum a_\nu(b - c)^\nu.$$

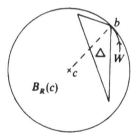

HARDY and LITTLEWOOD showed in 1912 (in this connection see also L. HOLZER, *Deutsche Math.* **4**(1939), 190-193) that for approach to b along a path W which lies in no such triangle Δ, so-called tangential approach, the function defined by the power series need not have a limit at all. □

There don't seem to be any "natural applications" of this generalized limit theorem. The interested reader will find a proof in KNESER [14], pp. 143-144 or in KNOPP [15]. In the latter and in [Lan] he will also find material concerning the converse of Abel's limit theorem (theorems of TAUBER, HARDY and LITTLEWOOD, and others).

Exercises

Exercise 1. The *hypergeometric series* determined by $a, b, c \in \mathbb{C}, -c \notin \mathbb{N}$, is given by

$$F(a, b, c, z) := 1 + \frac{ab}{c} z + \frac{a(a+1)b(b+1)}{2c(c+1)} z^2 + \cdots +$$

$$+ \frac{a(a+1) \cdots (a+n-1)b(b+1) \cdots (b+n-1)}{n! c(c+1) \cdots (c+n-1)} z^n + \cdots$$

Show that in case $-a, -b \notin \mathbb{N}$:

a) The hypergeometric series determined by a, b, c has radius of convergence 1.

b) If $\Re(a + b - c) < 0$, the series converges absolutely for $|z| = 1$.

Hint to b): Exercise 3, §4, Chapter 0.

Exercise 2. a) Let a_n be real and decrease to 0. Show that the power series $\sum a_n z^n$ converges compactly in $\overline{\mathbb{E}} \setminus \{1\}$. *Hint:* Investigate $(1 - z) \sum a_n z^n$.

b) Show that the logarithmic series $\lambda(z) = \sum_{\nu \geq 1} \frac{(-1)^{\nu-1}}{\nu} z^\nu$ converges compactly in $\overline{\mathbb{E}} \setminus \{1\}$.

Exercise 3. This exercise is concerned with proving a theorem of TAUBER which in a special case furnishes a converse of ABEL's continuity theorem.

Let $a_n \in \mathbb{C}$ satisfy $\lim_n n a_n = 0$. Then the power series $f(z) := \sum_{n=0}^{\infty} a_n z^n$ converges at least in $\mathbb{E}$. Suppose there is an $a \in \mathbb{C}$ such that $\lim_m f(x_m) = a$ for every sequence $\{x_m\}$ in $\mathbb{E}$ which approaches 1 "from the left", that is, for every sequence $\{x_m\} \subset (-1, 1)$ with $\lim_m x_m = 1$. Then $\sum_{n=0}^{\infty} a_n = a$.

Carry out a proof via the following steps:

a) Show that for $0 \leq x < 1$ and every integer $m > 0$

$$\left| \sum_{n=0}^{m} a_n - f(x) \right| \leq (1 - x) \sum_{n=1}^{m} n |a_n| + \sum_{n=m+1}^{\infty} |a_n| x^n, \text{ and}$$

$$\sum_{n=m+1}^{\infty} |a_n| x^n \leq \frac{1}{m(1-x)} \max\{n |a_n| : n > m\}.$$

b) Now consider the special sequence $x_m := 1 - \frac{1}{m}$ and use Exercise 3b) of §3, Chapter 0.

Exercise 4. Show that if almost all the a_n are real and non-negative, then the first hypothesis in Tauber's theorem above can be weakened to $\lim a_n = 0$ and the conclusion still follows.

§3   Holomorphy of power series

Except for the examples in 1.2.3, where transcendental real functions were put together to form holomorphic functions, we really don't know any holomorphic functions besides polynomials and rational functions. But from the *differential calculus* of real functions we know that *convergent real power series* represent functions which are *real-differentiable as often as we please* and that the summation and differentiation may be interchanged. Analogously, it turns out that in the *complex differential calculus* every *convergent complex power series* is *arbitrarily often complex-differentiable* and is consequently holomorphic; and here too the theorem on interchangeability of summation and differentiation holds.

1. Formal term-wise differentiation and integration. *If the series* $\sum a_\nu (z-c)^\nu$ *has radius of convergence* R, *then the series* $\sum \nu a_\nu (z-c)^{\nu-1}$ *and* $\sum (\nu+1)^{-1} a_\nu (z-c)^{\nu+1}$ *arising from it by term-wise differentiation and integration, respectively, each also have radius of convergence* R.

Proof. a) For the radius of convergence R' of the differentiated series we have

$$R' = \sup\{t \geq 0 : \text{ the sequence } \nu |a_\nu| t^{\nu-1} \text{ is bounded}\}.$$

Since the boundedness of the sequence $\nu |a_\nu| t^{\nu-1}$ certainly implies that of the sequence $|a_\nu| t^\nu$, the inequality $R' \leq R$ is clear.

In order to conclude that $R \leq R'$, it suffices to show that every $r < R$ satisfies $r \leq R'$. Given such an r, pick s with $r < s < R$. Then the series $|a_\nu| s^\nu$ is bounded. Moreover, for $q := r/s$, $\nu |a_\nu| r^{\nu-1} = (r^{-1}|a_\nu| s^\nu) \nu q^\nu$. Now for $\delta := q^{-1} - 1 > 0$ and $\nu \geq 2$ the binomial formula gives $q^{-\nu} = (1+\delta)^\nu \geq \binom{\nu}{2} \delta^2$, and so $0 < \nu q^\nu \leq \dfrac{2}{(\nu-1)\delta^2}$. It follows that νq^ν is a null sequence, and so too is its product with the bounded sequence $r^{-1}|a_\nu| s^\nu$, namely the sequence $\nu |a_\nu| r^{\nu-1}$. Then certainly $r \leq R'$. In summary, $R' = R$.

b) Let R^* denote the radius of convergence of the integrated series. According to what was proved in a), R^* is then also equal to the radius of convergence of the series gotten from $\sum (\nu+1)^{-1} a_\nu (z-c)^{\nu+1}$ by term-wise differentiation, namely the series $\sum a_\nu (z-c)^\nu$. That is, $R^* = R$.   □

2. Holomorphy of power series. The interchange theorem. In the theorem just proved there is nothing to the effect that the continuous limit function f of the power series $\sum a_\nu (z-c)^\nu$ is holomorphic in $B_R(c)$, but we shall now prove this. We claim

Theorem (Interchangeability of differentiation and summation in power series). *If the power series* $\sum a_\nu (z-c)^\nu$ *has radius of convergence*

$R > 0$, *then its limit function f is arbitrarily often complex-differentiable, and in particular holomorphic, in $B_R(c)$. Moreover,*

$$f^{(k)}(z) = \sum_{\nu \geq k} k! \binom{\nu}{k} (z - c)^{\nu - k} \, , \quad z \in B_R(c) \, , \, k \in \mathbb{N}$$

and in particular, for all $k \in \mathbb{N}$,

$$\frac{f^{(k)}(c)}{k!} = a_k \quad (\text{Taylor-coefficient formulas}).$$

Proof. It suffices to treat the case $k = 1$; the general result follows by iterating this conclusion. Set $B := B_R(c)$. First of all, theorem 1 makes it clear that $g(z) := \sum_{\nu \geq 1} \nu a_\nu (z - c)^{\nu - 1}$ is a well-defined complex-valued function on B. Our claim is that $f' = g$. We may and do assume that $c = 0$. Let $b \in B$ be fixed. In order to show that $f'(b) = g(b)$, we set

$$q_\nu(z) := z^{\nu - 1} + z^{\nu - 2} b + \cdots + z^{\nu - j} b^{j - 1} + \cdots + b^{\nu - 1}$$

$z \in \mathbb{C}$, $\nu = 1, 2, \ldots$. Then $z^\nu - b^\nu = (z - b) q_\nu(z)$ for all z and so

$$f(z) - f(b) = \sum_{\nu \geq 1} a_\nu (z^\nu - b^\nu) = (z - b) \sum_{\nu \geq 1} a_\nu q_\nu(z) \, , \, z \in B.$$

Now write $f_1(z)$ for $\sum_{\nu \geq 1} a_\nu q_\nu(z)$. It follows, upon noting the fact that $q_\nu(b) = \nu b^{\nu - 1}$, that

$$f(z) = f(b) + (z - b) f_1(z) \, , \, z \in B, \quad \text{and} \quad f_1(b) = \sum_{\nu \geq 1} \nu a_\nu b^{\nu - 1} = g(b).$$

All that remains therefore is to show that f_1 is *continuous at b*. This is accomplished by proving that the defining series $\sum_{\nu \geq 1} a_\nu q_\nu(z)$ for f_1 is *normally convergent in B*. But this latter fact is clear: for every $|b| < r < R$

$$|a_\nu q_\nu|_{B_r} \leq |a_\nu| \nu r^{\nu - 1}, \text{ so } \sum_{\nu \geq 1} |a_\nu q_\nu|_{B_r} \leq \sum_{\nu \geq 1} \nu |a_\nu| r^{\nu - 1} < \infty,$$

by theorem 1.   □

The above proof is valid (word for word) if instead of $\mathbb{C}$ *any complete valued field*, e.g., $\mathbb{R}$, underlies the series. (Of course for fields of non-zero characteristic one writes $\nu(\nu - 1) \cdots (\nu - k + 1)$ instead of $k! \binom{\nu}{k}$.)

3. Historical remarks on term-wise differentiation of series. For

EULER it went without saying that term-wise differentiation of power series and function series resulted in the derivative of the limit function. ABEL was the first to point out, in the 1826 letter to HOLMBOE already cited in 3.1.4, that the theorem on interchange of differentiation and summation is not generally valid for convergent series of differentiable functions. ABEL, who had just found his way in Berlin from the mathematics of the Euler period to the critical logical rigor of the Gauss period, writes with all the enthusiasm of a neophyte (*loc. cit.*, p. 258): "La théorie des séries infinies en général est jusqu'à présent très mal fondée. On applique aux séries infinies toutes les opèrations, comme si elles étaient finies; mais cela est-il bien permis? Je crois que non. Où est-il démontré qu'on obtient la différentielle d'une série infinie en prenant la différentielle de chaque terme? Rien n'est plus facile que de donner des exemples où cela n'est pas juste; par exemple

$$(1) \qquad \frac{x}{2} = \sin x - \frac{1}{2}\sin 2x + \frac{1}{3}\sin 3x - \text{ etc.}$$

En différentiant on obtient

$$(2) \qquad \frac{1}{2} = \cos x - \cos 2x + \cos 3x - \text{ etc.},$$

résultat tout faux, car cette série est divergente. (Until now the theory of infinite series in general has been very badly grounded. One applies all the operations to infinite series as if they were finite; but is that permissible? I think not. Where is it demonstrated that one obtains the differential of an infinite series by taking the differential of each term? Nothing is easier than to give instances where this is not so; for example (1) holds but in differentiating it one obtains (2), a result which is wholly false because the series there is divergent.)"

The correct function-theoretic generalization of the theorem on termwise differentiation of power series is the famous theorem of WEIERSTRASS on the term-wise differentiation of compactly convergent series of holomorphic functions. We will prove this theorem in 8.4.2 by means of the Cauchy estimates.

4. Examples of holomorphic functions. 1) From the *geometric* series $\frac{1}{1-z} = \sum_0^\infty z^\nu$, which converges throughout the open unit disc $\mathbb{E}$, after k-fold differentiation we get

$$\frac{1}{(1-z)^{k+1}} = \sum_{\nu \geq k} \binom{\nu}{k} z^{\nu-k}, \qquad z \in \mathbb{E}.$$

2) The *exponential function* $\exp z = \sum \frac{z^\nu}{\nu!}$ is holomorphic throughout $\mathbb{C}$:

$$\exp'(z) = \exp z, \qquad z \in \mathbb{C}.$$

This important differential equation can be made the starting point for the theory of the exponential function (cf. 5.1.1).

3) The *cosine function* and the *sine function*

$$\cos z = \sum \frac{(-1)^\nu}{(2\nu)!} z^{2\nu} \ , \qquad \sin z = \sum \frac{(-1)^\nu}{(2\nu+1)!} z^{2\nu+1} \ , \qquad z \in \mathbb{C}$$

are holomorphic in $\mathbb{C}$:

$$\cos'(z) = -\sin z \ , \quad \sin'(z) = \cos z \ , \ z \in \mathbb{C}.$$

These equations also follow immediately if one is willing to use the fact already established that $\exp' = \exp$, together with the Euler representations

$$\cos z = \frac{1}{2}[\exp(iz) + \exp(-iz)] \ , \quad \sin z = \frac{1}{2i}[\exp(iz) - \exp(-iz)].$$

4) The *logarithmic series* $\lambda(z) = z - \frac{z^2}{2} + \frac{z^3}{3} - \cdots$ is holomorphic in $\mathbb{E}$:

$$\lambda'(z) = 1 - z + z^2 - \cdots = \frac{1}{1+z} \ , \qquad z \in \mathbb{E}.$$

5) The *arctangent series* $a(z) = z - \frac{z^3}{3} + \frac{z^5}{5} - \cdots$ is holomorphic in $\mathbb{E}$ with derivative $a'(z) = \frac{1}{1+z^2}$. The designation "arctangent" will be justified in 5.2.5 and 5.5.2.

6) The *binomial series* $b_\sigma(z) = \sum \binom{\sigma}{\nu} z^\nu \ , \ \sigma \in \mathbb{C}$, is holomorphic in $\mathbb{E}$:

$$b'_\sigma(z) = \sigma b_{\sigma-1}(z) = \frac{\sigma}{1+z} b_\sigma(z) \ , \qquad z \in \mathbb{E}.$$

To see the first of these equalities, use the fact that $\nu\binom{\sigma}{\nu} = \sigma\binom{\sigma-1}{\nu-1}$ to get

$$b'_\sigma(z) = \sum_1^\infty \nu \binom{\sigma}{\nu} z^{\nu-1} = \sigma \sum_1^\infty \binom{\sigma-1}{\nu-1} z^{\nu-1} = \sigma b_{\sigma-1}(z).$$

The second equality is now a consequence of the multiplication formula 2.3.

□

The exponential series, the logarithmic series and the binomial series are connected by the important equation

$$(*) \qquad\qquad b_\sigma(z) = \exp(\sigma\lambda(z)) \ , \qquad z \in \mathbb{E},$$

a special case of which is

$$1 + z = \exp\lambda(z) \qquad \text{for all } z \in \mathbb{E}.$$

Proof. The function $f(z) := b_\sigma(z)\exp(-\sigma\lambda(z))$ is holomorphic in $\mathbb{E}$ and by the chain and product rules for differentiation

$$f'(z) = [b'_\sigma(z) - \sigma b_\sigma(z)\lambda'(z)]\exp(-\sigma\lambda(z)) = 0,$$

since by 4) and 6) $b'_\sigma = \sigma b_\sigma \lambda'$. Consequently, according to 1.3.3 f is constant in $\mathbb{E}$. That constant is $f(0) = 1$, so $(*)$ follows if we anticipate 5.1.1, according to which $\exp(-w) = (\exp(w))^{-1}$ for all w.

Exercises

Exercise 1. Show that the hypergeometric function $F(a, b, c, z)$ introduced in Exercise 1 of §2 satisfies the differential equation

$$z(1-z)F''(a,b,c,z) + \{c - (1+a+b)z\}F'(a,b,c,z) - abF(a,b,c,z) = 0$$

for all $z \in \mathbb{E}$.

Exercise 2. Show that for $c \in \mathbb{C}^{\times}$, $d \in \mathbb{C} \setminus \{c\}$, $k \in \mathbb{N}$

$$\frac{1}{(c-z)^{k+1}} = \frac{1}{(c-d)^{k+1}} \sum_{\nu \geq k} \binom{\nu}{k} \left(\frac{z-d}{c-d}\right)^{\nu-k}, \qquad z \in B_{|c-d|}(d).$$

Exercise 3. (Partial fraction decomposition of rational functions). Let $f(z) := \frac{p(z)}{q(z)}$ be a rational function, in which the degree of the polynomial p is less than that of q and q has the factorization $q(z) = c(z - c_1)^{\nu_1}(z - c_2)^{\nu_2} \cdots (z - c_m)^{\nu_m}$ with $c \in \mathbb{C}^{\times}$ and distinct complex numbers c_j.

a) Show that $f(z)$ has a representation of the form

$$f(z) = \frac{a_{11}}{z - c_1} + \frac{a_{12}}{(z - c_1)^2} + \cdots + \frac{a_{1\nu_1}}{(z - c_1)^{\nu_1}} + \frac{a_{21}}{z - c_2} + \cdots$$

$$+ \frac{a_{m1}}{z - c_m} + \frac{a_{m2}}{(z - c_m)^2} + \cdots + \frac{a_{m\nu_m}}{(z - c_m)^{\nu_m}}$$

for certain complex numbers a_{jk}.

b) Show that the particular coefficients $a_{k\nu_k}$, $1 \leq k \leq m$, from a) are determined by

$$a_{k\nu_k} = \frac{p(c_k)}{c(c_k - c_1)^{\nu_1} \cdots (c_k - c_{k-1})^{\nu_{k-1}}(c_k - c_{k+1})^{\nu_{k+1}} \cdots (c_k - c_m)^{\nu_m}}$$

Hint to a): Induction on $n := $ degree $q = \nu_1 + \nu_2 + \cdots + \nu_m$.

Exercises 2 and 3, in connection with the Fundamental Theorem of Algebra (cf. 9.1) insure that any rational function can be developed in a power series about each point in its domain of definition.

Exercise 4. Develop each of the given rational functions into power series about each of the given points and specify the radii of convergence of these series:

a) $\dfrac{1}{z^3 - iz^2 - z + i}$ about 0 and about 2.

b) $\dfrac{z^4 - z^3 - 8z^2 + 14z - 3}{z^3 - 4z^2 + 5z - 2}$ about 0 and about i.

Hint to b): First use long division to get a proper fraction.

Exercise 5. The sequence defined recursively by $c_0 := c_1 := 1$, $c_n := c_{n-1} + c_{n-2}$ for $n \geq 2$ is called the *Fibonacci numbers*. Determine these numbers explicitly. *Hint:* Consider the power series development about 0 of the rational function $\frac{1}{z^2+z-1}$.

§4   Structure of the algebra of convergent power series

The set $\bar{A}$ of all (formal) power series centered at 0 is (with the Cauchy multiplication) a commutative $\mathbb{C}$-algebra with 1. We will denote by A the set of all *convergent* power series centered at 0. Then

  A *is a* $\mathbb{C}$-*subalgebra of* $\bar{A}$ *which is characterized by*

(1) $A = \{f = \sum a_\nu z^\nu \in \bar{A} : \exists$ *positive real* s, M *such that* $|a_\nu| s^\nu \leq M$
      $\forall \nu \in \mathbb{N}\}$.

This latter is clear by virtue of the convergence lemma 1.1.

In what follows the structure of the ring A will be exhaustively described; in doing so we will consistently use the language of modern algebra. The tools are the order function $v : A \to \mathbb{N} \cup \{\infty\}$ and theorem 2 on units, which is not completely trivial for A. Since these tools are trivially also available in $\bar{A}$, all propositions of this section hold, *mutatis mutandis*, for the ring of formal power series as well. These results are, moreover, valid with any *complete valued* field k taking over the role of $\mathbb{C}$.

1. The order function. For every power series $f = \sum a_\nu z^\nu$ the *order* $v(f)$ of f is defined by

$$v(f) := \begin{cases} \min \{\nu \in \mathbb{N} : a_\nu \neq 0\} & \text{, in case } f \neq 0, \\ \infty & \text{, in case } f = 0. \end{cases}$$

For example, $v(z^n) = n$.

Rules of Computation for the order function. *The mapping* $v : \bar{A} \to \mathbb{N} \cup \{\infty\}$ *is a non-archimedean valuation of* $\bar{A}$; *that is, for all* $f, g \in \bar{A}$
   1) $v(fg) = v(f) + v(g)$ (*product rule*)
   2) $v(f + g) \geq \min\{v(f), v(g)\}$ (*sum rule*).

The reader can easily carry out the proof, if he recalls the conventions $n + \infty = n$, $\min\{n, \infty\} = n$ for $n \in \mathbb{R} \cup \{\infty\}$. Because the range $\mathbb{N} \cup \{\infty\}$ of v is "discrete" in $\mathbb{R} \cup \{\infty\}$, the valuation v is also called *discrete*.

The product rule immediately implies that:

The algebra $\bar{A}$ and so also its subalgebra A is an integral domain, that is, it contains no non-zero zero-divisors: from $fg = 0$ follows either $f = 0$ or $g = 0$.

The sum rule can be sharpened: in general, for any non-archimedean valuation, $v(f + g) = \min\{v(f), v(g)\}$ whenever $v(f) \neq v(g)$.

2. The theorem on units. An element e of a commutative ring R with 1 is called a *unit in R* if there exists an $\hat{e} \in R$ with $e\hat{e} = 1$. The units in R form a multiplicative group, the so-called *group of units of R*. To characterize the units of A we need

Lemma on units. *Every convergent power series $e = 1 - b_1 z - b_2 z^2 - b_3 z^3 - \cdots$ is a unit in A.*

Proof. We will have $e\hat{e} = 1$ for $\hat{e} := 1 + k_1 z + k_2 z^2 + k_3 z^3 + \cdots \in \bar{A}$ if we define

$$(*) \quad k_1 := b_1 \ , \quad k_n := b_1 k_{n-1} + b_2 k_{n-2} + \cdots + b_{n-1} k_1 + b_n \quad \text{for all } n \geq 2.$$

It remains then to show that $\hat{e} \in A$. Because $e \in A$ there exists a real $s > 0$ such that $|b_n| \leq s^n$ for all $n \geq 1$. From this and induction we get

$$|k_n| \leq \frac{1}{2}(2s)^n \ , \qquad n = 1, 2, \ldots.$$

Indeed, this is clear for $n = 1$ and the passage from $n - 1$ to n proceeds via $(*)$ as follows:

$$|k_n| \leq \sum_1^{n-1} |b_\nu||k_{n-\nu}| + |b_n| \leq \frac{1}{2} \sum_1^{n-1} s^\nu (2s)^{n-\nu} + s^n = \frac{1}{2}(2s)^n.$$

Therefore for the positive number $t := (2s)^{-1}$ we have $|k_n| t^n \leq 1/2$ for all $n \geq 1$ and this means (recall the defining equation (1) in the introduction to this section) that $\hat{e} \in A$. □

The preceding proof was given by HURWITZ in [12], pp. 28, 29; it probably goes back to WEIERSTRASS. In 7.4.1 we will be able to give a, wholly different, "two-line proof"; for the polynomial ring $\mathbb{C}[z]$ there is however no analog of this lemma on units. □

Theorem on units. *An element $f \in A$ is a unit of A if and only if $f(0) \neq 0$.*

Proof. a) The condition is obviously necessary: if $\hat{f} \in A$ and $f\hat{f} = 1$, then $f(0)\hat{f}(0) = 1$ and so $f(0) \neq 0$.

b) Let $f = \sum a_\nu z^\nu \in A$ with $a_0 = f(0) \neq 0$. For the series $e := a_0^{-1} f = 1 + a_0^{-1} a_1 z + a_0^{-1} a_2 z^2 + \cdots \in A$ the preceding lemma furnishes an $\hat{e} \in A$ with $e\hat{e} = 1$. It follows that $f \cdot (a_0^{-1}\hat{e}) = 1$.   □

The theorem on units can also be formulated thus:

$$f \in A \text{ is a unit of } A \Leftrightarrow v(f) = 0.$$

The lemma and the theorem on units both naturally hold as well for formal power series; in this case the proof of the lemma just reduces to the first two lines of the above proof.

3. Normal form of a convergent power series. *Every $f \in A \setminus \{0\}$ has the form*

(1) $f = ez^n$ *for some unit e of A and some $n \in \mathbb{N}$.*

The representation (1) is unique and $n = v(f)$.

Proof. a) Let $n = v(f)$, so that f has the form $f = a_n z^n + a_{n+1} z^{n+1} + \cdots$ with $a_n \neq 0$. Then $f = ez^n$, where $e := a_n + a_{n+1} z + \cdots$ is, by the theorem on units, a unit of A.

b) Let $f = \tilde{e} z^m$ be another representation of f with $m \in \mathbb{N}$ and $\tilde{e}$ a unit of A. By the theorem on units, $v(e) = v(\tilde{e}) = 0$. Therefore from $ez^n = \tilde{e}z^m$ and the product rule for the order function it follows that

$$n = v(e) + v(z^n) = v(ez^n) = v(\tilde{e}z^m) = m$$

and then $e = \tilde{e}$ also follows.   □

We call (1) *the normal form* of f.

An element p of an integral domain R is called a *prime element* if it is not a unit of R and if whenever $f, g \in R$ and p divides the product fg, then p divides one of the factors f or g. An integral domain R each of whose non-zero elements is a product of finitely many primes is called a *unique factorization domain* (in Bourbaki, simply *factorial*).

From the normal form (1) we see immediately that

Corollary. *The ring $\mathcal{A}$ is factorial and, up to multiplication by units, the element z is the only prime in $\mathcal{A}$.*

In contrast to $\mathcal{A}$, its subring $\mathbb{C}[z]$, which is itself a unique factorization domain, has the continuum-many prime elements $z - c$, $c \in \mathbb{C}$.

In the foregoing the prime z played a distinguished role. But the theorem and its corollary remain valid if instead of z any other fixed element $\tau \in \mathcal{A}$ with $v(\tau) = 1$ is considered. Every such τ is a prime element of $\mathcal{A}$, on an equal footing with z, and is known in the classical terminology as a *uniformizer of $\mathcal{A}$*.

Every integral domain R possesses a *quotient field $Q(R)$*. It is immediate from the corollary that:

The quotient field $Q(\mathcal{A})$ consists of all series of the form $\sum_{\nu \geq n} a_\nu z^\nu$, $n \in \mathbb{Z}$, where $\sum_0^\infty a_\nu z^\nu$ is a convergent power series.

Series of this kind are called "Laurent series with finite principal part" (cf. 12.1.3). The reader is encouraged to supply a proof for the above statement about $Q(\mathcal{A})$, as well as the following easy exercise:

The function $v : \mathcal{A} \to \mathbb{N} \cup \{\infty\}$ can be extended in exactly one way to a non-archimedean valuation $v : Q(\mathcal{A}) \to \mathbb{Z} \cup \{\infty\}$. This valuation is given by

$$v(f) = n \quad if \quad f = \sum_{\nu \geq n} a_\nu z^\nu \quad with\ a_n \neq 0.$$

4. Determination of all ideals. A commutative ring R is called a *principal ideal domain* if every one of its ideals in *principal*, that is, has the form Rf for some $f \in R$.

Theorem. *$\mathcal{A}$ is a principal ideal domain: the ideals $\mathcal{A}z^n$, $n \in \mathbb{N}$, comprise all the non-zero ideals of $\mathcal{A}$.*

Proof. Let $\mathfrak{a}$ be any non-zero ideal of $\mathcal{A}$. Choose an element $f \in \mathfrak{a}$ of minimal order $n \in \mathbb{N}$. According to theorem 3, $z^n = \hat{e}f$, with $\hat{e} \in \mathcal{A}$. It follows that $\mathcal{A}z^n \subset \mathfrak{a}$. If g is an arbitrary non-zero element of $\mathfrak{a}$, then $g = \tilde{e}z^m$, for some unit $\tilde{e} \in \mathcal{A}$ and $m = v(g)$. By minimality of n, $m \geq n$. It follows that $g = (\tilde{e}z^{m-n})z^n \in \mathcal{A}z^n$. This shows that conversely $\mathfrak{a} \subset \mathcal{A}z^n$ and gives the equality of the two. $\qquad\square$

An ideal $\mathfrak{p}$ of a ring R is called a *prime ideal* if $fg \in \mathfrak{p}$ always implies that one or the other of the factors f, g lies in $\mathfrak{p}$. An ideal $\mathfrak{m} \neq R$ in R

is called *maximal* if the only ideal which properly contains it is R itself. *Maximal ideals are prime ideals.*

Theorem. *The set* $\mathrm{m}(\mathcal{A})$ *of all non-units of* $\mathcal{A}$ *is a maximal ideal of* $\mathcal{A}$. *It coincides with* $\mathcal{A}z$ *and is the unique non-zero prime ideal of* $\mathcal{A}$.

Proof. According to the lemma on units $\mathrm{m}(\mathcal{A}) = \{f \in \mathcal{A} : v(f) \geq 1\}$; therefore $\mathrm{m}(\mathcal{A})$ is an ideal. From the preceding theorem it follows that $\mathrm{m}(\mathcal{A}) = \mathcal{A}z$.

If $\mathfrak{a}$ is an ideal which properly contains $\mathrm{m}(\mathcal{A})$, then $\mathfrak{a}$ contains a unit and so necessarily contains 1 and therewith is all of $\mathcal{A}$. Therefore $\mathrm{m}(\mathcal{A})$ is a maximal ideal, and in particular a prime ideal of $\mathcal{A}$.

If $n \geq 2$, $z \cdot z^{n-1} \in \mathcal{A}z^n$ but $z \notin \mathcal{A}z^n$ and $z^{n-1} \notin \mathcal{A}z^n$. Consequently $\mathcal{A}z^n$ is not a prime ideal and we see that $\mathcal{A}z$ is the unique non-zero prime ideal of $\mathcal{A}$. □

In modern algebra an integral domain is called a *discrete valuation ring* if it is a principal ideal domain which possesses a unique non-zero prime ideal. Thus we have shown that

The ring $\mathcal{A}$ *of convergent power series is a discrete valuation ring.*

A ring is called *local* if it has exactly one maximal ideal. All discrete valuation rings are local, so in particular $\mathcal{A}$ is a local ring.

The reader might check that the ring $\bar{\mathcal{A}}$ of formal power series is also a discrete valuation ring and, in particular, a local ring.

Chapter 5

Elementary Transcendental Functions

Post quantitates exponentiales considerari debent arcus circulares eorumque sinus et cosinus, quia ex ipsis exponentialibus, quando imaginariis quantitatibus involuntur, proveniunt. (After exponential quantities the circular functions, sine and cosine, should be considered because they arise when imaginary quantities are involved in the exponential.) - L. EULER, *Introductio*.

In this chapter the classical transcendental functions, already treated by EULER in his *Introductio* [E], will be discussed. At the center stands the exponential function, which is determined (§1) both by its differential equation and its addition theorem. In Section 2 we will prove directly, using differences and the logarithmic series and *without borrowing any facts from real analysis*, that the exponential function defines a homomorphism of the additive group $\mathbb{C}$ *onto* the multiplicative group $\mathbb{C}^{\times}$. This epimorphism theorem is basic for everything else; for example, it leads immediately to the realization that there is a *uniquely determined* positive real number π such that $\exp z = 1$ precisely when z is one of the *numbers* $2n\pi i$, $n \in \mathbb{Z}$. This famous constant thereby "occurs naturally among the complex numbers".

Following Euler's recommendation, all the important properties of the trigonometric functions are derived from the exponential function via the identities

$$\cos z = \frac{1}{2}(e^{iz} + e^{-iz}) \ , \ \sin z = \frac{1}{2i}(e^{iz} - e^{-iz}).$$

In particular, we will see that π is the *smallest positive zero* of the sine function and $\frac{\pi}{2}$ is the smallest positive zero of the cosine function, just as we learned in the infinitesimal calculus. In connection with §1-3 compare

133

also the presentation in *Numbers* [19], where among other things a completely elementary approach to the equation $e^{2\pi i} = 1$ will be found.

Logarithm functions will be treated in detail in §4 and §5, where the general power function and the Riemann zeta function will also be introduced.

§1   The exponential and trigonometric functions

The most important holomorphic function which is not a rational function is the one defined by the power series $\sum \frac{z^\nu}{\nu!}$ and designated $\exp z$. Its dominant role in the complex theatre is due to the Euler formulas and to its invariance under differentiation: $\exp' = \exp$. This latter property, along with $\exp 0 = 1$, characterizes the exponential function; it makes possible a very elegant derivation of the basic properties of this function. Decisive in many arguments is the elemental fact that a holomorphic function f with $f' = 0$ is necessarily constant.

1. Characterization of $\exp z$ by its differential equation. First let us note that

The exponential function is zero-free in $\mathbb{C}$ and

$$(\exp z)^{-1} = \exp(-z) \qquad \text{for all } z \in \mathbb{C}.$$

Proof. The holomorphic function $h(z) := \exp z \cdot \exp(-z)$ satisfies $h' = h - h = 0$ throughout the connected set $\mathbb{C}$. Consequently (1.3.3) it is constant in $\mathbb{C}$. Since $h(0) = 1$, it follows that $\exp(z) \cdot \exp(-z) = 1$ for all $z \in \mathbb{C}$, an equation which contains both of the claims.

Theorem. *Let G be a region in $\mathbb{C}$. Then the following statements about a holomorphic function f in G are equivalent:*

i) $f(z) = a \exp(bz)$ *in G, for constants $a, b \in \mathbb{C}$.*

ii) $f'(z) = bf(z)$ *in G.*

Proof. The implication i) $\Rightarrow$ ii) is trivial. To prove ii) $\Rightarrow$ i) we consider the holomorphic function $h(z) := f(z) \exp(-bz)$ in G. It satisfies $h' = bh - bh = 0$ throughout G. Consequently 1.3.3 furnishes an $a \in \mathbb{C}$ such that $h(z) = a$ for all $z \in G$. Because of the product form of h and the previously noted form for the reciprocal of an exponential, the equation $h(z) = a$ yields $f(z) = a \exp(bz)$. $\square$

Contained in this theorem is the fact that

If f is holomorphic in $\mathbb{C}$ and satisfies $f' = f$, $f(0) = 1$, then necessarily $f = \exp$.

It follows in particular that the function $\tilde{e}$ which came up in 1.2.3, 2) is in fact exp, and so we obtain

$$\exp z = e^x \cos y + ie^x \sin y \qquad \text{for } z = x + iy.$$

2. The addition theorem of the exponential function. *For all z and w in $\mathbb{C}$*

$$(\exp w) \cdot (\exp z) = \exp(w + z).$$

Proof. Fix w in $\mathbb{C}$. The function $f(z) := \exp(w + z)$ is holomorphic in $\mathbb{C}$ and it satisfies $f' = f$. So by theorem 1, $f(z) = a \exp z$ for an appropriate constant $a \in \mathbb{C}$. This constant is found, by considering $z = 0$, to be $a = f(0) = \exp w$. □

A second (less sophisticated) proof of the addition theorem consists of just calculating the Cauchy product (cf. 3.3.1) of the power series for $\exp w$ and $\exp z$. Since

$$p_\lambda = \sum_{\mu+\nu=\lambda} \frac{1}{\mu!} w^\mu \frac{1}{\nu!} z^\nu = \frac{1}{\lambda!} \sum_{\nu=0}^{\lambda} \binom{\lambda}{\nu} w^{\lambda-\nu} z^\nu = \frac{1}{\lambda!}(w + z)^\lambda,$$

we have

$$(\exp w)(\exp z) = \sum_0^\infty p_\lambda = \sum_0^\infty \frac{1}{\lambda!}(w + z)^\lambda = \exp(w + z).$$

WEIERSTRASS preferred to put the addition theorem into the form

$$(\exp w)(\exp z) = [\exp \frac{1}{2}(w + z)]^2,$$

where it reads: the function value at the arithmetic mean of two arguments is the geometric mean of the function values at the two respective arguments.

The addition theorem also characterizes the exponential function.

Theorem. *Let G be a region which contains 0, $e : G \to \mathbb{C}$ a function which is complex-differentiable at 0 and satisfies $e(0) \neq 0$ and the functional equation*

$$(*) \qquad e(w + z) = e(w)e(z) \quad \text{whenever } w, z \text{ and } w + z \text{ all lie in } G.$$

Then for $b := e'(0)$ we have

$$e(z) = \exp(bz) \qquad for\ all\ z \in G.$$

Proof. Since $e(0) \neq 0$ and $e(0) = e(0)e(0)$ by $(*)$, it follows that $e(0) = 1$. Given $z \in G$, equation $(*)$ holds for all sufficiently small w, because G is a neighborhood of both 0 and z. For such w then

$$\frac{e(z+w) - e(z)}{w} = \frac{e(w) - 1}{w} \cdot e(z) = \frac{e(w) - e(0)}{w - 0} \cdot e(z).$$

The existence of $e'(0)$ therefore implies the existence of $e'(z)$ and the identity $e'(z) = e'(0)e(z) = be(z)$. As this is the case for each $z \in G$, the function e is holomorphic in G and the present theorem is a consequence of the preceding one. □

3. Remarks on the addition theorem. The addition theorem is really a "power rule". In order to see this most clearly and also to clarify the term "exponential function", let us write $e := \exp(1) = 1 + \frac{1}{1!} + \frac{1}{2!} + \cdots$ and $e^z := \exp z$ for complex z just as is usually done for real z. With this definition of the symbol "e^z" the addition theorem reads like a *power rule* or *law of exponents*:

$$e^w \cdot e^z = e^{w+z}.$$

Remark. The symbol "e" was introduced by Euler; in a letter to GOLDBACH of November 25, 1731 we read "e denotat hic numerum, cujus logarithmus hyperbolicus est $= 1$ (e denotes here the number whose hyperbolic logarithm is equal to 1)". Cf. the "Correspondance entre Leonhard Euler et Chr. Goldbach 1729-1763," in *Correspondance mathématique et physique de quelques célèbres géomètres du* XVIII[iéme] *siècle*, ed. P. H. FUSS, St. Pétersbourg 1843, vol. 1, p. 58. □

The addition theorem contains the identity $(\exp z)^{-1} = \exp(-z)$. Also from the addition theorem and the Euler formula we can (without recourse to 1.2.3, 2)) get the *decomposition of the exponential function into real and imaginary parts*:

$$\exp z = e^x e^{iy} = e^x \cos y + ie^x \sin y \qquad for\ z = x + iy.$$

As further applications of the addition theorem we note

(1) $\exp x > 0$ *for* $x \in \mathbb{R}$; $\exp x = 1$ *for* $x \in \mathbb{R} \Leftrightarrow x = 0$;
(2) $|\exp z| = \exp(\Re z)$ *for* $z \in \mathbb{C}$.

Proof. Ignoring non-negative terms in the power series shows that $\exp x \geq 1 + x$ for $x \geq 0$ and then $e^{-x} = (e^x)^{-1}$ gives (1). Now (2) follows from this positivity and the calculation

$$\begin{aligned}|\exp z|^2 &= \exp z \cdot \overline{\exp z} = \exp z \cdot \exp \overline{z} = \exp(z + \overline{z}) \\ &= \exp(2\Re z) = (\exp(\Re z))^2.\end{aligned} \qquad \square$$

We see in particular that

(3) $$|\exp w| = 1 \Leftrightarrow w \in \mathbb{R}i.$$

As an amusing application of the addition theorem we derive a

Trigonometric summation formula. *For all $z \in \mathbb{C}$ such that $\sin \frac{1}{2}z \neq 0$*

(∗) $$\frac{1}{2} + \cos z + \cos 2z + \cdots + \cos nz = \frac{\sin(n + \frac{1}{2}z)}{2 \sin \frac{1}{2}z}, \qquad n \in \mathbb{N}.$$

Proof. Since $\cos \nu z = \frac{1}{2}(e^{i\nu z} + e^{-i\nu z})$, the addition theorem gives

$$\frac{1}{2} + \sum_1^n \cos \nu z = \frac{1}{2}\sum_{-n}^n e^{i\nu z} = \frac{1}{2}e^{-inz}\sum_0^{2n} e^{i\nu z}.$$

Use of the usual formula for the sum of a finite geometric series and another application of the addition theorem lead to

$$\frac{1}{2} + \sum_1^n \cos \nu z = \frac{1}{2}e^{-inz}\frac{e^{i(2n+1)z} - 1}{e^{iz} - 1} = \frac{e^{i(n+1/2)z} - e^{-i(n+1/2)z}}{2(e^{iz/2} - e^{-iz/2})}.$$

Since $2i \sin w = e^{iw} - e^{-iw}$, this is the claimed equality. $\qquad \square$

For $z = x \in \mathbb{R}$ equation (∗) is a formula in the real domain. It was derived here by a calculation in $\mathbb{C}$. Conclusions of this kind aroused quite a bit of admiration in Euler's time; HADAMARD is supposed to have said of this phenomenon: "*Le plus court chemin entre deux énoncés réels passe par le complexe* (The shortest path between two assertions about the reals passes through the complexes)."

4. Addition theorems for cos z and sin z. *For all $w, z \in \mathbb{C}$*

$$\cos(w + z) = \cos w \cos z - \sin w \sin z \ , \ \sin(w + z) = \sin w \cos z + \cos w \sin z.$$

Proof. The point of departure is the identity

$$\begin{aligned}e^{i(w+z)} &= e^{iw} \cdot e^{iz} = (\cos w + i\sin w)(\cos z + i\sin z) \\ &= (\cos w \cos z - \sin w \sin z) + i(\sin w \cos z + \cos w \sin z).\end{aligned}$$

Replacing w with $-w$, z with $-z$ yields the companion

$$e^{-i(w+z)} = (\cos w \cos z - \sin w \sin z) - i(\sin w \cos z + \cos w \sin z)$$

Addition and subtraction of these two identities give the claimed addition formulas for sine and cosine.   □

As in the case of the corresponding addition theorems for real arguments, innumerable further identities flow from these two. E.g., the useful formulas

$$\cos w - \cos z = -2 \sin \frac{1}{2}(w + z) \sin \frac{1}{2}(w - z)$$

$$\sin w - \sin z = 2 \cos \frac{1}{2}(w + z) \sin \frac{1}{2}(w - z)$$

follow easily for all complex w, z. From the plethora of other possible formulas we note explicitly only that

$$1 = \cos^2 z + \sin^2 z \ , \quad \cos 2z = \cos^2 z - \sin^2 z \ , \quad \sin 2z = 2 \sin z \cos z$$

for all $z \in \mathbb{C}$.

5. Historical remarks on cos z and sin z. These functions were discovered by geometers long before the advent of the exponential function; a formula closely related to the addition theorem for $\sin(\alpha + \beta)$ and $\sin(\alpha - \beta)$ was known already to ARCHIMEDES. In PTOLEMY the addition theorem can be found implicitly in the form of a proposition about circularly inscribed quadrilaterals (cf. 3.4.5 in the book *Numbers* [19]). Toward the end of the 16th century — before the discovery of logarithms — formulas like

$$\cos x \cos y = \frac{1}{2} \cos(x + y) + \frac{1}{2} \cos(x - y)$$

were used (for purposes of astronomy and navigation) to multiply two numbers A and B: in a table of sines (which of course is also a table of cosines) the angles x and y were found for which $\cos x = A$, $\cos y = B$. Then $x + y$ and $x - y$ were formed and the table again consulted for the values of $\cos(x + y)$ and $\cos(x - y)$. Half the sum of the latter then gave AB.

The power series developments of the functions $\cos x$ and $\sin x$ were known to NEWTON around 1665; e.g., he found the sine series by *inversion* of the series

$$\arcsin x = x + \frac{1}{6}x^3 + \frac{3}{40}x^5 + \frac{5}{112}x^7 + \cdots$$

which in turn he had arrived at by geometric considerations. But a systematic exposition of the theory is not to be found until the 8th chapter "Von den transcendenten Zahlgrössen, welche aus dem Kreise entspringen" of Euler's *Introductio* [E]. Here for the first time the trigonometric functions are defined on the unit circle in the way that has since become standard. Besides the addition theorems EULER presents an abundance of formulas, in particular in §138 his famous (cf. 4.2.1)

$$\cos z = \frac{1}{2}(e^{iz} + e^{-iz}) \; , \; \sin z = \frac{1}{2i}(e^{iz} - e^{-iz}) \; , \qquad z \in \mathbb{C}.$$

6. Hyperbolic functions. As is done for real arguments, the *hyperbolic cosine* and *hyperbolic sine* functions are defined by

$$\cosh z := \frac{1}{2}(e^{z} + e^{-z}) \; , \; \sinh z := \frac{1}{2}(e^{z} - e^{-z}) \; , \qquad z \in \mathbb{C}.$$

These functions are holomorphic in $\mathbb{C}$ and

$$(\cosh)'(z) = \sinh z \; , \quad (\sinh)'(z) = \cosh z;$$
$$\cosh z = \cos(iz) \; , \quad \sinh z = -i\sin(iz) \; , \; z \in \mathbb{C}.$$

All the important properties of these function are derivable from these equations. Thus we have

$$\cosh z = \sum \frac{z^{2\nu}}{(2\nu)!} \; , \quad \sinh z = \sum \frac{z^{2\nu+1}}{(2\nu+1)!} \qquad \text{for } z \in \mathbb{C}$$

and the addition theorems have the form

$$\cosh(w + z) = \cosh w \cosh z + \sinh w \sinh z,$$
$$\sinh(w + z) = \sinh w \cosh z + \cosh w \sinh z,$$

while $1 = \cos^2(iz) + \sin^2(iz)$ yields $1 = \cosh^2 z - \sinh^2 z$, an identity which clarifies somewhat the adjective "*hyperbolic*", if one remembers that $x^2 - y^2 = 1$ is the equation of a hyperbola in $\mathbb{R}^2$.

The hyperbolic functions provide a convenient means for decomposing the complex sine and cosine functions into real and imaginary parts:

$$\cos(x + iy) = \cos x \cosh y - i \sin x \sinh y,$$
$$\sin(x + iy) = \sin x \cosh y + i \cos x \sinh y.$$

Exercises

Exercise 1. Show that for $z \in \mathbb{C}$

$$
\begin{aligned}
\sin 3z &= 3\sin z - 4\sin^3 z \\
\sin 4z &= 8\cos^3 z \sin z - 4\cos z \sin z \\
\cos 3z &= 4\cos^3 z - 3\cos z \\
\cos 4z &= 8\cos^4 z - 8\cos^2 z + 1.
\end{aligned}
$$

Prove similar formulas for $\cosh 3z$, $\cosh 4z$, $\sinh 3z$, $\sinh 4z$.

Exercise 2. Let $k \in \mathbb{N}$, $k \geq 1$ and let Γ denote the boundary of the square in $\mathbb{C}$ which has the vertices $\pi k(\pm 1 \pm i)$. Show that $|\cos z| \geq 1$ for all $z \in \Gamma$.

Exercise 3. Let f, g be holomorphic in a region G which contains 0. Suppose that

$$
f(w + z) = f(z)f(w) - g(z)g(w) \quad \text{and} \quad g(w + z) = g(z)f(w) + g(w)f(z)
$$

hold whenever w, z and $w + z$ all belong to G. Show that if $f(0) = 1$ and $f'(0) = 0$, then there is a disc B centered at 0 and lying in G such that

$$
f(z) = \cos bz \quad \text{and} \quad g(z) = \sin bz \qquad \text{for all } z \in B,
$$

b being the number $g'(0)$.

Exercise 4. The power series $f(z) := \sum_{\nu \geq 0} a_\nu z^\nu$ converges in a disc B centered at 0. For every $z \in B$ such that $2z$ is also in B, f satisfies $f(2z) = (f(z))^2$. Show that if $f(0) \neq 0$, then $f(z) = \exp bz$, with $b := f'(0) = a_1$.

Exercise 5. Show that the sequence of functions $f_k(z) := F(1, k, 1, z/k)$, $k \geq 1$, $z \in \mathbb{E}$, converges compactly in the unit disc $\mathbb{E}$ to the exponential function. (For the definition of $F(a, b, c, z)$ see exercise 1 in Chapter 4, §2.)

Exercise 6. Show that for every $R > 0$ there is an $N \in \mathbb{N}$ such that none of the polynomials $1 + \frac{z}{1!} + \frac{z^2}{2!} + \cdots + \frac{z^n}{n!}$ with $n \geq N$ has any zeros in $B_R(0)$.

§2 The epimorphism theorem for $\exp z$ and its consequences

Because $\exp z$ is zero-free, $\exp$ is a holomorphic mapping of $\mathbb{C}$ into $\mathbb{C}^{\times}$. The addition theorem can be formulated in this context as

The mapping $\exp : \mathbb{C} \to \mathbb{C}^{\times}$ *is a group homomorphism of the additive group* $\mathbb{C}$ *of all complex numbers into the multiplicative group* $\mathbb{C}^{\times}$ *of all non-zero complex numbers.*

Whenever mathematicians see group homomorphisms $\psi : G \to H$ they instinctively inquire about the subgroups $\ker \psi := \{g \in G : \psi(g) = \text{the identity element of } H\}$ and $\psi(G)$ of G and H, respectively. For the *exponential homomorphism* these groups can be determined explicitly and this leads to a simple definition of the circular ratio π. Decisive for this undertaking is the

1. Epimorphism theorem. *The exponential homomorphism*

$$\exp : \mathbb{C} \to \mathbb{C}^{\times}$$

is an epimorphism (that is, it is surjective).

First we prove a

Lemma. *The subgroup* $\exp(\mathbb{C})$ *of* $\mathbb{C}^{\times}$ *is an open subset of* $\mathbb{C}^{\times}$.

Proof. According to 4.3.4, $\exp \lambda(z) = 1 + z$ for all $z \in \mathbb{E}$. From this follows first of all that $B_1(1) \subset \exp(\mathbb{C})$, because if $c \in B_1(1)$, then $c - 1 \in \mathbb{E}$, the domain of λ, so $b := \lambda(c - 1)$ exists in $\mathbb{C}$ and $\exp b = c$.

Now consider an arbitrary $a \in \mathbb{C}^{\times}$. Evidently $aB_1(1) = B_{|a|}(a)$. Moreover, by the group property of $\exp(\mathbb{C})$, $a \exp(\mathbb{C}) = \exp(\mathbb{C})$ if $a \in \exp(\mathbb{C})$. Therefore, in general

$$B_{|a|}(a) = aB_1(1) \subset a \exp(\mathbb{C}) = \exp(\mathbb{C}) \qquad \text{for all } a \in \exp(\mathbb{C}).$$

This says that for each $a \in \exp(\mathbb{C})$ the whole open disc $B_{|a|}(a)$ of radius $|a| > 0$ about a lies in $\exp(\mathbb{C})$; that is, $\exp(\mathbb{C})$ is open in $\mathbb{C}^{\times}$.   □

The epimorphism theorem follows now from a purely topological argument. This argument is quite general and establishes the following theorem about topological groups: an *open* subgroup A of a *connected* topological group G must exhaust G. So here we are considering $A := \exp(\mathbb{C})$, $G := \mathbb{C}^{\times}$. Consider also $B := \mathbb{C}^{\times} \setminus A$. Every set bA, $b \in B$, is also open in $\mathbb{C}^\times$. Therefore $\bigcup_{b\in B} bA$ is an open subset of $\mathbb{C}^\times$. But from elementary group theory we know, since A is a subgroup, that

$$B = \bigcup_{b\in B} bA = \text{ the union of all cosets different from } A.$$

Consequently, $\mathbb{C}^\times = A \cup B$ displays $\mathbb{C}^\times$ as the disjoint union of open sets one of which, A, is not empty $(1 = \exp(0) \in A)$. Since $\mathbb{C}^\times$ is connected (cf. 0.6), it follows that $B = \emptyset$. That is, $A = \mathbb{C}^\times$.

2. The equation $\ker(\exp) = 2\pi i\mathbb{Z}$. The kernel

$$K := \ker(\exp) = \{w \in \mathbb{C} : e^w = 1\}$$

is an additive subgroup of $\mathbb{C}$. From the epimorphism theorem we can easily infer the following preliminary result:

(1)   *K is not the trivial group : $K \neq \{0\}$.*

Proof. Since $\exp(\mathbb{C}) = \mathbb{C}^\times$ there is an $a \in \mathbb{C}$ with $e^a = -1 \neq 1 = e^0$. Then $a \neq 0$, so $c := 2a \neq 0$, yet $e^c = (e^a)^2 = 1$. That is, $c \in K$.   □

All the remaining steps in the characterization of K are quite elementary. Since, by 1.3(3), $|e^w| = 1$ is possible only for $w \in \mathbb{R}i$, we learn next that

(2) $K \subset \mathbb{R}i.$

Thirdly, we will show that

(3)   *There is a neighborhood U of $0 \in \mathbb{C}$ such that $U \cap K = \{0\}$.*

Proof. If this were not the case, then there would be a null sequence h_n in $\mathbb{C}$, all $h_n \neq 0$, with $\exp h_n = 1$. This would lead to the contradiction

$$1 = \exp(0) = \exp'(0) = \lim_{n\to\infty} \frac{\exp(h_n) - \exp(0)}{h_n} = 0.$$ □

Our goal is now within a few lines' reach:

Theorem. *There is a unique positive real number π such that*

(4) $\ker(\exp) = 2\pi i\mathbb{Z}.$

Proof. The continuity of $\exp$ and (1)–(3) ensure that there is a *smallest positive real number* π with $2\pi i \in K$ (note that $-K = K$). Since K is a group, $2\pi i\mathbb{Z} \subset K$ follows. If conversely $r \in \mathbb{R}$ and $ri \in K$, then $\pi \neq 0$ means that $2n\pi \leq r < 2(n+1)\pi$ for an appropriate $n \in \mathbb{Z}$. Since $ri - 2n\pi i \in K$

and $0 \leq r - 2n\pi < 2\pi$, the minimality of π implies that $r - 2n\pi = 0$, i.e., $r \in 2\pi\mathbb{Z}$. This proves that $K \subset 2\pi i\mathbb{Z}$ and completes the proof of (4). The uniqueness of π is clear.   □

As far as this book is concerned, the content of the above theorem is the *definition* of π. It follows directly that

$$(5) \qquad\qquad e^{i\pi} = -1$$

and from this that $e^{i\pi/2} = \pm i$. With only the results presently at hand the minus sign *cannot* be precluded: that will be accomplished in subsection 6 with the help of the Intermediate Value Theorem.

3. Periodicity of exp z. A function $f : \mathbb{C} \to \mathbb{C}$ is called *periodic* if there is a complex number $\omega \neq 0$ such that $f(z + \omega) = f(z)$ for all $z \in \mathbb{C}$. Such a number ω is called a *period* of f. If f is periodic, the set

$$\operatorname{per}(f) := \{\omega \in \mathbb{C} : \omega \text{ is a period of } f\} \cup \{0\}$$

of all periods of f together with 0 constitutes *an additive (abelian) subgroup of $\mathbb{C}$*.

The Periodicity Theorem. *The function* exp *is periodic and*

$$\operatorname{per}(\exp) = \ker(\exp) = 2\pi i\mathbb{Z}.$$

Proof. For a number $\omega \in \mathbb{C}$ the number $\exp(z + \omega) = \exp z \exp \omega$ coincides with $\exp z \in \mathbb{C}^{\times}$ for every $z \in \mathbb{C}$ exactly when $\exp \omega = 1$; which proves that $\operatorname{per}(\exp) = \ker(\exp)$.   □

The equation $\ker(\exp) = \operatorname{per}(\exp) = 2\pi i\mathbb{Z}$ describes the essential difference in the behavior of the e-function on the reals and on the complexes: on the real line, because $\ker(\exp) \cap \mathbb{R} = \{0\}$, it *takes on every positive real value exactly once*; on the complex plane by contrast it has the purely imaginary (hence unbeknown to its real restriction) *minimal period $2\pi i$* and it takes on *every* non-zero complex as well as real value *countably infinitely often*.

On the basis of this discussion a simple visualization of the exponential function is possible: divide the z-plane into the infinitely many contiguous strips

$$S_n := \{z \in \mathbb{C} : 2n\pi \leq \Im z < 2(n+1)\pi\}, \qquad n \in \mathbb{Z}.$$

Every strip S_n is mapped *bijectively* onto the set $\mathbb{C}^{\times}$ in the w-plane and in the process the "*orthogonal cartesian x-y-system of the z-plane*" goes over, with preservation of angles, into the "*orthogonal polar coordinate* system of the w-plane*."

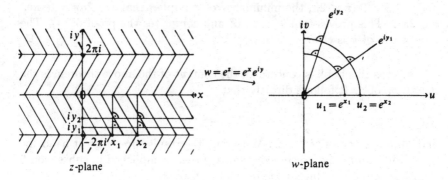

z-plane   w-plane

Remark. The difficulties which the complex exponential function caused mathematicians are well illustrated by the following exercise, which TH. CLAUSEN (known from the Clausen-von Staudt formula for the Bernoulli numbers) set in 1827 and which CRELLE saw fit to reproduce in his famous journal (*Journal für die Reine und Angewandte Mathematik* Vol. 2, pp. 286/287): "If e denotes the base of the hyperbolic logarithms, π half the circumference of a unit circle, and n a positive or negative whole number, then, as is known, $e^{2n\pi i} = 1$, $e^{1+2n\pi i} = e$; consequently also $e = (e^{1+2n\pi i})^{1+2n\pi i} = e = e^{(1+2n\pi i)^2} = e^{1+4n\pi i - 4n^2\pi^2}$. Since however $e^{1+4n\pi i} = e$, it would follow that $e^{-4n^2\pi^2} = 1$, which is absurd. Find the error in the derivation of this result".

The reader for his part might give the matter some thought too.

4. Course of values, zeros, and periodicity of cos z and sin z.

The exponential function assumes every complex value except zero. The trigonometric functions have no exceptional values:

$\cos z$ *and* $\sin z$ *assume every value* $c \in \mathbb{C}$ *countably often.*

Proof. Solving the equations $e^{iz} + e^{-iz} = 2c$ and $e^{iz} - e^{-iz} = 2ic$ for e^{iz} leads to $e^{iz} = c \pm \sqrt{c^2 - 1}$, respectively, $e^{iz} = ic \pm \sqrt{1 - c^2}$. Moreover, for no $c \in \mathbb{C}$ are the numbers $c \pm \sqrt{c^2 - 1}$ or $ic \pm \sqrt{1 - c^2}$ equal to 0. Therefore, since $\exp(\mathbb{C}) = \mathbb{C}^\times$ and $\ker(\exp) = 2\pi i\mathbb{Z}$, there are countably many z satisfying each of the latter two exponential equations, and hence satisfying $\cos z = c$, $\sin z = c$, respectively.   □

Because $\cos(\mathbb{C}) = \sin(\mathbb{C}) = \mathbb{C}$, cos and sin are *unbounded in the complex plane* (in contrast to their behavior on the real line, where both are real-valued and the identity $\cos^2 x + \sin^2 x = 1$ requires that $|\cos x| \le 1$ and $|\sin x| \le 1$): For example, on the imaginary axis, for $y > 0$,

$$\cos iy = \frac{1}{2}(e^y + e^{-y}) > 1 + \frac{1}{2}y^2 \,, \qquad i\sin iy = \frac{1}{2}(e^{-y} - e^y) < -y.$$

In contrast to $\exp z$, $\cos z$ and $\sin z$ have zeros. With π denoting the constant introduced in subsection 2 we will show

The Theorem on Zeros. *The zeros of $\sin z$ in $\mathbb{C}$ are precisely all the real numbers $n\pi$, $n \in \mathbb{Z}$. The zeros in $\mathbb{C}$ of $\cos z$ are precisely all the real numbers $\frac{1}{2}\pi + n\pi$, $n \in \mathbb{Z}$.*

Proof. Taking note of the fact that $e^{i\pi} = -1$, we have

$$2i \sin z = e^{-iz}(e^{2iz} - 1) \ , \ 2\cos z = e^{i(\pi - z)}(e^{2i(z - \frac{1}{2}\pi)} - 1),$$

from which we read off that

$$\begin{aligned}
\sin w = 0 \ &\Leftrightarrow \ 2iw \in \ker(\exp) = 2\pi i\mathbb{Z} \\
&\Leftrightarrow \ w = n\pi \ , \ n \in \mathbb{Z}; \\
\cos w = 0 \ &\Leftrightarrow \ 2i\left(w - \tfrac{1}{2}\pi\right) \in 2\pi i\mathbb{Z} \\
&\Leftrightarrow \ w = \tfrac{1}{2}\pi + n\pi.
\end{aligned}$$

$\square$

Remark. We see that π (respectively, $\frac{1}{2}\pi$) is indeed the smallest positive zero of sin (respectively, cos). Even if all the real zeros of sin and cos are known from the real theory, it would still have had to be shown that the extension of the domain of these functions to $\mathbb{C}$ introduces no new, non-real, complex zeros.

Next we show that sin and cos are also periodic on $\mathbb{C}$ and have the same periods there as on $\mathbb{R}$.

Periodicity Theorem. $\operatorname{per}(\cos) = \operatorname{per}(\sin) = 2\pi\mathbb{Z}$.

Proof. Since $\cos(z + w) - \cos z = -2\sin(z + \frac{1}{2}w)\sin\frac{1}{2}w$, by 1.4(1), the number w is in $\operatorname{per}(\cos)$ if and only if $\sin\frac{1}{2}w = 0$, that is, if and only if $w \in 2\pi\mathbb{Z}$. The claim about the sine function follows by the same reasoning from the identity $\sin(z + w) - \sin z = 2\cos(z + \frac{1}{2}w)\sin\frac{1}{2}w$.

Remark. Again, even if one knows that cos and sin each have the minimal period 2π on $\mathbb{R}$, he would still have had to show that the extensions to $\mathbb{C}$ also have 2π as a period and that moreover no new non-real periods are introduced.

5. Cotangent and tangent functions. Arctangent series. By means of the equations

$$\begin{aligned}
\cot z \ &:= \ \frac{\cos z}{\sin z} , \qquad z \in \mathbb{C} \setminus \pi\mathbb{Z}, \\
\tan z \ &:= \ \frac{1}{\cot z} = \frac{\sin z}{\cos z} , \qquad z \in \mathbb{C} \setminus (\tfrac{1}{2}\pi + \pi\mathbb{Z})
\end{aligned}$$

the known cotangent and tangent functions are extended into the complex plane. Their zero-sets are $\frac{1}{2}\pi + \pi\mathbb{Z}$ and $\pi\mathbb{Z}$, respectively. Both functions are holomorphic in these new domains:

$$\cot'(z) = \frac{-1}{\sin^2 z}, \qquad \tan'(z) = \frac{1}{\cos^2 z}.$$

In classical analysis the cotangent seems to enjoy a more important role than the tangent (cf. e.g., 11.2). From the Euler formulas for cos and sin follow

$$\cot z = i\frac{e^{2iz} + 1}{e^{2iz} - 1} = i\left(1 - \frac{2}{1 - e^{2iz}}\right),$$

$$\tan z = i\frac{1 - e^{2iz}}{1 + e^{2iz}} = i\left(1 - \frac{1}{1 + e^{-2iz}}\right).$$

Since the kernel of e^{2iz} is $\pi\mathbb{Z}$, we see immediately that

The functions $\cot z$ *and* $\tan z$ *are periodic, and*

$$\mathrm{per}(\cot) = \mathrm{per}(\tan) = \pi\mathbb{Z}.$$

We also take note of some computationally verifiable formulas which will be used later:

$$\frac{1}{\sin z} = \cot z + \tan \tfrac{1}{2}z, \qquad \cot'(z) + (\cot z)^2 + 1 = 0,$$

$$2\cot 2z = \cot z + \cot(z + \tfrac{1}{2}\pi) \qquad (\textit{ double-angle formula }).$$

From the addition theorems for $\cos z$ and $\sin z$ we get addition theorems for $\cot z$ and $\tan z$. E.g.,

$$\cot(w + z) = \frac{\cot w \cot z - 1}{\cot w + \cot z}; \text{ in particular, } \cot(z + \tfrac{1}{2}\pi) = -\tan z.$$

There is an especially elegant "cyclic" way to write the addition theorem:

$$\cot u \cot v + \cot v \cot w + \cot w \cot u = 1$$

in case $u + v + w = 0$. (Proof!)

In 4.3.4, 5) we introduced the arctangent series, which is defined and holomorphic in the unit disc $\mathbb{E}$:

$$a(z) = z - \frac{z^3}{3} + \frac{z^5}{5} - \cdots + (-1)^n\frac{z^{2n+1}}{2n + 1} + \cdots , \text{ with } a'(z) = \frac{1}{1 + z^2}.$$

Since $\tan 0 = 0$, the function $a(\tan z)$ is defined and holomorphic in some open disc B centered at 0. We claim that

$$a(\tan z) = z \qquad in \ B.$$

Proof. The function $F(z) := a(\tan z) - z$ satisfies

$$F'(z) = \frac{1}{1 + \tan^2 z} \cdot \frac{1}{\cos^2 z} - 1 = 0 \qquad in \ B.$$

Consequently F is constant in B and since $F(0) = 0$ that constant is 0 and the claim follows. $\square$

The identity $a(\tan z) = z$ explains the designation "arctangent". It is customary to write $\arctan z$ for the function $a(z)$; in 5.2 we will see among other things that $\tan(\arctan z) = z$ holds as well as $\arctan(\tan z) = z$.

6. The equation $e^{i\frac{\pi}{2}} = i$. From $e^{i\pi} = -1$ it follows that $e^{i\frac{\pi}{2}} = \pm i$. In order to determine the sign we will show, with the help of the Intermediate Value Theorem that

$$(1) \qquad\qquad \sin x > 0 \qquad \text{for } 0 < x < \pi.$$

Proof. Doing a little grouping and factoring of terms in the power series for sin gives $\sin z = z\left(1 - \frac{z^2}{6}\right) + \frac{z^5}{5!}\left(1 - \frac{z^2}{6 \cdot 7}\right) + \cdots$; from which we see that $\sin x$ is positive for $x \in (0, \sqrt{6})$. Were $\sin x$ to be negative anywhere in $(0, \pi)$, then by the Intermediate Value Theorem it would have to have a zero somewhere in $(0, \pi)$, contrary to the theorem 4 on zeros. $\square$

From (1) it follows that $\sin \frac{1}{2}\pi = 1$, because $\cos^2 x + \sin^2 x = 1$ and $\cos \frac{1}{2}\pi = 0$. Then from $e^{ix} = \cos x + i \sin x$ it is clear that

$$(2) \qquad e^{i\frac{\pi}{2}} = i \qquad \text{(equation of Johann BERNOULLI 1702).}$$

Because $\left(e^{i\frac{\pi}{4}}\right)^2 = i$ and $\Im e^{i\frac{\pi}{4}} = \sin \frac{\pi}{4} > 0$, it further follows that

$$e^{i\frac{\pi}{4}} = \frac{1}{2}\sqrt{2}(1 + i).$$

Thus for the functions $\cos z$ and $\sin z$ we have

$$\cos \frac{\pi}{2} = 0 \ , \ \sin \frac{\pi}{2} = 1 \ ; \ \cos \frac{\pi}{4} = \sin \frac{\pi}{4} = \frac{\sqrt{2}}{2} \ , \ \text{whence} \ \cot \frac{\pi}{4} = \tan \frac{\pi}{4} = 1.$$

The reader should determine $e^{i\frac{\pi}{8}}$.

Exercises

Exercise 1. a) For what $z \in \mathbb{C}$ is $\cos z$ real?

b) For what $z \in \mathbb{C}$ does $\cos z$ lie in $[-1, 1]$?

Exercise 2. For $z \in \mathbb{C}$ show that

a) $\cos z = 1$ if and only if $z \in 2\pi\mathbb{Z}$;

b) $\sin z = 1$ if and only if $z \in \frac{1}{2}\pi + 2\pi\mathbb{Z}$.

Exercise 3. Show that for $z, w \in \mathbb{C}$:

a) $\cos z = \cos w$ if and only if either $z + w \in 2\pi\mathbb{Z}$ or $z - w \in 2\pi\mathbb{Z}$;

b) $\sin z = \sin w$ if and only if either $z + w \in \pi + 2\pi\mathbb{Z}$ or $z - w \in 2\pi\mathbb{Z}$;

c) $\tan z = 1$ if and only if $z \in \frac{\pi}{4} + \pi\mathbb{Z}$;

d) $\tan z = \tan w$ if and only if $z - w \in \pi\mathbb{Z}$.

Exercise 4. Determine the largest possible regions in which each of $\cos z$, $\sin z$, $\tan z$ is injective.

§3   Polar coordinates, roots of unity and natural boundaries

In the plane $\mathbb{R}^2 = \mathbb{C}$ polar coordinates are introduced by writing every point $z = x + iy \neq 0$ in the form

$$z = |z|(\cos \varphi + i \sin \varphi)$$

where φ measures the angle between the x-axis and the vector z (cf. the figure).

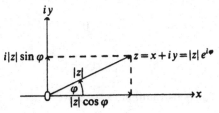

To make this intuitively clear idea precise is not so trivial; it will be done later. We will discuss roots of unity further and, as a consequence of these considerations, present examples of power series which have the unit circle as natural boundary.

1. Polar coordinates. The circle $S_1 := \partial \mathbb{E} = \{z \in \mathbb{C} : |z| = 1\}$ is a group with respect to multiplication and we have

The Epimorphism Theorem. *The mapping $p : \mathbb{R} \to S_1$, $\varphi \mapsto e^{i\varphi}$, is an epimorphism of the additive group $\mathbb{R}$ onto the multiplicative group S_1 and its kernel is $2\pi \mathbb{Z}$.*

This follows immediately from the epimorphism theorem 2.1, since by 1.3(3), $\exp w \in S_1 \Leftrightarrow w \in \mathbb{R}i$. □

The process of wrapping the number line around the circle of circumference 2π is accomplished analytically by the "polar-coordinate epimorphism" p. It now follows easily that

Every complex number $z \in \mathbb{C}^\times$ can be uniquely written in the form

$$z = |z|e^{i\varphi} = |z|(\cos \varphi + i \sin \varphi) \qquad with \ \varphi \in [0, 2\pi).$$

Proof. Since $z|z|^{-1} \in S_1$, the range of p, there is a $\varphi \in \mathbb{R}$ with $z = |z|e^{i\varphi}$; and since $\ker p = 2\pi\mathbb{Z}$ there is such a $\varphi \in [0, 2\pi)$. The latter requirement uniquely determines φ. If $|z|e^{i\varphi} = |z|e^{i\psi}$ with $\psi \in [0, 2\pi)$, then $e^{i(\psi-\varphi)} = 1$, so $\psi - \varphi \in 2\pi\mathbb{Z}$. Since $0 \leq |\psi - \varphi| < 2\pi$, it follows from $\psi - \varphi \in 2\pi\mathbb{Z}$ that $\psi - \varphi = 0$. □

The real numbers $|z|$, φ are called *the polar coordinates* of z; the number φ is called *an argument* of z. Our normalization of φ to the interval $[0, 2\pi)$ was *arbitrary*; in general any half-open interval of length 2π is suitable, and in subsequent sections it often proves advantageous to use the interval $(-\pi, \pi]$.

The multiplication of two complex numbers is especially easy when they are given in polar coordinates: for $w = |w|e^{i\psi}$, $z = |z|e^{i\varphi}$

$$wz = |w||z|e^{i(\psi+\varphi)}.$$

Since $(e^{iy})^n = e^{iny}$ this observation contains

de Moivre's formula. *For every number $\cos \varphi + i \sin \varphi \in \mathbb{C}$*

$$(\cos \varphi + i \sin \varphi)^n = \cos n\varphi + i \sin n\varphi , \qquad n \in \mathbb{Z}.$$

By expanding the left side (binomial expansion) and passing to real and imaginary parts we get, for every $n \geq 1$, representations of $\cos n\varphi$ and $\sin n\varphi$ as polynomials in $\cos \varphi$ and $\sin \varphi$. For example,

$$\cos 3\varphi = \cos^3 \varphi - 3 \cos \varphi \sin^2 \varphi \ , \ \sin 3\varphi = 3 \cos^2 \varphi \sin \varphi - \sin^3 \varphi.$$

The above method of deriving these identities is a further illustration of Hadamard's "principle of the shortest path through the complexes" — cf. 1.3.

Historical note. In 1707 Abraham DEMOIVRE indicated the discovery of his "magic" formula with numerical examples. By 1730 he appears to have known the general formula

$$\cos\varphi = \frac{1}{2}\sqrt[n]{\cos n\varphi + i\sin n\varphi} + \frac{1}{2}\sqrt[n]{\cos n\varphi - i\sin n\varphi}\,, \qquad n > 0.$$

In 1738 he described a (complicated) process for finding roots of the form $\sqrt[n]{a+ib}$; his prescription amounts in the end to the formula which is now named after him. The present-day point of view first emerged with EULER in 1748 (cf. [E], Chap. VIII) and the first rigorous proof for all $n \in \mathbb{Z}$ was also given by EULER 1749, with the help of differential calculus. For biographical particulars on DEMOIVRE see 12.4.6.

2. Roots of unity. *For every natural number n there are exactly n different complex numbers z with $z^n = 1$, namely*

$$\zeta_\nu := \zeta^\nu\,, 0 \le \nu < n\,, \qquad \text{where } \zeta := \exp\frac{2\pi i}{n} = \cos\frac{2\pi}{n} + i\sin\frac{2\pi}{n}.$$

Proof. $\zeta_\nu^n = 1$ for each ν by deMoivre's formula. Since $\zeta_\nu\zeta_\mu^{-1} = \exp\frac{2\pi i}{n}(\nu-\mu)$ and $\ker(\exp) = 2\pi i\mathbb{Z}$, it is clear that $\zeta_\nu = \zeta_\mu$ exactly if $\frac{1}{n}(\nu-\mu) \in \mathbb{Z}$. And since $|\nu - \mu| < n$ it follows that $\zeta_\nu = \zeta_\mu \Leftrightarrow \nu = \mu$; that is, $\zeta_0, \zeta_1, \ldots, \zeta_{n-1}$ are all different. Since the nth degree polynomial $z^n - 1$ has at most n different zeros, the claim is fully established. $\square$

Every number $w \in \mathbb{C}$ with $w^n = 1$ is called an nth *root of unity* and the specific one $\zeta = \exp(2\pi i/n)$ is called a *primitive* nth root of unity. The set G_n of all nth roots of unity is a *cyclic* subgroup of S_1 of order n. The sets

$$G := \bigcup_{n=0}^{\infty} G_n \quad \text{and } H := \bigcup_{n=0}^{\infty} G_{2^n}$$

are also subgroups of S_1, with $H \subset G$. We claim (for the concept of a dense set, see 0.2.3):

Density Theorem. *The groups H and G are dense in S_1.*

Proof. Since $G \supset H$ we only need to consider H. The set of all fractions having denominators which are powers of 2 is dense in $\mathbb{Q}$ hence in $\mathbb{R}$. Then $M := \{2\pi m 2^{-n} : m \in \mathbb{Z}\,, n \in \mathbb{N}\}$ is also dense in $\mathbb{R}$. The polar-coordinate epimorphism p maps M onto H. Now generally a continuous map $f : X \to$

Y sends every dense subset A of X onto a dense subset $f(A)$ of $f(X)$ (since $f(\overline{A}) \subset \overline{f(A)}$). Consequently, $H = p(M)$ is dense in $S_1 = p(\mathbb{R})$.   □

In the next subsection we will give an interesting application of this density theorem.

3. Singular points and natural boundaries. If $B = B_R(c)$, $0 < R < \infty$, is the convergence disc of the power series $f(z) = \sum a_\nu(z - c)^\nu$, then a boundary point $w \in \partial B$ is called a *singular point of f* if there is no holomorphic function $\hat{f}$ in any neighborhood U of w which satisfies $\hat{f}|U \cap B = f|U \cap B$. The set of singular points of f on ∂B is always closed and can be empty (on this point see however 8.1.5). If *every* point of ∂B is a singular point of f, ∂B is called the *natural boundary* of f and B is called the *region of holomorphy* of f.

Now let us consider the series $g(z) := \sum z^{2^\nu} = z + z^2 + z^4 + z^8 + \cdots$. It has radius of convergence 1 (why?) and for any $z \in \mathbb{E}$, $n \in \mathbb{N}$

$$(*) \quad g(z^{2^n}) = g(z) - (z + z^2 + \cdots + z^{2^{n-1}}) \, , \text{ so } |g(z^{2^n})| \leq |g(z)| + n.$$

The latter appraisal of $g(z^{2^n})$ has the following consequences:

For every 2^nth root of unity ζ, $n \in \mathbb{N}$, we have $\lim_{t\uparrow 1} |g(t\zeta)| = \infty$.

Proof. Since $g(t) > \sum_0^q t^{2^\nu} > (q+1)t^{2^q} > \frac{1}{2}(q+1)$ for all $q \in \mathbb{N}$ and all t satisfying $(\sqrt[2^n]{2})^{-1} < t < 1$, it follows that $\lim_{t\uparrow 1} g(t) = \infty$. From this and $(*)$ we get, in case $\zeta^{2^n} = 1$,

$$|g(t\zeta)| \geq g(t^{2^n}) - n, \quad \text{hence} \quad \lim_{t\uparrow 1} |g(t\zeta)| = \infty. \qquad □$$

Using the density theorem 2 there now follows quickly the surprising

Theorem. *The boundary of the unit disc is the natural boundary of $g(z)$.*

Proof. Every 2^nth root of unity $\zeta \in H$ is a singular point of g because $\lim_{t\uparrow 1} |g(t\zeta)| = \infty$. Since H is dense in $\partial \mathbb{E}$ and the singular points constitute a closed set, the claim follows.   □

Corollary. *The unit disc* $\mathbb{E}$ *is the region of holomorphy of the function* $h(z) := \sum 2^{-\nu} z^{2^{\nu}}$ *and this function (unlike the earlier g) is continuous on* $\overline{\mathbb{E}} = \mathbb{E} \cup \partial\mathbb{E}$.

Proof. Were $\partial\mathbb{E}$ not the natural boundary of h, that would also be true of h' and of $zh'(z) = g(z)$ (the function h' being holomorphic wherever h is — cf. 7.4.1). Since the series for h is normally convergent in $\overline{\mathbb{E}}$, h is continuous in $\mathbb{E}$ together with its boundary $\partial\mathbb{E}$. □

Likewise it is elementary to show that $\partial\mathbb{E}$ is the natural boundary of $\sum z^{\nu!}$. Also the famous theta series $1 + 2 \sum_1^\infty z^{\nu^2}$ has $\partial\mathbb{E}$ as a natural boundary; the proof of this will be given in volume 2 (cf. also subsection 4 below).

We mention here without proof a beautiful, charming and, at first glance paradoxical, theorem which was conjectured in 1906 by P. FATOU (French mathematician, 1878-1929) and elegantly proved in 1916 by A. HURWITZ (Swiss-German mathematician in Zürich, 1859-1919) — see his *Math. Werke* 1, p.733:

Let $\mathbb{E}$ *be the convergence disc of the power series* $\sum a_\nu z^\nu$. *Then there is a sequence* $\varepsilon_0, \varepsilon_1, \varepsilon_2, \ldots$, *each being* -1 *or* 1, *such that the unit disc is the region of holomorphy of the function* $\sum \varepsilon_\nu a_\nu z^\nu$.

4. Historical remarks about natural boundaries. KRONECKER and WEIERSTRASS knew from the theory of the elliptic modular functions that $\partial\mathbb{E}$ is the natural boundary of the theta series $1 + 2 \sum_1^\infty z^{\nu^2}$ (cf. [Kr], p. 182, as well as [W$_4$], p. 227). In 1880 WEIERSTRASS showed that the boundary of $\mathbb{E}$ is the natural boundary of every series

$$\sum b_\nu z^{a^\nu}, \qquad a \in \mathbb{N} \text{ odd} \neq 1 \,;\, b \text{ real}, \, ab > 1 + \frac{3}{2}\pi$$

([W$_4$], p. 223); he writes there: "Es ist leicht, unzählige andere Potenzreihen von derselben Beschaffenheit ... anzugeben, und selbst für einen beliebig begrenzten Bereich der Veränderlichen x die Existenz der Functionen derselben, die über diesen Bereich hinaus nicht fortgesetzt werden können, nachzuweisen." (It is easy to give innumerable other power series of the same nature and, even for an arbitrarily bounded domain of the variable x, to prove the existence of functions of this kind which cannot be continued beyond that domain.) Thus it is already being maintained here that *every region in* $\mathbb{C}$ *is a region of holomorphy*. But this general theorem will not be proved until the second volume.

In 1891 the Swedish mathematician I. FREDHOLM, known for his contributions to the theory of integral equations, showed that $\mathbb{E}$ is the region of holomorphy of every power series $\sum a^\nu z^{\nu^2}$, $0 < |a| < 1$ and that such functions are even *infinitely often differentiable in* $\overline{\mathbb{E}}$ (*Acta Math.* **15**, 279-280). Cf. also [G$_2$], vol. II, part 1, §88 of the English translation.

The phenomenon of the existence of power series with natural boundaries was somewhat clarified in 1892 by J. HADAMARD. In the important and oft-cited work "Essai sur l'étude des fonctions données par leur développement de Taylor" (*Jour. math. pures et appl.* (4) **8**(1892), 101-186), where he also re-discovered Cauchy's limit-superior formula, he proved (pp. 116ff.) the famous

Gap Theorem. *Let the power series* $f(z) = \sum_{\nu=0}^{\infty} b_\nu z^{\lambda_\nu}$, $0 \leq \lambda_0 < \lambda_1 < \cdots$, *have radius of convergence* $R < \infty$ *and suppose there is a fixed positive number* δ *such that for almost all* ν

$$\lambda_{\nu+1} - \lambda_\nu \geq \delta\lambda_\nu \qquad \text{(lacunarity condition)}.$$

Then the disc $B_R(0)$ *of convergence is the region of holomorphy of* f.

The literature on the gap theorem and its generalizations is vast; we refer the reader to GAIER's notes pp. 140-142 of the 3rd edition of [Lan] for a guide to it and to pp. 76-86 and 168-174 of that book for proofs of one of the deeper generalizations, that of E. FABRY. The simplest proof of Hadamard's gap theorem is the one given in 1927 by L. J. MORDELL, "On power series with the circle of convergence as a line of essential singularities," *Jour. London Math. Soc.* **2** (1927), 146-148. We will return to this matter in volume 2; cf. also H. KNESER [14], pp. 152ff.

Exercises

Exercise 1. For $a, b \in \mathbb{R}$, $b > 0$ determine the image under the exponential map of the rectangle $\{x + iy : x, y \in \mathbb{R}, |x - a| \leq b, |y| \leq b\}$.

Exercise 2 (Cf.4.2.5*). Define $b_n := 3^n$ for odd n and $b_n := 2 \cdot 3^n$ for even n. The power series $f(z) := \sum_{n \geq 1} \frac{(-1)^n}{n} z^n$ then has radius of convergence 1 and converges at $z = 1$ (proof!). In this exercise you are asked to construct a sequence $z_m \in \mathbb{E}$ such that $\lim z_m = 1$ and $\lim |f(z_m)| = \infty$. *Hint.* Choose a strictly increasing sequence of natural numbers $k_1 < k_2 < \cdots < k_n < \cdots$ such that for all $n \in \mathbb{N}$, $n \geq 1$, we have $\sum_{\nu=n}^{k_n} \frac{1}{\nu} > 3 \sum_{\nu=1}^{n-1} \frac{1}{\nu}$. Then set $z_m := 2^{-(b_{k_m})^{-1}} e^{i\pi 3^{-m}}$.

Exercise 3. Show that if $b \in \mathbb{R}$, $d \in \mathbb{N}$ and $b > 0$, $d \geq 2$, then the unit disc is the region of holomorphy of the series $\sum_{\nu=1}^{\infty} b^\nu z^{d^\nu}$.

Exercise 4. Show that $\mathbb{E}$ is the region of holomorphy of the series $\sum z^{\nu!}$.

§4   Logarithm functions

Logarithm functions are *holomorphic* functions ℓ which satisfy the equation $\exp \circ \ell = \text{id}$ throughout their domains of definition. Characteristic of such functions is the differential equation $\ell'(z) = 1/z$. Examples of logarithm functions are

1) in $B_1(1)$ the power series $\sum_1^\infty \frac{(-1)^{\nu-1}}{\nu}(z-1)^\nu$,

2) in the "slit plane" $\{z = re^{i\varphi} : r > 0 , \alpha < \varphi < \alpha + 2\pi\}$, $\alpha \in \mathbb{R}$ fixed, the function defined by

$$\ell(z) := \log r + i\varphi.$$

1. Definition and elementary properties. Just as in the reals, a number $b \in \mathbb{C}$ is called *a logarithm* of a number $a \in \mathbb{C}$, in symbols (fraught with danger) $b = \log a$, if $e^b = a$ holds. The properties of the e-function at once yield:

The number 0 has no logarithm. Every positive real number $r > 0$ has exactly one real logarithm $\log r$. Every complex number $c = re^{i\varphi} \in \mathbb{C}^\times$ has as logarithms precisely the countably many numbers

$$\log r + i\varphi + i2\pi n , n \in \mathbb{Z} , \quad \text{in which } \log r \in \mathbb{R}.$$

One is less interested in the logarithms of individual numbers than in logarithm functions. The discussion of such functions however demands considerable care, owing to the multiplicity of logarithms that each number possesses. We formally define

A holomorphic function $\ell : G \to \mathbb{C}$ on a region $G \subset \mathbb{C}$ is called a logarithm function on G if $\exp(\ell(z)) = z$ for all $z \in G$.

If $\ell : G \to \mathbb{C}$ is a logarithm function, then certainly G does not contain 0. If we know at least one logarithm function on G, then all other such functions can be written down; specifically,

Let $\ell : G \to \mathbb{C}$ be a logarithm function of G. Then the following assertions about a function $\hat{\ell} : G \to \mathbb{C}$ are equivalent:

i) $\hat{\ell}$ *is a logarithm function on G.*

ii) $\hat{\ell} = \ell + 2\pi i\hat{n}$ *for some $\hat{n} \in \mathbb{Z}$.*

Proof. i) $\Rightarrow$ ii) We have $\exp(\hat{\ell}(z)) = \exp(\ell(z))$, that is, $\exp(\hat{\ell}(z) - \ell(z)) = 1$, for all $z \in G$. This has the consequence that $\hat{\ell}(z) - \ell(z) \in 2\pi i\mathbb{Z}$ for all $z \in G$. In other words $\frac{1}{2\pi i}(\hat{\ell} - \ell)$ is a *continuous*, integer-valued function on G. The image of the *connected* set G is therefore a connected subset of $\mathbb{Z}$, viz., a single point, $\hat{n}$, say. So ii) holds.

ii) $\Rightarrow$ i) $\hat{\ell}$ is holomorphic in G and satisfies

$$\exp(\hat{\ell}(z)) = \exp(\ell(z)) \cdot \exp(2\pi i\hat{n}) = \exp(\ell(z)) = z$$

for all $z \in G$.   □

Logarithm functions are characterized by their first derivatives.

The following assertions about a function $\ell \in \mathcal{O}(G)$ are equivalent:

i) *ℓ is a logarithm function on G.*

ii) *$\ell'(z) = 1/z$ throughout G and $\exp(\ell(a)) = a$ holds for at least one $a \in G$.*

Proof. i) $\Rightarrow$ ii) From $\exp(\ell(z)) = z$ and the chain rule follows $1 = \ell'(z) \cdot \exp'(\ell(z)) = \ell'(z) \exp(\ell(z)) = \ell'(z) \cdot z$ and so $\ell'(z) = 1/z$.

ii) $\Rightarrow$ i) The function $g(z) := z \exp(-\ell(z))$, $z \in G$, is holomorphic in G and satisfies

$$g'(z) = \exp(-\ell(z)) - z\ell'(z) \exp(-\ell(z)) = 0$$

for all $z \in G$. Since G is a region it follows (cf. 1.3.3) that $g = c \in \mathbb{C}^\times$; thus $c \exp(\ell(z)) = z$ throughout G. Since $\exp(\ell(a)) = a$, $c = 1$ follows and therewith i).

2. Existence of logarithm functions. It is easy to write down some logarithm functions explicitly.

Existence Theorem. *The function $\log z := \sum_1^\infty \frac{(-1)^{\nu-1}}{\nu}(z - 1)^\nu$ is a logarithm function in $B_1(1)$.*

Proof. By 4.3.4, 4) $\lambda(z) = \sum_1^\infty \frac{(-1)^{\nu-1}}{\nu} z^\nu$ is holomorphic in the unit disc $\mathbb{E}$ and satisfies $\lambda'(z) = (z + 1)^{-1}$ there. Since $\log z$ as defined here equals $\lambda(z - 1)$, it follows that $\log \in \mathcal{O}(B_1(1))$ with $\log'(z) = z^{-1}$. Since $\log(1) = 0$, $e^{\log 1} = 1$ holds and so, by the derivative characterization of the preceding subsection, $\log$ is a logarithm function on $B_1(1)$.   □

The expression "logarithmic series" used in 4.2.2 for the series $\lambda(z)$ defining $\log(1 + z)$ is now justified by the above existence theorem. We also note that

(1) $$|\log(1 + w) - w| \le \frac{1}{2} \frac{|w|^2}{1 - |w|} \qquad \text{for all } w \in \mathbb{E}.$$

Proof. Since $\log(1 + w) - w = -\frac{w^2}{2} + \frac{w^3}{3} - + \cdots$, we have for all $w \in \mathbb{E}$

$$|\log(1 + w) - w| \leq \frac{1}{2}|w|^2(1 + |w| + |w|^2 + \cdots) \leq \frac{1}{2}\frac{|w|^2}{1 - |w|}. \qquad \square$$

Existence is easily transferred from $B_1(1)$ to other discs:

Let $a \in \mathbb{C}^\times$ and $b \in \mathbb{C}$ be a logarithm of a. Then $\log(za^{-1}) + b$ is a logarithm function in $B_{|a|}(a)$.

This is clear because $\ell_a(z) := \log(za^{-1}) + b$ is holomorphic in $B_{|a|}(a)$ and for $z \in B_{|a|}(a)$

$$
\begin{aligned}
\exp(\ell_a(z)) = \exp(\log(za^{-1}) + b) &= \exp(\log(za^{-1}))\exp(b) \\
&= za^{-1}\exp b = za^{-1}a = z. \qquad \square
\end{aligned}
$$

Now logarithm functions are holomorphic *per definitionem*. But in fact holomorphy follows just from continuity!

Let $\ell : G \to \mathbb{C}$ be continuous and satisfy $\exp \circ \ell = \mathrm{id}$ in G. Then ℓ is holomorphic in G and consequently it is a logarithm function in G.

Proof. Fix $a \in G$. Of course $a \neq 0$. Let ℓ_a denote the holomorphic logarithm function $\log(za^{-1}) + b$ in $B_{|a|}(a)$, where b is any logarithm of a. Then $\exp(\ell(z) - \ell_a(z)) = 1$, whence $\ell(z) - \ell_a(z) \in 2\pi i\mathbb{Z}$ for all $z \in G \cap B_{|a|}(a)$. The continuous function $\frac{1}{2\pi i}(\ell - \ell_a)$ is therefore integer-valued, hence locally constant. ℓ is therefore holomorphic in a neighborhood of a. $\square$

We emphasize: *"continuous" logarithm functions are already holomorphic.*

3. The Euler sequence $(1 + z/n)^n$. Motivated by, among other things, questions of compound interest, EULER considered in [E] the polynomial sequence

$$\left(1 + \frac{z}{n}\right)^n, \; n \geq 1$$

and, via the binomial expansion and a passage to the limit which was insufficiently justified in view of the subtleties present, he showed

Theorem. *The sequence $(1 + z/n)^n$ converges compactly in $\mathbb{C}$ to $\exp z$.*

We base the proof on the following

Composition Lemma. *Let X be a metric space, f_n a sequence in $\mathcal{C}(X)$ compactly convergent to $f \in \mathcal{C}(X)$. Then the sequence $\exp \circ f_n$ converges compactly in X to $\exp \circ f$.*

Proof. All the functions $\exp \circ f_n$, $\exp \circ f$ are continuous on X. Let K be a compact subset of X. Since

$$\exp \circ f_n - \exp \circ f = [\exp \circ (f_n - f) - 1] \exp \circ f$$

and $|\exp w - w| \leq 2|w|$ for $|w| \leq 1/2$ (cf. 4.2.1), it follows that

$$|\exp \circ f_n - \exp \circ f|_K \leq 2|\exp \circ f|_K |f_n - f|_K$$

whenever $|f_n - f|_K \leq 1/2$. Since $\lim |f_n - f|_K = 0$, the assertion follows.
□

Remark. The assertion of the Composition Lemma also follows directly from the Composition Theorem in 3.1.5*.

We can now proceed with the proof of Euler's convergence theorem. Let a compact set $K \subset \mathbb{C}$ be given. There is an $m \in \mathbb{N}$ such that $|z/n| \leq 1/2$ for all $n \geq m$ and all $z \in K$. Since by 2.(1), $|\log(1 + w) - w| \leq |w|^2$ for $|w| \leq 1/2$, it follows that

$$\log\left(1 + \frac{z}{n}\right) \in \mathcal{C}(K) \quad \text{and} \quad \left|n\log\left(1 + \frac{z}{n}\right) - z\right| \leq \frac{1}{n}|z|_K^2 \quad \text{for } n \geq m.$$

Consequently, $n\log(1 + \frac{z}{n})$ converges compactly in $\mathbb{C}$ to z. Because of the identity

$$\exp\left(n\log\left(1 + \frac{z}{n}\right)\right) = \left(1 + \frac{z}{n}\right)^n,$$

the composition lemma ensures that $(1 + \frac{z}{n})^n$ converges compactly in $\mathbb{C}$ to $\exp z$.

4. Principal branch of the logarithm. Next we will introduce a *logarithm function* in the "slit plane" $\mathbb{C}^-$ (cf. 2.2.4). Our point of departure is the real function

$$\log : \mathbb{R}^+ \to \mathbb{R} \ , \ r \mapsto \log r \qquad (\text{where } \mathbb{R}^+ := \{x \in \mathbb{R} : x > 0\}).$$

We will assume known from the infinitesimal calculus that this function is *continuous on* $\mathbb{R}^+$ (although this is easily inferred from its being the inverse of the strictly increasing continuous function $x \mapsto e^x$ on $\mathbb{R}$). We "continue this function into the complexes": every number $z \in \mathbb{C}^-$ is uniquely representable in the form $z = |z|e^{i\varphi}$, where $|z| > 0$ and $-\pi < \varphi < \pi$. We claim that:

The function $\log : \mathbb{C}^- \to \mathbb{C}$, $z = |z|e^{i\varphi} \mapsto \log|z| + i\varphi$, *is a logarithm function in* $\mathbb{C}^-$. *In* $B_1(1) \subset \mathbb{C}^-$ *it coincides with the function defined by the power series* $\sum_1^\infty \frac{(-1)^{\nu-1}}{\nu}(z-1)^\nu$.

Proof. First we want to see that the function is continuous. This comes down to checking that $z \mapsto \varphi$ is continuous. Thus suppose $z_n = |z_n|e^{i\varphi_n} \in \mathbb{C}^-$ and $z_n \to z = |z|e^{i\varphi} \in \mathbb{C}^-$, with $\varphi_n, \varphi \in (-\pi, \pi)$, and yet φ_n does not converge to φ. We use a compactness argument to reach a contradiction. A subsequence φ_{n_j} converges to some $\theta \in [-\pi, \pi]$ different from φ. But by continuity of $\exp$, $z_{n_j} = |z_{n_j}|e^{i\varphi_{n_j}} \to |z|e^{i\theta}$, so $z = |z|e^{i\theta}$, whence $e^{i(\varphi-\theta)} = 1$. Since $0 < |\varphi - \theta| < 2\pi$, this contradicts the periodicity theorem 5.3. Next notice that

$$\exp(\log z) = \exp(\log|z| + i\varphi) = e^{\log|z|} \cdot e^{i\varphi} = |z|e^{i\varphi} = z$$

for all $z \in \mathbb{C}^-$. Therefore, by 2, $\log z$ is a logarithm function in $\mathbb{C}^-$. By the existence theorem 2, the function $\sum_1^\infty \frac{(-1)^{\nu-1}}{\nu}(z-1)^\nu$ is a logarithm function in $B_1(1) \subset \mathbb{C}^-$, hence it can differ from $\log z$ in the connected set $B_1(1)$ only by a constant, and that constant is 0 because both functions vanish at $z = 1$. $\square$

The logarithm function in the slit plane $\mathbb{C}^-$ just introduced is called *the principal branch of the logarithm*; for it $\log i = \frac{1}{2}\pi i$. The infinitely many other logarithm functions $\log z + 2\pi i n$, $n \in \mathbb{Z}$, $z \in \mathbb{C}^-$, are called *secondary branches* or just simply *branches*; since $\mathbb{C}^-$ is a region (for each point $z \in \mathbb{C}^-$ the line segment $[1, z]$ from 1 to z lies wholly in $\mathbb{C}^-$), these branches are *all* the logarithm functions in $\mathbb{C}^-$.

The function $\tilde{\ell}(z) = \frac{1}{2}\log(x^2 + y^2) + i\arctan(y/x)$ considered in 1.2.3, 3), coincides in the right half-plane $\{z \in \mathbb{C} : \Re z > 0\}$ with the principal branch $\log z$, since $x^2 + y^2 = |z|^2$ and $\arctan(y/x) = \varphi$ if $x > 0$. By contrast, however, $\tilde{\ell}$ is *not* a logarithm function at all in the left half-plane; since there evidently $\exp(\tilde{\ell}(z)) = -z$.

From now on $\log$ will *always* mean the principal branch of the logarithm. In our definition the plane $\mathbb{C}$ was slit along the negative real axis, and a certain arbitrariness was involved in this. One could as easily remove any other half-line starting at 0 and by procedures analogous to the above define a logarithm function in the resulting region. However, there is *no logarithm function in the whole of* $\mathbb{C}^\times$; for any such function would have to coincide in $\mathbb{C}^-$ with some branch $\log z + 2\pi i n$ and consequently would fail to be continuous at each point of the negative real axis.

5. Historical remarks on logarithm functions in the complex domain. The extension of the real logarithm function to complex arguments led at first to a phenomenon in analysis that was unknown in the real domain: a function defined by natural properties becomes *multiple-valued* in the complex domain. On the basis of the permanence principle, according to which all functional relations that hold in the real domain should continue to hold in the complex domain, people believed, as late as the beginning of the 18th century, in the existence of a (unique) function $\log z$ which satisfies the equations

$$\exp(\log z) = z \quad \text{and} \quad \frac{d \log z}{dz} = \frac{1}{z}.$$

From 1700 until 1716 LEIBNIZ and Johann BERNOULLI were involved in a controversy over the true values of the logarithms of -1 and i; they got entangled in irresolvable contradictions. At any rate, in 1702 BERNOULLI already knew the remarkable equation (see also 2.6):

$$\log i = \frac{1}{2}\pi i \quad , \text{that is, } i \log i = -\frac{1}{2}\pi \qquad (\text{EULER, 1728}).$$

EULER was the first to call the permanence principle into question; in his 1749 work "De la controverse entre Messrs. Leibniz et Bernoulli sur les logarithmes des nombres négatifs et imaginaires" (*Opera Omnia* (1) **17**, pp.195-232) he says quite clearly (p. 229) that every number has *infinitely many* logarithms: "Nous voyons donc qu'il est essentiel à la nature des logarithmes que chaque nombre ait une infinité de logarithmes, et que tous ces logarithmes soient differens [sic] non seulement entr'eux, mais aussi de tous les logarithmes de tout autre nombre. (We see therefore that it is essential to the nature of logarithms that each number have an infinity of logarithms and that all these logarithms be different, not only from one another, but also [different] from all the logarithms of every other number.)"

Exercises

For $n \in \mathbb{N}$, $n \geq 2$ and a region $G \subset \mathbb{C}$, a holomorphic function $w : G \to \mathbb{C}$ is called a *holomorphic nth-root* if $w^n(z) = z$ for all $z \in G$.

Exercise 1. Show that if ℓ is a logarithm function in G and $w : G \to \mathbb{C}$ is *continuous* and satisfies $w^n(z) = z$ for all $z \in G$, then $w(z) = e^{2\pi i k/n} e^{\frac{1}{n}\ell(z)}$ for all $z \in G$ and some $k \in \{0, 1, \ldots, n-1\}$.

Exercise 2. There is no holomorphic nth-root in any region G which contains 0.

§5   Discussion of logarithm functions

The real logarithm function is frequently introduced as the inverse of the (real) exponential function, and so $\log(\exp x) = x$ for $x \in \mathbb{R}$. In the complex domain this equation no longer enjoys unrestricted validity, since $\exp : \mathbb{C} \to \mathbb{C}^\times$ is not injective. The latter fact is also the reason why the real addition theorem

$$\log(xy) = \log x + \log y \qquad \text{for all positive real } x \text{ and } y$$

is not unrestrictedly valid in the complex domain. First we will discuss how these formulas have to be modified. Then we will study general power functions, and finally we will show that

$$\zeta(z) := \sum_{1}^{\infty} \frac{1}{n^z}$$

is normally convergent in the half-plane $\{z \in \mathbb{C} : \Re z > 1\}$.

1. On the identities $\log(wz) = \log w + \log z$ and $\log(\exp z) = z$.
For numbers w, z and wz in $\mathbb{C}^-$, with $w = |w|e^{i\varphi}$, $z = |z|e^{i\psi}$, $wz = |wz|e^{i\chi}$, where $\varphi, \psi, \chi \in (-\pi, \pi)$, there is an $\eta \in \{-2\pi, 0, 2\pi\}$ such that $\chi = \varphi + \psi + \eta$. From this it follows that

$$
\begin{aligned}
\log(wz) &= \log(|w||z|) + i\chi = (\log|w| + i\varphi) + (\log|z| + i\psi) + i\eta \\
&= \log w + \log z + i\eta.
\end{aligned}
$$

We see in particular that

$$\log(wz) = \log w + \log z \Leftrightarrow \varphi + \psi \in (-\pi, \pi).$$

Since the condition $-\pi < \varphi + \psi < \pi$ is met whenever $\Re w > 0$ and $\Re z > 0$, a special case of the above is

$$\log(wz) = \log w + \log z \qquad \text{for all } w, z \in \mathbb{C} \text{ with } \Re w > 0, \Re z > 0. \qquad \square$$

The number $\log(\exp z)$ lacks definition precisely for those $z = x + iy$ for which $e^z = e^x \cos y + ie^x \sin y$ falls into $\mathbb{C} \setminus \mathbb{C}^-$. This happens exactly when $e^x \cos y \le 0$ and $e^x \sin y = 0$, that is, when $y = (2n+1)\pi$ for some $n \in \mathbb{Z}$. Therefore $\log \circ \exp$ is well-defined in the domain

$$B := \mathbb{C} \setminus \{z : \Im z = (2n+1)\pi, n \in \mathbb{Z}\} = \bigcup_{n \in \mathbb{Z}} G_n,$$

where for each $n \in \mathbb{Z}$

$$G_n := \{z \in \mathbb{C} : (2n-1)\pi < \Im z < (2n+1)\pi\},$$

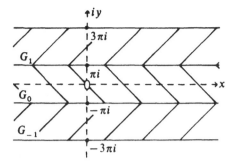

a strip of width 2π parallel to the x-axis (cf. the figure above).

For $z = x + iy \in G_n$ we have $e^z = e^x e^{i(y-2n\pi)}$ and $y - 2n\pi \in (-\pi, \pi)$. It follows that

$$\log(\exp z) = \log e^x + i(y - 2n\pi)$$

and so

$$\log(\exp z) = z - 2\pi in \qquad \text{for all } z \in G_n, \quad n \in \mathbb{Z}.$$

Only in the strip G_0 does $\log(\exp z) = z$ prevail. Since however $\exp(\log z) = z$ always holds, we have

The strip $G_0 = \{z \in \mathbb{C} : -\pi < \Im z < \pi\}$ is mapped biholomorphically (and so certainly topologically) onto the slit-plane $\mathbb{C}^-$ by the exponential function, and the inverse mapping is the principal branch of the logarithm.

2. Logarithm and arctangent. The arctangent function, defined in the unit disc in 2.5 satisfies

$$(1) \qquad\qquad \arctan z = \frac{1}{2i} \log \frac{1 + iz}{1 - iz}, \qquad z \in \mathbb{E}.$$

Proof. The function $h(z) := \frac{1+z}{1-z} \in \mathcal{O}(\mathbb{C} \setminus \{1\})$ is, to within the factor i, the Cayley mapping $h_{C'}$ from 2.2.2; therefore $h(\mathbb{E}) = \{z \in \mathbb{C} : \Re z > 0\}$. Accordingly, the function $H(z) := \log h(z)$ is well-defined in $\mathbb{E}$ and lies in $\mathcal{O}(\mathbb{E})$. It satisfies $H'(z) = \frac{h'(z)}{h(z)} = \frac{2}{1-z^2}$. Therefore $G(z) := H(iz) - 2\arctan z \in \mathcal{O}(\mathbb{E})$ satisfies

$$G'(z) = iH'(iz) - 2i(\arctan)'(z) = \frac{2i}{1 + z^2} - \frac{2i}{1 + z^2} = 0$$

for all $z \in \mathbb{E}$. Since $G(0) = 0$, it follows that $G \equiv 0$. $\qquad\square$

On the basis of the identity $\arctan(\tan z) = z$ proved in 2.5, equation (1) yields

(2) $$z = \frac{1}{2i} \log \frac{1 + i \tan z}{1 - i \tan z} \qquad \text{for all } z \text{ near } 0.$$

We further infer that

(3) $$\tan(\arctan z) = z \quad \text{for } z \in \mathbb{E}.$$

Proof. $w := \arctan z$ satisfies $e^{2iw} = \dfrac{1 + iz}{1 - iz}$, on account of (1). Since $\tan w = i\dfrac{1 - e^{2iw}}{1 + e^{2iw}}$ (cf. 2.5), it follows that $\tan w = z$.

3. Power functions. The NEWTON–ABEL formula. As soon as a logarithm function is available, general power functions can be introduced. If $\ell : G \to \mathbb{C}$ is a logarithm function, we consider the function

$$p_\sigma : G \to \mathbb{C} , \; z \mapsto \exp(\sigma \ell(z)).$$

for each complex number σ. We call p_σ the *power function with exponent σ* based on ℓ. This terminology is motivated by the following easily verified assertions:

Every function p_σ is holomorphic in G and satisfies $p_\sigma' = \sigma p_{\sigma-1}$. For all $\sigma, \tau \in \mathbb{C}$, $p_\sigma p_\tau = p_{\sigma+\tau}$ and for $n \in \mathbb{N}$, $p_n(z) = z^n$ throughout G.

In the slit-plane $\mathbb{C}^-$ a power function with exponent σ is defined by $\exp(\sigma \log z)$. Except for a brief interlude in 14.2.2, we reserve the (sometimes dangerously seductive) symbol z^σ for this power function. For whole numbers $\sigma \in \mathbb{Z}$ this agrees with the usual notation and, as remarked above, is consistent with the prior meaning of that notation. We have, for example,

$$1^\sigma = 1 , \; i^i = e^{-\frac{\pi}{2}} \cong 0.2078795763...$$

Remark. That i^i is real was remarked by EULER at the end of a letter to GOLDBACH of June 14, 1746: "Letztens habe gefunden, daß diese expressio $(\sqrt{-1})^{\sqrt{-1}}$ einen valorem realem habe, welcher in fractionibus decimalibus $= 0{,}2078795763$, welches mir merkwürdig zu seyn scheinet. (Recently I have found that this expression $(\sqrt{-1})^{\sqrt{-1}}$ has a real value, which in decimal fraction form $= 0.2078795763$; this seems remarkable to me.)" Cf. p. 383 of the "Correspondence entre Leonard Euler et Chr. Goldbach" cited in 1.3.

The rules already noted above can be suggestively written in the new notation thus:

$$(z^\sigma)' = \sigma z^{\sigma-1} , \qquad z^\sigma z^\tau = z^{\sigma+\tau} , \qquad z \in \mathbb{C}^-.$$

From the defining equation $z^\sigma = e^{\sigma \log z}$, $z \in \mathbb{C}^-$, follows:

For $z = re^{i\varphi}$, $\varphi \in (-\pi, \pi)$, and $\sigma = s + it$, we have $|z^\sigma| = |z|^{\Re\sigma} e^{-\varphi\Im\sigma}$, and so $|z^\sigma| \leq |z|^{\Re\sigma} e^{\pi|\Im\sigma|}$.

Proof. All is clear because $|e^w| = e^{\Re w}$ and $|e^{-\varphi\Im\sigma}| \leq e^{\pi|\Im\sigma|}$.   □

The function $(1 + z)^\sigma$ is, in particular, well defined in the unit disc $\mathbb{E}$. Since $(1 + z)^\sigma = \exp(\sigma \log(1 + z)) = b_\sigma(z)$ according to 4.3.4($*$), we have the following

NEWTON-ABEL formula:

$$(1 + z)^\sigma = \sum_0^\infty \binom{\sigma}{\nu} z^\nu \qquad \text{for all } \sigma \in \mathbb{C} \text{ , } z \in \mathbb{E}.$$

By means of this formula the value of the binomial series can be explicitly calculated. Setting $\sigma = s + it$ and $1 + z = re^{i\varphi}$, we have

$$b_\sigma(z) = \exp(\sigma \log(1 + z)) = e^{(s+it)(\log r + i\varphi)} = r^s e^{-t\varphi} e^{i(t \log r + s\varphi)}.$$

If you write $z = x + iy$, comparison of real and imaginary parts on both sides of the equation $1 + x + iy = re^{i\varphi}$ yields $r = ((1 + x)^2 + y^2)^{1/2}$, $\varphi = \arctan \frac{y}{1+x}$, and consequently

$$\begin{aligned}
(1 + z)^\sigma =\; & ((1+x)^2 + y^2)^{\frac{1}{2}s} e^{-t \arctan \frac{y}{1+x}} \times \\
& \times \left[\cos\left(s \arctan \frac{y}{1+x} + \frac{1}{2} t \log((1+x)^2 + y^2) \right) \right. \\
& \left. + i \sin\left(s \arctan \frac{y}{1+x} + \frac{1}{2} t \log((1+x)^2 + y^2) \right) \right]
\end{aligned}$$

for all $z \in \mathbb{E}$. This monstrous formula occurs just like this on p.329 of ABEL's 1826 work [A].

4.   The Riemann ζ-function. For all $n \in \mathbb{N}$, $n^z = \exp(z \log n)$ is holomorphic in $\mathbb{C}$ and $|n^z| = n^{\Re z}$, according to the foregoing.

Theorem. *The series $\sum_1^\infty n^{-z}$ converges uniformly in every half-plane $\{z \in \mathbb{C} : \Re z \geq 1 + \varepsilon\}$, $\varepsilon > 0$ and converges normally in $\{z \in \mathbb{C} : \Re z > 1\}$.*

Proof. For $\varepsilon > 0$, $|n^z| = n^{\Re z} \geq n^{1+\varepsilon}$ if $\Re z \geq 1 + \varepsilon$. It follows from this that

$$\sum_1^\infty \left| \frac{1}{n^z} \right| \leq \sum_1^\infty \frac{1}{n^{1+\varepsilon}} \qquad \text{for all } z \text{ with } \Re z \geq 1 + \varepsilon.$$

But it is well known that the series on the right converges whenever $\varepsilon > 0$ (see, e.g., KNOPP [15], p. 115), so the theorem follows from the majorant criterion 3.2.2 and the definition of normal convergence.   □

The function

$$\zeta(z) := \sum_1^\infty \frac{1}{n^z} , \qquad \Re z > 1$$

is therefore well defined and at least continuous in its domain of definition. In 8.4.2 we will see that $\zeta(z)$ is actually holomorphic. Although EULER had already studied this function, today it is called the *Riemann* zeta-function. At this point we cannot go more deeply into this famous function and its history, but in 11.3.1 we will determine the values $\zeta(2), \zeta(4), \ldots, \zeta(2n), \ldots$

Exercises

In the first three exercises $G := \mathbb{C} \setminus \{x \in \mathbb{R} : |x| \geq 1\}$.

Exercise 1. Find a function $f \in \mathcal{O}(G)$ which satisfies $f(0) = i$ and $f^2(z) = z^2 - 1$ for all $z \in G$. *Hints.* Set $f_1(z) := \exp(\frac{1}{2} \log(z+1))$ for $z \in \mathbb{C} \setminus \{x \in \mathbb{R} : x \leq -1\}$ and $f_2(z) := \exp(\frac{1}{2}\ell(z-1))$ for $z \in \mathbb{C} \setminus \{x \in \mathbb{R} : x \geq 1\}$, where ℓ is an appropriate branch of the logarithm in $\mathbb{C} \setminus \{x \in \mathbb{R} : x \geq 0\}$. Then consider $f_1(z)f_2(z)$ for $z \in G$.

Exercise 2. Show that $q : \mathbb{C}^\times \to \mathbb{C}^\times$ given by $z \mapsto \frac{1}{2}(z + z^{-1})$ maps the upper half-plane $\mathbb{H}$ biholomorphically onto the region G and determine the inverse mapping. *Hints.* In Exercise 3 to Chapter 2, §1 it was shown that q maps the upper half-plane bijectively onto G. Letting $u : G \to \mathbb{H}$ denote the inverse mapping, show that u is related to the function f constructed in Exercise 1 above by

$(*)$ $(u(z) - z)^2 = z^2 - 1 = f^2(z) , \quad z \in G.$

Check that u is continuous and then infer from $(*)$ that $u(z) = z + f(z)$.

Exercise 3. Show that $z \mapsto \cos z$ maps the strip $S := \{z \in \mathbb{C} : 0 < \Re z < \pi\}$ biholomorphically onto the region G. The inverse mapping $\arccos : G \to S$ is given by

$$\arccos(w) = -i \log(w + \sqrt{w^2 - 1})$$

where $\sqrt{w^2 - 1}$ suggestively denotes the function $f(w)$ from Exercise 1 above.

Exercise 4. For $G' := \mathbb{C} \setminus [-1, 0]$ find a function $g \in \mathcal{O}(G')$ such that $g(1) = \sqrt{2}$ and $g^2(z) = z(z+1)$ for all $z \in G'$.

Exercise 5. Determine the image G of $G' := \{z \in \mathbb{C} : \Re z < \Im z < \Re z + 2\pi\}$ under the exponential mapping. Show that $\exp$ maps G' biholomorphically onto G. Finally, determine the values of the inverse mapping $\ell : G \to G'$ on the connected components of $G \cap \mathbb{R}^+$.

Chapter 6

Complex Integral Calculus

Du kannst im Großen nichts verrichten
Und fängst es nun im Kleinen an
(Nothing is brought about large-scale
But is begun small-scale).
J. W. GOETHE

Calculus integralis est methodus, ex data differential-
ium relatione inveniendi relationem ipsarum quantita-
tum (Integral calculus is the method for finding, from a
given relation of differentials, the relation of the quantities themselves). L. EULER

1. GAUSS wrote to BESSEL on December 18, 1811: "What should we make
of $\int \varphi x \cdot dx$ for $x = a + bi$? Obviously, if we're to proceed from clear
concepts, we have to assume that x passes, via infinitely small increments
(each of the form $\alpha + i\beta$), from that value at which the integral is supposed
to be 0, to $x = a + bi$ and that then all the $\varphi x \cdot dx$ are summed up. In
this way the meaning is made precise. But the progression of x values can
take place in infinitely many ways: Just as we think of the realm of all
real magnitudes as an infinite straight line, so we can envision the realm
of all magnitudes, real and imaginary, as an infinite plane wherein every
point which is determined by an abscissa a and an ordinate b represents
as well the magnitude $a + bi$. The continuous passage from one value of
x to another $a + bi$ accordingly occurs along a curve and is consequently

167

possible in infinitely many ways. But I maintain that the integral $\int \varphi x \cdot dx$ computed via two different such passages always gets the same value as long as $\varphi x = \infty$ never occurs in the region of the plane enclosed by the curves describing these two passages. This is a very beautiful theorem, whose not-so-difficult proof I will give when an appropriate occasion comes up. It is closely related to other beautiful truths having to do with developing functions in series. The passage from point to point can always be carried out without ever touching one where $\varphi x = \infty$. However, I demand that those points be avoided lest the original basic conception of $\int \varphi x \cdot dx$ lose its clarity and lead to contradictions. Moreover it is also clear from this how a function generated by $\int \varphi x \cdot dx$ could have several values for the same values of x, depending on whether a point where $\varphi x = \infty$ is gone around not at all, once, or several times. If, for example, we define $\log x$ via $\int \frac{1}{x} dx$ starting at $x = 1$, then arrive at $\log x$ having gone around the point $x = 0$ one or more times or not at all, every circuit adding the constant $+2\pi i$ or $-2\pi i$; thus the fact that every number has multiple logarithms becomes quite clear." (*Werke* **8**, 90-92).

This famous letter shows that already in 1811 GAUSS knew about contour integrals and the Cauchy integral theorem and had a clear notion of periods of integrals. Yet GAUSS did not publish his discoveries before 1831.

2. In this chapter the foundations of the theory of complex contour-integration are presented. We reduce such integrals to integrals along real intervals; alternatively, one could naturally define them by means of Riemann sums taken along paths. Complex contour integrals are introduced in two steps: First we will integrate over *continuously differentiable paths,* then *integrals along piecewise continuously differentiable paths will be introduced* (Section 1). *The latter are adequate to all the needs of classical function theory.*

In Section 3 criteria for the *path independence* of contour integrals will be derived; for *star-shaped regions* a particularly simple *integrability criterion* is found. The primary tool in these investigations is the Fundamental Theorem of the Differential and Integral Calculus on real intervals (cf. 0.2).

§0   Integration over real intervals

The theory of integration of *real-valued continuous functions* on real intervals should be known to the reader. We plan to carry this theory over to *complex-valued continuous functions*, to the extent necessary for the needs of function theory. $I = [a, b]$, with $a \leq b$ will designate a compact interval in $\mathbb{R}$.

1. The integral concept. Rules of calculation and the standard estimate.

For every continuous function $f : I \to \mathbb{C}$ the definition

$$\int_r^s f(t)dt := \int_r^s (\Re f)(t)dt + i \int_r^s (\Im f)(t)dt \in \mathbb{C}$$

makes sense for any $r, s \in I$ because $\Re f$ and $\Im f$ are real-valued continuous, consequently integrable, functions. We have the following simple

Rules of calculation. *For all* $f, g \in \mathcal{C}(I)$, *all* $r, s \in I$ *and all* $c \in \mathbb{C}$

$$(1) \quad \int_r^s (f + g)(t)dt = \int_r^s f(t)dt + \int_r^s g(t)dt \ , \quad \int_r^s cf(t)dt = c \int_r^s f(t)dt,$$

$$(2) \quad \int_r^x f(t)dt + \int_x^s f(t)dt = \int_r^s f(t)dt \qquad \text{for every } x \in I,$$

$$(3) \quad \int_s^r f(t)dt = - \int_r^s f(t)dt \qquad (\text{reversal rule})$$

$$(4) \quad \Re \int_r^s f(t)dt = \int_r^s \Re f(t)dt, \quad \Im \int_r^s f(t)dt = \int_r^s \Im f(t)dt.$$

The mapping $\mathcal{C}(I) \to \mathbb{C}$, $f \mapsto \int_a^b f(t)dt$ is thus in particular a *complex-linear form* on the $\mathbb{C}$-vector space $\mathcal{C}(I)$. We call $\int_a^b f(t)dt$ the *integral of* f *along the (real) interval* $[a, b]$. For real-valued functions $f, g \in \mathcal{C}(I)$ there is a

Monotonicity rule: $\int_a^b f(t)dt \leq \int_a^b g(t)dt$ *in case* $f(t) \leq g(t)$ *for all* $t \in I$.

For complex-valued functions the appropriate analog of this rule is the

Standard estimate: $\left| \int_a^b f(t)dt \right| \leq \int_a^b |f(t)|dt \qquad$ *for all* $f \in \mathcal{C}(I)$.

Proof. For real-valued f this follows at once from the monotonicity rule and the inequalities $-|f(t)| \leq f(t) \leq |f(t)|$. The general case is reduced to this one as follows: There is a complex number c of modulus 1 such that $c \int_a^b f(t)dt \in \mathbb{R}$. From linearity and calculation rule 4) it follows that $c \int_a^b f(t)dt = \int_a^b \Re(cf(t))dt$. Then $|\Re(cf(t))| \le |c||f(t)| = |f(t)|$ for all $t \in I$ and the monotonicity rule finish the proof:

$$
\begin{aligned}
\left| \int_a^b f(t)dt \right| &= \left| c \int_a^b f(t)dt \right| = \left| \int_a^b \Re(cf(t))dt \right| \\
&\le \int_a^b |\Re(cf(t))|dt \qquad \text{(the case already discussed)} \\
&\le \int_a^b |f(t)|dt.
\end{aligned}
$$

The standard estimate is occasionally also referred to as a "triangle inequality". This usage is suggested by thinking of the definition of the integral in terms of Riemann sums. From that point of view the inequality just established does indeed generalize the Δ-inequality $|w + z| \le |w| + |z|$ for complex numbers.

2. The fundamental theorem of the differential and integral calculus. For calculating integrals the Fundamental Theorem of Calculus is indispensable. To formulate it, we first consider differentiable functions $f : I \to \mathbb{C}$. A function $f \in \mathcal{C}(I)$ is called (*continuously*) *differentiable* on I, if both $\Re f$ and $\Im f$ are (continuously) differentiable on I. We set

$$
\frac{d}{dt}f(t) := f'(t) := (\Re f)'(t) + i(\Im f)'(t)
$$

(called the *first derivative*) and verify painlessly that the sum, product and quotient rules retain their customary form; the chain rule says (cf. the figure) among other things that:

If f is holomorphic in the domain D and if $\gamma : I \to D$ is differentiable on I, then

$$
(f \circ \gamma)'(t) = f'(\gamma(t))\gamma'(t).
$$

A function $F \in \mathcal{C}(I)$ is called a *primitive* (or an *antiderivative*) of $f \in \mathcal{C}(I)$ on I, if F is differentiable on I and $F' = f$. Just as for real-valued functions, we have the

Fundamental Theorem of the Differential and Integral Calculus for Intervals. *Let $f \in \mathcal{C}(I)$. Then (existence theorem) $x \mapsto \int_a^x f(t)dt$, $x \in I$, is a primitive of f on I. If $F \in \mathcal{C}(I)$ is any primitive of f on I, then*

$$\int_r^s f(t)dt = F(s) - F(r) \qquad \text{for all } r, s \in I.$$

The proof consists in going over to real and imaginary parts, applying the fundamental theorem of calculus for real-valued functions and reassembling the pieces. A direct consequence of the fundamental theorem is:

If $F, \hat{F} \in \mathcal{C}(I)$ are primitives of f on I, then $\hat{F} - F$ is constant on I.

Here are two other useful applications of the fundamental theorem:

Substitution rule. *If J is an interval in $\mathbb{R}$ and $\varphi : J \to I$ is continuously differentiable, then for every function $f \in \mathcal{C}(I)$*

$$\int_r^s f(\varphi(t))\varphi'(t)dt = \int_{\varphi(r)}^{\varphi(s)} f(t)dt \qquad \text{for all } r, s \in J.$$

Proof. Let F be a primitive of f on I. Then

$$\int_{\varphi(r)}^{\varphi(s)} f(t)dt = F(\varphi(s)) - F(\varphi(r)).$$

Because $(F \circ \varphi)' = (F' \circ \varphi) \cdot \varphi'$, $F \circ \varphi$ is a primitive of $(f \circ \varphi) \cdot \varphi'$ on J. According to the fundamental theorem we then have

$$\int_r^s f(\varphi(t)) \cdot \varphi'(t)dt = (F \circ \varphi)(s) - (F \circ \varphi)(r).$$

Integration by parts rule. *For all continuously differentiable functions $f, g \in \mathcal{C}(I)$*

$$\int_a^b f(t)g'(t)dt = f(b)g(b) - f(a)g(a) - \int_a^b f'(t)g(t)dt.$$

Proof. If F is a primitive of $f'g$, then $fg - F$ is a primitive of fg'.

§1 Path integrals in ℂ

We first define complex contour integrals $\int_\gamma f dz$ along continuously differentiable paths in ℂ. But this class of integrals isn't adequate for function theory, where we must often integrate along paths with "corners". In all important applications on the other hand, we only need to integrate over paths comprised of line segments and circular arcs strung together. If in spite of this we consider the broader class of all *piecewise continuously differentiable paths*, the reason is not the all too frequent and pedagogically dangerous striving for generality at any price, but rather the realization that the formulation in terms of curves built from segments and circular arcs is not any simpler and in fact notationally is often more complicated.

In (older) textbooks on function theory complex integrals along arbitrary "rectifiable" curves are frequently considered. There was a time when it was fashionable to sacrifice valuable lecture time to developing the most general theory of line integrals. Nowadays it is more customary in lectures on basic function theory to restrict oneself to integration along piecewise continuously differentiable curves and get on with the main business of the theory.

By I we again denote a real interval $[a, b]$, where $a \leq b$.

1. Continuous and piecewise continuously differentiable paths.

According to 0.6.2 every continuous mapping $\gamma : I \to \mathbb{C}$ is called a *path* or a *curve*, with *initial point* $\gamma(a)$ and *terminal point* $\gamma(b)$. Instead of $\gamma(t)$ the more suggestive notation $z(t)$, or occasionally $\zeta(t)$, is also used. The path is called *continuously differentiable* or *smooth* if the function γ is continuously differentiable on I.

Examples. 0) A path γ is called a *null path* if the function γ is constant; such paths are of course continuously differentiable.

1) The *segment* $[z_0, z_1]$ from z_0 to z_1 is the continuously differentiable curve

$$z(t) := (1 - t)z_0 + tz_1 , \quad t \in [0, 1].$$

2) Let $c \in \mathbb{C}$, $r > 0$. The function

$$z(t) := c + re^{it} = \Re c + r \cos t + i(\Im c + r \sin t) , \quad t \in [a, b],$$

where $0 \leq a < b \leq 2\pi$, is continuously differentiable. The corresponding curve γ is called, as intuition dictates, a *circular arc* on the boundary of the disc $B_r(c)$. In case $a = 0$, $b = 2\pi$, γ is the circle of radius r around the center c. This curve is *closed* (meaning that initial point = terminal point); we designate this curve by $S_r(c)$ or sometimes simply by S_r and it is often convenient to identify S_r with the boundary $\partial B_r(c)$ of the disc $B_r(c)$.   □

If $\gamma_1, \ldots, \gamma_m$ are paths in $\mathbb{C}$ and the terminal point of γ_μ coincides with the initial point of $\gamma_{\mu+1}$ for each $1 \leq \mu < m$, then the path-sum $\gamma := \gamma_1 + \gamma_2 + \cdots + \gamma_m$ was defined in 0.6.2; its initial point is the initial point of γ_1 and its terminal point is the terminal point of γ_m. A path γ is called *piecewise continuously differentiable* (or *piecewise smooth*) if it has the form $\gamma = \gamma_1 + \cdots + \gamma_m$ with each γ_μ continuously differentiable. The figure shows such a path whose components γ_μ consist of segments and circular arcs.

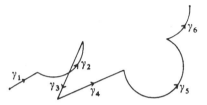

Every polygon is piecewise continuously differentiable.

In the sequel we will be working exclusively with piecewise continuously differentiable paths and so *we will agree that, from now on, the term "path" will be understood to mean piecewise continuously differentiable path*. Paths are then always *piecewise continuously differentiable functions* $\gamma : [a, b] \to$ ℂ, that is, γ is continuous and there are points $a_1, \ldots, a_{m+1}$ with $a = a_1 < a_2 < \cdots < a_m < a_{m+1} = b$ such that the restrictions $\gamma_\mu := \gamma|[a_\mu, a_{\mu+1}]$, $1 \leq \mu \leq m$ are continuously differentiable.

2. Integration along paths. As is 0.6.2, $|\gamma| = \gamma(I)$ designates the (compact) trace of the path γ. The trace of the circle $S_r(c)$ is, e.g., the boundary of the disc $B_r(c)$ (cf. 0.6.5); we also write $\partial B_r(c)$ instead of $S_r(c)$.

If γ is continuously differentiable, then $f(z(t)) \cdot z'(t) \in \mathcal{C}(I)$ for every function $f \in \mathcal{C}(|\gamma|)$; therefore according to 0.1 the complex number

$$\int_\gamma f \, dz := \int_\gamma f(z) \, dz := \int_a^b f(z(t)) z'(t) \, dt$$

exists. It is called the *path integral* or the *contour integral* or the *curvilinear integral* of $f \in \mathcal{C}(|\gamma|)$ *along the continuously differentiable path* γ. Instead of $\int_\gamma f \, dz$ we sometimes write $\int_\gamma f \, d\zeta = \int_\gamma f(\zeta) \, d\zeta$. In the special case where γ is the real interval $[a, b]$, described via $z(t) := t$, $a \leq t \leq b$, we obviously have

$$\int_\gamma f \, dz = \int_a^b f(t) \, dt.$$

It follows that the integrals discussed in section 0 are themselves path integrals.

It is now easy to define the path integral $\int_\gamma f \, dz$ for every path $\gamma = \gamma_1 + \gamma_2 + \cdots + \gamma_m$ for which the γ_μ are continuously differentiable paths and for every function $f \in \mathcal{C}(|\gamma|)$. We simply set

$$(*) \qquad\qquad \int_\gamma f dz := \sum_{\mu=1}^{m} \int_{\gamma_\mu} f dz.$$

Note that each summand on the right side is well-defined because each γ_μ is a continuously differentiable path with $|\gamma_\mu| \subset |\gamma|$.

Our contour integral concept is in an obvious sense independent of the particular way γ is expressed as a sum; but to make this precise we would have to introduce some rather unwieldy terminology and talk about refinements of representations as sums. We will leave to the interested reader the task of formulating the appropriate notion of equivalence among piecewise continuously differentiable paths and proving that contour integrals depend only on the equivalence class of the paths involved.

3. The integrals $\int_{\partial B} (\zeta - c)^n d\zeta$. Fundamental to function theory is the following

Theorem. *For $n \in \mathbb{Z}$ and all discs $B = B_r(c)$, $r > 0$,*

$$\int_{\partial B} (\zeta - c)^n d\zeta = \begin{cases} 0 & \text{for } n \neq -1, \\ 2\pi i & \text{for } n = -1. \end{cases}$$

Proof. Parameterize the boundary ∂B of B by $\zeta(t) := c + re^{it}$, with $t \in [0, 2\pi]$. Then $\zeta'(t) = ire^{it}$ and

$$\int_{\partial B} (\zeta - c)^n d\zeta = \int_0^{2\pi} (re^{it})^n ire^{it} dt = r^{n+1} \int_0^{2\pi} ie^{i(n+1)t} dt.$$

Since $\frac{1}{n+1} e^{i(n+1)t}$ is a primitive of $ie^{i(n+1)t}$ if $n \neq -1$, the claimed equality follows. $\square$

Much of function theory depends on the fact that $\int_{\partial B} (\zeta - c)^{-1} d\zeta \neq 0$.

The theorem shows that integrals along closed curves do not always vanish. The calculation involved also shows (*mutatis mutandis*) that integrals along curves which each have the same initial point and each have the same terminal point need not be equal. Thus, e.g., (see the left-hand figure below)

$$\int_{\gamma^+} \frac{d\zeta}{\zeta - c} = \pi i , \qquad \int_{\gamma^-} \frac{d\zeta}{\zeta - c} = -\pi i.$$

In 1841 in his proof of LAURENT's theorem WEIERSTRASS determined the value of the integral $\int_{\partial B} \zeta^{-1} d\zeta$ using a "rational parameterization" (cf. [W₁], p.52):

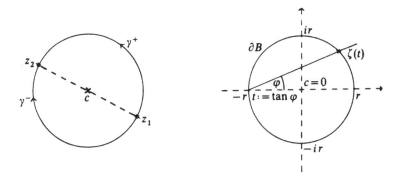

He describes the boundary ∂B of the disc (with $c = 0$) by means of $\zeta(t) := r\frac{1+ti}{1-ti}$, $-\infty < t < +\infty$. Evidently — as everyone used to learn in school — $\zeta(t)$ is the second point of intersection of ∂B with the half-line starting at $-r$ and having slope $t := \tan\varphi$ (cf. the right-hand figure). Because

$$\zeta'(t) = r\frac{2i}{(1-ti)^2}, \qquad \frac{\zeta'(t)}{\zeta(t)} = \frac{2i}{1+t^2}$$

and it follows that

$$\int_{\partial B} \frac{d\zeta}{\zeta} = 2i \int_{-\infty}^{\infty} \frac{dt}{1+t^2} = 4i \int_{0}^{\infty} \frac{dt}{1+t^2}.$$

WEIERSTRASS now defines (!) (which we proved above)

$$\pi := \int_{-\infty}^{\infty} \frac{dt}{1+t^2} = 4 \int_{0}^{1} \frac{dt}{1+t^2},$$

the reduction to a proper integral being accomplished via the substitution $t := \frac{1}{\tau}$ in $\int_{1}^{\infty} \frac{dt}{1+t^2}$. WEIERSTRASS remarked that all he really needed to know in his further deliberations was that this integral has a finite non-zero value. Cf. also 5.4.5 of the book *Numbers* [19].

4. On the history of integration in the complex plane. The first integrations through imaginary regions were published in 1813 by S. D. POISSON (French mathematician, professor at the École Polytechnique). Nevertheless the first systematic investigations of integral calculus in the complex plane were made by CAUCHY in the two treatises $[C_1]$ and $[C_2]$ already cited in the introduction to Chapter 1. The work $[C_1]$ was presented to the Paris Academy on August 22, 1814 but only submitted for printing in the "Mémoires présentés par divers Savants à l'Académie royale des Sciences de l'Institut de France" on September 14, 1825 and published in 1827. The second, essentially shorter work $[C_2]$ appeared as a special document (*magistral mémoire*) in Paris in 1825. *This document already contains the Cauchy integral theorem and is considered to be the first exposition of classical function theory*; it is customary and just (GAUSS' letter to BESSEL notwithstanding) to begin the history of function theory with CAUCHY's treatise. Repeated reference will be made to it as we progress. CAUCHY was only gradually led to study integrals in the complex plane. His works make clear that he thought a long time about this circle of questions: only *after* he'd solved his problems by separation of the functions into real and imaginary parts did he recognize that it is better not only not to make such a separation at all, but also to combine the *two* integrals

$$\int (udx - vdy) \, , \qquad \int (vdx + udy),$$

which come up in mathematical physics in the study of two-dimensional flows of incompressible fluids, into a *single* integral

$$\int f dz \qquad \text{with } f := u + iv \, , \qquad dz := dx + idy.$$

A good exposition of the development of the integral calculus in the complex plane along with detailed literature references is to be found in P. STÄCKEL's "Integration durch imaginäres Gebiet. Ein Beitrag zur Geschichte der Funktionentheorie," *Biblio. Math.* (3) **1**(1900), 109-128 and the supplement to it by the same author: "Beiträge zur Geschichte der Funktionentheorie im achtzehnten Jahrhundert," *Biblio. Math.* (3) **2**(1901), 111-121.

5. Independence of parameterization. Paths are mappings $\gamma : I \to \mathbb{C}$. You can think of γ as a "parameterization" of the trace or impression. Then it is clear that this parameterization is somewhat accidental: one is inclined to regard as the same curves which are merely traversed in the same direction but in a different time interval or with different speeds. This can all be made precise rather easily:

Two continuously differentiable paths $\gamma : I \to \mathbb{C}$, $\tilde{\gamma} : \tilde{I} = [\tilde{a}, \tilde{b}] \to \mathbb{C}$ *are called* equivalent *if there is a continuously differentiable bijection* $\varphi : \tilde{I} \to I$ *with everywhere strictly positive derivative* φ', *such that* $\tilde{\gamma} = \gamma \circ \varphi$.

The mapping φ is called a "parameter transformation" and is, because $\varphi' > 0$, a strictly increasing function with a differentiable inverse. It then follows that $\varphi(\tilde{a}) = a$, $\varphi(\tilde{b}) = b$. The inequality $\varphi' > 0$ means intuitively that in parameter transformations the direction of progression along the curve does not change (no time-reversal!).

We immediately confirm that the equivalence concept thus introduced is a genuine *equivalence relation* in the totality of continuously differentiable paths. *Equivalent paths have the same trace.* We prove the important

Independence theorem. *If* γ, $\tilde{\gamma}$ *are equivalent continuously differentiable paths, then*

$$\int_\gamma f dz = \int_{\tilde{\gamma}} f dz \qquad \text{for every function } f \in \mathcal{C}(|\gamma|).$$

Proof. In the foregoing notation, $\tilde{\gamma}(t) = \gamma(\varphi(t))$ and so $\tilde{\gamma}'(t) = \gamma'(\varphi(t))\varphi'(t)$, $t \in \tilde{I}$. It therefore follows that

$$\int_{\tilde{\gamma}} f dz = \int_{\tilde{a}}^{\tilde{b}} f(\tilde{\gamma}(t))\tilde{\gamma}'(t)dt = \int_{\tilde{a}}^{\tilde{b}} f(\gamma(\varphi(t)))\gamma'(\varphi(t))\varphi'(t)dt.$$

According to Substitution Rule 0.2, applied to $f(\gamma(t))\gamma'(t)$, the integral on the right coincides with $\int_{\varphi(\tilde{a})}^{\varphi(\tilde{b})} f(\gamma(t))\gamma'(t)dt$. Because $\varphi(\tilde{a}) = a$, $\varphi(\tilde{b}) = b$, we consequently have $\int_{\tilde{\gamma}} f dz = \int_a^b f(\gamma(t))\gamma'(t)dt = \int_\gamma f dz$. □

Thus the value of a path integral does not depend on the accidental parameterization of the path; and so, e.g., the Weierstrass parameterization (cf. 1.3) and the standard parameterization of $\partial B_r(0)$ both give the same values to integrals. Ideally we should from this point onward consider only equivalence classes of parameterized paths, even extending this idea in the natural way to piecewise continuously differentiable paths. But then every time we make a new definition (like sums of paths, the negative of a path, the length of a path) we would be obliged to show that it is independent of the class representative used in making it; the exposition would be considerably more unwieldy and complicated. For this reason we will work throughout with the mappings themselves and not with their equivalence classes.

6. Connection with real curvilinear integrals. As is well known, for a continuously differentiable path γ presented as $z(t) = x(t) + iy(t)$, $a \leq t \leq b$, and real-valued continuous functions p, q on $|\gamma|$, a real path integral is defined by

$$(*) \quad \int_\gamma (p dx + q dy) := \int_a^b p(x(t), y(t))x'(t)dt + \int_a^b q(x(t), y(t))y'(t)dt \in \mathbb{R}.$$

Theorem. *Every function $f \in C(|\gamma|)$, with $u := \Re f$, $v := \Im f$, satisfies*

$$\int_\gamma f dz = \int_\gamma (u dx - v dy) + i \int_\gamma (v dx + u dy).$$

Proof. $f = u + iv$ and $z'(t) = x'(t) + iy'(t)$, so

$$f(z(t))z'(t) = [u(x(t), y(t)) + iv(x(t), y(t))][x'(t) + iy'(t)]$$

and the claim follows upon multiplying everything out and integrating. □

The formula in the statement of the theorem is gotten by writing $dz = dx + idy$, $f dz = (u + iv)(dx + idy)$ and "formally" multiplying out the terms; cf. also subsection 4.

One could just as easily have begun the complex integral calculus by using $(*)$ to *define* the real integrals $\int_\gamma (p dx + q dy)$. It is entirely a matter of taste which avenue is preferred. □

One can also introduce general complex path integrals of the form $\int_\gamma f dx$, $\int_\gamma f dy$, $\int_\gamma f d\bar{z}$ for continuously differentiable paths γ and arbitrary $f \in \mathcal{C}(|\gamma|)$, understanding by them the respective complex numbers

$$\int_a^b f(z(t))x'(t)dt \,, \qquad \int_a^b f(z(t))y'(t)dt \,, \qquad \int_a^b f(z(t))\overline{z'(t)}dt.$$

Then the identities

$$\int_\gamma f dx = \frac{1}{2}\left(\int_\gamma f dz + \int_\gamma f d\bar{z}\right) \,, \quad \int_\gamma f dy = \frac{1}{2i}\left(\int_\gamma f dz - \int_\gamma f d\bar{z}\right) \,, \quad \int_\gamma f d\bar{z} = \overline{\int_\gamma \bar{f} dz}$$

are immediate.

Exercises

Exercise 1. Consider the rectangle $R := \{z \in \mathbb{C} : -r < \Re z < r \,, \, -s < \Im z < s\}$, whose boundary is the polygonal path

$$[-r - is, r - is] + [r - is, r + is] + [r + is, -r + is] + [-r + is, -r - is],$$

where $r > 0$ and $s > 0$. Calculate $\int_{\partial R} \zeta^{-1} d\zeta$.

Exercise 2. Let $\gamma : [0, 2\pi] \to \mathbb{C}$ be $\gamma(t) := e^{it}$ and let $g : |\gamma| \to \mathbb{C}$ be continuous. Show that

$$\overline{\int_\gamma g(\zeta)d\zeta} = -\int_\gamma \overline{g(\zeta)}\zeta^2 d\zeta.$$

Exercise 3. For $a, b \in \mathbb{R}$ define $\gamma : [0, 2\pi] \to \mathbb{C}$ by $\gamma(t) := a\cos t + ib\sin t$ and compute $\int_\gamma |\zeta|^2 d\zeta$.

§2   Properties of complex path integrals

The calculation rules from 0.1 carry over to path integrals; this will be our first order of business. With the help of the notion of the euclidean length of a curve we then derive (subsection 2) the standard estimate for path integrals, which is indispensable for applications. From it, for example, follow immediately (subsection 3) theorems dealing with interchange of limit and path integration.

1. Rules of Calculation. *For all* $f, g \in \mathcal{C}(|\gamma|)$ *,* $c \in \mathbb{C}$

1) $\int_\gamma (f + g)dz = \int_\gamma f dz + \int_\gamma g dz \,, \quad \int_\gamma cf dz = c\int_\gamma f dz.$

2) *If $\gamma*$ is a path whose initial point is the terminal point of γ, then*

$$\int_{\gamma+\gamma*} f\,dz = \int_{\gamma} f\,dz + \int_{\gamma*} f\,dz \qquad \text{for all } f \in \mathcal{C}(|\gamma+\gamma*|).$$

Proof. Because of definition $(*)$ in 1.2 it suffices to verify 1) for continuously differentiable paths. In that case however 1) follows immediately from the corresponding rule in 0.1; thus, for example, if γ is defined on $[a,b]$

$$\int_{\gamma} cf\,dz = \int_{a}^{b} cf(\gamma(t))\gamma'(t)dt = c\int_{a}^{b} f(\gamma(t))\gamma'(t)dt = c\int_{\gamma} f\,dz.$$

Rule 2) is immediate from definition $(*)$ in 1.2. $\qquad\qquad\qquad\qquad\square$

In order to get an analog of the reversal of limits in 0.1 (Rule 3), we assign to every path $\gamma : I \to \mathbb{C}$ its *reversed path* $-\gamma$ defined as $\gamma \circ \varphi$ where $\varphi : I \to I$ is given by $\varphi(t) := a + b - t$. Intuitively $-\gamma$ consists of "running over γ in the opposite direction." The sum path $\gamma + (-\gamma)$ is always defined; for every sum path $\gamma + \gamma*$ we have $-(\gamma + \gamma*) = -\gamma* + (-\gamma)$.

γ and $-\gamma$ have the same trace and $-\gamma$ is piecewise continuously differentiable if γ is. Note however that φ does *not* effect an equivalence between the paths γ and $-\gamma$ because $\varphi'(t) = -1 < 0$. Integrals along $-\gamma$ can be determined easily by means of the

Reversal rule. $\int_{-\gamma} f\,dz = -\int_{\gamma} f\,dz \qquad$ *for all $f \in \mathcal{C}(|\gamma|)$.*

Proof. We need only consider continuously differentiable paths γ. Since $\varphi(a) = b$, $\varphi(b) = a$, application of the substitution rule to $f(\gamma(t))\gamma'(t)$ gives

$$\int_{-\gamma} f\,dz = \int_{a}^{b} f(\gamma(\varphi(t)))\gamma'(\varphi(t))\varphi'(t)dt = \int_{b}^{a} f(\gamma(t))\gamma'(t)dt.$$

The reversal rule 3) from 0.1 now gives

$$\int_{-\gamma} f\,dz = -\int_{a}^{b} f(\gamma(t))\gamma'(t)dt = -\int_{\gamma} f\,dz. \qquad\qquad \square$$

Rule 4) from 0.1 does not carry over: in general $\Re \int_{\gamma} f\,dz \neq \int_{\gamma} \Re f\,dz$. Thus, e.g., $\int_{\gamma} dz = i$ for the path $\gamma := [0,i]$ and so $0 = \Re \int_{\gamma} f\,dz$, but $\int_{\gamma} \Re f\,dz = i$ for the function $f := 1$.

Important is the

Transformation rule. *Let $g : \hat{D} \to D$ be a holomorphic mapping with continuous derivative g'; let $\hat{\gamma}$ be a path in $\hat{D}$ and $\gamma := g \circ \hat{\gamma}$ the image path in D. Then*

$$\int_{\gamma} f(z)dz = \int_{\hat{\gamma}} f(g(\zeta))g'(\zeta)d\zeta \qquad \text{for all } f \in C(|\gamma|).$$

Proof. We may assume that $\hat{\gamma}$ is continuously differentiable. Then it follows that

$$\int_{\gamma} f(z)dz = \int_{a}^{b} f(g(\hat{\gamma}(t)))g'(\hat{\gamma}(t))\hat{\gamma}'(t)dt = \int_{\hat{\gamma}} f(g(\zeta))g'(\zeta)d\zeta.$$

2. The standard estimate. For every continuously differentiable path $\gamma : [a, b] \to \mathbb{C}$, $t \mapsto z(t) = x(t) + iy(t)$, the (real) integral

$$L(\gamma) := \int_{a}^{b} |z'(t)|dt = \int_{a}^{b} \sqrt{x'(t)^2 + y'(t)^2}dt$$

is called the (*euclidean*) *length of* γ. (It can be shown rather easily that $L(\gamma)$ is independent of the parameterization of γ.) We motivate this choice of language by means of two

Examples. 1) The line segment $[z_0, z_1]$ given by $z(t) = (1 - t)z_0 + tz_1$, $t \in [0, 1]$, has $z'(t) = z_1 - z_0$ and hence has length

$$L([z_0, z_1]) = \int_{0}^{1} |z_1 - z_0|dt = |z_1 - z_0|,$$

as we feel it should.

2) The circular arc γ on the disc $B_r(c)$, given by $z(t) = c + re^{it}$, $t \in [a, b]$, has length $L(\gamma) = r(b - a)$ since $|z'(t)| = |rie^{it}| = r$. The length of the whole circular periphery $\partial B_r(c)$, corresponding to $a := 0$, $b := 2\pi$, is $L(\partial B_r(c)) = 2r\pi$, in accordance with elementary geometry. □

If $\gamma = \gamma_1 + \gamma_2 + \cdots + \gamma_m$ is a path with continuously differentiable constituent paths γ_μ, then we call

$$L(\gamma) := L(\gamma_1) + L(\gamma_2) + \cdots + L(\gamma_m)$$

the (*euclidean*) *length of* γ. We can now prove the fundamental

Standard estimate for path integrals. *For every (piecewise continuously differentiable) path γ in $\mathbb{C}$ and every function $f \in C(|\gamma|)$*

$$\left| \int_{\gamma} fdz \right| \leq |f|_{\gamma} L(\gamma) \,, \quad \text{where} \quad |f|_{\gamma} := \max_{t \in [a,b]} |f(z(t))|.$$

Proof. First let γ be continuously differentiable. Then the inequality

$$\left| \int_\gamma f\, dz \right| = \left| \int_a^b f(z(t)) z'(t)\, dt \right| \leq \int_a^b |f(z(t))||z'(t)|\, dt$$

follows from the standard estimate 0.1. Then our claim follows from the fact that $|f(z(t))| \leq |f|_\gamma$ for all $t \in [a, b]$ and the monotonicity of real integrals.

Now let $\gamma = \gamma_1 + \gamma_2 + \cdots + \gamma_m$ be an arbitrary path. Since $\int_\gamma f\, dz = \sum_1^m \int_{\gamma_\mu} f\, dz$ and $|f|_{\gamma_\mu} \leq |f|_\gamma$ (due to $|\gamma_\mu| \subset |\gamma|$), it follows from what has already been proven that

$$\left| \int_\gamma f\, dz \right| \leq \sum_1^m \left| \int_{\gamma_\mu} f\, dz \right| \leq \sum_1^m |f|_{\gamma_\mu} L(\gamma_\mu) \leq |f|_\gamma \sum_1^m L(\gamma_\mu) = |f|_\gamma L(\gamma). \quad \square$$

In the standard estimate strict inequality prevails whenever there is at least one point $c \in |\gamma|$ where $|f(c)| < |f|_\gamma$. (Why?) This sharper version of the result won't really be used in this book, but as an application of it we will show now that

$$|e^z - 1| < |z| \qquad \text{for all } z \in \mathbb{C} \text{ with } \Re z < 0.$$

Proof. Let $\gamma := [0, z]$, $f(\zeta) := e^\zeta$. Then $\int_\gamma f\, dz = e^z - 1$ and $|f(\zeta)| = |e^\zeta| = e^{\Re \zeta} < 1$ for all $\zeta \in \mathbb{C}$ with $\Re \zeta < 0$, in particular for all ζ on γ except its initial point. Therefore from the sharp form of the standard estimate

$$|e^z - 1| = \left| \int_\gamma e^\zeta dz \right| < L(\gamma) = |z|.$$

3. Interchange theorems. With the help of the standard estimate it follows easily that integration and convergence of functions are interchangeable.

Interchange theorem for sequences. *Let γ be a path and $f_n \in \mathcal{C}(|\gamma|)$ a sequence of functions which converges uniformly on $|\gamma|$ to a function $f : |\gamma| \to \mathbb{C}$. Then*

$$\lim \int_\gamma f_n dz = \int_\gamma (\lim f_n) dz = \int_\gamma f\, dz.$$

Proof. According to the continuity theorem 3.1.2, $f \in \mathcal{C}(|\gamma|)$. Therefore $\int_\gamma f\, dz$ exists. From the standard estimate and the fact that $\lim |f_n - f|_\gamma = 0$ we then get

$$\left| \int_\gamma f_n dz - \int_\gamma f\, dz \right| = \left| \int_\gamma (f_n - f) dz \right| \leq |f_n - f|_\gamma L(\gamma) \to 0. \qquad \square$$

Applying this result to partial sums gives us

Interchange theorem for series. *Let γ be a path, $f_\nu \in C(|\gamma|)$ functions for which the series $\sum f_\nu$ converges uniformly on $|\gamma|$ to a function $f : |\gamma| \to \mathbb{C}$. Then*

$$\sum \int_\gamma f_\nu dz = \int_\gamma \left(\sum f_\nu \right) dz = \int_\gamma f dz.$$

The significance of the interchange theorems for function theory will only gradually become clear; e.g., from them follows Weierstrass' theorem concerning the holomorphy of the limit of a compactly convergent sequence of holomorphic functions (cf. 8.4.1). In the next section we give our first application; in it (as almost always) the series involved is normally convergent.

4. The integral $\frac{1}{2\pi i} \int_{\partial B} \frac{d\zeta}{\zeta - z}$. Because the boundary ∂B of the disc $B := B_r(c)$ is given by $\zeta(t) = c + re^{it}$, $t \in [0, 2\pi]$, we have

$$\frac{1}{2\pi i} \int_{\partial B} \frac{d\zeta}{\zeta - z} = \frac{r}{2\pi} \int_0^{2\pi} \frac{e^{it} dt}{re^{it} + (c - z)} \qquad \text{for all } z \in \mathbb{C} \setminus \partial B.$$

The straight-forward calculation of this integral is difficult if $z \neq c$. Therefore we don't attempt to evaluate it directly, but (by means of a trick) reduce to the case $z = c$. This is done via the geometric series.

Lemma. *The following equations hold:*

(1) $\quad \dfrac{1}{\zeta - z} = \dfrac{1}{\zeta - c} \displaystyle\sum_0^\infty \left(\dfrac{z - c}{\zeta - c} \right)^\nu \qquad$ *for all ζ, z with $|z - c| < |\zeta - c|$,*

(2) $\quad \dfrac{1}{\zeta - z} = \dfrac{-1}{z - c} \displaystyle\sum_0^\infty \left(\dfrac{\zeta - c}{z - c} \right)^\nu \qquad$ *for all ζ, z with $|z - c| > |\zeta - c|$.*

For fixed c, r and $z \in \mathbb{C} \setminus \partial B_r(c)$ these are normally convergent series in the variable $\zeta \in \partial B_r(c)$.

For the proof of (1) we set $w := (z - c)(\zeta - c)^{-1}$ and write

$$\frac{1}{\zeta - z} = \frac{1}{\zeta - c} \frac{1}{1 - w} = \frac{1}{\zeta - c} \sum_0^\infty w^\nu \qquad \text{for all } w \in \mathbb{E}.$$

Correspondingly one verifies (2) by developing $(\zeta - z)^{-1} = -[(z - c)(1 - w)]^{-1}$, with $w := (\zeta - c)(z - c)^{-1}$, into a geometric series. Actually (2) is just (1) with the roles of z and ζ interchanged, so it does not require separate proof.

If we set $q := |z - c|r^{-1}$, then for fixed $z \in B := B_r(c)$, $0 \le q < 1$ and $\max_{\zeta \in \partial B} \left| \left(\frac{z-c}{\zeta-c} \right)^n \right| = q^n$, $n \in \mathbb{N}$. Therefore the series (1) is normally convergent in $\zeta \in \partial B$. The normal convergence of the series (2) is treated in a similar manner.

Theorem. $\frac{1}{2\pi i} \int_{\partial B} \frac{d\zeta}{\zeta - z} = \begin{cases} 1 & \text{for } z \in B \\ 0 & \text{for } z \in \mathbb{C} \setminus \overline{B}. \end{cases}$

Proof. a) In case $z \in B$, (1) and the interchange theorem for series give

$$\int_{\partial B} \frac{d\zeta}{\zeta - z} = \sum_0^\infty (z - c)^\nu \int_{\partial B} \frac{d\zeta}{(\zeta - c)^{\nu+1}}.$$

According to 1.3 all the integrals on the right vanish except when $\nu = 0$ and that integral has the value $2\pi i$.

b) In case $z \in \mathbb{C} \setminus \overline{B}$, (2) and the interchange theorem for series give

$$\int_{\partial B} \frac{d\zeta}{\zeta - z} = -\sum_0^\infty \frac{1}{(z - c)^{\nu+1}} \int_{\partial B} (\zeta - z)^\nu d\zeta.$$

Now 1.3 insures that without exception all the integrals on the right vanish.

□

The reader will find another proof of a) via the Cauchy Integral Theorem in 7.1.2.

The trick of developing $1/(\zeta - z)$ into a geometric series around c used above will be exploited again in developing holomorphic functions into power series (cf. also 7.3.1).

In the *theory of the index* the integral $\frac{1}{2\pi i} \int_\gamma \frac{d\zeta}{\zeta-z}$ will be studied for every *closed* path γ as a function of $z \in \mathbb{C} \setminus |\gamma|$. Then we will see that the function defined by this integral is locally constant, has values only in $\mathbb{Z}$ and vanishes for all "large values" of z (cf. 9.5.1).

Exercises

Exercise 1. Let $G := \{z \in \mathbb{E} : \Re z + \Im z > 1\}$. Find a convenient parameterization γ of ∂G and compute $\int_\gamma \Im \zeta d\zeta$ as well as $\int_\gamma \frac{\zeta}{|\zeta|} d\zeta$.

Exercise 2. For any polynomial $p(z)$, any $c \in \mathbb{C}$, $r \in \mathbb{R}^+$

$$\int_{\partial B_r(c)} \overline{p(\zeta)}\, d\zeta = 2\pi i r^2 \overline{p'(c)}.$$

Exercise 3. Let $\gamma : [a, b] \to \mathbb{C}$ be a continuously differentiable path with $\gamma'(t) \neq 0$ for all $t \in [a, b]$. Then there is a path $\tilde{\gamma} : [\tilde{a}, \tilde{b}] \to \mathbb{C}$ which is equivalent to γ and satisfies $|\tilde{\gamma}'(t)| = 1$ for all $t \in [\tilde{a}, \tilde{b}]$.

Exercise 4 (Sharpened standard estimate). Let γ be a path in $\mathbb{C}$ and $f \in \mathcal{C}(|\gamma|)$. If there exists $c \in |\gamma|$ such that $|f(c)| < |f|_\gamma := \max_{\zeta \in |\gamma|} |f(\zeta)|$, then

$$\left| \int_\gamma f(\zeta)\, d\zeta \right| < |f|_\gamma \cdot L(\gamma).$$

Exercise 5. Let $t_n(z) := 1 + z + \frac{1}{2!}z^2 + \cdots + \frac{1}{n!}z^n$ be the nth Taylor polynomial approximant to e^z. Show that $|e^z - t_n(z)| < |z|^{n+1}$ for all $n \in \mathbb{N}$ and all $z \in \mathbb{C}$ with $\Re z < 0$.

§3 Path independence of integrals. Primitives.

The path integral $\int_\gamma f d\zeta$ is, for fixed $f \in \mathcal{C}(D)$, a *function of the path* γ *in* D. Two points $z_I, z_T \in D$ can be joined, if at all, by a multitude of paths γ in D. We saw in 1.3 that, even in the case of a holomorphic function f in D, the integral $\int_\gamma f d\zeta$ in general depends not just on the initial point z_I and the terminal point z_T but on the whole course of the path γ. Here we will discuss conditions which guarantee the *path independence* of the integral $\int_\gamma f d\zeta$, in the sense that its value is determined solely by the initial and terminal points of the path.

1. Primitives. We want to generalize the concept of primitive (function) introduced in 0.2. Fundamental here is the following

Theorem. *If f is continuous in D, the following assertions about a function $F : D \to \mathbb{C}$ are equivalent:*

i) *F is holomorphic in D and satisfies $F' = f$.*

ii) *For every pair $w, z \in D$ and every path γ in D with initial point w and terminal point z*

$$\int_\gamma f d\zeta = F(z) - F(w).$$

Proof i) $\Rightarrow$ ii) If $\gamma : [a, b] \to D$, $t \mapsto \zeta(t)$ is continuously differentiable, then

$$\int_\gamma f d\zeta = \int_a^b f(\zeta(t))\zeta'(t) dt = \int_a^b F'(\zeta(t))\zeta'(t) dt.$$

By the chain rule (cf. 0.2), $F'(\zeta(t))\zeta'(t) = \frac{d}{dt} F(\zeta(t))$. Since $w = \zeta(a)$ and $z = \zeta(b)$, it follows from this and the fundamental theorem of 0.2 that

$$\int_\gamma f d\zeta = \int_a^b \frac{d}{dt} F(\zeta(t)) dt = F(\zeta(b)) - F(\zeta(a)) = F(z) - F(w).$$

If more generally $\gamma = \gamma_1 + \cdots + \gamma_m$ is an arbitrary path in D from w to z and $z_I(\gamma_\mu)$, $z_T(\gamma_\mu)$ designate respectively the initial and terminal points of γ_μ, $1 \leq \mu \leq m$, then $w = z_I(\gamma_1)$, $z_T(\gamma_\mu) = z_I(\gamma_{\mu+1})$ for $\mu = 1, 2, \cdots, m-1$ and $z_T(\gamma_m) = z$. Therefore, according to what has already been proved

$$\int_\gamma f d\zeta = \sum_{\mu=1}^m \int_{\gamma_\mu} f d\zeta = \sum_{\mu=1}^m (F(z_T(\gamma_\mu)) - F(z_I(\gamma_\mu))) = F(z) - F(w).$$

ii) $\Rightarrow$ i) We show that at every point $c \in D$, $F'(c)$ exists and equals $f(c)$. Let $\overline{B} \subset D$ be a disc centered at c. By the hypothesis of ii)

$$F(z) = F(c) + \int_{[c,z]} f d\zeta \qquad \text{for all } z \in B.$$

Set

$$F_1(z) := \frac{1}{z-c} \int_{[c,z]} f d\zeta \quad \text{for } z \in B \setminus \{c\} \quad \text{and } F_1(c) := f(c).$$

Then $F(z) = F(c) + (z-c)F_1(z)$ holds for all $z \in B$. If we can show that F_1 is continuous at c, then $F'(c) = F_1(c) = f(c)$ will follow. For $z \in B \setminus \{c\}$

$$F_1(z) - F_1(c) = \frac{1}{z-c} \int_{[c,z]} (f(\zeta) - f(c)) d\zeta,$$

due to the fact that $\int_{[c,z]} d\zeta = z - c$. Because the segment $[c, z]$ has length $|z - c|$, the standard estimate yields

$$|F_1(z) - F_1(c)| \leq \frac{1}{|z-c|} \cdot |f - f(c)|_{[c,z]} \cdot |z - c| \leq |f - f(c)|_B,$$

for all $z \in B$. The continuity of F_1 at c thus follows from the continuity of f at c.   □

From now on a function $F : D \to \mathbb{C}$ will be called a *primitive of* $f \in \mathcal{C}(D)$, if F satisfies i) and ii) of the preceding theorem.

2. Remarks about primitives. An integrability criterion. Theorem 1 furnishes an important method for calculating complex path integrals. As soon as we know a primitive F of f, we need never parameterize any paths but just determine the difference of two values of F. And primitives can often be found directly by using i). Thus, e.g., for every integer $n \neq -1$ the function z^n in $\mathbb{C}^\times$ (or in $\mathbb{C}$ when $n \geq 0$) has the function $\frac{z^{n+1}}{n+1}$ as a primitive; therefore $\int_\gamma \zeta^n d\zeta = 0$ for *every* closed path γ in $\mathbb{C}^\times$ and every $n \in \mathbb{Z}$, $n \neq -1$.

Because $\int_{\partial B}(\zeta - c)^{-1} d\zeta = 2\pi i$ for every disc B centered at c (by theorem 2.4), it is now clear that

For no $c \in \mathbb{C}$ is there a neighborhood U of c such that the function $(z - c)^{-1} \in \mathcal{O}(\mathbb{C} \setminus \{c\})$ has a primitive in $U \setminus \{c\}$.

For $c := 0$ this reflects the fact, already realized in 5.4.4, that there is no logarithm function in $\mathbb{C}^\times$.

Within its disc of convergence every power series $f(z) = \sum a_\nu(z-c)^\nu$ has the convergent series $F(z) = \sum \frac{a_\nu}{\nu+1}(z - c)^{\nu+1}$ as a primitive; this follows immediately from theorem 4.3.2. $\square$

If F is holomorphic in D and $F' = 0$ throughout D, then F is locally constant in D.

Proof. The hypothesis says that F is a primitive of the constant 0 function. Because in every disc $B \subset D$ each point $z \in B$ can be joined radially to the center c, it follows that

$$F(z) - F(c) = \int_\gamma 0 d\zeta = 0,$$

that is, $F(z) = F(c)$ for all $z \in B$. $\square$

As promised in 1.3.3, we have found another proof for the important theorem 1.3.3. Another immediate consequence is:

If both F and $\hat{F}$, elements of $\mathcal{O}(D)$, are primitives of f, then $\hat{F} - F$ is locally constant in D. $\square$

Every function $f \in \mathcal{C}(I)$ has primitives (the existence assertion of the fundamental theorem 0.2). If we pass from intervals I to domains D in $\mathbb{C}$, this statement is not true without further qualification. Possessing a primitive is a special property that a function $f \in \mathcal{C}(D)$ may or may not have; those that do are called *integrable in D*. It is clear that, for a function

f which is integrable in D, the integral $\int_\gamma f d\zeta$ along any closed path γ in D must vanish (according to theorem 1). This property is characteristic of integrability; we thus get as the appropriate analog of the existence assertion in the fundamental theorem 0.2 the

Integrability Criterion. *The following statements about a continuous function f in a domain D are equivalent:*

i) *f is integrable in D.*

ii) *$\int_\gamma f d\zeta = 0$ for every closed path γ in D.*

If ii) holds and if D is a region, then a primitive F of f can be obtained as follows: Fix a point $z_1 \in D$ and to each point $z \in D$ "somehow" associate a path γ_z in D from z_1 to z; finally set

$$F(z) := \int_{\gamma_z} f d\zeta \,, \qquad z \in D.$$

Proof. Only the implication ii) $\Rightarrow$ i) needs verification. Because a path must remain within a single connected component of D (cf. 0.6.4), we can assume that D is a region. In order to show that F is a primitive of f, consider an arbitrary path γ in D from w to z. Choose paths γ_w, γ_z in D from z_1 to w and z, respectively. Then $\gamma_w + \gamma - \gamma_z$ is a closed path in D and therefore

$$0 = \int_{\gamma_w + \gamma - \gamma_z} f d\zeta = \int_{\gamma_w} f d\zeta + \int_\gamma f d\zeta - \int_{\gamma_z} f d\zeta = F(w) + \int_\gamma f d\zeta - F(z).$$

Consequently F has property ii) in theorem 1.   □

We will see in 8.2.1 that integrable functions are always holomorphic. If we write $f = u + iv$, then on the basis of theorem 1.6 the single complex equation $\int_\gamma f d\zeta = 0$ goes over into the two real equations

$$\int_\gamma (u dx - v dy) = 0 \qquad \text{and} \qquad \int_\gamma (v dx + u dy) = 0.$$

(On this point cf. also 1.4.)

3. Integrability criterion for star-shaped regions. The condition that *all* integrals $\int_\gamma f d\zeta$ along *all* closed paths γ in G vanish is not verifiable in practice (it is a so-called academic point); consequently the integrability criterion 2 is to a large extent useless in applications. It is of fundamental importance for the Cauchy theory that this condition can be significantly weakened in certain special regions in $\mathbb{C}$.

A set $M \subset \mathbb{C}$ is called *star-shaped* or *star-like* if there is a point $z_1 \in M$ such that the segment $[z_1, z]$ lies wholly in M whenever $z \in M$. Any such point is called a (*star-*) *center* of M (cf. the left-hand figure below). It is clear that every star-shaped domain D in $\mathbb{C}$ is a region; such regions will be called *star regions*.

A set $M \subset \mathbb{C}$ is called *convex* if the segment $[w, z]$ lies in M whenever its endpoints w and z do; such sets were introduced already in antiquity, by ARCHIMEDES, occasioned by his investigations of surface area. Every convex set is star-shaped and each of its points is a star-center. In particular, every convex domain, e.g., every open disc, is a star region. The plane slit along the negative real axis (cf. 2.2.4) is a star region $\mathbb{C}^-$ (which is not convex), all points $x \in \mathbb{R}$, $x > 0$ (and only these) being star- centers of $\mathbb{C}^-$. The punctured plane $\mathbb{C}$ is not a star region.

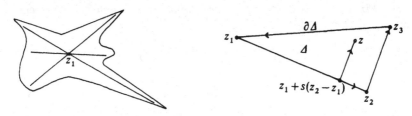

We want to show that in the study of the integrability problem in star regions it suffices to consider, instead of all closed paths, only those which are boundaries of triangles. Whenever z_1, z_2, z_3 are three points in $\mathbb{C}$, the compact set

$$\Delta = \{z \in \mathbb{C} : z = z_1 + s(z_2 - z_1) + t(z_3 - z_1), s \geq 0, t \geq 0, s + t \leq 1\}$$

$$= \{z \in \mathbb{C} : z = t_1 z_1 + t_2 z_2 + t_3 z_3, t_1 \geq 0, t_2 \geq 0, t_3 \geq 0, t_1 + t_2 + t_3 = 1\}$$

is called the (*compact*) *triangle* with *vertices* z_1, z_2, z_3 (*barycentric representation*).

The closed polygonal path

$$\partial\Delta := [z_1, z_2] + [z_2, z_3] + [z_3, z_1]$$

is called the *boundary* of Δ (with initial and terminal point z_1); the trace $|\partial\Delta|$ is actually the set-theoretic boundary of Δ (see the right-hand figure above). We claim

Integrability criterion for star regions. *Let G be a star region with center z_1. Let $f \in \mathcal{C}(G)$ satisfy $\int_{\partial\Delta} f d\zeta = 0$ for the boundary $\partial\Delta$ of each triangle $\Delta \subset G$ which has z_1 as a vertex.*

Then f is integrable in G, and the function

$$F(z) := \int_{[z_1, z]} f d\zeta \,, \qquad z \in G,$$

is a primitive of f in G. In particular, $\int_\gamma f d\zeta = 0$ for every closed path γ in G.

Proof. Because G is star-shaped with star-center z_1, $[z_1, z] \subset G$ for every $z \in G$, and so F is well defined. Let $c \in G$ be fixed. If z is near enough to c, then the triangle Δ with vertices z_1, c, z lies wholly in G.

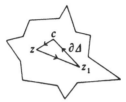

Because, by hypothesis, the integral of f along $\partial\Delta = [z_1, c] + [c, z] + [z, z_1]$ vanishes, we have

$$F(z) = F(c) + \int_{[c,z]} f d\zeta, \qquad z \in G \text{ near } c.$$

From this it follows literally by the same argument used in the proof of the implication ii) $\Rightarrow$ i) of theorem 1, that F is complex-differentiable at c and that $F'(c) = f(c)$. □

In the next chapter we will see that the condition $\int_{\partial\Delta} f d\zeta = 0$ in the integrability criterion just proved (in contrast to the more general condition in the criterion 2) is actually verifiable in important and non-trivial cases.

Exercises

Exercise 1. Let $G := \mathbb{C} \setminus [0, 1]$, $f : G \to \mathbb{C}$ the function $f(z) := \dfrac{1}{z(z - 1)}$. Show that for every closed path γ in G

$$\int_\gamma f(\zeta) d\zeta = 0$$

Exercise 2. Let D be a domain in $\mathbb{C}$, $f_n : D \to \mathbb{C}$ a sequence of continuous, integrable functions which converge compactly to $f : D \to \mathbb{C}$. Show that f is also integrable.

Exercise 3. Let G_1, G_2 be regions in $\mathbb{C}$ such that $G_1 \cap G_2$ is connected. Suppose $f : G_1 \cup G_2 \to \mathbb{C}$ is continuous and $\int_\gamma f(\zeta) d\zeta = 0$ for every closed path γ in G_1 and for every closed path γ in G_2. Show that then this equality holds for every closed path γ in $G_1 \cup G_2$ as well.

Chapter 7

The Integral Theorem, Integral Formula and Power Series Development

> Integralsatz und Integralformel sind zusammen von solcher Tragweite, dass man ohne Uebertreibung sagen kann, in diesen beiden Integralen liege die ganze jetzige Functionentheorie conzentrirt vor (The integral theorem and the integral formula together are of such scope that one can say without exaggeration: the whole of contemporary function theory is concentrated in these two integrals) — L. KRONECKER.

The era of complex integration begins with CAUCHY. It is consequently condign that his name is associated with practically every major result of this theory. In this chapter the principal Cauchy theorems will be derived in their simplest forms and extensively discussed (sections 1 and 2). We show in section 3 the most important application which is that holomorphic functions may be locally developed into power series. "Ceci marque un des plus grands progrès qui aient jamais été réalisés dans l'Analyse. (This marks one of the greatest advances that have ever been realized in analysis.)" – [Lin], pp. 9,10. As a consequence of the CAUCHY–TAYLOR development of a function we immediately prove (in 3.4) the Riemann continuation theorem, which is indispensable in many subsequent consid-

191

erations. In section 4 we discuss further consequences of the power series theorem. In a closing section we consider the Taylor series of the special functions $z \cot z$, $\tan z$ and $z/\sin z$ around 0; the coefficients of these series are determined by the so-called Bernoulli numbers. "Le développement de Taylor rend d'importants services aux mathématiciens. (The Taylor development renders great services to mathematicians.)" – J. HADAMARD, 1892

§1   The Cauchy Integral Theorem for star regions

The main result of this section is theorem 2. In order to prove it we will need in addition to integrability criterion 6.3.3, the

1. Integral lemma of GOURSAT. *Let f be holomorphic in the domain D. Then for the boundary $\partial\Delta$ of every triangle $\Delta \subset D$ we have*

$$\int_{\partial\Delta} f \, d\zeta = 0.$$

For the proof we require two elementary facts about perimeters of triangles:

1) $\max_{w,z \in \Delta} |w - z| \leq L(\partial\Delta)$.

2) $L(\partial\Delta') = \frac{1}{2}L(\partial\Delta)$ for each of the four congruent sub-triangles Δ' arising from connecting the midpoints of the three sides of Δ (cf. the left-hand figure below).

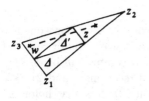
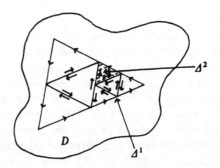

We now prove the integral lemma. As a handy abbreviation we use $a(\Delta) := \int_{\partial\Delta} f \, d\zeta$. By connecting with straight line segments the midpoints of the sides of Δ we divide Δ into four congruent sub-triangles Δ_ν, $1 \leq \nu \leq 4$; and then

$$a(\Delta) = \sum_{1}^{4} \int_{\partial \Delta_\nu} f d\zeta =: \sum_{1}^{4} a(\Delta_\nu),$$

because the segments connecting the midpoints of the sides of Δ are each traversed twice, in opposite directions (as one sees clearly in the right-hand figure above), causing the corresponding integrals to cancel each other (the reversal rule), while the union of the remaining sides of the Δ_ν is just $\partial \Delta$.

From among the four integrals $a(\Delta_\nu)$ we select one with the largest absolute value and label the corresponding triangle Δ^1. Thus

$$|a(\Delta)| \leq 4|a(\Delta^1)|.$$

Apply the same subdivision and selection process to Δ^1 to get a triangle Δ^2 for which $|a(\Delta)| \leq 4|a(\Delta^1)| \leq 4^2|a(\Delta^2)|$. Continuation of this procedure generates a descending sequence $\Delta^1 \supset \Delta^2 \supset \cdots \supset \Delta^n \supset \cdots$ of compact triangles satisfying

(1) $$|a(\Delta)| \leq 4^n|a(\Delta^n)|\ ,\qquad n = 1, 2, \ldots$$

From preliminary remark 2) follows moreover that

(2) $$L(\partial\Delta^n) = \frac{1}{2^n} L(\partial\Delta)\ ,\qquad n = 1, 2, \ldots$$

The intersection $\bigcap_{1}^{\infty} \Delta^\nu$ consists of *precisely one* point $c \in \Delta$ (*nested interval principle*). Because f lies in $\mathcal{O}(D)$, there is a function $g \in \mathcal{C}(D)$ such that

$$f(\zeta) = f(c) + f'(c)(\zeta - c) + (\zeta - c)g(\zeta)\ ,\qquad \zeta \in D,$$

and $g(c) = 0$. Then from the equations (which are valid on trivial grounds or can be justified by the evident existence of primitives)

$$\int_{\partial\Delta^n} f(c)d\zeta = 0 \qquad \text{and} \qquad \int_{\partial\Delta^n} f'(c)(\zeta - c)d\zeta = 0 \quad \text{for all } n \geq 1,$$

it follows that

$$a(\Delta^n) = \int_{\partial\Delta^n} (\zeta - c)g(\zeta)d\zeta\ ,\qquad n = 1, 2, \ldots$$

From the standard estimate for curvilinear integrals together with the first preliminary remark, we get the inequality

$$|a(\Delta^n)| \leq \max_{\zeta \in \partial\Delta^n} (|\zeta - c||g(\zeta)|)L(\partial\Delta^n) \leq L(\partial\Delta^n)^2|g|_{\partial\Delta^n}\ ,\qquad n = 1, 2, \ldots$$

From (1) and (2) it also follows that

$$|a(\Delta)| \leq 4^n |a(\Delta^n)| \leq L(\partial\Delta)^2 |g|_{\partial\Delta^n}, \qquad n = 1, 2, \ldots$$

Because $g(c) = 0$ and g is continuous at c, for every $\varepsilon > 0$, there is a $\delta > 0$ such that $|g|_{B_\delta(c)} \leq \varepsilon$. To this δ corresponds an n_0 such that $\Delta^n \subset B_\delta(c)$ for all $n \geq n_0$. Accordingly, $|g|_{\partial\Delta^n} \leq \varepsilon$ for $n \geq n_0$, and so

$$|a(\Delta)| \leq L(\partial\Delta)^2 \varepsilon.$$

Since $L(\partial\Delta)$ is a fixed number and arbitrary positive values can be chosen for ε, $a(\Delta) = 0$.   □

It is frequently and correctly maintained that all of Cauchy's theory of functions can be developed, by and large without any additional calculation, from Goursat's integral lemma.

2. The Cauchy Integral Theorem for star regions. *Let G be a star region with center c and let $f : G \to \mathbb{C}$ be holomorphic in G. Then f is integrable in G and the function $F(z) := \int_{[c,z]} f\,d\zeta$, $z \in G$, is a primitive of f in G. In particular,*

$$\int_\gamma f\,d\zeta = 0 \qquad \text{for every closed path } \gamma \text{ in } G.$$

Proof. Since $f \in \mathcal{O}(G)$, we have $\int_{\partial\Delta} f\,d\zeta = 0$ for the boundary of every triangle $\Delta \subset G$, on the basis of Goursat's integral lemma. The present claim therefore follows from the integrability criterion 6.3.3 for star regions.

□

Applications. 1) In the star region $\mathbb{C}^-$ with center 1, $\int_{[1,z]} \frac{d\zeta}{\zeta}$ is a primitive of $\frac{1}{z}$. If we next choose as our path, from 1 to $z = re^{i\varphi} \in \mathbb{C}^-$, the segment $[1, r]$ followed by the circular arc W from r to z (cf. the figure below), then because of path independence we get

$$\int_{[1,z]} \frac{d\zeta}{\zeta} = \int_1^r \frac{dt}{t} + \int_0^\varphi \frac{ire^{it}}{re^{it}}\,dt = \log r + j\varphi.$$

The original primitive thus turns out to be the principal branch of the logarithm function in $\mathbb{C}^-$ and we have another proof of the existence of the principal branch.

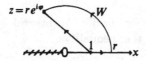

2) We give a second, direct proof of theorem 6.2.4 in the case where $z \in B$. Let $\varepsilon > 0$ be small enough that $\overline{B_\varepsilon}(z) \subset B$. We introduce two intermediate paths defined by (see the left-hand figure below)

$$\Gamma_1 := \gamma_1 + \alpha + \gamma_3 + \beta , \qquad \Gamma_2 := \gamma_2 - \beta + \gamma_4 - \alpha$$

Since the function $h(\zeta) := \frac{1}{\zeta - z}$ is holomorphic in the slit plane shown (right-hand figure below) and Γ_1 is a closed path in this star region, we have $\int_{\Gamma_1} h(\zeta)d\zeta = 0$. A similar argument shows that $\int_{\Gamma_2} h(\zeta)d\zeta = 0$. It follows (cf. the figures) that

$$0 = \int_{\Gamma_1} hd\zeta - \int_{\Gamma_2} hd\zeta = \int_{\Gamma_1 - \Gamma_2} hd\zeta = \int_{\partial B} hd\zeta - \int_{\partial B_\epsilon(z)} hd\zeta$$

and so $\int_{\partial B} hd\zeta = \int_{\partial B_\epsilon(z)} hd\zeta$. This reduces the integral to be determined to one in which z is the center of the circular path of integration, a (quite simple) case already dealt with.

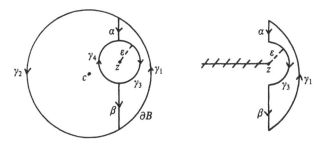

3) *If R is an open triangle or rectangle in $\mathbb{C}$, then*

$$\int_{\partial R} \frac{d\zeta}{\zeta - z} = 2\pi i \qquad \text{for all } z \in R.$$

The proof is carried out just as in the preceding text, with R now replacing B in the arguments and the figures.

3. On the history of the Integral Theorem. CAUCHY stated his theorem in 1825 in [C₂]. The publication of this classical work occurred in a very strange way. It went out of print soon and not until 1874/75 – long after RIEMANN and WEIERSTRASS had created their own theory of functions – was it reprinted as "Mélanges" in *Bull. Sci. Math. Astron.* 7(1874), 265-304, along with two continuations in Volume 8(1875), 43-55 and 148-159. P. STÄCKEL made a German translation "Abhandlung über bestimmte Integrale zwischen imaginären Grenzen" in 1900; it is the 65-page Volume 112 in the well-known series *Ostwald's Klassiker der exakten Wissenschaften.*

In his book *La vie et les travaux du baron Cauchy* (Paris, 1868 in 2 volumes; reprint: Paris, Blanchard, 1970) Cauchy's pupil and biographer VALSON enthusiastically praised this work, which indeed by 1868 was already epochal: "Ce Mémoire peut être considéré comme le plus important des travaux de Cauchy, et les hommes compétents n'hésitent pas à le comparer à tout ce que l'ésprit humain a jamais produit de plus beau dans la domaine des sciences. (This memoir may be considered the most important work of Cauchy and knowledgeable people don't hesitate to compare it to any of the most beautiful achievements of the human mind in the domain of science.)" It seems all the more astonishing that in the 27 volumes of the *Œuvres complètes d'Augustin Cauchy* which the French Academy of Science published between 1882 and 1974 (1st series with 12 volumes, 2nd series with 15 volumes) this particular work of Cauchy first showed up in a shortened form in 1958 (pp. 57-65, Vol. 2 of 2nd series) and in full only in 1974 in the last volume of the 2nd series (pp. 41-89).

CAUCHY formulated his theorem for the boundaries of rectangles (p. 7 of STÄCKEL's translation):

"We now think of the function $f(x + iy)$ as finite and continuous as long as x remains between the bounds x_0 and X and y between the bounds y_0 and Y. Then one proves easily that the value of the integral

$$\int_{x_0+iy_0}^{X+iY} f(z)dz = \int_{t_0}^{T} [\varphi'(t) + i\chi'(t)]f[\varphi(t) + i\chi(t)]dt$$

is independent of the nature of the functions $x = \varphi(t)$, $y = \chi(t)$."

This is for rectangular regions precisely the independence of the integral from the path $\varphi(t) + i\chi(t)$, $t \in [t_0, T]$. One is surprised to read that CAUCHY only hypothesizes that the function f be finite and continuous but in the proof uses, without further consideration, the existence and continuity of f'. This reflects the conviction, going back to the Euler tradition and also held by CAUCHY – at least in the early years of his work – that continuous functions are perforce given by analytic expressions and are therefore differentiable according to the rules of the differential calculus.

CAUCHY proves the Integral Theorem by methods of the calculus of variations: he replaces the functions $\varphi(t)$, $\chi(t)$ by "neighboring" functions $\varphi(t) + \varepsilon u(t)$, $\chi(t) + \varepsilon v(t)$, where $u(t_0) = v(t_0) = u(T) = v(T) = 0$, and determines the "variation of the integral" as follows (pp. 7,8 of STÄCKEL): "The integral will experience a corresponding change, which can be developed into ascending powers of ε. In this way one gets a series in which the infinitely small term of first order is the product of ε with the integral [1]

$$(*) \qquad \int_{t_0}^{T} [(u + iv)(x' + iy')f'(x + iy) + (u' + iv')f(x + iy)]dt$$

Now by partial integration we find

---

[1]If we abbreviate $\zeta := \varphi + i\chi$, $\eta := u + iv$ and put $f(\zeta + \varepsilon\eta)$ in the form $f(\zeta) + f'(\zeta) \cdot \varepsilon\eta$ + higher order terms in η, then the various ε-dependent integrands have the form $f(\zeta + \varepsilon\eta)(\zeta' + \varepsilon\eta') = f(\zeta)\zeta' + [\eta\zeta'f'(\zeta) + \eta'f(\zeta)]\varepsilon$ + higher order terms in ε, and this confirms the claim $(*)$.

$$\int_{t_0}^T (u' + iv')f(x + iy)dt = -\int_{t_0}^T (u + iv)(x' + iy')f'(x + iy)dt.$$

Consequently the integral (∗) reduces to zero." Thus did CAUCHY determine that the variation of the integral vanishes; which for him established the correctness of his theorem. This method of proof can be made rigorous but is almost forgotten in modern function theory.

The sentence "Ich behaupte nun, dass das Integral $\int \varphi x \cdot dx$ nach zweien verschiednen Übergängen immer einerlei Werth erhalte, ...", quoted in the introduction to Chapter 6, from GAUSS' letter of Dec. 18, 1811 to BESSEL shows that GAUSS knew the integral theorem that early. "Aber es ist doch ein grosser Unterschied; ob Jemand eine mathematische Wahrheit mit vollem Beweise und der Darlegung ihrer ganzen Tragweite veröffentlicht, oder ob ein Anderer sie nur so nebenher einem Freund unter Discretion mittheilt. Deshalb können wir den Satz mit Recht als das *Cauchy'sche Theorem* bezeichnen (But there is nevertheless a big difference between someone who publishes a mathematical truth with a full proof and an indication of its complete scope, and another who only incidentally communicates it privately to a friend. Therefore the theorem can rightly be designated as the *Cauchy Theorem*)" - KRONECKER, on p. 52 of [Kr].

4. On the history of the integral lemma. Édouard GOURSAT (1858-1936, French mathematician, member of the Académie des Sciences) communicated his proof to HERMITE in an 1883 letter (published as "Démonstration du Théorème de Cauchy," *Acta Math.* 4(1884), 197-200); he employs rectangles instead of triangles and explicitly used the continuity of the derivative (see the bottom of his page 199). But he must have soon become aware of the superfluousness of this continuity hypothesis, as he begins his 1899 work [G_1] with the sentence: "J'ai reconnu depuis longtemps que la démonstration du théorème de Cauchy, que j'ai donnée en 1883, ne supposait pas la continuité de la derivée. (I have recognized for a long time that the demonstration of Cauchy's theorem which I gave in 1883 didn't really presuppose the continuity of the derivative.)" And in the last sentence of this work he says: "On voit qu'en se plaçant au point de vue de Cauchy il *suffit*, pour édifier la théorie des fonctions analytiques, de supposer la *continuité* de $f(z)$ et l'*existence* de la dérivée. (One sees that, from Cauchy's point of view, it *suffices*, for purposes of erecting the theory of analytic functions, to hypothesize the *continuity* of $f(z)$ and the *existence* of the derivative.)"

GOURSAT considered regions G with quite general boundaries and applied his bisection method also to rectangles which partly protruded outside of G. The technical difficulties which this occasioned were noted as early as 1901 by Alfred PRINGSHEIM (1850-1941, German mathematician in Munich; doctorate 1872 in Heidelberg; 1877 failed attempt at *Habili-*

tation in Bonn "on account of the great ignorance of the candidate" (allegedly PRINGSHEIM refused to explain to the august faculty how one solves quadratic equations); successful *Habilitation* in Munich in 1877; by his own testimony "one of the most prominent exponents of the specifically Weierstrassian 'elementary' function theory"; owner of a coal mine in Silesia; friend of Richard WAGNER; father-in-law of Thomas MANN, who in 1905 had to withdraw his already published short story *Wälsungenblut* under pressure from the Pringsheim family) in his paper "Über den Goursatschen Beweis des Cauchyschen Integralsatzes" (*Trans. Amer. Math. Soc.* **2**(1901), 413-421). PRINGSHEIM proceeded from triangles, saying on p.418: "Der wahre *Kern* jenes Integralsatzes liegt in seiner Gültigkeit für irgend einen *Special*-Bereich *einfachster* Art z.B. ein *Dreieck* · · · Die Möglichkeit, ihn auf *krummlinig* begrenzte Bereiche zu übertragen, beruht dagegen lediglich auf *Stetigkeits*-Eigenschaften, welche den Integralen *jeder stetigen* Function zukommen. (The real *kernel* of that [Goursat's] integral theorem lies in its validity for any *special* domain of the *simplest* kind, e.g., a *triangle* · · · The possibility of extending it to domains with curved boundaries rests, by contrast, merely on *continuity* properties which are to be found in the integrals of *every continuous* function.)"

By means of his "triangle" proof PRINGSHEIM essentially simplified Goursat's method of proof and gave it the elegant, final form that it has had to this day. The triangle variant also has the economic advantage that it yields the integral theory for star regions immediately, whereas the rectangle version cannot do this.

5*. Real analysis proof of the integral lemma. From the point of view of real analysis one likes to think of Goursat's lemma as a special case of STOKES' formula. For triangles in $\mathbb{R}^2$ this reads

> Let p, q be real-valued and continuously differentiable functions in a domain $D \subset \mathbb{R}^2$. Then for the boundary $\partial\Delta$ of every triangle $\Delta \subset D$ we have

$$\int_{\partial\Delta} (p\,dx + q\,dy) = \int\int_{\Delta} \left(\frac{\partial q}{\partial x} - \frac{\partial p}{\partial y} \right) dx\,dy,$$

where $\int\int \cdots dx\,dy$ indicates the area integral over Δ.

From this the integral lemma follows immediately but only *under the supplemental hypothesis that the derivative f' of f is continuous in D*: For then $u = \Re f$ and $v = \Im f$ are *continuously real-differentiable*, so that there follows (cf. 6.1.6)

$$\int_{\partial\Delta} f\,dz \;=\; \int_{\partial\Delta} (u\,dx - v\,dy) + i \int_{\partial\Delta} (v\,dx + u\,dy)$$

$$= -\int\int_\Delta \left(\frac{\partial v}{\partial x} + \frac{\partial u}{\partial y}\right) dx dy + i \int\int_\Delta \left(\frac{\partial u}{\partial x} - \frac{\partial v}{\partial y}\right) dx dy.$$

In these double integrals both integrands are 0 by virtue of the Cauchy-Riemann differential equations; therefore $\int_{\partial\Delta} f dz = 0$ follows.   □

The foregoing proof was known to CAUCHY in 1846, as his *Comptes Rendus* note "Sur les intégrales qui s'étendent à tous les points d'une courbe fermée" (*Œuvres* (1) **10**, pp. 70-74) shows. It is possible that this proof, which moreover is supposed to have been known to WEIERSTRASS as early as 1842, was suggested to CAUCHY by GREEN's work from the year 1828. In 1851 RIEMANN discussed in detail and utilized Stokes' formula ([R], article 7 ff.). Cauchy's name is never mentioned in Riemann's work.   □

We have repeatedly emphasized that for the construction of the Cauchy theory of functions – in contrast to real analysis – only the *existence of the first derivative*, not however its continuity, need be hypothesized. Function theorists are occasionally reproached for making too much of an issue of this fine point in their theory, all the more because it is meaningless as far as applications are concerned. It does seem that for all the holomorphic functions f which occur in (mathematical) nature, the continuity of the derivative f' is known *a priori* (and in most cases it is even known in advance that f is arbitrarily often complex-differentiable!) Nevertheless, it remains a surprising and deep discovery that the continuity of f' does not need to be postulated. Moreover, the Goursat proof is less "imposing" than the real variable proof based on Stokes' formula, which in the final analysis doesn't just fall into our hands from heaven. There is naturally also a Goursat lemma for real integrals, where instead of complex-differentiability conditions of real-integrability are imposed.

Discussions about the value of and best proof for a mathematical proposition will (perforce) come up again and again as long as mathematics is done by human beings, and yet to many mathematicians the ensuing polemics are as hard to understand as the disputations of the Byzantines about the genders of angels.

6*. The Fresnel integrals $\int_0^\infty \cos t^2 dt$, $\int_0^\infty \sin t^2 dt$ have played an important role in the theory of light diffraction since A. J. FRESNEL (1788-1827, French engineer and physicist). With the help of Cauchy's integral theorem we reduce the evaluation of these integrals to the "*error integral*"

$$\lim_{R\to\infty} \int_0^R e^{-t^2} dt = \int_0^\infty e^{-t^2} dt = \frac{1}{2}\sqrt{\pi}.$$

(In turn this formula will be derived later in various different ways, among others via the residue calculus, and it will be generalized; cf. 12.4.3, 12.4.6, as well as 14.3.2 and 14.3.3.)

Theorem. *For every $a \in \mathbb{R}$ with $|a| \leq 1$*

(1) $$\int_0^\infty e^{-(1+ia)^2 t^2} dt = \frac{1}{2} \frac{1}{1+ia} \sqrt{\pi}.$$

Proof. The case $a = 0$ reduces to the error integral, which, as promised will be proved later. The case $a < 0$ reduces to the case $a > 0$ by conjugating both sides of (1). Thus we only have to deal with $0 < a \leq 1$. Define $f(z) := e^{-z^2}$. For this function the integral theorem gives

(*) $$\int_{\gamma_3} f d\zeta = \int_{\gamma_1} f d\zeta + \int_{\gamma_2} f d\zeta,$$

since (see the accompanying figure) $\gamma_1 + \gamma_2 - \gamma_3$ is a closed path and f is holomorphic throughout $\mathbb{C}$. Now $\gamma_2(t) = r + it$, $0 \leq t \leq ar$, so

$$|f(\gamma_2(t))| = e^{-r^2 + t^2} \leq e^{-r^2} e^{rt}, \qquad \text{if } 0 \leq t \leq r$$

and from this, the fact that $\gamma_2'(t) = i$ and $a \leq 1$, follows

$$\left| \int_{\gamma_2} f dz \right| \leq \int_0^{ar} |f(\gamma_2(t))| dt \leq e^{-r^2} \int_0^r e^{rt} dt \leq \frac{1}{r}, \quad \text{that is,} \quad \lim_{r \to \infty} \int_{\gamma_2} f d\zeta = 0.$$

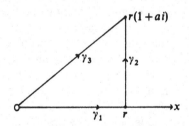

Because $\gamma_3(t) = (1 + ia)t$, $0 \leq t \leq r$, and $\gamma_3'(t) = 1 + ia$, it now follows from (*) that

$$(1+ia) \int_0^\infty e^{-(1+ia)^2 t^2} dt = \lim_{r \to \infty} \int_{\gamma_3} f d\zeta = \lim_{r \to \infty} \int_{\gamma_1} f d\zeta = \int_0^\infty e^{-t^2} dt = \frac{\sqrt{\pi}}{2}.$$

$\square$

Splitting (1) into real and imaginary parts gives

(2)
$$\int_0^\infty e^{(a^2-1)t^2} \cos 2at^2 dt = \frac{1}{2(1+a^2)} \sqrt{\pi}, \qquad -1 \leq a \leq 1,$$

$$\int_0^\infty e^{(a^2-1)t^2} \sin 2at^2 dt = \frac{a}{2(1+a^2)} \sqrt{\pi}, \qquad -1 \leq a \leq 1.$$

Taking $a := 1$ and using t in the role of $\sqrt{2}t$,

$$(3) \qquad \int_0^\infty \cos t^2 dt = \int_0^\infty \sin t^2 dt = \frac{1}{2}\sqrt{\frac{\pi}{2}}.$$

FRESNEL knew these formulas in 1819. The equations (2) were familiar to EULER by 1781; in his work "De valoribus integralium a termino variabilis $x = 0$ usque ad $x = \infty$ extensorum" (*Opera Omnia* (1) **19**, 217-227) he gets, with the help of his Gamma function $\Gamma(z)$ (whose theory will be developed in the second volume of this book) the following (p. 225):

For $p, q \in \mathbb{R}$ with $p \geq 0$ and $f := \sqrt{p^2 + q^2} \neq 0$

$$\int_0^\infty e^{-px}\frac{\cos qx}{\sqrt{x}}dx = \frac{\sqrt{\pi}}{f}\sqrt{\frac{f+p}{2}} \, , \qquad \int_0^\infty e^{-px}\frac{\sin qx}{\sqrt{x}}dx = \frac{\sqrt{\pi}}{f}\sqrt{\frac{f-p}{2}}.$$

If we substitute $t = \sqrt{x}$ and set $p := 1 - a^2$, $q := 2a$, then equations (2) result.

The method of computing the Fresnel integrals described above was well known in the 19th century, appearing, for example, in H. LAURENT's *Traité d'analyse*, Paris, 1888, Vol. 3, pp. 257-260.

Exercise

Exercise. Show that $\int_{-\infty}^\infty e^{-u^2x^2}dx = \sqrt{\pi}/u$ for all $u \in \mathbb{C}^\times$ with $|\Im u| \leq \Re u$. (In other words, the evaluation can be achieved by acting as if the substitution $t := ux$ were permissible.)

§2   Cauchy's Integral Formula for discs

The integral theorem 1.2 is inadequate for deriving the Cauchy integral formula. A sharper version, which we discuss next, is needed. Cauchy's integral formula itself will then follow in a few lines.

1. A sharper version of Cauchy's Integral Theorem for star regions. *Let G be a star region with star-center c. Let $f : G \to \mathbb{C}$ be continuous in G and holomorphic in the punctured region $G\backslash\{c\}$. Then f is integrable in G.*

The proof of this is a verbatim transcription of the proof of theorem 1.2 except that now we invoke the

Sharpened version of GOURSAT's integral lemma. *Let D be a domain, $c \in D$. Let $f : D \to \mathbb{C}$ be continuous in D and holomorphic in $D \setminus \{c\}$. Then $\int_{\partial \Delta} f d\zeta = 0$ for every triangle $\Delta \subset D$ which has a vertex at c.*

The proof of this sharper version consists of reducing to the original version: on the sides of Δ of which c is an endpoint we select any two points and consider the sub-triangles Δ_1 (containing c as a vertex), Δ_2 and Δ_3 thereby created. (See the figure below.)

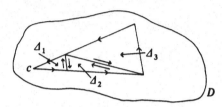

The integrals over the "interior paths" cancel each other out and, because $\Delta_2 \cup \Delta_3 \subset D \setminus \{c\}$ where f is holomorphic, the original Goursat integral lemma affirms that the integrals over $\partial\Delta_2$ and $\partial\Delta_3$ vanish. It follows that

$$\int_{\partial\Delta} f d\zeta = \int_{\partial\Delta_1} f d\zeta \, , \text{ and so } \left| \int_{\partial\Delta} f d\zeta \right| \leq |f|_\Delta L(\partial\Delta_1).$$

Since $L(\partial\Delta_1)$ can be made arbitrarily small, it follows that $\int_{\partial\Delta} f d\zeta = 0$.

Remark. The propositions of this section are preliminaries to Riemann's continuation theorem, which, among other things, asserts that every function which is continuous in D and holomorphic in $D \setminus \{c\}$ is in fact holomorphic throughout D (cf. 3.4). The above sharpening of the integral theorem is therefore really no sharpening at all; but at this point in the development of the theory we are not in any position to see this (cf. also 3.5).

2. The Cauchy Integral Formula for discs. *Let f be holomorphic in the domain D and let $B := B_r(c)$, $r > 0$, be an open disc which together with its boundary ∂B lies wholly in D. Then for all $z \in B$*

$$\boxed{f(z) = \frac{1}{2\pi i} \int_{\partial B} \frac{f(\zeta)}{\zeta - z} d\zeta.}$$

Proof. Let $z \in B$ be given and fixed, and consider the function

$$g(\zeta) := \frac{f(\zeta) - f(z)}{\zeta - z} \quad \text{for } \zeta \in D \setminus \{z\}, \ g(z) := f'(z).$$

Since $f \in \acute{\mathcal{O}}(D)$, g is holomorphic in $D \setminus \{z\}$ and continuous in D. Since $\overline{B} \subset D$, there is an $s > r$ close enough to r that $B' := B_s(c) \subset D$. Because

B' is convex, the sharpened version of the Integral Theorem 1 says that $g|B'$ is integrable, and so, in particular, $\int_{\partial B} g(\zeta)d\zeta = 0$. By theorem 6.2.4, the definition of g on ∂B gives

$$0 = \int_{\partial B} g(\zeta)d\zeta = \int_{\partial B} \frac{f(\zeta)}{\zeta - z}d\zeta - f(z) \int_{\partial B} \frac{d\zeta}{\zeta - z} = \int_{\partial B} \frac{f(\zeta)}{\zeta - z}d\zeta - 2\pi i f(z).$$

$\square$

The enormous significance of Cauchy's Integral Formula for function theory will manifest itself over and over again. What immediately attracts our attention is that it allows every value $f(z)$, $z \in B$, to be computed merely from knowledge of the values of f on the boundary ∂B. There is no analog of this in real analysis. It foreshadows the Identity Theorem and is the first indication of the (*sit venia verbo*) "analytic mortar" between the values of a holomorphic function. In the integrand of the Cauchy Integral Formula z *appears explicitly only as a parameter in the denominator, no longer tied to the function* f! We will be able to glean a lot of information about holomorphic functions generally from the simple structure of the special function $(\zeta - z)^{-1}$; among other things, the power series development of f and the Cauchy estimates for the higher derivatives of f. The function $(\zeta - z)^{-1}$ is often called the *Cauchy kernel (of the integral formula)*.

The special instance of the integral formula where z is the center c of the disc and ∂B is parameterized by $c + re^{i\varphi}$, $\varphi \in [0, 2\pi]$, is known as the

Mean value equality. *Under the hypotheses of theorem 2*

$$f(c) = \frac{1}{2\pi} \int_0^{2\pi} f(c + re^{i\varphi})d\varphi.$$

From which, e.g., immediately follows, using the standard estimate 6.2.2, the

Mean value inequality:

$$|f(c)| \leq |f|_{\partial B},$$

which however is only a special case of the general Cauchy inequalities for the Taylor coefficients (cf. 8.3.1).

Remark. By means of a beautiful trick of LANDAU's (see *Acta Math.* **40**(1916), 340, footnote 1)) much more than the mean value inequality can be immediately inferred from the Cauchy integral formula:

(#) $\qquad\qquad |f(z)| \leq |f|_{\partial B} \qquad$ *for all* $z \in B$.

Proof. First we note that the integral formula and the standard estimate show that

$$|f(z)| \leq a_z |f|_{\partial B} \text{ , where } a_z := r|1/(\zeta - z)|_{\partial B} \qquad \text{for } z \in B.$$

This estimate is of course valid as well for every positive integer power f^k of f, since $f^k \in \mathcal{O}(D)$. It then gives $|f(z)| \leq \sqrt[k]{a_z}|f|_{\partial B}$, for all $z \in B$, and since $\lim_{k \to \infty} \sqrt[k]{a_z} = 1$, the claim follows. — The inequality (#) is a forerunner of the maximum principle for bounded regions; cf. 8.5.2.

The Cauchy integral formula holds for other configurations besides discs. We will be content here to cite two such:

Let $f \in \mathcal{O}(D)$ and let R be an (open) triangle or rectangle which together with its boundary lies in D. Then

$$f(z) = \frac{1}{2\pi i} \int_{\partial R} \frac{f(\zeta)}{\zeta - z} d\zeta \qquad \text{for all } z \in R.$$

Proof. Since there is a convex region G with $R \subset G \subset D$, it follows, as above for discs, that

$$0 = \int_{\partial R} \frac{f(\zeta)}{\zeta - z} d\zeta - f(z) \int_{\partial R} \frac{d\zeta}{\zeta - z}. \qquad \qquad \square$$

According to 1.2 3) the last integral has the value $2\pi i$.

In 13.1.1 below we will see a considerable generalization of this.

3. Historical remarks on the Integral Formula. CAUCHY discovered his famous formula in 1831 during his exile in Turin. Its first publication was in a lithographed treatise *Sur la mécanique céleste et sur un nouveau calcul appelé calcul des limites*, lu à l'Académie de Turin le 11 octobre 1831. The integral formula first became generally accessible in 1841, when CAUCHY, back in Paris, published it in the 2nd volume of his *Exercices d'analyse et de physique mathématique* (*Œuvres* (2), **12**, 58-112). CAUCHY writes his formula for $c = 0$ in the following fashion (*loc. cit.*, p.61)

$$f(x) = \frac{1}{2\pi} \int_{-\pi}^{\pi} \frac{\overline{x} f(\overline{x})}{\overline{x} - x} dp,$$

where he is denoting by $\overline{x}$ the variable of integration (and not the complex conjugate of x) and is writing $\overline{x} = X e^{p\sqrt{-1}}$ (so that $X = |\overline{x}|$). This formula coincides, naturally, with ours if the boundary circle is described by $\zeta = re^{i\varphi}$, $-\pi \leq \varphi \leq \pi$:

$$\frac{1}{2\pi i} \int_{\partial B} \frac{f(\zeta)}{\zeta - z} d\zeta = \frac{1}{2\pi i} \int_{-\pi}^{\pi} \frac{f(\zeta)}{\zeta - z} i r e^{i\varphi} d\varphi = \frac{1}{2\pi} \int_{-\pi}^{\pi} \frac{\zeta f(\zeta)}{\zeta - z} d\varphi. \qquad \square$$

The mean value equality $f(c) = \frac{1}{2\pi} \int_0^{2\pi} f(c + re^{i\varphi}) d\varphi$ is to be found as early as 1823 in the work of POISSON: in "Suite du mémoire sur les intégrales définies et sur la sommation des séries," *Journ. de l'École Poly-technique*, Cahier 19, 404-509 (esp. 498). POISSON did not recognize the full scope of his formula; it was "von gränzenlosen Zauberformeln dünenartig zugedeckt" (cf. p.120 of the article "Integration durch imaginäres Gebiet" by STÄCKEL; this quotation – "covered by limitless sand dune-like magic formulae" –, used in another connection, is from GOETHE *Über Mathematik und deren Mißbrauch* 2nd Abt., volume **11**, p.85 of the Weimar edition of 1893, Verlag Hermann Böhlau.)

4*. The Cauchy integral formula for continuously real-differenti-able functions. Under the supplemental hypothesis of the continuity of f' the Cauchy integral theorem is a special case of the theorem of STOKES (cf. 1.5). Therefore it is not surprising that, under the same additional hypothesis, the Cauchy integral formula is a special case of a general integral formula for continuously real-differentiable functions.

Theorem. *Let $f : D \to \mathbb{C}$ be continuously real-differentiable in the domain D. Let $B := B_r(c)$, $r > 0$, be an open disc which together with its boundary ∂B lies in D. Then for every $z \in B$*

$$f(z) = \frac{1}{2\pi i} \int_{\partial B} \frac{f(\zeta)}{\zeta - z} d\zeta + \frac{1}{2\pi i} \int \int_B \frac{\partial f}{\partial \bar{\zeta}} \frac{1}{\zeta - z} d\zeta \wedge d\bar{\zeta}.$$

The area integral on the right is the "*correction term*" which vanishes in case f is holomorphic. In the general case part of what has to be proven is the existence of this integral. Note that to compute it we must know the values of f throughout B. We won't go into these matters any further because we don't plan to make any use of this so-called *inhomogeneous Cauchy integral formula*. (For a proof of it, see, e.g., [10].) Anyway, this generalized integral formula was unknown in classical function theory; it apparently first appeared in 1912 in a work of D. POMPEIU "Sur une classe de fonctions d'une variable complexe ...," *Rend. Circ. Mat. Palermo* **35** (1913), 277-281. Not until the 1950's was it put to use in the theory of functions of several variables by DOLBEAULT and GROTHENDIECK.

5*. Schwarz' integral formula . From the Cauchy theorems countless other integral formulas can be derived by skillful manipulation; for example:

If f is holomorphic in a neighborhood of the closure of the disc $B = B_s(0)$, then

(1) $\bar{f}(0) = \dfrac{1}{2\pi i} \int_{\partial B} \dfrac{\bar{f}(\zeta)}{\zeta - z} d\zeta$ *for all $z \in B$.*

Proof. For $z \in B$ the function $h(w) := \dfrac{\bar{z} f(w)}{s^2 - \bar{z}w}$ is holomorphic in a neighborhood of the closure of B. Therefore $\int_{\partial B} h(\zeta) d\zeta = 0$ by the integral theorem. For $\zeta \in \partial B$ we have

$$\frac{f(\zeta)}{\zeta} + h(\zeta) = \frac{f(\zeta)}{\zeta} \frac{\bar{\zeta}}{\bar{\zeta} - \bar{z}} \qquad \text{(on account of } \zeta\bar{\zeta} = s^2\text{)}.$$

From this and the Cauchy integral formula we get, for $z \in B$,

$$2\pi i f(0) = \int_{\partial B} \frac{f(\zeta)}{\zeta} d\zeta = \int_{\partial B} \left(\frac{f(\zeta)}{\zeta} + h(\zeta) \right) d\zeta = \int_{\partial B} \frac{f(\zeta)}{\zeta} \frac{\bar{\zeta}}{\bar{\zeta} - \bar{z}} d\zeta.$$

Because $\zeta d\bar{\zeta} + \bar{\zeta} d\zeta = 0$ for $\zeta\bar{\zeta} = s^2$, it follows by conjugating the above that

$$-2\pi i \bar{f}(0) = \int_{\partial B} \frac{\bar{f}(\zeta)}{\bar{\zeta}} \frac{\zeta}{\zeta - z} d\zeta = \int_{\partial B} \frac{\bar{f}(\zeta)}{\bar{\zeta}} \frac{\zeta}{\zeta - z} d\bar{\zeta} = -\int_{\partial B} \frac{\bar{f}(\zeta)}{\zeta - z} d\zeta. \quad \square$$

The last line of the foregoing proof can also be gotten directly, if less elegantly, without recourse to the "differential" equation $\zeta d\bar{\zeta} = -\bar{\zeta} d\zeta$, by introducing the parameterization $\zeta = se^{i\varphi}$ for ∂B, conjugating the resulting integral $\int_0^{2\pi} \cdots d\varphi$, and then reverting to the unparameterized integral. The careful reader should carry through the details of this little calculation.

In his 1870 work "Zur Integration der partiellen Differentialgleichung $\frac{\partial^2 u}{\partial x^2} + \frac{\partial^2 u}{\partial y^2} = 0$" (*Gesammelte mathematische Abhandlungen*, **2**, 175-210) Hermann Amandus SCHWARZ (1843-1921, German mathematician at Halle, Zürich and Göttingen and, from 1892 at Berlin as WEIERSTRASS' successor) presented an integral formula for holomorphic functions in which only the real part of f entered into the integrand. He showed (p. 186)

Schwarz' integral formula for discs centered at 0. *If f is holomorphic in a neighborhood of the closure of the disc $B := B_s(0)$, then*

$$f(z) = \frac{1}{2\pi i} \int_{\partial B} \frac{\Re f(\zeta)}{\zeta} \frac{\zeta + z}{\zeta - z} d\zeta + i\Im f(0) \qquad \text{for all } z \in B.$$

Proof. On account of $\frac{1}{\zeta} \frac{\zeta + z}{\zeta - z} = \frac{2}{\zeta - z} - \frac{1}{\zeta}$, $2\Re f = f + \bar{f}$, and the Cauchy integral formula, the integral in the equation to be proved has the value

$$\int_{\partial B} \frac{f(\zeta) + \bar{f}(\zeta)}{\zeta - z} d\zeta - \frac{1}{2} \int_{\partial B} \frac{f(\zeta) + \bar{f}(\zeta)}{\zeta} d\zeta =$$

$$= 2\pi i f(z) + \int_{\partial B} \frac{\bar{f}(\zeta)}{\zeta - z} d\zeta - \pi i f(0) - \frac{1}{2} \int_{\partial B} \frac{\bar{f}(\zeta)}{\zeta} d\zeta.$$

According to (1) both integrals on the right here have the value $2\pi i \bar{f}(0)$. Therefore

$$\frac{1}{2\pi i}\int_{\partial B}\frac{\Re f(\zeta)}{\zeta}\frac{\zeta+z}{\zeta-z}d\zeta = f(z)+\bar{f}(0)-\frac{1}{2}f(0)-\frac{1}{2}\bar{f}(0) = f(z)-i\Im f(0).$$

$\square$

The Schwarz formula shows that every value $f(z)$, $z \in B$, is already determined by the values of the real part of f on ∂B and by the number $\Im f(0)$. Upon setting $u := \Re f$, $\zeta = se^{i\psi}$ we get

$$f(z) = \frac{1}{2\pi}\int_0^{2\pi} u(se^{i\psi})\frac{\zeta+z}{\zeta-z}d\psi + i\Im f(0), \qquad z \in B.$$

Since $\Re\frac{\zeta+z}{\zeta-z} = \frac{|\zeta|^2-|z|^2}{|\zeta-z|^2}$, consideration of real parts gives

$$u(z) = \frac{1}{2\pi}\int_0^{2\pi} u(se^{i\psi})\frac{s^2-|z|^2}{|\zeta-z|^2}d\psi.$$

This is the famous *Poisson integral formula for harmonic functions*: if we further set $z = re^{i\varphi}$, then $|\zeta-z|^2 = s^2-2rs\cos(\psi-\varphi)+r^2$ and the formula takes its classical form

$$u(re^{i\varphi}) = \frac{1}{2\pi}\int_0^{2\pi} u(se^{i\psi})\frac{s^2-r^2}{s^2-2rs\cos(\psi-\varphi)+r^2}d\psi.$$

Exercises

Exercise 1. Using the Cauchy integral formula calculate

a) $\displaystyle\int_{\partial B_2(0)}\frac{e^z dz}{(z+1)(z-3)^2}$,

c) $\displaystyle\int_{\partial B_2(-2i)}\frac{dz}{z^2+1}$,

b) $\displaystyle\int_{\partial B_2(0)}\frac{\sin z}{z+i}dz$,

d) $\displaystyle\int_{\partial B_1(0)}\frac{e^z}{(z-2)^3}dz$.

Exercise 2. Let $r > 0$, D an open neighborhood of $\overline{B_r(0)}$, $f : D \to \mathbb{C}$ a holomorphic function and a_1, a_2 distinct points of $B_r(0)$.

a) Express $\int_{\partial B_r(0)}\frac{f(\zeta)d\zeta}{(\zeta-a_1)(\zeta-a_2)}$ in terms of $\int_{\partial B_r(0)}\frac{f(\zeta)}{(\zeta-a_j)}d\zeta$ ($j = 1,2$).

b) Use a) to deduce the "theorem of LIOUVILLE": Every bounded function in $\mathcal{O}(\mathbb{C})$ is constant. (Cf. also 8.3.3.)

Exercise 3. Let $r > 0$ and $f : \overline{B_r(0)} \to \mathbb{C}$ be a continuous function which is holomorphic in $B_r(0)$. Show that

$$f(z) = \frac{1}{2\pi i}\int_{\partial B_r(0)}\frac{f(\zeta)}{\zeta-z}d\zeta \qquad \text{for all } z \in B_r(0).$$

§3   The development of holomorphic functions into power series

A function $f : D \to \mathbb{C}$ is said to be *developable into a power series around* $c \in D$, if for some $r > 0$ with $B := B_r(c) \subset D$ there is a power series $\sum a_\nu(z - c)^\nu$ which converges in B to $f|B$. From the commutativity of differentiation and summation in a power series (Theorem 4.3.2) it follows at once that:

If f is developable around c in $B \subset D$ into a power series $\sum a_\nu(z - c)^\nu$, then f is infinitely often complex-differentiable, and $a_\nu = \frac{f^{(\nu)}(c)}{\nu!}$ for all $\nu \in \mathbb{N}$.

A power series development of a function f around c is, whatever the radius r of the disc B, *uniquely determined by the derivatives of f at c* and always has the form

$$f(z) = \sum \frac{f^{(\nu)}(c)}{\nu!}(z - c)^\nu.$$

This series is called (as for functions on $\mathbb{R}$) *the Taylor series of f around c*; it converges compactly in B.

The most important consequence of the Cauchy integral formula is the acquisition for every holomorphic function of a power series development about every point in its domain of definition. This development leads easily to the Riemann continuation theorem. The point of departure of our considerations here is a simple

1. Lemma on developability. If γ is a piecewise continuously differentiable path in $\mathbb{C}$, then to every continuous function $f : |\gamma| \to \mathbb{C}$ we associate the function

$$(1) \qquad F(z) := \frac{1}{2\pi i} \int_\gamma \frac{f(\zeta)}{\zeta - z} d\zeta , \qquad z \in \mathbb{C} \setminus |\gamma|$$

and claim:

Lemma on developability. *The function F is holomorphic in $\mathbb{C} \setminus |\gamma|$. For each $c \in \mathbb{C} \setminus |\gamma|$ the power series*

$$\sum_0^\infty a_\nu(z - c)^\nu \quad \text{with } a_\nu := \frac{1}{2\pi i} \int_\gamma \frac{f(\zeta)}{(\zeta - c)^{\nu+1}} d\zeta$$

converges in every open disc centered at c which does not touch $|\gamma|$ and in fact converges to F. The function F is infinitely often complex-differentiable in $\mathbb{C} \setminus |\gamma|$ and it satisfies

(2) $F^{(k)}(z) = \dfrac{k!}{2\pi i}\displaystyle\int_\gamma \dfrac{f(\zeta)}{(\zeta - z)^{k+1}}d\zeta$ for all $z \in \mathbb{C}\setminus|\gamma|$ and all $k \in \mathbb{N}$.

Proof. Fix $B := B_r(c)$ with $r > 0$ and $B \cap |\gamma| = \emptyset$. The series $\frac{1}{(1-w)^{k+1}} = \sum_{\nu \geq k} \binom{\nu}{k} w^{\nu-k}$, which converges throughout $\mathbb{E}$, is transformed by $w := (z - c)(\zeta - c)^{-1}$ into

$$\frac{1}{(\zeta - z)^{k+1}} = \sum_{\nu \geq k} \binom{\nu}{k} \frac{1}{(\zeta - c)^{\nu+1}}(z - c)^{\nu-k}$$

which converges for all $z \in B$, $\zeta \in |\gamma|$, $k \in \mathbb{N}$.

Set $g_\nu(\zeta) := f(\zeta)/(\zeta - c)^{\nu+1}$ for $\zeta \in |\gamma|$. It follows that for $z \in B$

(∗) $\dfrac{k!}{2\pi i}\displaystyle\int_\gamma \dfrac{f(\zeta)}{(\zeta - z)^{k+1}}d\zeta = \dfrac{1}{2\pi i}\int_\gamma \left[\sum_{\nu \geq k} k!\binom{\nu}{k}g_\nu(\zeta)(z - c)^{\nu-k}\right]d\zeta.$

Because $|\zeta - c| \geq r$ for all $\zeta \in |\gamma|$, it follows from the definition of g_ν that $|g_\nu|_\gamma \leq r^{-(\nu+1)}|f|_\gamma$ and therefore, with $q := |z - c|/r$,

$$\max_{\zeta \in |\gamma|}|g_\nu(\zeta)(z - c)^{\nu-k}| \leq \frac{1}{r^{k+1}}|f|_\gamma q^{\nu-k}.$$

Since $0 \leq q < 1$ for every $z \in B$ and since $\sum_{\nu \geq k}\binom{\nu}{k}q^{\nu-k} = \frac{1}{(1-q)^{k+1}}$, the series in (∗) converges normally in $\zeta \in |\gamma|$ for each fixed $z \in B$. Consequently, according to the interchange theorem for series (6.2.3),

$$\frac{k!}{2\pi i}\int_\gamma \frac{f(\zeta)}{(\zeta - z)^{k+1}}d\zeta = \sum_{\nu \geq k} k!\binom{\nu}{k}a_\nu(z - c)^{\nu-k},$$

with $a_\nu := \frac{1}{2\pi i}\int_\gamma \frac{f(\zeta)}{(\zeta-c)^{\nu+1}}d\zeta$. Thus it has been established that the function F defined by (1) is representable in the disc B by the power series $\sum a_\nu(z - c)^\nu$ ($k = 0$) and further, on account of theorem 4.3.2, that F is complex-differentiable in B and satisfies

$$F^{(k)}(z) = \sum_{\nu \geq k} k!\binom{\nu}{k}a_\nu(z - c)^{\nu-k}, \qquad z \in B, \ k \in \mathbb{N}.$$

Since B is an arbitrary open disc in $\mathbb{C}\setminus|\gamma|$, (2) is proved and in particular $F \in \mathcal{O}(\mathbb{C}\setminus|\gamma|)$.   □

The trick of developing the kernel $1/(\zeta - z)^{k+1}$ into a power series about c, then inverting the order of integration and summation was used in the case $k = 0$ by CAUCHY as early as 1831 – cf. *Œuvres* (2) **12**, p.61 (pp. 37-40 of G. BIRKHOFF, *A Source Book in Classical Analysis*, Harvard University Press (1973), for an English translation).

2. The CAUCHY-TAYLOR representation theorem.

For every point c in the domain D we will denote by $d := d_c(D)$ the distance from c to the boundary of D; cf. 0.6.5. Thus $B_d(c)$ is the *largest* disc centered at c which lies wholly in D.

CAUCHY-TAYLOR Representation Theorem. *Every holomorphic function f in D is developable around each point $c \in D$ into a Taylor series $\sum a_\nu(z - c)^\nu$ which compactly converges to it in $B_d(c)$. The Taylor coefficients a_ν are given by the integrals*

$$(1) \qquad a_\nu = \frac{f^{(\nu)}(c)}{\nu!} = \frac{1}{2\pi i} \int_{\partial B} \frac{f(\zeta)d\zeta}{(\zeta - c)^{\nu+1}}$$

whenever $B := B_r(c)$, $\quad 0 < r < d$.

In particular, f is infinitely often complex-differentiable in D and in every disc B of the above kind the Cauchy integral formulas hold:

$$(2) \qquad f^{(k)}(z) = \frac{k!}{2\pi i} \int_{\partial B} \frac{f(\zeta)d\zeta}{(\zeta - z)^{k+1}} , \qquad z \in B , \text{ for all } k \in \mathbb{N}.$$

Proof. Since $f \in \mathcal{O}(D)$, the Cauchy formula

$$f(z) = \frac{1}{2\pi i} \int_{\partial B} \frac{f(\zeta)}{\zeta - z} d\zeta , \qquad z \in B$$

holds for every disc $B = B_r(c)$, $0 < r < d$. Therefore according to the lemma on power series developability (with $F := f$, $\gamma := \partial B$), f has a Taylor development around c which converges in $B_r(c)$ and whose Taylor coefficients are given by (1). As every choice of $r < d$ generates the same series, the convergence to f occurs throughout $B_d(c)$.

Likewise the identities (2) follow directly from the lemma on developability. $\qquad \square$

Remark 1. The integral formulas (2) for the derivatives $f^{(k)}(z)$ flow from the Cauchy integral formula for $f(z)$ and the trivial identities

$$\frac{d^k}{dz^k}\left(\frac{1}{\zeta - z}\right) = \frac{k!}{(\zeta - z)^{k+1}} , \qquad k \in \mathbb{N}$$

as soon as one knows that this differentiation operation is permutable with the integration. In the above proof such a permutation was not used (but instead *summation* was permuted with *differentiation* and *integration*.)

Remark 2. The lemma on developability describes a simple process for passing from certain integrals to power series. What makes this process applicable in the above proof is the Cauchy integral formula.

Because power series are holomorphic, the lemma on developability (with $\gamma :=\partial B$) leads to a $\mathbb{C}$-vector space homomorphism $C(\partial B) \to \mathcal{O}(B)$, $f \mapsto F$. If the function f is not, in addition, holomorphic in B and continuous on $\overline{B}$, then in general for approach to points of ∂B, F does *not* realize the values of f as *boundary values*: For $B := \mathbb{E}$ and $f(\zeta) := \zeta^{-1}$ on $\partial\mathbb{E}$, for example, we have $F \equiv 0$ in $\mathbb{E}$ because for $z \in \mathbb{E} \setminus \{0\}$

$$F(z) = \frac{1}{2\pi i} \int_{\partial\mathbb{E}} \frac{d\zeta}{\zeta(\zeta - z)} = -\frac{1}{2\pi i z} \left[\int_{\partial\mathbb{E}} \frac{d\zeta}{\zeta} - \int_{\partial\mathbb{E}} \frac{d\zeta}{\zeta - z} \right] = 0.$$

This example also shows that the homomorphism $C(\partial B) \to \mathcal{O}(B)$ is *not injective*.

3. Historical remarks on the representation theorem. In Brook TAYLOR's *Methodus incrementorum directa et inversa*, Londini 1715, on pp. 21 ff. we find the first formulation and derivation of the theorem in the domain of the real numbers. An exhaustive analysis was given by A. PRINGSHEIM in "Zur Geschichte des Taylorschen Lehrsatzes," *Biblio. Math.* (3) 1(1900), 433-479; Cauchy's contributions are also gone into in detail there.

Cauchy immediately recognized that his integral formula implied, via development of its kernel $(\zeta - z)^{-1}$ into a geometric series, the representation theorem. He expressed it thus in 1841 (*Œuvres* (2) **12**, p.61, as well as Théorème I on p.64):

"La fonction $f(x)$ sera développable par la formule de Maclaurin en une série convergente ordonnée suivant les puissances ascendantes de x, si le module de la variable réelle ou imaginaire x conserve une valeur inférieure à celle pour laquelle la fonction (ou sa dérivée du premier ordre) cesse d'être finie et continue. (The function will be developable according to the formula of Maclaurin in a series of ascending powers of x, if the value of the modulus of this variable x, real or imaginary, is kept below that for which the function (or its derivative of the first order) ceases to be finite or continuous.)" In order to understand the last line of Cauchy's text we have to realize that the only singularities which were accepted in Cauchy's time were poles.

Moreover CAUCHY also gave a representation for the remainder term, following the model for functions on $\mathbb{R}$. And he described just *how well* the remainder term converges to 0. CAUCHY called his method the "calcul des limites". KRONECKER wrote this about the integral formula ([Kr], p.176): "in diese[r] hat man das Prius, in ih[r] liegt implicite schon die Reihenentwicklung, wie alle Eigenschaften der Functionen, wohl darum, weil in [ihrer] Geltung alle die höchst verwickelten Bedingungen, die für die Function $f(z)$ bestehen müssen, zusammengefaßt sind (in this we have what is absolutely primary; in it lies implicitly the power series development and indeed all the properties of functions – probably for the reason that in its validity the most complicated conditions bearing on the function are united)".

4. The Riemann continuation theorem. It was already emphasized in 2.1 that the sharpening of the Cauchy integral theorem there is really not a genuine sharpening; as a matter of fact, every function which is continuous in D and holomorphic everywhere in D with the possible exception of one point $c \in D$, is automatically holomorphic throughout D. This statement in turn is a special case of a general theorem about extending a holomorphic function over a *discrete and closed exceptional set*.

If A is a closed set lying in D and f is holomorphic in $D \setminus A$, then f is said to be *continuously*, respectively, *holomorphically extendable over A* if $f = F|(D \setminus A)$ for some function $F : D \to \mathbb{C}$ which is continuous, respectively, holomorphic in D. It is appropriate to introduce here the concept of a discrete set. If A is a subset of a metric space X, then a point $p \in A$ is called an *isolated point of A* if there is a neighborhood U of p such that $U \cap A = \{p\}$. The set A is called *discrete in X* if each point of A is an isolated point of A. The set A is discrete exactly when there is no cluster point of A in X which belongs to A.

The Riemann continuation theorem. *If A is discrete and closed in D, then the following assertions about a holomorphic function f in $D \setminus A$ are equivalent:*

i) *f is holomorphically extendable over A.*

ii) *f is continuously extendable over A.*

iii) *f is bounded in a neighborhood of each point of A.*

iv) *$\lim_{z \to c}(z - c)f(z) = 0$ for each point $c \in A$.*

Proof. We may assume that A consists of just one point, $c = 0$. The chain i) $\Rightarrow$ ii) $\Rightarrow$ iii) $\Rightarrow$ iv) is trivial. For proving iv) $\Rightarrow$ i) we introduce the functions

$$g(z) := zf(z) \quad \text{for } z \in D \setminus \{0\}\ , \ g(0) := 0 \quad \text{and } h(z) := zg(z).$$

By assumption g is *continuous at 0*. Therefore the identity $h(z) = h(0) + zg(z)$ shows that h is *complex-differentiable at the point 0*, with $h'(0) = g(0) = 0$. Of course $f \in \mathcal{O}(D \setminus \{0\})$ entails $h \in \mathcal{O}(D \setminus \{0\})$. In summary, h is holomorphic throughout D. Therefore according to the representation theorem 2, h admits a Taylor development $a_0 + a_1 z + a_2 z^2 + a_3 z^3 + \cdots$ around 0. Because $h(0) = h'(0) = 0$, it follows that $h(z) = z^2(a_2 + a_3 z + \cdots)$. Since $h(z) = z^2 f(z)$ for $z \neq 0$, we see that $F(z) := a_2 + a_3 z + \cdots$ is the desired holomorphic extension of f to D. $\qquad\square$

5. Historical remarks on the Riemann continuation theorem. In 1851 in his dissertation RIEMANN derived the implication ii) ⇒ i), more generally for "lines" of exceptional points ([R], Lehrsatz p. 23). There were lengthy discussions in the last century concerning correct and incorrect proofs of this implication. The American mathematician William Fogg OSGOOD (1864-1943, professor at Harvard and Peking; doctorate in 1890 from Erlangen under Max NOETHER; wrote his well-known textbook [Os] in 1906) reported on this in 1896 in an interesting article entitled "Some points in the elements of the theory of functions" (*Bull. Amer. Math. Soc.* **2**, 296-302). In 1905 E. LANDAU joined the discussion with the short note "On a familiar theorem of the theory of functions" (*Bull. Amer. Math. Soc.* **12**, 155-156; *Collected Works* **2**, pp. 204-5) in which he proved the implication iii) ⇒ i) by means of Cauchy's integral formula for the first derivative. This had already been done in 1841 by WEIERSTRASS (cf. [W₁], p.63); he also used in addition the theorem about the Laurent series development in circular regions, which he proved at that time, before LAURENT (cf. in this connection chapter 12.1.4). This work of Weierstrass was however not published until 1894.

By 1916 the situation was completely clear. At that time Friedrich Hermann SCHOTTKY (German mathematician, 1851-1935, professor at Marburg and from 1902 in Berlin), in a work still worth reading today entitled "Über das Cauchysche Integral" (cf. [Sch]), sketched the path from the (Riemann) definition of holomorphy to the continuation theorem that we currently follow. SCHOTTKY emphasized that essentially everything can be reduced to the *sharpened version* of *Goursat's integral lemma*.

Exercises

Exercise 1. Develop each of the following functions into power series about 0:

a) $f(z) = \exp(z + \pi i)$,

b) $f(z) = \sin^2 z$,

c) $f(z) = \cos(z^2 - 1)$,

d) $f(z) = \frac{2z+1}{(z^2+1)(z+1)^2}$.

Exercise 2. For $a, b \in \mathbb{C}$, $|a| < 1 < |b|$, and $m, n \in \mathbb{N}$ determine the value of

$$\int_{\partial B_1(0)} \frac{d\zeta}{(\zeta - a)^m (\zeta - b)^n}.$$

Exercise 3. Determine all entire functions f which satisfy the differential equation $f'' + f = 0$.

Exercise 4. Let f be holomorphic in $B_r(0)$, $r > 1$. Calculate the integrals $\int_{\partial \mathbb{E}} (2 \pm (\zeta + \zeta^{-1})) \frac{f(\zeta)}{\zeta} d\zeta$ in two different ways and thereby deduce that

$$\pi^{-1} \int_0^{2\pi} f(e^{it}) \cos^2(\tfrac{1}{2}t) dt = f(0) + \tfrac{1}{2} f'(0),$$

$$\pi^{-1} \int_0^{2\pi} f(e^{it}) \sin^2(\tfrac{1}{2}t) dt = f(0) - \tfrac{1}{2} f'(0).$$

Exercise 5. Let $r > 0$, $\overline{B_r}(0) \subset D$ open $\subset \mathbb{C}$ and $f, g \in \mathcal{O}(D)$. Suppose there is an $a \in \partial B_r(0)$ such that $g(a) = 0$, $g'(a) \neq 0$, $f(a) \neq 0$ and g is zero-free in $\overline{B_r}(0) \setminus \{a\}$. Let $\sum_{n \geq 0} a_n z^n$ be the power series development of f/g around 0. Show that $\lim_{n \to \infty} \frac{a_n}{a_{n+1}} = a$. *Hint.* Use the geometric series.

Exercise 6. Let D open $\subset \mathbb{C}$, $a \in D$, $f : D \setminus \{a\} \to \mathbb{C}$ holomorphic. Show that if f' has a holomorphic extension to D, then so does f.

§4   Discussion of the representation theorem

We will now draw some immediate conclusions from the power series representation theorem; among other things we will discuss the rearrangement and product theorems for power series. We will briefly go into the principle of analytic continuation and also show how to determine the radii of convergence of power series "directly".

1. Holomorphy and complex-differentiability of every order. From the representation theorem together with theorem 4.3.2 we immediately get

Every function which is holomorphic in D is arbitrarily often complex-differentiable in D.

This statement demonstrates especially clearly how strong a difference there is between real- and complex-differentiability: on the real line the derivative of a differentiable function in general need not even be continuous; for example, the function defined by $f(x) := x^2 \sin(1/x)$ for $x \in \mathbb{R} \setminus \{0\}$ and $f(0) := 0$ is differentiable on $\mathbb{R}$ but the derivative is discontinuous at the origin.

The representation theorem has no analog on the real line: there are *infinitely-often-differentiable* functions $f : \mathbb{R} \to \mathbb{R}$, which are not developable into a power series in *any* neighborhood of the origin. The standard example

$$f(x) := \exp(-x^{-2}) \qquad \text{for } x \neq 0 \text{ , } f(0) := 0,$$

can be found in CAUCHY's 1823 *Calcul Infinitésimal* (*Œuvres* (2) **4**, p. 230; pp. 7,8 of G. BIRKHOFF, *A Source Book in Classical Analysis*, Harvard University Press (1973) for an English translation). For this function $f^{(n)}(0) = 0$ for all $n \in \mathbb{N}$. On the real line it is possible to prescribe arbitrarily the values of all the derivatives at one point. The French mathematician Émile BOREL (1871-1956) proved this in his 1895 thesis; he showed (*Ann. Scient. École Norm. Sup.* (3) **12**, p. 44; also *Œuvres* **1**, p. 274) that

For every sequence $(r_n)_{n \geq 0}$ of real numbers there is an infinitely-often-differentiable function $f : \mathbb{R} \to \mathbb{R}$ having $f^{(n)}(0) = r_n$ for every n.

We will prove this theorem and more in 9.5.5.

The representation theorem makes possible the "two-line proof" of the lemma on units promised in 4.4.2: If $e = 1 - b_1 z - b_2 z^2 - \cdots$ is a convergent power series, then e is holomorphic near 0. Because $e(0) \neq 0$, so is $1/e$; consequently $1/e$ is also given by a convergent power series.

2. The rearrangement theorem. *If $f(z) = \sum a_\nu (z - c)^\nu$ is a power series which converges in $B_R(c)$, then f is developable into a power series $\sum b_\nu (z - z_1)^\nu$ about each point $z_1 \in B_R(c)$; the radius of convergence of this new power series is at least $R - |z_1 - c|$, and its coefficients are given by*

$$b_\nu = \sum_{j=\nu}^{\infty} \binom{j}{\nu} a_j (z_1 - c)^{j-\nu} , \qquad \nu \in \mathbb{N}.$$

Proof. In the disc about z_1 of radius $R - |z_1 - c|$ the representation theorem says that $f(z) = \sum \frac{f^{(\nu)}(z_1)}{\nu!} (z - z_1)^\nu$, while by theorem 4.3.2 $\frac{f^{(\nu)}(z_1)}{\nu!} = \sum_{j=\nu}^{\infty} \binom{j}{\nu} a_j (z_1 - c)^{j-\nu}$. □

The name "rearrangement theorem" is based on the following: in the situation described in the theorem, the equation

$$f(z) = \sum_{j=0}^{\infty} a_j [(z_1 - c) + (z - z_1)]^j = \sum_{j=0}^{\infty} \left[\sum_{\nu=0}^{j} a_j \binom{j}{\nu} (z_1 - c)^{j-\nu} (z - z_1)^\nu \right]$$

holds. If we uncritically rearrange the double sum on the right as though only a finite number of summands were involved, we get the double sum

$$\sum_{\nu=0}^{\infty}\left[\sum_{j=\nu}^{\infty}\binom{j}{\nu}a_j(z_1-c)^{j-\nu}\right](z-z_1)^{\nu}$$

which is precisely the development of f around z_1 which the theorem affirms. The formal rearrangement process can be rigorously justified without recourse to the Cauchy-Taylor theorem. To see this, recall that an absolutely convergent series can be divided up into infinitely many infinite subseries, the sum of which coincides with that of the original series. The proof thus carried out remains wholly within the Weierstrassian framework and is valid as well when the coefficients come from any complete valued field of characteristic 0.

3. Analytic continuation. If f is holomorphic in the region G and we develop f into a power series around $c \in G$ according to theorem 3.2, then the radius of convergence R of this series is not less than the distance $d_c(G)$ of c from the boundary of G. It can actually be greater (cf. the left-hand figure). In this case we say that f is "analytically continued" beyond G (but it would be more accurate to speak of a "holomorphic" continuation). For example, the geometric series $\sum z^{\nu} \in \mathcal{O}(\mathbb{E})$ has the Taylor series $\sum (z-c)^{\nu}/(1-c)^{\nu+1}$ with radius of convergence $|1-c|$ around the point $c \in \mathbb{E}$; in case $|1-c| > 1$ we have an analytic continuation. (In this example the function $(1-z)^{-1} \in \mathcal{O}(\mathbb{C} \setminus \{1\})$ is naturally the largest possible analytic continuation.)

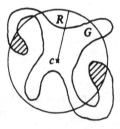
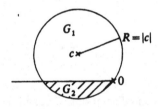

The principle of analytic continuation plays a significant role in (Weierstrassian) function theory. At this point we can't go into analytic continuation any deeper; we wish, nevertheless, to at least bring to the reader's attention the problem of multi-valuedness. This can occur if $B_R(c) \cap G$ is disconnected; in such situations the Taylor series around c doesn't always represent the original function f in the connected components of $B_R(c) \cap G$ which do not contain c. (The latter are the shaded regions of the left-hand figure above.) The holomorphic logarithm $\log z$ in the slit plane $\mathbb{C}^-$ serves to illustrate this phenomenon. Around each $c \in \mathbb{C}^-$ this function has the Taylor series

$$\log c + \sum_1^\infty \frac{(-1)^{\nu-1}}{\nu} \frac{1}{c^\nu} (z-c)^\nu,$$

whose radius of convergence is $|c|$. In case $\Re c < 0$, $\mathbb{C}^- \cap B_{|c|}(c)$ consists of two connected components G_1, G_2 (see the right-hand figure above). In G_1 the series represents the principal branch of log, but not in G_2. To see this, note that the negative real axis separates G_1 from G_2 and the principal branch "jumps by $\pm 2\pi i$" in passing over this axis, whereas the series is continuous at each point of this boundary line.

4. The product theorem for power series. *Let* $f(z) = \sum a_\mu z^\mu$ *and* $g(z) = \sum b_\nu z^\nu$ *be convergent in the respective discs* B_s, B_t. *Then for* $r := \min\{s,t\}$ *the product function* $f \cdot g$ *has the power series representation*

$$(f \cdot g)(z) = \sum p_\lambda z^\lambda \quad \text{with } p_\lambda := \sum_{\mu+\nu=\lambda} a_\mu b_\nu \quad \text{(Cauchy product, cf. 0.4.6)}$$

in the disc B_r.

Proof. As a product of holomorphic functions, $f \cdot g$ is holomorphic in B_r and so according to the representation theorem it is developable in this disc into the Taylor series

$$\sum \frac{(f \cdot g)^{(\lambda)}(0)}{\lambda!} z^\lambda.$$

Since $f^{(\mu)}(0) = \mu! a_\mu$ and $g^{(\nu)}(0) = \nu! b_\nu$, the Leibniz rule

$$(f \cdot g)^{(\lambda)}(0) = \sum_{\mu+\nu=\lambda} \frac{\lambda!}{\mu!\nu!} f^{(\mu)}(0) g^{(\nu)}(0)$$

for higher derivatives shows that this series is the one described in the conclusion of the theorem.

With the help of Abel's limit theorem there follows immediately the

Series multiplication theorem of ABEL. *If the series* $\sum_0^\infty a_\mu$, $\sum_0^\infty b_\nu$ *and* $\sum_0^\infty p_\lambda$, *with* $p_\lambda := a_0 b_\lambda + \cdots + a_\lambda b_0$ *for every* λ, *converge to the respective sums* a, b *and* p, *then* $ab = p$.

Proof. The series $f(z) := \sum a_\mu z^\mu$, $g(z) := \sum b_\nu z^\nu$ converge in the unit disc $\mathbb{E}$; therefore we also have $(f \cdot g)(z) = \sum p_\lambda z^\lambda$ in $\mathbb{E}$. The convergence of the three series $\sum a_\mu$, $\sum b_\nu$, and $\sum p_\lambda$ implies (cf. 4.2.5):

$$\lim_{x \to 1-0} f(x) = a, \quad \lim_{x \to 1-0} g(x) = b, \quad \lim_{x \to 1-0} (f \cdot g)(x) = p.$$

Since $\lim_{x \to 1-0}(f \cdot g)(x) = [\lim_{x \to 1-0} f(x)][\lim_{x \to 1-0} g(x)]$, the claim follows. $\square$

The series multiplication theorem of ABEL is to be found in [A], p. 318. The reader should compare this theorem with the product theorem of CAUCHY (0.4.6). The series multiplication theorem of ABEL is susceptible to the following direct proof (cf. CÉSARO, *Bull. Sci. Math.* (2) **14**(1890), p. 114): We set $s_n := a_0 + \cdots + a_n$, $t_n := b_0 + \cdots + b_n$, $q_n := p_0 + \cdots + p_n$ and verify that $q_n = a_0 t_n + a_1 t_{n-1} + \cdots + a_n t_0$ and that further

$$q_0 + q_1 + \cdots + q_n = s_0 t_n + s_1 t_{n-1} + \cdots + s_n t_0.$$

On the basis of Exercise 0.3.3 and the preceding equation, it follows that

$$p = \lim \frac{s_0 t_n + s_1 t_{n-1} + \cdots + s_n t_0}{n+1} = ab.$$

5. Determination of radii of convergence.

The radius of convergence R of the Taylor series $\sum a_\nu (z - c)^\nu$ is determined by the coefficients (via the Cauchy-Hadamard formula 4.1.3 or the ratio criterion 4.1.4). The representation theorem frequently allows the number R to be read off at a glance from the properties of the holomorphic function so represented, without knowledge of the coefficients. Thus, e.g.,

Let f and g be holomorphic in $\mathbb{C}$ and have no common zero in $\mathbb{C}^\times$. Let $c \in \mathbb{C}^\times$ be a "smallest" non-zero zero of g (that is, $|w| \geq |c|$ for every zero $w \neq 0$ of g). If the function f/g, which is then holomorphic in $B_{|c|}(0) \setminus \{0\}$, is holomorphically continuable over the point 0, then its Taylor series around 0 has radius of convergence $|c|$.

Proof. This is clear from the representation theorem 3.1, since $f(c) \neq 0$ means that f/g tends to ∞ as c is approached.

Examples. 1) The Taylor series around 0 of $\tan z = \sin z / \cos z$ has radius of convergence $\frac{1}{2}\pi$, because $\frac{1}{2}\pi$ is a "smallest" zero of $\cos z$.

2) The functions $z \cot z = z \cos z / \sin z$, $z / \sin z$ and $z/(e^z - 1)$ are holomorphically continuable over zero (each receiving value 1 there, since the power series around 0 of each denominator begins with the term z). Since π, respectively, $2\pi i$, is a "smallest" zero of $\sin z$, respectively, $e^z - 1$, it follows that π is the radius of convergence of $z \cot z$ and of $z/\sin z$ around 0, whereas the Taylor series of $z/(e^z - 1)$ around 0 has radius of convergence 2π. $\square$

The determination of the radii of convergence of these real series by means of the Cauchy-Hadamard formula or the ratio criterion is rather tedious (cf. 11.3.1). The elegant route through the complexes is especially impressive for the function $z/(e^z - 1)$, because its denominator has no real zeros except 0. Here we have a beautiful example illustrating GAUSS's prophetic words (cf. the Historical Introduction), according to which complete knowledge of the nature of an analytic function in $\mathbb{C}$ is often indispensable for a correct assessment of its behavior on $\mathbb{R}$.

The technique described here for determining radii of convergence can be converted into a *method of approximating* zeros. If, say, g is a polynomial with only real zeros, all non-zero, we can develop $1/g$ into its Taylor series $\sum a_\nu z^\nu$ around 0 and consider the sequence $a_\nu/a_{\nu+1}$: if it has a finite limit r, then either r or $-r$ is the smallest zero of g. This technique was developed in 1732 and 1738 by Daniel BERNOULLI (1700-1782) and is extensively discussed by EULER in §335 ff. of [E]. Concerning BERNOULLI's original works, with commentary by L. P. BOUCKAERT, see *Die Werke von Daniel Bernoulli*, vol. 2, Birkhäuser Verlag, Basel-Boston-Stuttgart, 1982. The BERNOULLI-EULER method may be generalized to polynomials with complex zeros. For this see, e.g., Problem 243 in Part 3 of the first volume of G. PÓLYA and G. SZEGÖ.

Exercises

Exercise 1. Develop f into a power series about 0 and determine its radius of convergence:

a) $f(z) = \dfrac{e^z}{1 - tz}$, $t \in \mathbb{C}$,

b) $f(z) = \dfrac{\sin^2 z}{z}$.

Exercise 2. Determine all entire functions f which satisfy $f(z^2) = (f(z))^2$ for all $z \in \mathbb{C}$.

Exercise 3. Let $R > 0$, $\overline{B_R(0)} \subset D$ open $\subset \mathbb{C}$, $c \in \partial B_R(0)$ and $f \in \mathcal{O}(D \setminus \{c\})$. Let $\sum_{k \geq 0} a_k z^k$ be the power series development of f around 0. Show that if $f(z)(z - c)$ is bounded near c, then for all sufficiently small $r > 0$ and all $k \in \mathbb{N}$

$$\int_{\partial B_r(c)} \frac{f(\zeta)d\zeta}{(\zeta - c)^k} = 2\pi i \sum_{\nu=k}^{\infty} \binom{\nu}{k} (a_{\nu-1} - ca_\nu)c^{\nu-k},$$

where $a_{-1} := 0$.

§5*   Special Taylor series. Bernoulli numbers

The power series developments given in 4.2.1 for $\exp z$, $\cos z$, $\sin z$, $\log(1 + z)$, etc. are the Taylor series of these functions around the origin. At the center of this section will be the Taylor series (around 0) of the holomorphic (near 0) function

$$g(z) := \frac{z}{e^z - 1} \qquad \text{for } z \neq 0 \,,\, g(0) := 1,$$

which plays a significant role in classical analysis. Because of 5.2.5 this function is linked with the cotangent and tangent functions through the equations

(1) $$\cot z = i + z^{-1} g(2iz),$$

(2) $$\tan z = \cot z - 2 \cot 2z.$$

Consequently from the Taylor series of $g(z)$ around 0 we can get those of $z \cot z$ and $\tan z$.

The Taylor coefficients of the power series of $g(z)$ around 0 are essentially the so-called *Bernoulli numbers*, which turn up in many analytic and number-theoretic problems. We will encounter them again in 11.2.4. It should be emphasized that the considerations of this section are quite elementary. The representation theorem is not even needed, because for our purposes knowledge of the exact radii of convergence of the series concerned is not particularly relevant.

1.   The Taylor series of $z(e^z - 1)^{-1}$. Bernoulli numbers. For historical reasons we write the Taylor series of $g(z) = z(e^z - 1)^{-1}$ around 0 in the form

$$\frac{z}{e^z - 1} = \sum_0^\infty \frac{B_\nu}{\nu!} z^\nu \,, \qquad B_\nu \in \mathbb{C}.$$

Since $\cot z$ is an odd function, $z \cot z$ is an even function. Since $g(z) + \frac{1}{2}z = \frac{1}{2i} z \cot(\frac{1}{2i} z)$, from equation (1) above, $g(z) + \frac{1}{2}z$ is an *even* function. Consequently,

$$B_1 = -\frac{1}{2} \quad \text{and} \quad B_{2\nu+1} = 0 \qquad \text{for all } \nu \geq 1.$$

Thus

(1) $$\frac{z}{e^z - 1} = 1 - \frac{z}{2} + \sum_1^\infty \frac{B_{2\nu}}{(2\nu)!} z^{2\nu}.$$

From $1 = \frac{e^z-1}{z} \cdot \frac{z}{e^z-1} = \left(\sum_1^\infty \frac{z^{\nu-1}}{\nu!}\right)\left(\sum_0^\infty \frac{B_\nu}{\nu!}z^\nu\right)$ we get, by multiplying everything out and comparing coefficients (uniqueness of the Taylor series (cf. 4.3.2) justifies this) we get the formula

$$\binom{n}{0}B_0 + \binom{n}{1}B_1 + \binom{n}{2}B_2 + \cdots + \binom{n}{n-1}B_{n-1} = 0.$$

The $B_{2\nu}$ are called the *Bernoulli numbers*[2]; they can be *recursively* determined from this last equation.

Every Bernoulli number $B_{2\nu}$ is rational, with

$$B_0 = 1 \ , \ B_2 = \frac{1}{6} \ , \ B_4 = -\frac{1}{30} \ , \ B_6 = \frac{1}{42}$$

(2)

$$B_8 = -\frac{1}{30} \ , \ B_{10} = \frac{5}{66} \ , \ B_{12} = -\frac{691}{2730} \ , \ B_{14} = \frac{7}{6}.$$

Since the radius of convergence of the series in (1) is finite (and in fact equal to 2π), we further see that

The sequence $B_{2\nu}$ of Bernoulli numbers is unbounded.

The explicit values of the first few Bernoulli numbers thus conduce to a false impression of the behavior of the other terms; thus $B_{26} = 8553103/6$ and B_{122} has a 107 decimal-place numerator but likewise the denominator 6.

2. The Taylor series of $z\cot z$, $\tan z$ and $\dfrac{z}{\sin z}$. From equations (1) and (2) of the introduction and from identity 1.(1) we immediately get

(1)
$$\cot z = \frac{1}{z} + \sum_1^\infty (-1)^\nu \frac{4^\nu}{(2\nu)!}B_{2\nu}z^{2\nu-1},$$

(2)
$$\tan z = \sum_1^\infty (-1)^{\nu-1}\frac{4^\nu(4^\nu-1)}{(2\nu)!}B_{2\nu}z^{2\nu-1}$$
$$= z + \frac{1}{3}z^3 + \frac{2}{15}z^5 + \frac{17}{315}z^7 + \cdots$$

Equation (1) is valid in a punctured disc $B_R(0) \setminus \{0\}$ (it isn't necessary to know that according to 4.5 $R = \pi$). Later we will see that $(-1)^{\nu-1}B_{2\nu}$ is always positive (cf. 11.2.4), and therefore all the series coefficients in (1) are negative and in (2) all are positive.

The series (1) and (2) originate with EULER, and will be found, e.g., in chapters 9 and 10 of his *Introductio* [E]. Equation (1) can also be put in the graceful form

---

[2]The enumeration of these numbers is not uniform throughout the literature. Frequently the vanishing ones $B_{2\nu+1}$ are not designated at all and instead of $B_{2\nu}$ the notation $(-1)^{\nu-1}B_\nu$ is used.

$$(1') \qquad \frac{1}{2}z \cot \frac{1}{2}z = 1 - B_2\frac{z^2}{2!} + B_4\frac{z^4}{4!} - B_6\frac{z^6}{6!} + - \cdots$$

Because $\cot z + \tan \frac{1}{2}z = \frac{1}{\sin z}$ (cf. 5.2.5), it further follows from (1) and (2) that

$$(3) \qquad \frac{z}{\sin z} = \sum_0^\infty (-1)^{\nu-1}\frac{(4^\nu - 2)}{(2\nu)!}B_{2\nu}z^{2\nu}.$$

3. Sums of powers and Bernoulli numbers. *For all* $n, k \in \mathbb{N} \setminus \{0\}$

$$1^k + 2^k + \cdots + n^k = \frac{1}{k+1}n^{k+1} + \frac{1}{2}n^k + \sum_2^k \frac{B_\mu}{\mu}\binom{k}{\mu-1}n^{k+1-\mu}.$$

Proof. Writing $S(n^k) := \sum_1^n \nu^k$ and $S(n^0) := n+1$, it follows that

$$E_n(w) := 1 + e^w + \cdots + e^{nw} = \sum_0^\infty \frac{1}{k!}S(n^k)w^k.$$

On the other hand we have

$$E_{n-1}(w) = \frac{w}{e^w - 1} \cdot \frac{e^{nw} - 1}{w} = \left(\sum_{\mu=0}^\infty \frac{B_\mu}{\mu!}w^\mu\right)\left(\sum_{\lambda=0}^\infty \frac{n^{\lambda+1}}{(\lambda+1)!}w^\lambda\right).$$

And therefore from the product theorem 4.4 and the facts that $B_0 = 1$, $B_1 = -\frac{1}{2}$ and $\frac{k!}{\mu!(k+1-\mu)!} = \frac{1}{\mu}\binom{k}{\mu-1}$ we get for $S((n-1)^k)$ the expression

$$\sum_{\mu+\lambda=k} \frac{k!}{\mu!(\lambda+1)!}B_\mu n^{\lambda+1} = \frac{1}{k+1}n^{k+1} - \frac{1}{2}n^k + \sum_{\mu=2}^k \frac{B_\mu}{\mu}\binom{k}{\mu-1}n^{k+1-\mu}.$$

Since $S(n^k) = S((n-1)^k) + n^k$, this is the claim we were trying to prove.   □

Remark. Jakob BERNOULLI (1665-1705) found the numbers now named after him while computing the sums of powers of successive integers. In his *Ars Conjectandi*, published posthumously in 1713, he wrote A, B, C, D for B_2, B_4, B_6, B_8 and gave the sums $\sum_1^n \nu^k$ explicitly for $1 \leq k \leq 10$ but offered no general proof. (Cf. *Die Werke von Jakob Bernoulli*, vol. 3, Birkhäuser, Basel (1975), pp. 166/167; see also W. WALTER: *Analysis I*, Grundwissen Mathematik, vol. 3, Springer-Verlag, Berlin, (1985).)

If we introduce the *rational* $(k+1)$th degree polynomial $\Phi_k(w) := \frac{1}{k+1}(w - 1)^{k+1} + \frac{1}{2}(w-1)^k + \sum_{\mu=2}^k \frac{B_\mu}{\mu}\binom{k}{\mu-1}(w-1)^{k+1-\mu} \in \mathbb{Q}[w]$, $k = 1, 2, \cdots$, then Bernoulli's theorem says

$$(*) \qquad 1^k + 2^k + \cdots + (n-1)^k = \Phi_k(n) , \ n = 1, 2, \cdots$$

One checks that $\Phi_k(w) = \frac{1}{k+1}w^{k+1} - \frac{1}{2}w^k +$ lower degree terms. E.g.,

$$\Phi_1(w) = \frac{1}{2}(w^2 - w) \,,\; \Phi_2(w) = \frac{1}{6}(2w^3 - 3w^2 + w) \,,\; \Phi_3(w) = \frac{1}{4}(w^4 - 2w^3 + w^2).$$

We have further

$$\Phi_4(w) = \frac{1}{30}(6w^5 - 15w^4 + 10w^3 - w) = \frac{1}{10}(w-1)w(2w-1)(w^2 - w - \frac{1}{3})$$

and so from $(*)$

$$1^4 + 2^4 + \cdots + (n-1)^4 = \frac{1}{10}(n-1)n(2n-1)(n^2 - n - \frac{1}{3}) \,,\; n = 2, 3, \cdots$$

which readers who enjoy calculating can also confirm by induction on n. The equations $(*)$ actually characterize the sequence of polynomials $\Phi_1, \Phi_2, \cdots$. Namely, if $\Phi(w)$ is any polynomial (over the complex field!) such $(*)$ holds for a fixed $k \geq 1$ and all $n \geq 1$, then the polynomial $\Phi(w) - \Phi_k(w)$ vanishes for all $w \in \mathbb{N}$ and must accordingly be the 0 polynomial, that is, $\Phi = \Phi_k$.

4. Bernoulli polynomials. For every complex number w the function $ze^{wz}/$ $(e^z - 1)$ is holomorphic in $\mathbb{C} \setminus \{\pm 2\nu\pi i : \nu = 1, 2, \cdots\}$. According to the representation theorem we have, for each $w \in \mathbb{C}$, a Taylor series development around 0

$$(1) \qquad\qquad F(w, z) := \frac{ze^{wz}}{e^z - 1} = \sum \frac{B_k(w)}{k!} z^k$$

for appropriate complex numbers $B_k(w)$. (The series representation is actually valid in the disc of radius 2π about 0.) The functions $B_k(w)$ admit explicit formulas:

Theorem. $B_k(w)$ *is a monic rational polynomial in w of degree k.*

$$(2) \qquad B_k(w) = \sum_{\nu=0}^{k} \binom{k}{\nu} B_\nu w^{k-\nu} = w^k - \frac{1}{2}kw^{k-1} + \cdots + B_k \,, \qquad k \in \mathbb{N}.$$

In particular, $B_k(0)$ is the kth Bernoulli number.

Proof. Since $F(w, z) = e^{wz} \cdot \frac{z}{e^z - 1}$ and $\frac{z}{e^z - 1} = \sum \frac{B_\nu}{\nu!} z^\nu$, (1) gives

$$\sum \frac{B_k(w)}{k!} z^k = \left(\sum \frac{1}{\mu!} w^\mu z^\mu \right) \left(\sum \frac{B_\nu}{\nu!} z^\nu \right)$$

for all z near 0. From this and the product theorem 4.4 it follows that

$$B_k(w) = \sum_{\mu+\nu=k} \frac{k!}{\mu!\nu!} B_\mu w^\mu = \sum_{\nu=0}^{k} \binom{k}{\nu} B_\nu w^{k-\nu}.$$

We call $B_k(w)$ the kth *Bernoulli polynomial*. We note three interesting formulas:

(3) $B'_k(w) = kB_{k-1}(w),$ $k \geq 1$ (*derivative formula*)

(4) $B_k(w+1) - B_k(w) = kw^{k-1},\ k \geq 1$ (*difference equation*)

(5) $B_k(1-w) = (-1)^k B_k(w),$ $k \geq 1$ (*complementarity formula*).

Equation (2) yields direct proofs if one is willing to calculate a bit. But it is more elegant to consider $F(w,z)$ as a holomorphic function of w and exploit the three obvious identities

$$\frac{\partial}{\partial w} F(w,z) = zF(w,z)\ ,\ F(w+1,z) - F(w,z) = ze^{wz}\ ,\ F(1-w,z) = F(w,-z).$$

The corresponding power series can be obtained from (1) by differentiation with respect to w, since that series converges normally in the variable $w \in \mathbb{C}$. Then (3) – (5) follow directly by comparing coefficients on the two sides of each of the three identities. – From (4) we get immediately

$$1^k + 2^k + \cdots + n^k = \frac{1}{k+1}[B_{k+1}(n+1) - B_{k+1}(1)].$$

A simple connection exists between the polynomials $\Phi_k(w)$ introduced in 3. and the Bernoulli polynomials. Since

$$\frac{\partial}{\partial w}\left(\frac{e^{wz} - e^z}{e^z - 1}\right) = \frac{ze^{wz}}{e^z - 1}\quad \text{and}\quad \frac{e^{wz} - e^z}{e^z - 1} = \frac{1}{z}[F(w,z) - F(1,z)],$$

it follows upon engaging exercise 2 that

(6) $B_k(w) = \Phi'_k(w)\ ,$ $k\Phi_{k-1}(w) = B_k(w) - B_k(1)\ ,$ $k = 1, 2, \cdots .$

One sees in particular that $\Phi_k(1) = 0$ for all $k \in \mathbb{N}$.

The first four Bernoulli polynomials are:

$$B_0(w) = 1\ ,\ B_1(w) = w - \frac{1}{2}\ ,\ B_2(w) = w^2 - w + \frac{1}{6}\ ,$$

$$B_3(w) = w^3 - \frac{3}{2}w^2 + \frac{1}{2}w\ ,\ B_4(w) = w^4 - 2w^3 + w^2 - \frac{1}{30}.$$

Exercises

Exercise 1. Derive the Taylor series of $\tan^2 z$ around 0 from 2.(2) by differentiation.

Exercise 2. Setting $\Phi_0(w) := w - 1$, show that

$$\frac{e^{wz} - e^z}{e^z - 1} = \sum_{\nu \geq 0} \frac{\Phi_\nu(w)}{\nu!} z^\nu.$$

Exercise 3. Show that for $k, n \in \mathbb{N}$

$$\sum_{j=1}^n j^k = \frac{1}{k+1} \sum_{\nu=0}^k \binom{k+1}{\nu}(n+1)^{k+1-\nu} B_\nu.$$

Chapter 8

Fundamental Theorems about Holomorphic Functions

Having led to the Cauchy integral formula and the Cauchy-Taylor representation theorem, the theory of integration in the complex plane will temporarily pass off of center-stage. The power of the two mentioned results has already become clear but this chapter will offer further convincing examples of this power. First off, in section 1 we prove and discuss the Identity Theorem, which makes a statement about the "cohesion among the values taken on by a holomorphic function." In the second section we illuminate the holomorphy concept from a variety of angles. In the third, the Cauchy estimates are discussed. As applications of them we get, among other things, LIOUVILLE's theorem and, in section 4, the convergence theorems of WEIERSTRASS. The Open Mapping Theorem and the Maximum Principle are proved in section 5.

§1 The Identity Theorem

A holomorphic function is *locally* represented by its Taylor series. An identity theorem is already contained in this observation; namely:

If f and g are holomorphic in D and there is a point c ∈ D together with a (possibly quite small) neighborhood U ⊂ D of c, such that f|U = g|U, then in fact f|B_d(c) = g|B_d(c), where d := d_c(D) is the distance from c to the boundary of D.

This is clear because f and g are represented throughout $B_d(c)$ by their

227

Taylor series around c and the coefficients of the one series coincide with those of the other due to $f|U = g|U$.

An identity theorem of a different kind is an immediate consequence of the integral formula:

If f and g are holomorphic in a neighborhood of the closed disc $\overline{B}$ and if $f|\partial B = g|\partial B$, then in fact $f|\overline{B} = g|\overline{B}$.

The identity theorem which we are about to become acquainted with contains these two as special cases. As an application of it we will show (in section 5) among other things that every power series must have a singularity at some point on its circle of convergence.

1. The Identity Theorem. *The following statements about a pair f, g of holomorphic functions in a region $G \subset \mathbb{C}$ are equivalent:*

  i) $f = g$.

  ii) *The coincidence set $\{w \in G : f(w) = g(w)\}$ has a cluster point in G.*

  iii) *There is a point $c \in G$ such that $f^{(n)}(c) = g^{(n)}(c)$ for all $n \in \mathbb{N}$.*

Proof. i) $\Rightarrow$ ii) is trivial.

  ii) $\Rightarrow$ iii) We set $h := f - g \in \mathcal{O}(G)$. Then the hypothesis says that the zero-set $M := \{w \in G : h(w) = 0\}$ of h has a cluster point $c \in G$. If there is an $m \in \mathbb{N}$ with $h^{(m)}(c) \neq 0$, then we consider the smallest such m. For it we have the factorization

$$h(z) = (z - c)^m h_m(z) \quad \text{with} \quad h_m(z) := \sum_{\mu \geq m} \frac{h^{(\mu)}(c)}{\mu!}(z - c)^{\mu - m} \in \mathcal{O}(B),$$

holding for every open disc $B \subset G$ centered at c, and $h_m(c) \neq 0$. This all follows from the representation theorem 7.3.2. Because of its continuity h_m is then zero-free in some neighborhood $U \subset B$ of c. It follows from the factorization above that $M \cap (U \setminus \{c\}) = \emptyset$; that is, c is not a cluster point of M after all. This contradiction shows that there is no such m, that is, $h^{(n)}(c) = 0$ for all $n \in \mathbb{N}$; i.e., $f^{(n)}(c) = g^{(n)}(c)$ for all $n \in \mathbb{N}$.

  iii) $\Rightarrow$ i) Again we set $h := f - g$. Each set $S_k := \{w \in G : h^{(k)}(w) = 0\}$ is (relatively) closed in G, on account of the continuity of $h^{(k)} \in \mathcal{O}(G)$. Therefore the intersection $S := \bigcap_0^\infty S_k$ is also (relatively) *closed* in G. However, this set is also *open in G*, because if $z_1 \in S$, then the Taylor series of h around z_1 is the zero series in any open disc B centered at z_1 which lies in G. This implies that $h^{(k)}|B = 0$ for every $k \in \mathbb{N}$, entailing that $B \subset S$. Since G is connected and S is not empty ($c \in S$ by hypothesis), it follows from 0.6.1 that $S = G$. That is, $f = g$. $\square$

The connectedness of G, used in the proof that iii) $\Rightarrow$ i), is essential to the validity of that implication. For example, if D is the union of two disjoint open discs B and B' and if we set

$$f := 0 \text{ in } D ; \qquad g := 0 \text{ in } B \qquad \text{and} \qquad g := 1 \text{ in } B',$$

then f and g are holomorphic in D and have properties ii) and iii), yet $f \neq g$ in D. On the other hand the equivalence of (ii) with (iii) is valid in any domain.

The conditions ii) and iii) of the theorem are fundamentally different in nature. The latter demands equality of *all* derivatives at a *single* point, whereas in the former no derivatives appear. Instead, we demand equality of the function values themselves at *sufficiently many* points.

The reader should prove the following variant of the implication iii) $\Rightarrow$ i) in the Identity Theorem:

If f and g are holomorphic in the region G and at some point of G all but finitely many of the derivatives of f coincide with the corresponding ones of g, then there is a polynomial $p \in \mathbb{C}[z]$ such that $f = g + p$ throughout G.

The Identity Theorem implies that a function f which is holomorphic in a region G is *completely determined* by its values on "very sparse" subsets of G, for example on very short lengths of curves W. Properties of f which are expressible as analytic identities therefore need only be verified on W. They then automatically "propagate themselves analytically from W to the whole of G." We can illustrate this *permanence principle* with the example of the power rule $e^{w+z} = e^w e^z$. If this identity is known for all real values of the arguments, then it follows for all complex values as well! First, for each real number $w := u$ the holomorphic functions of z expressed by e^{u+z} and $e^u e^z$ coincide for all $z \in \mathbb{R}$, hence for all $z \in \mathbb{C}$. Consequently, for each fixed $z \in \mathbb{C}$ the holomorphic functions of w expressed by e^{w+z} and $e^w e^z$ coincide for all $w \in \mathbb{R}$, and therefore coincide for all $w \in \mathbb{C}$.   □

We are also now in a position to see that the definition of the functions exp, cos and sin via their real power series represents the *only* possible way to extend these functions from the real line to holomorphic functions in the complex plane. In general

If $I := \{x \in \mathbb{R} : a < x < b\}$ is a real interval, $f : I \to \mathbb{R}$ a function defined on I and G is a region in $\mathbb{C}$ which contains I, then there is at most one holomorphic function $F : G \to \mathbb{C}$ which satisfies $F|I = f$.   □

An important consequence of the Identity Theorem is a characterization of connectedness in terms of the absence of zero-divisors.

The following assertions about a non-empty domain D in $\mathbb{C}$ are equivalent:

i) D is connected (and is thus a region).

ii) The algebra $\mathcal{O}(D)$ is an integral domain (that is, it has no non-zero zero-divisors).

Proof. i) $\Rightarrow$ ii) Suppose $f, g \in \mathcal{O}(D)$, f is not the function $0 \in \mathcal{O}(D)$ but $f \cdot g$ is, that is, $f(z)g(z) = 0$ for all $z \in D$. There is some $c \in D$ where $f(c) \neq 0$ and then on continuity grounds a neighborhood $U \subset D$ of c throughout which f is zero-free. Then $g(U) = 0$ and since D is connected this entails $g(D) = 0$ by the Identity Theorem. That is, $g = 0$, the zero element of the algebra $\mathcal{O}(D)$.

ii) $\Rightarrow$ i) If D were not connected, it would be expressible as the disjoint union of two non-empty domains D_1, D_2. The functions f, g defined in D by

$$f(z) := \begin{cases} 0 & for \quad z \in D_1, \\ 1 & for \quad z \in D_2, \end{cases}$$

$$g(z) := \begin{cases} 1 & for \quad z \in D_1, \\ 0 & for \quad z \in D_2, \end{cases}$$

are then holomorphic in D, neither is the zero function in $\mathcal{O}(D)$ and yet $f \cdot g$ is the zero function. This contradicts the hypothesis ii) that $\mathcal{O}(D)$ is free of non-zero zero-divisors. $\qquad\square$

The reader should compare the statement proved above with theorem 0.6.1.

In real analysis an important role is played by the *functions with compact support*, by which is meant real functions which are infinitely often real-differentiable and whose supports are compact, the support of a function being the *closure* of the complement of its zero-set. Such functions are used in a well-known construction of *infinitely differentiable partitions of unity*. But there is no comparable holomorphic partition of unity, because the support of a holomorphic function in a region G is either void or – by virtue of the Identity Theorem – all of G.

2. On the history of the Identity Theorem. On page 286 of the
second volume of Crelle's Journal, which appeared in 1827, we find

Aufgaben und Lehrsätze,
erstere   aufzulösen,   letztere   zu   beweisen.

1.
(Von Herrn *N. H. Abel.*)

49. $\mathbf{T}$heorème. Si la somme de la série infinie
$$a_0 + a_1 x + a_2 x^2 + a_3 x^3 + \ldots + a_m x^m + \ldots$$
est égale à zéro pour toutes les valeurs de x entre deux limites réelles
α et β; on aura nécessairement
$$a_0 = 0, \quad a_1 = 0, \quad a_2 = 0 \ldots a_m = 0 \ldots$$
en vertu de ce que la somme de la série s'évanouira pour une valeur
quelconque de x.

This is an embryonic form of the Identity Theorem. Only in the case $\alpha \leq$
$0 < \beta$ is the claim immediately evident, the series in that case representing
the zero function around 0, so that all of its Taylor coefficients consequently
must vanish. That case had already been treated in 1748 by EULER ([E],
§214). In an 1840 article entitled "Allgemeine Lehrsätze in Beziehung auf
die im verkehrten Verhältnisse des Quadrats der Entfernung wirkenden
Anziehungs- und Abstossungs-Kräfte" (*Werke* 5, pp. 197–242; English
translation on pp. 153–196, part 10, vol. 3(1843) of *Scientific Memoirs*,
edited by Richard TAYLOR, Johnson Reprint Corp. (1966), New York),
GAUSS enunciated an identity theorem for potentials of masses (p.223),
which RIEMANN incorporated into function theory in 1851 and expressed as
follows ([R], p. 28): "Eine Function $w = u + iv$ von z kann nicht längs einer
Linie constant sein, wenn sie nicht überall constant ist (A function $w =$
$u + iv$ of z cannot be constant along a line unless it is constant everywhere)."
The proofs of GAUSS and RIEMANN employ integral formulas and are not
really sound; like it or not, one has to call on continuity arguments and
facts about series developments.

An identity theorem by CAUCHY also shows up in 1845. He expresses
the matter as follows (*Œuvres* (1) **9**, p. 39): "Supposons que deux fonctions
de x soient toujours égales entre elles pour des valeurs de x très voisines
d'une valeur donnée. Si l'on vient à faire varier x par degrés insensibles, ces
deux fonctions seront encore égales tant qu'elles resteront l'une et l'autre
fonctions continues de x (Suppose that two [holomorphic] functions of x are
always equal to each other for the values of x very near a given value. If x
is varied by imperceptible degrees, the two functions will still be equal as
long as they remain continuous [= holomorphic] functions of x)." CAUCHY
made no applications of the identity theorem; its significance was first
recognized by RIEMANN and WEIERSTRASS.

From lecture notes transcribed by the Italian mathematician PINCHERLE ("Saggio di una introduzione alla teoria delle funzioni analitiche secondo i principii del Prof. C. Weierstrass, compilato dal Dott. S. Pincherle," *Giorn. Mat.* **18**(1880), 178-254 and 317-357) one can infer (pp. 343/44) that in his 1877/78 lectures WEIERSTRASS derived the implication ii) ⇒ i) for power series using Cauchy's inequalities for the Taylor coefficients. One would be inclined to doubt that WEIERSTRASS really argued in as roundabout a way as PINCHERLE records.

3. Discreteness and countability of the a-places. *Let $f : G \to \mathbb{C}$ be holomorphic and not constant. Then for every number $a \in \mathbb{C}$ the set*

$$f^{-1}(a) := \{z \in G : f(z) = a\},$$

comprising the so-called a-places of f, is discrete and relatively closed (possibly empty too) in G. In particular, for every compact set $K \subset G$, each set $f^{-1}(a) \cap K$, $a \in \mathbb{C}$, is finite and consequently $f^{-1}(a)$ is at most countable; that is, f has at most countably many a-places in G.

Proof. Because f is continuous, each fiber $f^{-1}(a)$ is relatively closed in G. If one of the fibers $f^{-1}(a')$ has a cluster point in G, then it would follow from theorem 1 that $f(z) \equiv a'$, which was excluded by hypothesis. If $K \cap f^{-1}(a)$ were infinite for some $a \in \mathbb{C}$ and some compact set $K \subset G$, then it would contain a sequence of pairwise distinct points. Since $K \cap f^{-1}(a)$ is *compact* such a sequence would have to have a cluster point in $K \cap f^{-1}(a)$, which is impossible since the points of $f^{-1}(a)$ are isolated from each other. Every domain in $\mathbb{C}$ being a countable union of compact sets (cf. 0.2.5), it further follows that $f^{-1}(a)$ is at most countably infinite.   □

The theorem just proved says in particular that

The zero-set of a function which is holomorphic but does not identically vanish in G is a discrete and relatively closed subset of G.

We should recall that the zeros of infinitely often real-differentiable functions needn't have this property. For example, the function defined by

$$f(x) := \exp(-1/x^2)\sin(1/x) \quad \text{for} \quad x \in \mathbb{R} \setminus \{0\}, \qquad f(0) := 0$$

is infinitely often differentiable everywhere in $\mathbb{R}$ (with $f^{(n)}(0) = 0$ for all $n \in \mathbb{N}$) and 0 is a cluster point of its other zeros $1/(\pi n)$, $n \in \mathbb{Z} \setminus \{0\}$.   □

The zeros of a holomorphic function $f \in \mathcal{O}(G)$ may very well cluster at a boundary point of G. Thus for example the function $\sin(\frac{z+1}{z-1})$ belongs to $\mathcal{O}(\mathbb{C} \setminus \{1\})$ and its zero-set $\{\frac{n\pi+1}{n\pi-1} : n \in \mathbb{Z}\}$ has 1 as a cluster point.

4. Order of a zero and multiplicity at a point. If f is holomorphic and not identically zero in a neighborhood of c, then from the Identity Theorem we know that there is a natural number m such that: $f(c) = f'(c) = \cdots = f^{(m-1)}(c) = 0$ and $f^{(m)}(c) \neq 0$. (This statement constitutes a strong generalization of the theorem that a holomorphic function with identically vanishing derivative is locally constant.) We set

$$o_c(f) := m = \min\{\nu \in \mathbb{N} : f^{(\nu)}(c) \neq 0\}.$$

This integer measures the degree to which f vanishes at c and is called the *order of the zero of f at c* or simply the *order of f at c*. Evidently

$$f(c) = 0 \Leftrightarrow o_c(f) > 0.$$

To complete the definition we set $o_c(f) := \infty$ for functions f which are identically 0 near c.

Examples. For $n \in \mathbb{N}$, $o_0(z^n) = n$ and $o_c(z^n) = 0$ for $c \neq 0$. The function $\sin \pi z$ has order 1 at each point of $\mathbb{Z}$.

With the usual agreement that $n + \infty = \infty$ and $\min(n, \infty) = n$ we verify directly two

Rules of computation. *For all functions f and g which are holomorphic near c*

1) $o_c(fg) = o_c(f) + o_c(g)$ (*the product rule*);

2) $o_c(f + g) \geq \min(o_c(f), o_c(g))$ *with equality whenever* $o_c(f) \neq o_c(g)$.

In 4.4.1 the order function $v : \mathcal{A} \to \mathbb{N} \cup \{\infty\}$ for the algebra $\mathcal{A}$ of convergent power series was introduced. If $f = \sum a_\nu (z - c)^\nu$ is holomorphic near c, then the function $f \circ \tau_c = \sum a_\nu z^\nu$, where $\tau_c(z) := z + c$, belongs to $\mathcal{A}$ and we evidently have

$$o_c(f) = v(f \circ \tau_c).$$

Besides the order we often consider the number

$$\nu(f, c) := o_c(f - f(c)).$$

We say that f assumes the value $f(c)$ with *multiplicity* $\nu(f, c)$ at the point c. Of course $\nu(f, c) \geq 1$ always. One immediately confirms the equivalence of the following statements:

i) *f has multiplicity $n < \infty$ at c.*

ii) $f(z) = f(c) + (z - c)^n F(z)$ *for some F which is holomorphic near c and satisfies $F(c) \neq 0$.*

In particular we see that: $\nu(f, c) = 1 \Leftrightarrow f'(c) \neq 0$.

5. Existence of singular points. *On the boundary of the disc of convergence of a power series $f(z) = \sum a_\nu (z - c)^\nu$ there is always at least one singular point of f.*

Proof (by *reductio ad absurdum*). Let $B := B_R(c)$ be the disc (assumed bounded) of convergence of f and suppose the claim is false for this f. Then for every $w \in \partial B$ there is a disc $B_r(w)$ of positive radius $r = r(w)$ and a function $g \in \mathcal{O}(B_r(w))$ such that f and g coincide in $B \cap B_r(w)$. Finitely many of the discs $B_r(w)$, say $K_1, \ldots, K_\ell$ suffice to cover the compact set ∂B. Let $g_j \in \mathcal{O}(K_j)$ be so chosen that $f|B \cap K_j = g_j|B \cap K_j$, $1 \leq j \leq \ell$. There is an $\tilde{R} > R$ such that $\tilde{B} := B_{\tilde{R}}(c) \subset B \cup K_1 \cup \cdots \cup K_\ell$. We define a function $\tilde{f}$ in $\tilde{B}$ as follows: For $z \in B$, $\tilde{f}(z)$ shall be $f(z)$. If on the other hand z lies in $\tilde{B} \setminus B$, then we choose one of the discs K_j which contains z and set $\tilde{f}(z) := g_j(z)$. This definition is independent of the choice of the disc K_j. For if K_k is another one containing z, then $K_k \cap K_j \cap B$ is not empty (shaded region in the figure)

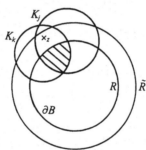

and in this open set g_k and g_j each coincide with f; from this and the Identity Theorem it follows that g_k and g_j coincide throughout the (convex) region $K_j \cap K_k$ and in particular at the point z.

The function $\tilde{f}$ being well-defined, is evidently holomorphic in $\tilde{B}$ and so according to the CAUCHY-TAYLOR theorem it is represented by a power series centered at c and convergent throughout $\tilde{B}$. Since this power series is also the power series at c which represents f, the smaller disc B could not have been the disc of convergence of f. We have reached a contradiction.

$\square$

HURWITZ ([12], p. 51) calls the result just proved a "fundamental theorem" and gives a direct proof of it which may well go back to WEIERSTRASS. His proof does not use the Cauchy-Taylor theorem. Without the use of integrals he gets:

Suppose that for every $w \in B_R(c)$ *the function* $f : B_R(c) \to \mathbb{C}$ *is developable into a convergent power series centered at* w *(i.e.,* f *is an "analytic" function in* WEIERSTRASS' *sense). Then the power series of* f *at* c *in fact converges throughout* $B_R(c)$.

This "global developability" theorem allows an *integral-free* construction of the WEIERSTRASS theory of analytic functions; on this point compare also SCHEEFFER's derivation of the Laurent expansion in 12.1.6.

The geometric series $\sum z^\nu = (1 - z)^{-1}$ has $z = 1$ as its only singular point. A substantial generalization of this example was described by G. VIVANTI in 1893 ("Sulle serie di potenze," *Rivista di Matematica* **3**, 111-114) and proved by PRINGSHEIM in 1894 ("Über Functionen, welche in gewissen Punkten endliche Differentialquotienten jeder endlichen Ordnung, aber keine Taylor'sche Reihenentwickelung besitzen," *Math. Annalen* **44**, 41–56):

Theorem. *Let the power series* $f(z) = \sum a_\nu z^\nu$ *have positive finite radius of convergence* R *and suppose that all but finitely many of its coefficients* a_ν *are real and non-negative. Then* $z := R$ *is a singular point of* f.

Proof. We may suppose that $R = 1$ and that $a_\nu \geq 0$ for all ν. If f were not singular at 1, its Taylor series centered at $1/2$ would be holomorphic at 1, that is, $\sum \frac{1}{\nu!} f^{(\nu)}(\frac{1}{2})(z - \frac{1}{2})^\nu$ would have radius of convergence $r > \frac{1}{2}$. Since, for every ζ with $|\zeta| = \frac{1}{2}$, we have

$$\left| \frac{1}{n!} f^{(n)}(\zeta) \right| = \left| \sum_n^\infty \binom{\nu}{n} a_\nu \zeta^{\nu-n} \right| \leq \sum_n^\infty \binom{\nu}{n} a_\nu \left(\frac{1}{2} \right)^{\nu-n} = \frac{1}{n!} f^{(n)} \left(\frac{1}{2} \right)$$

due to the fact that $a_\nu \geq 0$ for all ν, the Taylor series $\sum \frac{1}{\nu!} f^{(\nu)}(\zeta)(z - \zeta)^\nu$ of f centered at each ζ with $|\zeta| = 1/2$ would have radius of convergence $\geq r > \frac{1}{2}$. As a result there would be no singular point of f on $\partial \mathbb{E}$, contrary to the preceding theorem. $\square$

On the basis of this theorem, for example, 1 is a singular point of the series $\sum_{\nu \geq 1} \nu^{-2} z^\nu$, which is normally convergent in the whole closed disc $\mathbb{E} \cup \partial \mathbb{E}$. For a further extension of this VIVANTI-PRINGSHEIM theorem see §17 of [Lan].

Exercises

Exercise 1. Let G be a region, B a non-empty open disc lying in G. When is the algebra homomorphism $\mathcal{O}(G) \to \mathcal{O}(B)$ given by $f \mapsto f|B$ injective? When is it surjective?

Exercise 2. Show that for a region G in $\mathbb{C}$ and an $f \in \mathcal{O}(G)$ the following assertions are equivalent:

i) f is a polynomial.

ii) There is a point $c \in G$ such that $f^{(n)}(c) = 0$ for almost all $n \in \mathbb{N}$.

Exercise 3. Let G be a region in $\mathbb{C}$, $f, g \in \mathcal{O}(G)$ zero-free. Show that if the set $M := \{z \in G : \frac{f'(z)}{f(z)} = \frac{g'(z)}{g(z)}\}$ is not discrete in G, then $f = \lambda g$ for some $\lambda \in \mathbb{C}^{\times}$.

Exercise 4. Let G be a region in $\mathbb{C}$ which is symmetric about $\mathbb{R}$, that is, satisfies $G = \{\bar{z} : z \in G\}$, and let $f \in \mathcal{O}(G)$. Prove the equivalence of the following assertions:

i) $f(G \cap \mathbb{R}) \subset \mathbb{R}$.

ii) $f(\bar{z}) = \bar{f}(z)$ for all $z \in G$.

Exercise 5. For each of the following four properties the reader is asked to either produce a function f which is holomorphic in a neighborhood of 0 and enjoys that property or prove that no such function exists:

i) $f(\frac{1}{n}) = (-1)^n \frac{1}{n}$ for almost all $n \in \mathbb{N} \setminus \{0\}$;

ii) $f(\frac{1}{n}) = (n^2 - 1)^{-1}$ for almost all $n \in \mathbb{N} \setminus \{0, 1\}$;

iii) $|f^{(n)}(0)| \geq (n!)^2$ for almost all $n \in \mathbb{N}$

iv) $|f(\frac{1}{n})| \leq e^{-n}$ for almost all $n \in \mathbb{N} \setminus \{0\}$ and $o_0(f) \neq \infty$.

§2   The Concept of Holomorphy

Holomorphy is, according to the definition we have adopted, the same thing as complex-differentiability throughout an open set. In this section we describe other possible ways of introducing the fundamental concept of holomorphy. Moreover our list of equivalences could be considerably expanded without much trouble. But we will only take up those characterizations of holomorphy which are especially important and historically significant and which we feel every student of the subject should definitely know.

1. Holomorphy, local integrability and convergent power series. A continuous function f in D is called *locally integrable* in D if D can be covered by open subsets U such that $f|U$ is integrable in U.

Theorem. *The following assertions about a continuous function $f : D \to \mathbb{C}$ are equivalent:*

i) f is holomorphic (= complex-differentiable) in D.

ii) For every (compact) triangle $\Delta \subset D$, $\int_{\partial\Delta} f(\zeta)d\zeta = 0$.

iii) f is locally integrable in D (the MORERA condition).

iv) For every open disc B with $\overline{B} \subset D$

$$f(z) = \frac{1}{2\pi i} \int_{\partial B} \frac{f(\zeta)}{\zeta - z} d\zeta \qquad \text{holds for all } z \in B.$$

v) f is developable into a convergent power series around each point $c \in D$.

Proof. In the following scheme each labelled implication is clear from its label, so only iii) $\Rightarrow$ i) calls for further comment.

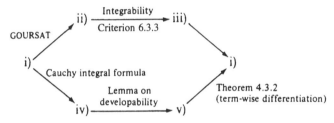

To wit, given $c \in D$ there is an open disc B such that $c \in B \subset D$ and $f|B$ has a primitive F in B, i.e., $F' = f|B$. Since according to 7.4.1 F is infinitely often complex-differentiable in B, it follows that f is holomorphic in B, whence throughout D.   □

We thus see that the validity of the Cauchy integral formula can serve as a characterization of holomorphic functions. Much more important however is the fact that to date no alternative proof of the implication i) $\Rightarrow$ v) has been found which is convincingly free of the use of integrals (cf. however the remarks in 5.1).

Among the various holomorphy criteria the equivalence of the concepts of "complex-differentiability" and "locally developable into power series" has played the principal role in the history of function theory. If you start from the differentiability concept, you are talking about the *Cauchy* or the *Riemann construction*, and if you give primacy to convergent power series, then you are talking about the *Weierstrass construction*. Supplemental remarks on this point will be found in subsection 4 of this section.

The implication iii) $\Rightarrow$ i) in theorem 1 is known in the literature as the

Theorem of Morera. *Every function which is locally integrable in D is holomorphic in D.*

The Italian mathematician Giacinto MORERA (1856-1909, professor of analytical mechanics at Genoa and after 1900 at Turin) proved this "converse of the Cauchy integral theorem" in 1886 in a work entitled "Un teorema fondamentale nella teorica delle funzioni di una variabile complessa," *Rend. Reale Istituto Lombardo di scienze e lettere* (2) **19**, 304-307. OSGOOD, in the 1896 work which was cited in 7.3.4, was probably the first to emphasize the equivalence i) ⇔ iii) and to define holomorphy via *local* integrability; at that time he did not yet know about Morera's work.

2. The holomorphy of integrals. The holomorphy property 1.ii) is frequently simple to verify for functions which are themselves defined by integrals. Here is a simple and very useful example. We designate by $\gamma : [a, b] \to \mathbb{C}$ a piecewise continuously differentiable path and claim

Theorem. *If $g(w, z)$ is a continuous function on $|\gamma| \times D$ and for every $w \in |\gamma|$, $g(w, z)$ is holomorphic in D, then the function*

$$h(z) := \int_\gamma g(\xi, z)d\xi \,, \qquad z \in D,$$

is holomorphic in D.

Proof. We verify 1.ii) for h. To this end let Δ be a triangle lying in D. Then

$$(*) \qquad \int_{\partial\Delta} h(\zeta)d\zeta = \int_{\partial\Delta} \left(\int_\gamma g(\xi, \zeta)d\xi \right) d\zeta = \int_\gamma \left(\int_{\partial\Delta} g(\xi, \zeta)d\zeta \right) d\xi,$$

because, on account of the continuity of the integrand g on $|\gamma| \times \Delta$, the order of integrations can be reversed. This is the well-known theorem of FUBINI, for which the reader can consult any text on the infinitesimal calculus. Since for each fixed ξ, $g(\xi, \zeta)$ is holomorphic in D, $\int_{\partial\Delta} g(\xi, \zeta)d\zeta = 0$ by 1.ii). This holds for each $\xi \in |\gamma|$, so from $(*)$ follows that $\int_{\partial\Delta} h(\zeta)d\zeta = 0$, as desired.   □

An application of this theorem will be made in 9.5.3.

3. Holomorphy, angle- and orientation-preservation (final formulation). In 2.1.2 holomorphy was characterized by the angle- and orientation-preservation properties. But the possible presence of zeros of the derivative, at which points angle-preservation is violated, necessitated certain precautions. The Identity Theorem and Riemann's continuation theorem combine to enable us to bring this result into a form in which the condition $f'(z) \neq 0$ no longer appears.

Theorem. *The following assertions about a continuously real-differentiable function* $f : D \rightarrow \mathbb{C}$ *are equivalent:*

i) f *is holomorphic and nowhere locally constant in* D.

ii) *There is a relatively closed and discrete subset* A *of* D *such that* f *is angle- and orientation-preserving in* $D \setminus A$.

Proof. i) $\Rightarrow$ ii) Since f' doesn't vanish identically in any open set, its zero-set A is discrete and relatively closed in D, according to 1.3. Theorem 2.1.2 insures that f is angle- and orientation-preserving throughout $D \setminus A$.

ii) $\Rightarrow$ i) f is holomorphic in $D \setminus A$ by theorem 2.1.2. But then Riemann's continuation theorem 7.3.4 insures that in fact the (continuous) function f is holomorphic in all of D. □

4. The Cauchy, Riemann and Weierstrass points of view. Weierstrass' creed. CAUCHY took over differentiability from the real domain without even commenting on it. For RIEMANN the deeper reason for studying complex-differentiable functions lay in the "Aehnlichkeit in den kleinsten Theilen," that is, in the angle- and orientation-preserving properties of these functions. WEIERSTRASS based everything on convergent power series. Operating out of a tradition in which the study of mathematics commences with real analysis, the CAUCHY-RIEMANN conception of holomorphy as complex-differentiability seems more natural to us than that of WEIERSTRASS, even though access to the latter requires only a single limit process, local uniform convergence, a fact which gives this theory great internal cohesion. A logically unimpeachable development of the CAUCHY-RIEMANN theory, in which contour integration occupies center-stage, only became possible after the infinitesimal calculus had been put on a firm foundation (by, among others, WEIERSTRASS himself).

One shouldn't overlook the fact that from youth WEIERSTRASS was thoroughly proficient with integration in the complex plane. He used it as early as 1841, long before RIEMANN and independently of CAUCHY, in a proof of the LAURENT's theorem (cf. [W$_1$]). The caution with which WEIERSTRASS at that time integrated in the complex plane shows that he had clearly perceived the difficulties of constructing a complex integral calculus; perhaps here lie the roots of his later phobia of the Cauchy theory.[1]

Nowadays all these inhibitions have disappeared. The integral concept and the relevant theorems about it have been grounded in a simple and satisfactory way. Consequently the Cauchy-Riemann point of departure seems the more natural. And it is in fact complex integration which has

---

[1] WEIERSTRASS is supposed to have scarcely cited CAUCHY. We even read in the highly interesting article "Eléments d'analyse de Karl Weierstrass," *Arch. Hist. Exact Sci.* **10**(1976), 41-176 by P. DUGAC that (p.61) in 1882 WEIERSTRASS did not even confirm to the French Academy the receipt of volume 1 of Cauchy's works which they had sent him.

furnished the most elegant methods and is even indispensable to a simple proof of the equivalence of the Cauchy-Riemann and Weierstrass theories. On the other hand, the Weierstrass definition is necessary if one wants to develop a function theory over more general complete valued fields K other than $\mathbb{R}$ or $\mathbb{C}$ (so-called p-adic function theory). There is no integral calculus over them because the fields K are *totally disconnected*, so only the Weierstrass point of view is fruitful. An Indian mathematician once wrote in this connection that WEIERSTRASS, the prince of analysis, was an algebraist.

D. HILBERT occasionally remarked that every mathematical discipline goes through three periods of development: the *naive*, the *formal* and the *critical*. In function theory the time before EULER was certainly the naive period; with EULER the formal period began, while the critical period began with CAUCHY, attaining its high point in 1860 with WEIERSTRASS' activity in Berlin.

F. KLEIN [H₈] (p.70 of the German original) said of CAUCHY that "mit seinen glänzenden Leistungen auf allen Gebieten der Mathematik fast neben Gauß stellen kann (with his brilliant achievements in all areas of mathematics he can almost be put alongside Gauss)". His assessment of RIEMANN and WEIERSTRASS is (p. 246): "Riemann ist der Mann der glänzenden Intuition. Wo sein Interesse geweckt ist, beginnt er neu, ohne sich durch Tradition beirren zu lassen und ohne den Zwang der Systematik anzuerkennen. Weierstraß ist in erster Linie Logiker; er geht langsam, systematisch, schrittweise vor. Wo er arbeitet, erstrebt er die abschließende Form. (Riemann is the man of brilliant intuition. Where his interest has been awakened he starts from scratch, without letting himself be misled by tradition and recognizing no compulsion to be systematic. Weierstrass is in the first place a logician; he proceeds slowly, systematically and stepwise. In his work he aims for the conclusive and definitive form.)" The reader should compare these sentences with POINCARÉ's words quoted in the Historical Introduction of this book.

In a letter of October 3, 1875 to SCHWARZ, WEIERSTRASS summarized his "Glaubensbekenntnis, in welchem ich besonders durch eingehendes Studium der Theorie der analytischen Functionen mehrerer Veränderlichen bekräftigt worden bin (creed in which I have been especially confirmed by a thorough study of the theory of analytic functions of several variables)" in the following sentences (*Mathematische Werke* 2, p. 235):

"Je mehr ich über die Principien der Functionen theorie nachdenke – und ich thue dies unablässig –, um so fester wird meine Überzeugung, dass diese auf dem Fundamente algebraischer Wahrheiten aufgebaut werden muss, und dass es deshalb nicht der richtige Weg ist, wenn umgekehrt zur Begründung einfacher und fundamentaler algebraischer Sätze das 'Transcendente', um mich kurz auszudrücken, in Anspruch genommen wird – so bestechend auch auf den ersten Anblick z.B. die Betrachtungen sein mögen, durch welche Riemann so viele der wichtigsten Eigenschaften algebraischer

Functionen entdeckt hat. [Dass dem Forscher, so lange er sucht, jeder Weg gestattet sein muss, versteht sich von selbst, es handelt sich nur um die systematische Begründung.] (The more I ponder the principles of function theory – and I do so unceasingly – the firmer becomes my conviction that they have to be built on a foundation of algebraic truths. It is therefore not correct to turn around and, expressing myself briefly, use "transcendental" notions as the basis of simple and fundamental algebraic propositions – however brilliant, e.g., the considerations may appear at first glance by which Riemann discovered so many of the most important properties of algebraic functions. [That every path should be permitted the researcher in the course of his investigations goes without saying; what is at issue here is merely the question of a systematic theoretical foundation.])"

Exercises

Exercise 1. Let D be an open subset of $\mathbb{C}$, L a straight line in $\mathbb{C}$, $f : D \to \mathbb{C}$ continuous. Show that if f is holomorphic in $D \setminus L$, then in fact f is holomorphic in D.

Exercise 2. Let D open $\subset \mathbb{C}$, γ a piecewise continuously differentiable path in $\mathbb{C}$, $g(w, z)$ a continuous complex-valued function on $|\gamma| \times D$. Suppose that for every $w \in |\gamma|$, $z \mapsto g(w, z)$ is holomorphic in D with derivative $\frac{\partial}{\partial z} g(w, z)$. Show that then the function $h(z) := \int_\gamma g(\zeta, z) d\zeta$, $z \in D$, is also holomorphic and that $h'(z) = \int_\gamma \frac{\partial}{\partial z} g(\zeta, z) d\zeta$ for all $z \in D$.

§3 The Cauchy estimates and inequalities for Taylor coefficients

According to 7.2.2 holomorphic functions satisfy the mean value inequality $|f(c)| \le |f|_{\partial B}$. This estimate can be significantly generalized.

1. The Cauchy estimates for derivatives in discs. *Let f be holomorphic in a neighborhood of the closed disc $\overline{B} = \overline{B}_r(c)$. Then for every $k \in \mathbb{N}$ and every $z \in B$, the estimate*

$$|f^{(k)}(z)| \le k! \frac{r}{d_z^{k+1}} |f|_{\partial B} \ , \ \text{with } d_z := d_z(B) = \min_{\zeta \in \partial B} |\zeta - z|,$$

holds.

Proof. The Cauchy integral formula 7.3.2(2) for derivatives says that

$$f^{(k)}(z) = \frac{k!}{2\pi i} \int_{\partial B} \frac{f(\zeta)}{(\zeta - z)^{k+1}} d\zeta , \qquad z \in B.$$

If we apply to this integral the standard estimate 6.2.2 for contour integrals, then the present claim follows from the facts that $L(\partial B) = 2\pi r$ and

$$\max_{\zeta \in \partial B} \frac{|f(\zeta)|}{|\zeta - z|^{k+1}} \leq |f|_{\partial B} \left(\min_{\zeta \in \partial B} |\zeta - z| \right)^{-k-1} = |f|_{\partial B} d_z^{-k-1}.$$

Corollary. *If f is holomorphic in a neighborhood of $\overline{B}$, then for every $k \in \mathbb{N}$ and every positive number $d < r$, the estimate*

$$|f^{(k)}(z)| \leq k! \frac{r}{d^{k+1}} |f|_{\partial B}$$

holds for all $z \in \overline{B}_{r-d}(c)$.

In the limiting case when d converges to r we get

Cauchy's inequalities for the Taylor coefficients.
Let $f(z) = \sum a_\nu (z-c)^\nu$ be a power series with radius of convergence greater than r, and set $M(r) := \max_{|z-c|=r} |f(z)|$. Then

$$|a_\nu| \leq \frac{M(r)}{r^\nu} , \qquad \text{for all } \nu \in \mathbb{N}. \qquad \square$$

A simple covering argument leads at once to

Cauchy's estimates for derivatives in compact sets. *Let D be a domain in $\mathbb{C}$, K a compact subset of D, L a compact neighborhood of K lying in D. For every $k \in \mathbb{N}$ there exists a finite constant M_k (which depends only on D, K and L) such that*

$$|f^{(k)}|_K \leq M_k |f|_L \qquad \text{for all } f \in \mathcal{O}(D).$$

We should note that the role of L cannot be taken over by K itself. For $f_n := z^n \in \mathcal{O}(\mathbb{C})$ and $K := \overline{\mathbb{E}}$ we have, e.g., $|f_n|_K = 1$ but $|f_n'|_K = n$ for each $n \in \mathbb{N}$.

2. The Gutzmer formula and the maximum principle.

The inequalities for the Taylor coefficeints can be derived directly, without invoking the integral formula for derivatives. They can even be refined, by observing that on the parameterized circle $z(\varphi) = c + re^{i\varphi}$, $0 \leq \varphi \leq 2\pi$, the power series $\sum a_\nu (z - c)^\nu$ are *trigonometric series* $\sum a_\nu r^\nu e^{i\nu\varphi}$. Because $\sum a_\nu r^\nu e^{i(\nu-n)\varphi}$ converges normally in $[0, 2\pi]$ to $f(c+re^{i\varphi})e^{-in\varphi}$, the "orthonormality relations"

$$\frac{1}{2\pi} \int_0^{2\pi} e^{i(m-n)\varphi} d\varphi = \begin{cases} 0 & \text{for } m \neq n \\ 1 & \text{for } m = n \end{cases}$$

immediately lead to the following representation of the Taylor coefficients:

If the series $f(z) = \sum a_\nu (z-c)^\nu$ has radius of convergence greater than r, then

$$(*) \qquad a_n r^n = \frac{1}{2\pi} \int_0^{2\pi} f(c + re^{i\varphi}) e^{-in\varphi} d\varphi , \qquad n \in \mathbb{N}.$$

From this follows immediately

The Gutzmer formula. Let $f(z) = \sum a_\nu (z-c)^\nu$ be a power series with radius of convergence greater than r, and set $M(r) := \max_{|z-c|=r} |f(z)|$. Then

$$\sum |a_\nu|^2 r^{2\nu} = \frac{1}{2\pi} \int_0^{2\pi} |f(c + re^{i\varphi})|^2 d\varphi \leq M(r)^2.$$

Proof. Since $\overline{f(c + re^{i\varphi})} = \sum \bar{a}_\nu r^\nu e^{-i\nu\varphi}$, we have

$$|f(c + re^{i\varphi})|^2 = \sum \bar{a}_\nu r^\nu f(c + re^{i\varphi}) e^{-i\nu\varphi},$$

with convergence normal on $[0, 2\pi]$. Consequently integration passes through the sum and $(*)$ yields

$$\int_0^{2\pi} |f(c + re^{i\varphi})|^2 d\varphi = \sum \bar{a}_\nu r^\nu \int_0^{2\pi} f(c + re^{i\varphi}) e^{-i\nu\varphi} d\varphi = 2\pi \sum |a_\nu|^2 r^{2\nu}.$$

On the other hand, the estimate $\int_0^{2\pi} |f(c + re^{i\varphi})|^2 d\varphi \leq 2\pi M(r)^2$ is trivial. □

The inequalities $|a_\nu| r^\nu \leq M(r)$, $\nu \in \mathbb{N}$, are naturally contained in the Gutzmer formula. Moreover, there follows directly the

Corollary. If $f(z) = \sum a_\nu (z-c)^\nu$ in $B_s(c)$ and if there is an $m \in \mathbb{N}$ and an r with $0 < r < s$ and $|a_m| r^m = M(r)$, then necessarily $f(z) = a_m (z-c)^m$.

Proof. From GUTZMER we have that $\sum_{\nu \neq m} |a_\nu|^2 r^{2\nu} \leq 0$, and so $a_\nu = 0$ for all $\nu \neq m$. □

This corollary together with the identity theorem implies the

Maximum principle. *Let f be holomorphic in the region G and have a local maximum at a point $c \in G$; that is, suppose that $|f(c)| = |f|_U$ for some neighborhood U of c in G. Then f is constant in G.*

Proof. If $\sum a_\nu (z - c)^\nu$ is the Taylor series of f about c, then by hypothesis $|a_0| = |f|_U$. For all sufficiently small positive r it then follows that $|a_0| \geq M(r)$. From the corollary (with $m = 0$) f is consequently constant in some disc centered at c. Then from the identity theorem f is in fact constant throughout G.   □

The maximum principle, which here is a somewhat incidental spin-off, will be set in a larger framework in 5.2.

The set of all power series with center at c and having a radius of convergence greater than r forms a complex vector space V. By means of the equation

$$\langle f, g \rangle := \frac{1}{2\pi} \int_0^{2\pi} f(c + re^{i\varphi})\overline{g(c + re^{i\varphi})}d\varphi \,, \qquad f, g \in V$$

a *hermitian bilinear form* can be introduced into V. The family $e_n := r^{-n}(z-c)^n$, $n \in \mathbb{N}$, forms an *orthonormal system* in V:

$$\langle e_m, e_n \rangle = \begin{cases} 0 & \text{for } m \neq n \\ 1 & \text{for } m = n. \end{cases}$$

Every $f = \sum a_\nu (z-c)^\nu \in V$ is an *orthogonal series* with the "Fourier coefficients" $\langle f, e_\nu \rangle = a_\nu r^\nu$; Gutzmer's equation is the *Parseval completeness relation*:

$$\| f \|^2 := \langle f, f \rangle = \sum_0^\infty |\langle f, e_\nu \rangle|^2.$$

And we have $\| f \| = 0 \Leftrightarrow \langle f, e_\nu \rangle = a_\nu r^\nu = 0$ for all $\nu \in \mathbb{N} \Leftrightarrow f = 0$. Then with respect to the form $\langle \, , \, \rangle$, V is a unitary *vector space*. But V is not complete, hence is only a pre-Hilbert *not* a *Hilbert space*. To see this (with $c := 0$, $r := 1$), consider the polynomials $p_n := \sum_1^n \frac{z^\nu}{\nu}$. Due to the equations

$$\| p_m - p_n \|^2 = \sum_{m+1}^n \frac{1}{\nu^2}, \qquad \text{for } m < n,$$

these polynomials constitute a Cauchy sequence in V with respect to $\| \, \|$. But this sequence has no limit in V, because the only limit candidate is the series $\sum_1^\infty \frac{z^\nu}{\nu}$, which has radius of convergence 1 and consequently does not lie in V.

3.   Entire functions.   LIOUVILLE's theorem.

Functions which are holomorphic everywhere in $\mathbb{C}$ were called *entire functions* by WEIERSTRASS ([W$_3$],p. 84). Every polynomial is of course entire; all other entire functions are called *transcendental*. Examples of the latter are $\exp z$, $\cos z$, $\sin z$.

The Cauchy inequalities immediately imply the famous

Theorem of Liouville. *Every bounded entire function is constant.*

Proof. The Taylor development $f(z) = \sum a_\nu z^\nu$ of f at 0 converges *throughout* $\mathbb{C}$ (by the representation theorem 7.3.2); according to subsection 1

$$r^\nu |a_\nu| \leq \max_{|z|=r} |f(z)| \qquad \text{holds for all } r > 0 \text{ and all } \nu \in \mathbb{N}.$$

Since f is bounded there is a finite M such that $|f(z)| \leq M$ for all $z \in \mathbb{C}$. It follows that $r^\nu |a_\nu| \leq M$ *for all* $r > 0$ *and all* $\nu \in \mathbb{N}$. Since r^ν can be made arbitrarily large if $\nu \neq 0$, it follows that for such ν we must have $a_\nu = 0$. Thus $f(z) = a_0$. □

Variant of the proof. Use the Cauchy inequality only for $\nu = 1$ but apply it at every point $c \in \mathbb{C}$, getting $|f'(c)| \leq Mr^{-1}$, for all $r > 0$, and consequently $f'(c) = 0$. That is, $f' \equiv 0$ and so $f \equiv \text{const.}$ □

We can also give a *second direct proof* by means of the Cauchy integral formula. Let $c \in \mathbb{C}$ be arbitrary. For $r > |c|$ and $S := \partial B_r(0)$ we have

$$f(c) - f(0) = \frac{1}{2\pi i} \int_S \left(\frac{1}{\zeta - c} - \frac{1}{\zeta} \right) f(\zeta) d\zeta = \frac{c}{2\pi i} \int_S \frac{f(\zeta) d\zeta}{(\zeta - c)\zeta}.$$

If we choose $r \geq 2|c|$, then $|\zeta - c| \geq \frac{1}{2}r$ for $\zeta \in S$ and it follows that

$$|f(c) - f(0)| \leq \frac{|c|}{2\pi} \max_{|\zeta|=r} \left| \frac{f(\zeta)}{(\zeta - c)\zeta} \right| 2\pi r \leq 2|c|Mr^{-1},$$

where, as before, M is a (finite) bound for f. If we let r increase indefinitely, it follows that $f(c) = f(0)$, for each $c \in \mathbb{C}$. □

LIOUVILLE's theorem also follows directly from the mean value equality in 7.2.2. An elegant proof of this kind was given by E. NELSON, "A proof of Liouville's theorem," *Proc. Amer. Math. Soc.* **12**(1961), p. 995.

We plan to use LIOUVILLE's theorem in 9.1.2 in a proof of the fundamental theorem of algebra. But as an immediate application we have

Every holomorphic mapping $f : \mathbb{C} \to \mathbb{E}$ is constant. In particular, there are no biholomorphic mappings of $\mathbb{E}$ onto $\mathbb{C}$ or of $\mathbb{H}$ onto $\mathbb{C}$.

It is however quite possible to map the plane *topologically*, even *real-analytically*, onto the open unit disc. Such a mapping from $\mathbb{C}$ to $\mathbb{E}$ is given, e.g., by $z \mapsto z/\sqrt{1+|z|^2}$, with inverse mapping from $\mathbb{E}$ to $\mathbb{C}$ given by $w \mapsto w/\sqrt{1-|w|^2}$.

Remark. The algebra $\mathcal{O}(\mathbb{C})$ of entire functions contains the algebra $\mathbb{C}[z]$ of polynomials. The greater abundance of functions in $\mathcal{O}(\mathbb{C})$ compared with $\mathbb{C}[z]$ is convincingly attested to by two approximation theorems: According to WEIER-STRASS (*Math. Werke* **3**, p.5), every continuous function $f : \mathbb{R} \to \mathbb{C}$ is uniformly approximable on each compact interval by polynomials. Uniform approximation on all of $\mathbb{R}$ is not generally possible; e.g., $\sin x$ is certainly not approximable uniformly on $\mathbb{R}$ by polynomials. But with entire functions, anything that's continuous on $\mathbb{R}$ can be approximated uniformly well on $\mathbb{R}$. Even more: In his paper "Sur un théorème de Weierstrass," *Ark. Mat. Astron. Fys.* **20B**, 1-5 (1927) T. CARLEMAN showed:

Let a continuous and strictly positive "error" function $\varepsilon : \mathbb{R} \to \mathbb{R}$ be given. Then for every continuous $f : \mathbb{R} \to \mathbb{C}$ there exists a $g \in \mathcal{O}(\mathbb{C})$ such that

$$|f(x) - g(x)| < \varepsilon(x) \qquad \text{for all } x \in \mathbb{R}.$$

The reader will find a proof in D. GAIER's book *Lectures on Complex Approximation*, Birkhäuser, Boston, (1987), p. 149.

4. Historical remarks on the Cauchy inequalities and the theorem of LIOUVILLE.

CAUCHY knew the inequalities for Taylor coefficients which bear his name by 1835 (cf. *Œuvres* (2) **11**, p. 434). WEIERSTRASS proved these inequalities in 1841 by an elementary method, which involved arithmetic means instead of integrals ([W$_2$], 67-74 and [W$_4$], 224-226). We will reproduce this beautiful proof in the next subsection. August GUTZMER (1860-1925, ordinarius professor at Halle, 1901-1921 sole editor of the high-level *Jahresberichte der Deutschen Mathematiker-Vereinigung*) published his formula in 1888 in the paper "Ein Satz über Potenzreihen," *Math. Annalen* **32**, 596-600.

Joseph LIOUVILLE (1809-1882, French mathematician and professor at the Collège de France) in 1847 put the theorem "Une fonction doublement périodique qui ne devient jamais infinie est impossible (a doubly periodic function which never becomes infinite is impossible)" at the beginning of his *Leçons sur les fonctions doublement périodiques*. Carl Wilhelm BOR-CHARDT (1817-1880, German mathematician at Berlin, pupil of JACOBI and close friend of WEIERSTRASS, from 1855-1880 CRELLE's successor as editor of the *Journal für die Reine und Angewandte Mathematik*) heard Liouville's lectures in 1847, published them in 1879 in the aforementioned journal, vol. 88, 277-310, and named the theorem after LIOUVILLE (cf. the footnore on p. 277). But the theorem originates with CAUCHY, who derived it in 1844 in his note "Mémoires sur les fonctions complémentaires" (*Œuvres* (1) **8**, pp. 378-385; see théorème II on page 378), via his residue calculus. The direct derivation from the Cauchy inequalities was given in 1883 by the French mathematician Camille JORDAN (1838-1921, professor at the École Polytechnique), in the second volume of his *Cours d'analyse*, théorème 312 on p. 312. (In the third edition of the second volume, which appeared in 1913 and which was reprinted in 1959 by Gauthier-Villars, this would be théorème 338 on p. 364.)

5*. Proof of the Cauchy inequalities following WEIERSTRASS. The kernel of the method is found in the proof of the following

Lemma. *Let* $m, n \in \mathbb{N}$ *and* $q(z) = \sum_{-m}^{n} a_\nu (z - c)^\nu$. *Then*

$$|a_0| \leq M(r) := \max_{|\zeta - c| = r} |q(\zeta)| \qquad \text{for every } r > 0.$$

Proof. We may assume that $c = 0$. Fix r and write simply M for $M(r)$. Choose $\lambda \in S_1$, the unit circle, so that for all $\nu \in \mathbb{Z} \setminus \{0\}$, $\lambda^\nu \neq 1$. For example, $\lambda := e^i$ or (more elementarily) $\lambda := (2 - i)(2 + i)^{-1}$. [For the verification in the latter case, suppose that $\lambda^n = 1$ for some $n \in \mathbb{N} \setminus \{0\}$. Then $(2 - i)^n = (2i + 2 - i)^n = (2i)^n + n(2i)^{n-1}(2 - i) + \cdots$, so $(2i)^n = (2 - i)(\alpha + i\beta)$ for certain $\alpha, \beta \in \mathbb{Z}$, whence $4^n = 5(\alpha^2 + \beta^2)$, which is absurd. Of course, instead of this calculation, one could just remark that in the *unique factorization domain* $\mathbb{Z}[i]$, $2 + i$ and $2 - i$ are prime elements which are not associates.] Let $\sum'$ denote summation over non-zero indices. It then follows that

$$\sum_{j=0}^{k-1} q(r\lambda^j) = ka_0 + \sum_{-m}^{n} {}' a_\nu r^\nu \frac{\lambda^{\nu k} - 1}{\lambda^\nu - 1}, \qquad k \geq 1.$$

Now $|q(r\lambda^j)| \leq M$, since $|r\lambda^j| = r$, and so we get

$$|a_0| \leq M + \frac{1}{k} \sum_{-m}^{n} {}' |a_\nu| \frac{2r^\nu}{|\lambda^\nu - 1|}, \qquad k \geq 1.$$

Since the value of the sum on the right is independent of k and k may be chosen as large as we like, it follows that $|a_0| \leq M$.

Theorem. *For some* $m \in \mathbb{N}$ *let* $f(z) = \sum_{-m}^{\infty} a_\nu (z - c)^\nu$ *be holomorphic in the punctured disc* $B_s(c) \setminus \{c\}$. *Then for every* r *with* $0 < r < s$, *setting* $M(r) := \max_{|z-c|=r} |f(z)|$, *we have*

$$|a_\mu| \leq \frac{M(r)}{r^\mu}, \qquad \text{for all } \mu \geq -m.$$

Proof. Again we may suppose $c = 0$. First look at $\mu = 0$. Let $\varepsilon > 0$ be given. Then choose $n \in \mathbb{N}$ so large that the remainder series $g(z) := \sum_{n+1}^{\infty} a_\nu z^\nu$ satisfies $\max_{|z|=r} |g(z)| \leq \varepsilon$. Then $q(z) := f(z) - g(z) = \sum_{-m}^{n} a_\nu z^\nu$ satisfies

$$\max_{|z|=r} |q(z)| \leq M(r) + \varepsilon.$$

It then follows from the lemma that $|a_0| \leq M(r) + \varepsilon$. Since $\varepsilon > 0$ is arbitrary, we have $|a_0| \leq M(r)$.

Finally, consider an arbitrary $\mu \geq -m$. The function $z^{-\mu}f(z)$ $= \sum_{-(m+\mu)}^{\infty} a_{\mu+\nu}z^{\nu}$ is, like f, holomorphic in $B_s(c) \setminus \{c\}$. Its constant term is a_μ and its relevant maximum is $\max_{|z|=r} |z^{-\mu}f(z)| = r^{-\mu}M(r)$. So from what was learned in the first paragraph of the proof, we have $|a_\mu| \leq r^{-\mu}M(r)$.

Exercises

Exercise 1. Generalize the Gutzmer equality to

$$\langle f, g \rangle = \sum_{\nu \geq 0} a_\nu \bar{b}_\nu r^{2\nu}$$

for $f(z) := \sum_{\nu \geq 0} a_\nu (z-c)^\nu$ and $g(z) := \sum_{\nu \geq 0} b_\nu (z-c)^\nu \in V$. [For the definition of V, see subsection 2.]

Exercise 2. For $f \in \mathcal{O}(\mathbb{C})$ prove the following sharper versions of Liouville's theorem:

  a) If for some $n \in \mathbb{N}$ and some finite constants R and M, the function f satisfies $|f(z)| \leq M|z|^n$ for all $|z| \geq R$, then f is a polynomial of degree no greater than n.

  b) If the function $\Re f$ is bounded in $\mathbb{C}$, then f is constant.

Exercise 3. Let $f : \mathbb{C} \to \mathbb{C}$ be entire. Show that the Taylor series of f at 0 converges to f *uniformly in* $\mathbb{C}$ if and only if f is a polynomial.

Exercise 4. Show that if the entire functions f and g satisfy $|f(z)| \leq |g(z)|$ for all $z \in \mathbb{C}$, then there is a $\lambda \in \mathbb{C}$ such that $f = \lambda g$.

§4   Convergence theorems of WEIERSTRASS

The focal point of this section is the theorem affirming that in function theory – as contrasted with real analysis – compact convergence commutes with differentiation and the sequence of derivatives converges compactly as well. As corollaries we get theorems about differentiating series.

1. **Weierstrass' convergence theorem.** *Let f_n be a sequence of holomorphic functions in a domain D which converges compactly there to f : $D \to \mathbb{C}$. Then f is holomorphic in D and for every $k \in \mathbb{N}$ the sequence $f_n^{(k)}$ of kth derivatives converges compactly in D to $f^{(k)}$.*

Proof. a) First of all, the limit function f is continuous in D (by the continuity theorem 3.1.2). For every triangle $\Delta \subset D$ the interchange theorem for sequences (6.2.3) then ensures that

$$\int_{\partial \Delta} f d\zeta = \lim_{n \to \infty} \int_{\partial \Delta} f_n d\zeta.$$

Since each $f_n \in \mathcal{O}(D)$, each integral on the right side vanishes (GOURSAT); consequently f is holomorphic in D, by the ii) $\Rightarrow$ i) implication in theorem 2.1.

b) Evidently it suffices to verify the convergence claim for $k = 1$. Let K be a compact subset of D. Cauchy's estimates for derivatives in compact sets furnish a compact $L \subset D$ and a finite constant M such that $|f_n' - f'|_K \leq M|f_n - f|_L$ holds for all n. Since $\lim |f_n - f|_L = 0$ by hypothesis, it follows that $\lim |f_n' - f'|_K = 0$. $\qquad\square$

The holomorphy of the limit function rests, in this proof, on the simple fact that a limit function under compact convergence inherits local integrability from the terms of the sequence. The theorem of MORERA then concludes the proof. In this kind of reasoning the developability of the functions into power series is irrelevant. For functions on the real line the convergence theorem is false for a variety of reasons: Limit functions of compactly convergent sequences of real-differentiable functions are in general not real-differentiable. Cf. paragraph 2. of the introductory material to chapter 3.

2. **Differentiation of series. Weierstrass' double series theorem.** Since the sequence $\{f_n\}$ and the series $\sum(f_\nu - f_{\nu-1})$ exhibit the same convergence behavior, from theorem 1 there follows at once

Weierstrass' differentiation theorem for compactly convergent series. *A series $\sum f_\nu$ of holomorphic functions in D which converges compactly in D has a holomorphic limit f in D. For every $k \in \mathbb{N}$ the k-times term-wise differentiated series $\sum f_\nu^{(k)}$ converges compactly in D to $f^{(k)}$:*

$$f^{(k)}(z) = \sum f_\nu^{(k)}(z) \, , \qquad z \in D.$$

This is a generalization of the theorem that convergent power series represent holomorphic functions and may be "differentiated term by term" (for f_ν take the monomial $a_\nu(z - c)^\nu$). In applications one frequently needs

Weierstrass' differentiation theorem for normally convergent series. *If for $f_\nu \in \mathcal{O}(D)$ the series $\sum f_\nu$ converges normally in D to $f \in \mathcal{O}(D)$, then for each $k \in \mathbb{N}$ the series $\sum f_\nu^{(k)}$ converges normally in D to $f^{(k)}$.*

Proof. Let $K \subset D$ be compact. The Cauchy estimates furnish a compact set L, $K \subset L \subset D$, and, for each $k \in \mathbb{N}$, a finite constant M_k, such that $|g^{(k)}|_K \leq M_k |g|_L$ for all $g \in \mathcal{O}(D)$. This implies that $\sum |f_\nu^{(k)}|_K \leq M_k \sum |f_\nu|_L < \infty$; that is, the series of derivatives converge normally in D. Since this entails compact convergence, the respective limits are the $f^{(k)}$. $\square$

Example. The Riemann zeta function $\zeta(z)$ introduced in 5.5.4 is holomorphic in the right half-plane $\{z \in \mathbb{C} : \Re z > 1\}$ because the ζ-series $\sum_{n \geq 1} n^{-z}$ converges normally there (theorem 5.5.4).

An immediate application of the first differentiation theorem is

Weierstrass' double series theorem. *Let $f_\nu(z) = \sum a_\mu^{(\nu)}(z - c)^\mu$ be power series convergent in a common disc B centered at c, for $\nu \in \mathbb{N}$. Suppose that the series $f(z) = \sum f_\nu(z)$ converges normally in B. Then for each $\mu \in \mathbb{N}$, $b_\mu := \sum_{\nu=0}^{\infty} a_\mu^{(\nu)}$ converges in $\mathbb{C}$, and f is represented in B by the convergent power series*

$$f(z) = \sum b_\mu(z - c)^\mu.$$

Proof. According to the differentiation theorem for compactly convergent series, $f \in \mathcal{O}(B)$ and $f^{(\mu)} = \sum_{\nu=0}^{\infty} f_\nu^{(\mu)}$ for every $\mu \in \mathbb{N}$. Then the representation theorem 7.3.2 says that f is represented in B by the Taylor series

$$\sum_0^\infty \frac{f^{(\mu)}(c)}{\mu!}(z - c)^\mu \text{ , where now } \frac{f^{(\mu)}(c)}{\mu!} = \sum_{\nu=0}^{\infty} \frac{f_\nu^{(\mu)}(c)}{\mu!} = \sum_{\nu=0}^{\infty} a_\mu^{(\nu)}. \quad \square$$

The designation "double series theorem" is almost self-explanatory: In B, f is given by the double series

$$f(z) = \sum_{\nu=0}^{\infty} \left(\sum_{\mu=0}^{\infty} a_\mu^{(\nu)}(z - c)^\mu \right),$$

and the theorem affirms that, as with polynomials, the summations may be interchanged without altering the convergence in B or the value of the limit:

$$f(z) = \sum_{\mu=0}^{\infty} \left(\sum_{\nu=0}^{\infty} a_{\mu}^{(\nu)} \right) (z-c)^{\mu}.$$

Finally, we mention the following corollary to the product theorem for convergent series 3.3.1:

Product theorem for normally convergent series of holomorphic functions. *If $f = \sum f_{\mu}$ and $g = \sum g_{\nu}$ are normally convergent series of holomorphic functions in D, then every product series $\sum h_{\lambda}$ in which $h_0, h_1, \ldots$ run through every product $f_{\mu}g_{\nu}$ exactly once, converges normally in D to $fg \in \mathcal{O}(D)$. In particular, $fg = \sum p_{\lambda}$ with $p_{\lambda} := \sum_{\mu+\nu=\lambda} f_{\mu}g_{\nu}$ (Cauchy product).*

This assertion becomes false if only compact convergence of the series for f and g is hypothesized; under this hypothesis it is even possible that $\sum p_{\lambda}$ diverge. This is illustrated by the example of the constant functions $f_{\nu} = g_{\nu} := \frac{(-1)^{\nu}}{\sqrt{\nu+1}}$ considered in 0.4.6.

3. On the history of the convergence theorems. In the 19th century series predominated over sequences, because series were thought of as "closed analytic expressions," a shibboleth until the function concept finally got put on a firm foundation. For WEIERSTRASS the double series theorem was the key to convergence theory. In 1841 in a work of his youth ([W$_2$], pp.70 ff.) he stated and proved this theorem, for power series in several complex variables even, knowing nothing about Cauchy's function theory. But it should be noted that besides the compact convergence of $\sum f_{\nu}(z)$, he hypothesized in addition the unconditional (= absolute) convergence of this series at each point of B. His proof is elementary, the only tool used being the Cauchy inequalities for Taylor coefficients, which he derived directly (as was done in 3.5) without employing integrals. In 1880 WEIERSTRASS came back once again to his convergence theorem for series [W$_4$]. This time he did not require the supplemental unconditional convergence.

From the double series theorem WEIERSTRASS easily obtained the theorem on differentiating compactly convergent series ([W$_2$], pp. 73/74) simply by developing each f_{ν} into its Taylor series around each point $c \in D$ and noting that, for fixed c, they all converge in some fixed disc B centered at c and lying in D.

In 1886 MORERA deduced the convergence theorem for compactly convergent series from the converse of Cauchy's integral theorem which he had discovered; see p. 306 of his work cited in 2.1, and also his "Sulla rappresentazione delle funzioni di una variabile complessa per mezzo di espressioni analitiche infinite," *Atti R. Accad. Sci. Torino* **21**(1886), pp. 894-897. It was Morera's technique that we used in part a) of the proof in subsection 1. In 1887 P. PAINLEVÉ (1863-1933, French mathematician; 1908 first airplane passenger of W. WRIGHT; 1915/16 minister of war, 1917 and again 1925 premier of France) proved Weierstrass' convergence theorem with the help of the Cauchy integral formula – in "Sur les lignes singulières des fonctions analytiques," *Ann. Fac. Sci. Toulouse* (1) **2**(1887), pp. 11-12.

In 1896 OSGOOD gave Morera's argument (pp. 297/298 of the work cited in 7.3.4), and wrote: "It is to be noticed that this proof belongs to the most elementary class of proofs, in that it calls for no explicit representation of the functions entering (e.g., by Cauchy's integral or by a power series)."

4. A convergence theorem for sequences of primitives. In convergence theorem 1 we inferred from the compact convergence of a sequence of functions, the convergence of the sequence of their derivatives. A supplemental hypothesis to control the otherwise arbitrary "constants of integration" enables us to also infer the compact convergence of certain sequences of primitives of the original functions.

Theorem. *Let G be a region, $f_0, f_1, \ldots$ a sequence in $\mathcal{O}(G)$ which converges compactly to $f \in \mathcal{O}(G)$. Let $F_n \in \mathcal{O}(G)$ be a primitive of f_n, for each $n \in \mathbb{N}$. Then the sequence $F_0, F_1, \ldots$ will converge compactly in G to a primitive F of f if there is a point $c \in G$ for which the numerical sequence $F_n(c)$ converges.*

Proof. We may assume that all $F_n(c)$ are 0 – otherwise we just pass over to primitives $F_n - F_n(c)$. Let $w \in G$ be arbitrary. We have, on the basis of theorem 6.3.1,

$$(1) \qquad F_n(w) = \int_\gamma f_n(\zeta)d\zeta$$

for every path γ in G from c to w because $F_n' = f_n$ and $F_n(c) = 0$. It therefore follows from the interchange theorem 6.2.3 that

$$(2) \qquad \lim F_n(w) \text{ exists and } F(w) := \lim F_n(w) = \int_\gamma f(\zeta)d\zeta.$$

A function $F : G \to \mathbb{C}$ is thereby defined which obviously satisfies condition ii) of theorem 6.3.1 and is therefore a primitive of f. Now consider an arbitrary compact disc $K = \overline{B_r(a)}$ in G. It follows directly (!) from (1) and (2) that

$$F(z) - F_n(z) = \int_{[a,z]} [f(\zeta) - f_n(\zeta)]d\zeta + F(a) - F_n(a) \text{ , for all } z \in K.$$

The standard estimate 6.2.2 for path integrals then gives

$$|F - F_n|_K \le |f - f_n|_K \cdot r + |F(a) - F_n(a)|.$$

Since by (2) $\lim F_n(a) = F(a)$, $\lim |F - F_n|_K = 0$ follows. Therewith the compact convergence in G of the sequence F_n to F is demonstrated.   □

Although Weierstrass' convergence theorem will be applied again and again in the sequel, this last theorem won't play any further role.

5*. A remark of WEIERSTRASS' on holomorphy. In [W$_4$] WEIER-STRASS studied the following problem thoroughly:

Let $\sum f_\nu$ be a series of *rational* functions which converges compactly to $f \in \mathcal{O}(D)$ on a domain D which decomposes into disjoint regions $G_1, G_2, \ldots$. What "analytic connections" subsist among the limit functions $f|G_1, f|G_2, \ldots$ on the various regions? (Naturally WEIERSTRASS was more precise in his formulation; he asked whether $f|G_1$ and $f|G_2$ are "branches" of one and the same "monogenic" holomorphic function, that is, whether they "arise from one another via analytic continuation.")

WEIERSTRASS discovered to his surprise (p. 216, *op. cit.*) that there need be *no* connection between $f|G_1$ and $f|G_2$; that there even exist disjoint regions G_1 and G_2 in $\mathbb{C}$ and a sequence of rational functions f_ν such that the series $\sum f_\nu$ converges compactly in $G_1 \cup G_2$ to $+1$ in G_1 and to -1 in G_2. Thus WEIERSTRASS discovered a special case of RUNGE's approximation theorem, which is so central to contemporary function theory. First of all we want to give a very simple example involving sequences.

The sequence $1/(1 - z^n)$ of rational functions are holomorphic in $\mathbb{C} \setminus \partial\mathbb{E}$ and compactly converge there to the function

$$h(z) := \begin{cases} 1 & for \quad |z| < 1 \\ 0 & for \quad |z| > 1. \end{cases}$$

From which we immediately obtain the following:

Let $f, g \in \mathcal{O}(\mathbb{C})$ be given but arbitrary. Then the sequence

$$f_n(z) := g(z) + \frac{f(z) - g(z)}{1 - z^n}$$

converges compactly in $\mathbb{C} \setminus \partial\mathbb{E}$ to the function

$$G(z) := \begin{cases} f(z) & for \quad |z| < 1 \\ g(z) & for \quad |z| > 1. \end{cases}$$

Proof. This is clear from the preceding, since $G = g + (f - g)h$. □

Simple examples involving series can also be adduced:

The series $\dfrac{1}{1 - z} + \dfrac{z}{z^2 - 1} + \dfrac{z^2}{z^4 - 1} + \dfrac{z^4}{z^8 - 1} + \dfrac{z^8}{z^{16} - 1} + \cdots$, whose terms are rational functions all of which are holomorphic in $\mathbb{C} \setminus \partial\mathbb{E}$, converges compactly in that domain to the function

$$F(z) := \begin{cases} 1 & for \quad |z| < 1 \\ 0 & for \quad |z| > 1. \end{cases}$$

Proof. This follows from the identities

$$\sum_1^n \frac{z^{2^{\nu-1}}}{z^{2^\nu} - 1} = \sum_1^n \left(\frac{1}{1 - z^{2^\nu}} - \frac{1}{1 - z^{2^{\nu-1}}} \right) = \frac{1}{1 - z^{2^n}} - \frac{1}{1 - z}. \qquad \square$$

Examples of series of this type go back to TANNERY, who used the series

$$\frac{1+z}{1-z} + \frac{2z}{z^2 - 1} + \frac{2z^2}{z^4 - 1} + \frac{2z^4}{z^8 - 1} + \frac{2z^8}{z^{16} - 1} + \cdots = \begin{cases} 1 & for \quad |z| < 1 \\ -1 & for \quad |z| > 1. \end{cases}$$

The reader is encouraged to provide a proof of this equality. (Compare also [W$_4$], 231/232.) WEIERSTRASS gave quite a bit more complicated an example, $\sum_1^\infty \frac{1}{z^n + z^{-n}}$. For him this convergence phenomenon was an inducement to express some of his criticisms of the concept of holomorphy. He wrote (*loc. cit.*, p. 210):

"$\cdots$ so ist damit bewiesen, dass der Begriff einer monogenen Function einer complexen Veränderlichen mit dem Begriff einer durch (arithmetische) Grössenoperationen ausdrückbaren Abhängigkeit sich nicht vollständig deckt. Daraus aber folgt dann, dass mehrere der wichtigsten Sätze der neueren Functionenlehre nicht ohne Weiteres auf Ausdrücke, welche im Sinne der älteren Analysten (Euler, Lagrange u.A.) Functionen einer complexen Veränderlichen sind, dürfen angewandt werden ($\cdots$ is thus proven that the concept of a monogenic function of a complex variable is not completely coextensive with that of dependence expressible by (arithmetic) operations on magnitudes. But from this it then follows that several of the most important theorems of contemporary function theory may not, without further justification, be applied to expressions which represent functions of a complex variable in the sense of the older analysts (Euler, Lagrange, *et al.*))."

WEIERSTRASS thereby represents another viewpoint from that of RIEMANN, who on p. 39 at the end of §20 of his dissertation [R] took a contrary position.

6*. A construction of WEIERSTRASS'. A number $\alpha \in \mathbb{C}$ is called *algebraic* if $p(\alpha) = 0$ for some non-zero polynomial p in $\mathbb{Z}[z]$. It is shown in algebra that the set K of all algebraic numbers is a countable field extension of $\mathbb{Q}$, and is consequently not all of $\mathbb{C}$. In an 1886 letter to L. KOENIGSBERGER (published in *Acta Math.* 39(1923), 238-239) WEIERSTRASS showed that

There exists a transcendental entire function $f(z) = \sum a_\nu z^\nu$ with $a_\nu \in \mathbb{Q}$ for all ν, such that $f(K) \subset K$ and $f(\mathbb{Q}) \subset \mathbb{Q}$.

Proof. Due to countability of $\mathbb{Z}[z]$, its non-zero elements can be enumerated $p_0, p_1, p_2, \ldots$. Set $q_n := p_0 p_1 \ldots p_n$ and let r_n denote its degree, $n \in \mathbb{N}$. A sequence $m_n \in \mathbb{N}$ is inductively defined by

$$m_0 := 0 \, , \, m_1 := m_0 + r_0 + 1, \ldots, m_{n+1} := m_n + r_n + 1.$$

For every non-zero rational number k_n the polynomial $k_n q_n(z) z^{m_n}$ involves only powers z^ℓ with $\ell \in \{m_n, \ldots, m_n + r_n\}$ and the term $z^{m_n + r_n}$ is present with a non-zero coefficient. Differently indexed polynomials therefore have no powers in common, so $f(z) := \sum k_n q_n(z) z^{m_n}$ is a formal power series with rational coefficients in which every summand $z^{m_n + r_n}$ actually occurs. Now if we choose the k_n so small that all the coefficients of $k_n q_n(z) z^{m_n}$ are smaller than $[(m_n + r_n)!]^{-1}$, then it will follow that $f \in \mathcal{O}(\mathbb{C})$ and $f \notin \mathbb{C}[z]$.

Given $\alpha \in K$, it is a zero of some p_s and so $q_n(\alpha) = 0$ for all $n \geq s$ and therefore

$$f(\alpha) = \sum_{0}^{s-1} k_n q_n(\alpha) \alpha^{m_n} \in K$$

with, moreover, $f(\alpha) \in \mathbb{Q}$ in case $\alpha \in \mathbb{Q}$.   □

This elegant construction of WEIERSTRASS' caused quite a sensation in its day, not the least because of the essential use it made of CANTOR's counting method, which at that time was by no means generally accepted. Following WEIERSTRASS, P. STÄCKEL wrote a paper in 1895 " Über arithmetische Eigenschaften analytischer Funktionen," *Math. Annalen* **46**, pp. 513-520, in which he showed that

If A is a countable and B a dense subset of $\mathbb{C}$, then there exists a transcendental entire function f with $f(A) \subset B$.

In particular, there is a transcendental entire function whose values at *every* algebraic argument are *transcendental* ($:=$ non-algebraic) numbers. A famous theorem of LINDEMANN, GELFOND and SCHNEIDER affirms that the exponential function takes transcendental values at every *non-zero* algebraic argument. In 1904 in *Math. Annalen* **58**, pp. 545-557 G. FABER constructed a transcendental entire function which together with all its derivatives takes algebraic values at algebraic arguments.

Weierstrass' construction laid to rest the idea that an entire function with rational coefficients which assumes rational values at every rational argument had to be itself rational, that is, a polynomial. Nevertheless under additional hypotheses this conclusion does follow. Already in 1892 HILBERT remarked, at the end of his paper "Über die Irreduzibilität ganzer rationaler Funktionen mit ganzzahligen Koeffizienten" (*Jour. für die Reine und Angew. Math.* **110**, pp. 104-129; also *Gesammelte Abhandlungen*, vol. II, pp. 264-286), that a power series $f(z)$ with positive radius of convergence is necessarily a polynomial if it is an *algebraic* function (meaning that $p(f(z), z) \equiv 0$ for some non-zero polynomial $p(w, z)$ of two variables) and if it assumes rational values for all rational arguments from some non-empty interval (however short) in $\mathbb{R}$.

Exercises

Exercise 1. Formulate and prove an inference from the normal convergence of a series $\sum f_\nu$, $f_\nu \in \mathcal{O}(G)$, to that of a series $\sum F_\nu$ of its primitives.

Exercise 2. Show that the series $\sum_{\nu \geq 1}(z^{2^n} - z^{-2^n})^{-1}$ converges compactly in $\mathbb{C}^\times \setminus \partial \mathbb{E}$ and determine the limit function.

Exercise 3. According to Exercise 3 in §2, Chapter 3, the series $\sum_{n \geq 1} \frac{(-1)^n}{z+n}$ converges compactly in $\mathbb{C} \setminus \{-1, -2, -3, \ldots\}$ to a holomorphic function f. Find the power series development of f around 0.

Exercise 4. Let f be holomorphic in some neighborhood of 0. Show that if the series

$$(*) \qquad\qquad \sum_{n \geq 1} f^{(n)}(z)$$

converges absolutely at $z = 0$, then f is an entire function and the series $(*)$ converges normally throughout $\mathbb{C}$.

Exercise 5. Let $0 \leq m_1 < m_2 < \cdots$ be a strictly increasing sequence of integers, p_ν complex polynomials satisfying degree $(z^{m_\nu} p_\nu(z)) < m_{\nu+1}$ for all ν. Suppose that the series $f(z) := \sum_{\nu \geq 0} z^{m_\nu} p_\nu(z)$ converges compactly in $\mathbb{E}$. Show that the Taylor series of f around 0 contains just those monomials $a_k z^k$ which occur in the summands $z^{m_\nu} p_\nu(z)$ (theorem on removal of parentheses).

§5 The open mapping theorem and the maximum principle

The fibers $f^{-1}(a)$ of a non-constant holomorphic function f consist of isolated points (cf. 1.3) and are thus very "thin". So f does not collapse open sets U too violently and correspondingly the image set $f(U)$ is "fat." This heuristic will now be made precise. To this end we introduce some convenient terminology.

A continuous mapping $f : X \to Y$ between topological spaces X and Y is called *open* if the image $f(U)$ of every open subset U of X is an open subset of Y. (By contrast continuity means that every open subset V of Y has an open pre-image $f^{-1}(V)$ in X.) Every topological mapping is open. The mapping $x \mapsto x^2$ of $\mathbb{R}$ into $\mathbb{R}$ is *not* open. But the latter phenomenon cannot occur among holomorphic mappings; we even have the

1. Open Mapping Theorem. *If f is holomorphic and nowhere locally constant in the domain D, then it is an open mapping of D into $\mathbb{C}$.*

The proof rests on a fact of some interest by itself:

Existence theorem for zeros. *Let V be an open disc centered at c, with $\overline{V} \subset D$. Let f be holomorphic in D and satisfy $\min_{z \in \partial V} |f(z)| > |f(c)|$. Then f has a zero in V.*

Proof. If f were zero-free in V, then it would be zero-free in an open neighborhood U of $\overline{V}$ lying in D. The function $g : U \to \mathbb{C}$, $z \mapsto 1/f(z)$ would then be holomorphic in U and the mean value inequality would imply that

$$|f(c)|^{-1} = |g(c)| \leq \max_{z \in \partial V} |g(z)| = \max_{z \in \partial V} \frac{1}{|f(z)|} = \left(\min_{z \in \partial V} |f(z)| \right)^{-1},$$

i.e., $|f(c)| \geq \min_{z \in \partial V} |f(z)|$, contrary to hypothesis. □

The existence theorem just proved delivers immediately a

Quantitative form of the open mapping theorem. *Let V be an open disc centered at c with $\overline{V} \subset D$ and let f be holomorphic in D and satisfy $2\delta := \min_{z \in \partial V} |f(z) - f(c)| > 0$. Then $f(V) \supset B_\delta(f(c))$.*

Proof. For every b with $|b - f(c)| < \delta$,

$$|f(z) - b| \geq |f(z) - f(c)| - |b - f(c)| > \delta \quad \text{for all } z \in \partial V.$$

It follows that $\min_{z \in \partial V} |f(z) - b| > |f(c) - b|$. The preceding existence theorem is therefore applicable to $f(z) - b$ and furnishes a $\tilde{z} \in V$ with $f(\tilde{z}) = b$. □

By now the proof of the open mapping theorem itself is trivial: let U be an open subset of D, $c \in U$. We have to show that $f(U)$ contains a disc about $f(c)$. Since f is not constant around c, there is a disc V centered at c with $\overline{V} \subset U$ and $f(c) \notin f(\partial V)$ (thanks to the Identity Theorem). Therefore the number $2\delta := \min_{z \in \partial V} |f(z) - f(c)|$ is positive. From this it follows that $B_\delta(f(c)) \subset f(V) \subset f(U)$. □

The open mapping theorem has important consequences. Thus for example it is immediately clear that a holomorphic function with constant real part, or constant imaginary part, or constant modulus is itself constant. More generally, the reader should satisfy himself that

If $P(X, Y) \in \mathbb{R}[X, Y]$ is a non-constant polynomial, then every function f which is holomorphic in a region G and for which $P(\Re f(z), \Im f(z))$ is constant, is itself constant.

The open mapping theorem is frequently also formulated as the

Theorem on the preservation of regions. *If f is holomorphic and non-constant in the region G, then $f(G)$ is also a region.*

Proof. According to the Identity Theorem f is nowhere locally constant in G, so that by the open mapping theorem the set $f(G)$ is open. As f is continuous, the image set inherits the connectedness of G as well.   □

Our proof of the open mapping theorem goes back to CARATHÉODORY ([5], pp. 139/140). The decisive tool is the mean value inequality hence, indirectly, the Cauchy integral formula. It is possible, but admittedly quite tedious, to carry through an integration-free proof; cf., say, G. T. WHY-BURN, *Topological Analysis*, Princeton University Press, 1964, p. 79. The fact that holomorphic functions are locally developable into power series admits an elementary proof by means of the open mapping theorem (see, e.g., P. PORCELLI and E. H. CONNELL, "A proof of the power series expansion without Cauchy's formula," *Bull. Amer. Math. Soc.* **67**(1961), 177-181).

2. The maximum principle. With the help of Gutzmer's formula the maximum principle was derived in 3.2 in the formulation: •

A function $f \in \mathcal{O}(G)$ whose modulus experiences a local maximum is constant in G.

This assertion is a special case of the open mapping theorem: namely, if there is a $c \in G$ and a neighborhood U of c in G with $|f(z)| \leq |f(c)|$ for all $z \in U$, then

$$f(U) \subset \{w \in \mathbb{C} : |w| \leq |f(c)|\}.$$

The set $f(U)$ is then certainly *not* a neighborhood of $f(c)$, that is, f is not an open mapping. Consequently, since G is connected, the open mapping theorem implies that f is constant.   □

If we interpret the real number $|f(z)|$ as altitude above the point z (measured perpendicularly to the z-plane), then we get a surface in $\mathbb{R}^3$ over $G \subset \mathbb{C} = \mathbb{R}^2$, which is occasionally designated as the *analytic landscape of* f. In terms of it the maximum principle may be stated in the following suggestive fashion:

In the analytic landscape of a holomorphic function there are no genuine peaks.

The maximum principle is often used in the following variant:

Maximum principle for bounded regions. *Let G be a bounded region, f a function which is holomorphic in G and continuous on $\overline{G} = G \cup \partial G$. Then the maximum of the function $|f|$ over $\overline{G}$ is assumed on the boundary ∂G of G:*

$$|f(z)| \leq |f|_{\partial G} \qquad \text{for all } z \in \overline{G}.$$

The reader should construct a proof for himself. The hypothesis that G be bounded is essential: the conclusion fails for the function $h(z) := \exp(\exp z)$ in the strip-like region $S := \{z \in \mathbb{C} : -\frac{1}{2}\pi < \Im z < \frac{1}{2}\pi\}$. Indeed in this example $|h|_{\partial S} = 1$, but $h(x) = \exp(e^x) \to \infty$ for $x \in \mathbb{R}$, $x \to +\infty$.

□

Application of the maximum principle to $1/f$ leads straightaway to the

Minimum principle. *Let f be holomorphic in G and let there be a point $c \in G$ at which $|f|$ experiences a local minimum, that is, at which $|f(c)| = \inf_{z \in U} |f(z)|$ for some neighborhood U of c in G. Then either $f(c) = 0$ or f is constant in G.*

Minimum principle for bounded regions. *Let G be a bounded region, f continuous in $\overline{G}$ and holomorphic in G. Then either f has zeros in G or else the minimum over $\overline{G}$ of $|f|$ is assumed on ∂G:*

$$|f(z)| \geq \min_{\zeta \in \partial G} |f(\zeta)| \qquad \text{for all } z \in G.$$

Evidently the minimum principle is a generalization of the existence theorem for zeros proved in subsection 1.

3. On the history of the maximum principle. RIEMANN wrote in 1851 (cf. [R], p. 22):

"*Eine harmonische Function u kann nicht in einem Punkt im Innern ein Minimum oder ein Maximum haben, wenn sie nicht überall constant ist.* (A harmonic function u cannot have either a minimum or a maximum at an interior point unless it is constant.)" BURKHARDT formulated this theorem for the real and imaginary parts of holomorphic functions on p.126 of his 1897 textbook [Bu] (p.197 of the English translation). In his 1906 work [Os] OSGOOD also treated the maximum and minimum principles only for harmonic functions (p. 652 of the 5th edition, 1928).

It seems to be difficult to find out when and where the theorem was first formulated for holomorphic functions and proved for them without a reduction to the harmonic case. Even experts in the history of function theory could not tell me whether the maximum principle occurs in Cauchy's work or not. In 1892 SCHOTTKY spoke of "a theorem of function theory" (the reader will find further details on this in 11.2.2). C. CARATHÉODORY

(German mathematician of Greek extraction, 1873-1950, originally an engineer, assistant to A. SOMMERFELD; at Munich from 1924 onward) gave his simple proof of the Schwarz lemma in 1912 (p. 110 of [Ca]) by means of the maximum principle for holomorphic functions (for this see 9.2.5), but in the course of doing so he said nothing about this important theorem. HURWITZ discussed the theorem in his *Vorlesungen über allgemeine Funktionentheorie und elliptische Funktionen* ([12], p. 107) which was first published in 1922 by Julius Springer, Berlin.

In 1915 L. BIEBERBACH (1886-1982) wrote in his little Göschen volume *Einführung in die konforme Abbildung* ([3], p. 8): "*Wenn f(z) im Inneren eines Gebietes G regulär und endlich ist, so besitzt |f(z)| kein Maximum im Inneren des Gebietes.* Die Behauptung (bekanntlich eine leichte Folgerung des Cauchyschen Integralsatzes) kann auch unmittelbar aus der *Gebietstreue* gefolgert werden. (If f(z) is regular and finite in the interior of a region G, then |f(z)| has no maximum there. This assertion (known to be an easy consequence of Cauchy's integral theorem) can also be immediately deduced from the open mapping theorem.)" In the second volume of his work [4], which appeared in 1927, BIEBERBACH spoke (p. 70) of the "principle of the maximum."

4. Sharpening the WEIERSTRASS convergence theorem. *Let G be a bounded region, f_n a sequence of functions which are continuous on $\overline{G}$ and holomorphic in G. If the sequence $f_n|\partial G$ converges uniformly on ∂G, then the sequence f_n converges uniformly in $\overline{G}$ to a limit function which is continuous on $\overline{G}$ and holomorphic in G.*

Proof. According to the maximum principle for bounded regions

$$|f_m - f_n|_{\overline{G}} = |f_m - f_n|_{\partial G}$$

holds for all m and n. Since $f_n|\partial G$ is a Cauchy sequence with respect to the supremum semi-norm $|\ |_{\partial G}$, this equality shows that f_n is a Cauchy sequence with respect to $|\ |_{\overline{G}}$. From the Cauchy convergence criterion 3.2.1, the continuity theorem 3.1.2 and Weierstrass' convergence theorem 4.1 the assertion follows.   □

WEIERSTRASS was aware of this phenomenon of the "*inward propagation of convergence.*" As simple consequences let us note

Corollary 1. *Let A be discrete in G, $f_n \in \mathcal{O}(G)$ a sequence which converges compactly in $G \backslash A$. Then in fact the sequence f_n converges compactly in the whole of G.*

And

Corollary 2. *Let there be associated with the sequence $f_n \in \mathcal{O}(G)$ a compactly convergent sequence $g_n \in \mathcal{O}(G)$ with a limit function which is not identically 0, such that the sequence $g_n f_n$ converges compactly in G. Then the original sequence f_n converges compactly in G.*

The reader will have no trouble carrying out the proofs of these, and is urged to do so.

5. The theorem of HURWITZ. This concerns the "preservation of zeros" under compact convergence. We first show

Lemma. *If the sequence $f_n \in \mathcal{O}(G)$ converges compactly in G to a non-constant f, then for every $c \in G$ there is an index $n_c \in \mathbb{N}$ and a sequence $c_n \in G$, $n \geq n_c$, such that*

$$\lim c_n = c \quad and \quad f_n(c_n) = f(c) \quad for \ all \quad n \geq n_c.$$

Proof. We may suppose $f(c) = 0$. By hypothesis $f \not\equiv 0$, so there exists (by the identity theorem) an open disc B centered at c with $\overline{B} \subset G$, such that f is zero-free in $\overline{B} \setminus \{c\}$. Since f_n converges to f uniformly on $\partial B \cup \{c\}$, there is an n_c such that $|f_n(c)| < \min\{|f_n(z)| : z \in \partial B\}$ for all $n \geq n_c$. According to the minimum principle (alternatively, the existence theorem for zeros in 1.), each f_n $(n \geq n_c)$ has a zero c_n in B. Necessarily $\lim_n c_n = c$. For otherwise there would be a subsequence c'_n convergent to a $c' \in \overline{B} \setminus \{c\}$ and then (continuous convergence) $0 = \lim f'_n(c_{n'}) = f(c')$, which cannot be. $\square$

This lemma is quantitatively sharpened by

Theorem of Hurwitz. *Suppose the sequence $f_n \in \mathcal{O}(G)$ converges compactly in G to $f \in \mathcal{O}(G)$. Let U be a bounded open subset of G with $\overline{U} \subset G$ such that f has no zeros on ∂U. Then there is an index $n_U \in \mathbb{N}$ such that for each $n \geq n_U$ the functions f and f_n have the same number of zeros in $\overline{U}$:*

$$\sum_{w \in \overline{U}} o_w(f) = \sum_{w \in \overline{U}} o_w(f_n) \quad for \ all \ n \geq n_U.$$

Proof. Since $f \not\equiv 0$ and $\overline{U}$ is compact, the number $m := \sum_{w \in \overline{U}} o_w(f)$ is finite (by the identity theorem). We carry out a proof by induction on m. In case $m = 0$, the number $\varepsilon := \min\{|f(z)| : z \in \overline{U}\}$ is positive. Since $|f_n - f|_{\overline{U}} < \varepsilon$ for almost all n, almost all f_n are zero-free in $\overline{U}$.

Now consider $m > 0$ and a zero $c \in U$ of f. According to the lemma there is an n_c and a sequence $c_n \in U$ for $n \geq n_c$ such that $f_n(c_n) = 0$ and $\lim c_n = c$. Then for these n there exist $h, h_n \in \mathcal{O}(G)$ such that

$$(*) \qquad f_n(z) = (z - c_n) h_n(z), \quad f(z) = (z - c) h(z).$$

Corollary 2 in 4. above and the fact that $\lim(z - c_n) = z - c$ insure that the sequence h_n converges compactly in G to h. But according to $(*)$, h has exactly $m - 1$ zeros in $\overline{U}$ and none on ∂U. Therefore the induction hypothesis furnishes an $n_U \geq n_c$ so that for each $n \geq n_U$ the function h_n has exactly $m - 1$ zeros in $\overline{U}$. Then from $(*)$ and the fact that $c_n \in U$ it follows that for each $n \geq n_U$ the function f_n has exactly m zeros in $\overline{U}$.

In 13.2.3 we will give a second proof of Hurwitz' theorem by means of a theorem of ROUCHÉ.

Hurwitz' theorem is always applicable if $f \neq 0$. In that case around each zero c of f there is a compact disc $\overline{B} \subset G$ such that f is zero-free in $\overline{B} \setminus \{c\}$. It is to be noted that there are sequences of zero-free functions, e.g., $f_n(z) := z/n$ in $\mathbb{C} \setminus \{0\}$, which converge compactly to the zero function.

Naturally a version of Hurwitz' theorem holds as well for a-places (consider the sequence $f_n(z) - a$). The theorem of HURWITZ (or even the lemma) contains as a special case

If f_n is a sequence of zero-free holomorphic functions in G which converges compactly in G to $f \in \mathcal{O}(G)$, then f is either identically zero or is zero-free in G.

This observation has the following consequence:

Let $f_n : G \to \mathbb{C}$ be injective holomorphic functions which converge compactly in G to $f : G \to \mathbb{C}$. Then f is either constant or injective.

Proof. Suppose that f is non-constant and consider any point $c \in G$. Each function $f_n - f_n(c)$ is zero-free in $G \setminus \{c\}$ because of the injectivity of f_n in G. The above observation, applied to the sequence $f_n - f_n(c)$ in the region $G \setminus \{c\}$, says that $f - f(c)$ is zero-free in $G \setminus \{c\}$; that is, $f(z) \neq f(c)$ for all $z \in G \setminus \{c\}$. Since c is any point whatsoever in G, this is just the assertion that f is injective in G. $\qquad\qquad \square$.

This assertion will play an important role in the proof, to be given in the second volume, of the Riemann mapping theorem.

Historical note. HURWITZ proved his theorem in 1889 with the aid of ROUCHÉ's theorem; cf. "Über die Nullstellen der Bessel'schen Funktion," (*Math. Werke* **1**, p.268). HURWITZ described his result in the following suggestive terms (p.269; here we maintain our earlier notation):

The zeros of f in G coincide with those places at which the roots of the equations $f_1(z) = 0, f_2(z) = 0, \ldots, f_\nu(z) = 0, \ldots$ "condense"..

Exercises

Exercise 1. Let $R > 0$, $f : B_R(0) \to \mathbb{C}$ holomorphic. For $p \in [0, R)$ define $M(p) := \sup\{|f(z)| : |z| = p\}$. Show that the mapping $p \mapsto M(p)$ of $[0, R)$ into $\mathbb{R}$ is non-decreasing and continuous. Show that this map is strictly increasing if f is not constant.

Exercise 2. Let G be a bounded region, f and g continuous and zero-free in $\overline{G}$ and holomorphic in G, with $|f(z)| = |g(z)|$ for all $z \in \partial G$. Show that then $f(z) = \lambda g(z)$ for some $\lambda \in \partial \mathbb{E}$ and all $z \in \overline{G}$.

Exercise 3. Let G be a region in $\mathbb{C}$, $f \in \mathcal{O}(G)$. Show that if $\Re f$ experiences a local maximum at some $c \in G$, then f is constant.

Chapter 9

Miscellany

> Wer vieles bringt, wird manchem etwas brin-
> gen (He who offers much will offer something
> to many).— J. W. von GOETHE

As soon as Cauchy's integral formula is available a plethora of themes
from classical function theory can be treated directly and independently
of each other. This freedom as to choice of themes forces one to impose
his own limits; in CARATHÉODORY ([5], p. viii) we read: "Die größte
Schwierigkeit bei der Planung eines Lehrbuches der Funktionentheorie liegt
in der Auswahl des Stoffes. Man muß sich von vornherein entschließen, alle
Fragen wegzulassen, deren Darstellung zu große Vorbereitungen verlangt.
(The greatest difficulty in planning a textbook on function theory lies in
the choice of material. You have to decide beforehand to leave aside all
questions whose treatment requires too much preparatory development.)"

The themes selected for this chapter, except for the theorem of RITT
on asymptotic power series developments, belong to the canonical material
of function theory. The theorem of RITT deserves to be rescued from
oblivion: its surprising statement generalizes an old theorem of É. BOREL
about arbitrarily specifying all the derivatives of an infinitely differentiable
real function at some point. This classical theorem of real analysis thus
finds a function-theoretic interpretation.

§1   The fundamental theorem of algebra

In chapter 4 (written by this author) of the book *Numbers* [19] we examined
the fundamental theorem of algebra and its history in some depth, giving
among others the proofs of ARGAND and LAPLACE. Here in what follows
we will present four function-theoretic proofs.

1. The fundamental theorem of algebra. *Every non-constant complex polynomial has at least one complex zero.*

This existence theorem was called by GAUSS the *Grundlehrsatz* of the theory of algebraic equations (cf. his *Werke 3*, p. 73). Since zeros always split off as linear factors (on this point cf. *Numbers*, chap. 4, §3), the theorem is equivalent to

Factorization theorem. *Every polynomial* $p(z) = a_0 + a_1 z + \cdots + a_n z^n \in \mathbb{C}[z]$ *of positive degree n (i.e., $a_n \neq 0$) is representable, uniquely up to the order of the factors, as a product*

$$p(z) = a_n(z - c_1)^{m_1}(z - c_2)^{m_2} \cdots (z - c_r)^{m_r},$$

where $c_1, \ldots, c_r \in \mathbb{C}$ are different, $m_1, \ldots, m_r \in \mathbb{N} \setminus \{0\}$ and $n = m_1 + \cdots + m_r$.

For real polynomials $p(z) \in \mathbb{R}[z]$, $\overline{p(z)} = p(\overline{z})$ and so whenever c is a zero, $\overline{c}$ is also a zero. Of course $(z - c)(z - \overline{c}) \in \mathbb{R}[z]$. Therefore

Every real polynomial $p(z)$ of degree $n \geq 1$ is uniquely expressible as a product of real linear factors and real quadratic polynomials.

By making use of the order function o_z we can formulate the fundamental theorem of algebra as an equation:

Every complex polynomial p of degree n satisfies $\sum_{z \in \mathbb{C}} o_z(p) = n$.

For transcendental entire functions there is no corresponding result. For example, we have both

$$\sum_{z \in \mathbb{C}} o_z(\exp) = 0 \quad \text{and} \quad \sum_{z \in \mathbb{C}} o_z(\sin) = \infty.$$

Almost all proofs of the fundamental theorem use the fact that polynomials of positive degree converge to ∞ uniformly as $|z|$ grows. We make this property more precise in the

Growth lemma. *Let $p(z) = \sum_0^n a_\nu z^\nu \in \mathbb{C}[z]$ be an nth degree polynomial. Then there exists a real $R > 0$ such that for all $z \in \mathbb{C}$ with $|z| \geq R$*

$$(1) \qquad \frac{1}{2}|a_n||z|^n \leq |p(z)| \leq 2|a_n||z|^n,$$

so that $\lim_{z \to \infty} \frac{|z|^\nu}{|p(z)|} = 0$ *for* $0 \leq \nu < n$.

Proof. We may suppose that $n \geq 1$. Set $r(z) := \sum_0^{n-1} |a_\nu||z|^\nu$. Then clearly for all z

$$|a_n||z|^n - r(z) \leq |p(z)| \leq |a_n||z|^n + r(z).$$

If $|z| \geq 1$ and $\nu < n$, then $|z|^\nu \leq |z|^{n-1}$ and so $r(z) \leq M|z|^{n-1}$, with $M := \sum_0^{n-1} |a_\nu|$. It follows that $R := \max\{1, 2M|a_n|^{-1}\}$ has the desired property. □

The proof just given is elementary in that only properties of the absolute value are used in it; consequently, the growth lemma is valid as well for polynomials over any *valued* field.

2. Four proofs of the fundamental theorem. The first three proofs involve *reductio ad absurdum*; so we assume that there is a polynomial $q(z) = \sum_0^n a_\nu z^\nu$ of degree $n \geq 1$ which has no complex zeros.

First proof (after R. P. BOAS: "Yet another proof of the fundamental theorem of algebra," *Amer. Math. Monthly* **71**(1964), 180; only the Cauchy integral theorem and the complex exponential function are used).

For $q^*(z) := \sum_0^n \bar{a}_\nu z^\nu \in \mathbb{C}[z]$, we have $q^*(\bar{c}) = \overline{q(c)}$ for every $c \in \mathbb{C}$. Therefore $g := qq^* \in \mathbb{C}[z]$ is zero-free and satisfies $g(x) = |q(x)|^2 > 0$ for all $x \in \mathbb{R}$. Writing $\zeta := e^{i\varphi}$, $0 \leq \varphi \leq 2\pi$, we have $\cos\varphi = \frac{1}{2}(\zeta + \zeta^{-1})$. It follows that

$$(\#) \quad 0 < \int_0^{2\pi} \frac{d\varphi}{g(2\cos\varphi)} = \frac{1}{i}\int_{\partial\mathbb{E}} \frac{d\zeta}{\zeta g(\zeta + \zeta^{-1})} = \frac{1}{i}\int_{\partial\mathbb{E}} \frac{\zeta^{2n-1}}{h(\zeta)}d\zeta,$$

with $h(z) := z^{2n}g(z + z^{-1})$; notice that $h(z)$ is a polynomial. Since g is zero-free in $\mathbb{C}$, h has no zeros in $\mathbb{C}^\times$. An easy calculation reveals that the constant term in h is $h(0) = |a_n|^2 \neq 0$. Therefore $1/h \in \mathcal{O}(\mathbb{C})$ and, since $n \geq 1$, so does $z^{2n-1}/h(z)$. Consequently according to Cauchy's integral theorem the integral on the right end of $(\#)$ vanishes, a contradiction! [In the first German edition of this book the integral $\int_{-r}^r \frac{dx}{g(x)}$ was considered instead of $(\#)$. By means of the growth lemma and the Cauchy integral theorem a contradiction in the form $0 = \lim_{r\to\infty} \int_{-r}^r \frac{dx}{g(x)}$ was reached. In BOAS's proof "there is no need to discuss the asymptotic behavior of any integrals."]

The remaining proofs all make use of the growth lemma.

Second proof (via the mean value inequality). Because q is zero-free, the function $f := 1/q$ is holomorphic in $\mathbb{C}$. Therefore $|f(0)| \leq |f|_{\partial B_r}$ for every circle ∂B_r of radius $r > 0$ centered at 0 (cf. 7.2.2). Since $\lim_{|z|\to\infty} |f(z)| =$

0 according to the growth lemma, we get $f(0) = 0$, in contradiction to $f(0) = q(0)^{-1} \neq 0$.

Third proof (via Liouville's theorem) . As in the second proof we use the facts that $f := 1/q \in \mathcal{O}(\mathbb{C})$ and $\lim_{|z| \to \infty} |f(z)| = 0$. But the latter fact implies that f is bounded in $\mathbb{C}$, hence by LIOUVILLE it is constant; that is, q is constant — which cannot be because $n \geq 1$.

Fourth proof (directly, from the minimum principle). Let $p(z) := \sum_0^n a_\nu z^\nu$ be a non-constant polynomial of degree n. Since $a_n \neq 0$, we can find an $s > 0$ so that $|p(0)| < \frac{1}{2}|a_n|s^n$. The growth lemma insures that if we increase s enough we can also have $|p(0)| < \min_{|z|=s} |p(z)|$. Therefore the minimum of $|p|$ over $\overline{B_s(0)}$ occurs at some point $a \in B_s(0)$. From the minimum principle it follows that $p(a) = 0$ (alternatively, we can also conclude this directly from the existence theorem for zeros in 8.5.1).

3. Theorem of GAUSS about the location of the zeros of derivatives. If $p(z)$ is a complex nth degree polynomial and $c_1, \ldots, c_n \in \mathbb{C}$ its (not necessarily distinct) zeros, then

(1) $$\frac{p'(z)}{p(z)} = \frac{1}{z - c_1} + \cdots + \frac{1}{z - c_n} = \sum_1^n \frac{\overline{z - c_\nu}}{|z - c_\nu|^2}.$$

This can be proved by induction on n, since from $p(z) = (z - c_n)q(z)$ follows $p'(z) = q(z) + (z - c_n)q'(z)$ and thus $p'(z)/p(z) = q'(z)/q(z) + 1/(z - c_n)$. With the help of (1) we quickly obtain:

Theorem (GAUSS, *Werke 3*, p.112). *If $c_1, \ldots, c_n$ are the (not necessarily distinct) zeros of the polynomial $p(z) \in \mathbb{C}[z]$, then for every zero $c \in \mathbb{C}$ of the derivative $p'(z)$ there are real numbers $\lambda_1, \ldots, \lambda_n$ such that*

$$c = \sum_1^n \lambda_\nu c_\nu \ , \ \lambda_1 \geq 0, \ldots, \lambda_n \geq 0 \ , \ \sum_1^n \lambda_\nu = 1.$$

Proof. If c is one of the zeros of p, say c_j, then set $\lambda_j := 1$ and $\lambda_\nu := 0$ for $\nu \neq j$. On the other hand, if $p(c) \neq 0$, then it follows from (1) that

$$0 = \frac{\overline{p'(c)}}{p(c)} = \sum_1^n \frac{c - c_\nu}{|c - c_\nu|^2}$$

and so, with $m_\nu := |c - c_\nu|^{-2} > 0$ and $m := \sum_1^n m_\nu$,

$$mc = \sum_1^n m_\nu c_\nu.$$

Consequently the numbers $\lambda_\nu := m_\nu/m$ have the required properties.   □

For any set $A \subset \mathbb{C}$ the intersection of all the *convex* sets which contain A is called the *convex hull* of A and denoted conv A. One checks readily that

$$\mathrm{conv}\{c_1, \ldots, c_n\} = \{z \in \mathbb{C} : z = \sum_1^n \lambda_\nu c_\nu \; ; \; \lambda_1 \geq 0, \ldots, \lambda_n \geq 0 \; , \; \sum_1^n \lambda_\nu = 1\}.$$

Consequently, the theorem of GAUSS may be expressed thus:

Every zero of $p'(z)$ lies in the convex hull of the set of zeros of $p(z)$.

Remark. From (1) it follows that all $z \in \mathbb{C}$ with $p'(z) \neq 0$ satisfy the inequality

$$\min_{1 \leq \nu \leq n} |z - c_\nu| \leq n|p(z)/p'(z)|,$$

which says that inside the circle of radius $n|p(z)/p'(z)|$ centered at z lies at least one zero of p. This information is successfully exploited in the numerical search for complex zeros of p via NEWTON's Method.

Exercises

Exercise 1. Let $p(z) \in \mathbb{C}[z]$ be a non-constant polynomial. Using the growth lemma and the open mapping theorem (but not the fundamental theorem of algebra) show that $p(\mathbb{C}) = \mathbb{C}$.

Exercise 2. Let f be an entire function with only finitely many zeros; let these be, each repeated as often as its multiplicity requires, $c_1, \ldots, c_n$.

a) Show that the equality $\frac{f'(z)}{f(z)} = \sum_{j=1}^n \frac{1}{z-c_j}$ for all $z \in \mathbb{C} \setminus \{c_1, \ldots, c_n\}$ holds if and only if f is a polynomial.

b) Show by means of an example that for an entire function f, the zeros of f' are not generally all to be found in the convex hull of those of f.

§2 Schwarz' lemma and the groups Aut 𝔼, Aut ℍ

The goal of this section is to prove that the automorphisms of the unit disc 𝔼 and of the upper half-plane ℍ described in 2.3.1-3 are *all* of the automorphisms of 𝔼 and ℍ, respectively. The tool used is a lemma concerning mappings of the unit disc which fix 0. The result goes back to H. A. SCHWARZ.

1. Schwarz' lemma. *Every holomorphic map* $f : \mathbb{E} \to \mathbb{E}$ *with* $f(0) = 0$ *satisfies*

$$|f(z)| \leq |z| \qquad \text{for all } z \in \mathbb{E}, \qquad \text{and } |f'(0)| \leq 1.$$

If there is at least one point $c \in \mathbb{E} \setminus \{0\}$ *with* $|f(c)| = |c|$, *or if* $|f'(0)| = 1$, *then* f *is a rotation about* 0, *that is, there is an* $a \in S_1$ *such that*

$$f(z) = az \qquad \text{for all } z \in \mathbb{E}.$$

Proof. $f(0) = 0$ means that

$$g(z) := f(z)/z \qquad \text{for } z \in \mathbb{E} \setminus \{0\} \, , \, g(0) := f'(0)$$

defines a holomorphic function g in $\mathbb{E}$. Since $|f(z)| < 1$ for every z in $\mathbb{E}$,

$$\max_{|z|=r} |g(z)| \leq \frac{1}{r} \qquad \text{for every positive } r < 1.$$

From the maximum principle follows then

$$|g(z)| \leq 1/r \qquad \text{for } z \in B_r(0) \, , \, 0 < r < 1.$$

Letting $r \to 1$, gives $|g(z)| \leq 1$, that is, $|f(z)| \leq |z|$, for all $z \in \mathbb{E}$ and $|f'(0)| = |g(0)| \leq 1$. In case either $|f'(0)| = 1$ or $|f(c)| = |c|$ for some $c \in \mathbb{E} \setminus \{0\}$, then $|g(0)| = 1$ or $|g(c)| = 1$; which says that $|g|$ attains a maximum in $\mathbb{E}$. According to the maximum principle, g is then a constant, of course of modulus 1. $\square$

2. Automorphisms of $\mathbb{E}$ fixing 0. The groups $\operatorname{Aut}\mathbb{E}$ and $\operatorname{Aut}\mathbb{H}$.
For every point c of a domain D in $\mathbb{C}$ and every subgroup L of $\operatorname{Aut} D$, the set of all automorphisms in L which fix c is a subgroup of L. It is called the *isotropy group of* c *with respect to* L. In case $L = \operatorname{Aut} D$ we denote this subgroup by $\operatorname{Aut}_c D$. For the group $\operatorname{Aut}_0\mathbb{E}$ of all *center-preserving* automorphisms of $\mathbb{E}$ we have

Theorem. *Every automorphism* $f : \mathbb{E} \to \mathbb{E}$ *with* $f(0) = 0$ *is a rotation:*

$$\operatorname{Aut}_0 \mathbb{E} = \{f : \mathbb{E} \to \mathbb{E}, z \mapsto f(z) = az : a \in S_1\}.$$

Proof. Certainly all rotations belong to $\operatorname{Aut}_0\mathbb{E}$. If conversely $f \in \operatorname{Aut}_0\mathbb{E}$, then also $f^{-1} \in \operatorname{Aut}_0\mathbb{E}$ and from Schwarz' lemma follow

$$|f(z)| \leq |z| \qquad \text{and} \qquad |z| = |f^{-1}(f(z))| \leq |f(z)| \qquad \text{for } z \in \mathbb{E},$$

that is, $|f(z)/z| = 1$ for all $z \in \mathbb{E} \setminus \{0\}$, and so $f(z)/z = a \in S_1$. $\square$

The explicit specification of all automorphisms of $\mathbb{E}$ is now simple. We base it on the following elementary

Lemma. *Let* J *be a subgroup of* $\operatorname{Aut} D$ *with the following properties:*

1) J acts transitively on D.

2) J contains, for some $c \in D$, the isotropy group $\mathrm{Aut}_c D$.

Then $J = \mathrm{Aut}\, D$.

Proof. Consider $h \in \mathrm{Aut}\, D$. On account of 1) there is a $g \in J$ with $g(h(c)) = c$. From 2) it follows that $f := g \circ h \in J$, so $h = g^{-1} \circ f \in J$.

Theorem.

$$\mathrm{Aut}\, \mathbb{E} \;=\; \left\{ \frac{az+b}{\bar{b}z+\bar{a}} : a, b \in \mathbb{C}\,,\ |a|^2 - |b|^2 = 1 \right\}$$

$$=\; \left\{ e^{i\varphi} \frac{z-w}{\bar{w}z-1} : w \in \mathbb{E}\,,\ 0 \le \varphi < 2\pi \right\}.$$

Proof. The two sets on the right are equal and constitute a subgroup J of $\mathrm{Aut}\, \mathbb{E}$ which acts transitively on $\mathbb{E}$ (cf. 2.3.2-4). On the basis of the preceding theorem $\mathrm{Aut}_0 \mathbb{E} = \{e^{i\varphi}z : 0 \le \varphi < 2\pi\} \subset J$ and so from the lemma it follows that $J = \mathrm{Aut}\, \mathbb{E}$.   □

According to 2.3.2 the mapping $h \mapsto h_{C'} \circ h \circ h_C$ of $\mathrm{Aut}\, \mathbb{E}$ into $\mathrm{Aut}\, \mathbb{H}$, where h_C, $h_{C'}$ designate the Cayley mappings, is a group isomorphism. Since we also have

$$\left\{ \frac{\alpha z + \beta}{\gamma z + \delta} : \begin{pmatrix} \alpha & \beta \\ \gamma & \delta \end{pmatrix} \in SL(2, \mathbb{R}) \right\}$$

$$=\; \left\{ h_{C'} \circ \frac{az+b}{\bar{b}z+\bar{a}} \circ h_C : a, b \in \mathbb{C}\,,\ |a|^2 - |b|^2 = 1 \right\}$$

as a result of 2.3.2, our theorem above has the

Corollary. $\mathrm{Aut}\, \mathbb{H} = \left\{ \dfrac{\alpha z + \beta}{\gamma z + \delta} : \begin{pmatrix} \alpha & \beta \\ \gamma & \delta \end{pmatrix} \in SL(2, \mathbb{R}) \right\}.$

3. Fixed points of automorphisms. Since the equation $\frac{az+b}{cz+d} = z$ has at most two solutions (unless $b = c = 0$ and $a = d$), the automorphisms of $\mathbb{E}$ and $\mathbb{H}$ other than the identity have at most two fixed points in $\mathbb{C}$. Here by *fixed point* of a mapping $f : D \to \mathbb{C}$ is meant any point $p \in D$ with $f(p) = p$.

Theorem. *Every automorphism h of $\mathbb{E}$ (respectively, $\mathbb{H}$) with two distinct fixed points in $\mathbb{E}$ (respectively, $\mathbb{H}$) is the identity.*

Proof. Because the groups $\mathrm{Aut}\, \mathbb{E}$ and $\mathrm{Aut}\, \mathbb{H}$ are isomorphic, it suffices to prove the result for $\mathbb{E}$. Since $\mathbb{E}$ is homogeneous, we may assume that one of the fixed points is 0. Then already we know that $h(z) = az$ for some $a \in S_1$, by Theorem 2. If there is another fixed point p, that is, one in $\mathbb{E} \setminus \{0\}$, then $h(p) = ap = p$, and so $a = 1$ and $h = \mathrm{id}$.   □

As in 2.2.1 let $h_A \in \mathrm{Aut}\,\mathbb{H}$ be the automorphism $z \mapsto \frac{\alpha z + \beta}{\gamma z + \delta}$ determined by the matrix $A = \begin{pmatrix} \alpha & \beta \\ \gamma & \delta \end{pmatrix} \in SL(2, \mathbb{R})$. The number $\mathrm{Tr}\,A := \alpha + \delta$ is called the *trace* of A. A direct verification shows

Theorem. *For $A \in SL(2, \mathbb{R}) \setminus \{\pm E\}$ the automorphism $h_A \in \mathrm{Aut}\,\mathbb{H}$ has a fixed point in $\mathbb{H}$, precisely when $|\mathrm{Tr}\,A| < 2$.*

All automorphisms $h_A : \mathbb{H} \to \mathbb{H}$ with $h_A \neq \mathrm{id}$ and $|\mathrm{Tr}\,A| \geq 2$ are therefore *fixed-point-free* in $\mathbb{H}$. Among such automorphisms are, in particular, all *translations* $z \mapsto z + 2\tau$, $\tau \in \mathbb{R} \setminus \{0\}$. To these translations correspond the automorphisms

$$z \mapsto \frac{(1 + i\tau)z - i\tau}{i\tau z + (1 - i\tau)}, \quad \tau \neq 0$$

of $\mathbb{E}$ (proof!), which are, of course, fixed-point-free in $\mathbb{E}$.

4. On the history of Schwarz' lemma. In a work entitled "Zur Theorie der Abbildung" (from the program of the Federal Polytechnical School in Zürich for the school-year 1869-70; vol. II, pp. 108-132 of his *Gesammelte Mathematische Abhandlungen*) Weierstrass' favorite student H. A. SCHWARZ stated a theorem, which for a long time attracted no attention, and used it, together with a convergence argument, in a proof of the Riemann mapping theorem. SCHWARZ formulated his proposition essentially as follows (cf. pp. 109-111):

Let $f : \mathbb{E} \to G$ be a biholomorphic mapping of the unit disc $\mathbb{E}$ onto a region G in $\mathbb{C}$ with $f(0) = 0 \in G$. Let ρ_1 denote the least and ρ_2 the greatest values of the distance function $|z|$, $z \in \partial G$ (cf. the figure below). Then

$$\rho_1 |z| \leq |f(z)| \leq \rho_2 |z| \quad \text{for all } z \in \mathbb{E}.$$

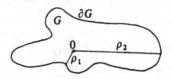

SCHWARZ proved this by examining the real part of the function

$\log[f(z)/z]$. His upper (respectively, lower) estimate can be gotten more simply by an application of the maximum principle to $f(z)/z$ (respectively, to $z/f(z)$).

In 1912 CARATHÉODORY [Ca] recognized the importance for function theory of the theorem SCHWARZ had used, and suggested (p. 110) that a particularly important variant of it be called Schwarz' lemma. The elegant proof offered in section 1, the one generally used nowadays, based on the maximum principle, is to be found in the work of CARATHÉODORY. It occurred earlier in his note "Sur quelques applications du théorème de Landau-Picard," *C.R. Acad. Sci. Paris* **144** (1907), 1203–1206 (*Gesammelte Math. Schriften* **3**, 6–9). There in a footnote CARATHÉODORY acknowledges his indebtedness for the proof to Erhard SCHMIDT: "Je dois cette démonstration si élégante d'un théorème connu de M. *Schwarz* (*Ges. Abh.* **2**, p. 108) à une communication orale de M. *E. Schmidt*. (I owe such an elegant proof of a known theorem of M. *Schwarz* to an oral communication from M. *E. Schmidt*)."

A beautiful application of Schwarz' lemma which is not so well known is

5. Theorem of STUDY. *Let* $f : \mathbb{E} \to G$ *be biholomorphic and let* $G_r := f(B_r(0))$ *denote the f-image of the open disc* $B_r(0)$, $0 < r < 1$. *Then*

a) *If* G *is convex, so is every* G_r, $0 < r < 1$.

b) *If* G *is star-shaped with center* $f(0)$, *so is every* G_r, $0 < r < 1$.

Proof. We assume $f(0) = 0$ (otherwise, replace $f(z)$ with $f(z) - f(0)$).

(a) Let $p, q \in G_r$, $p \neq q$, be given. We must show that every point $v = (1 - t)p + tq$, $0 \leq t \leq 1$, on the line segment joining p and q also belongs to G_r. Let $a, b \in B_r(0)$ be the f-preimages of p, q. We may suppose the notation such that $|a| \leq |b|$, and then $b \neq 0$. Then too $zab^{-1} \in \mathbb{E}$ for all $z \in \mathbb{E}$, and so the function

$$g(z) := (1 - t)f(zab^{-1}) + tf(z) , \; z \in \mathbb{E}$$

is well-defined. Because G is convex, $g(z)$ lies in G for every $z \in \mathbb{E}$. Therefore a holomorphic mapping $h : \mathbb{E} \to \mathbb{E}$ is well-defined by

$$h(z) := f^{-1}(g(z)) , \quad z \in \mathbb{E}.$$

Since $f(0) = 0$, we have $g(0) = 0$, hence $h(0) = 0$. From Schwarz' lemma therefore $|h(z)| \leq |z|$ for all $z \in \mathbb{E}$, and in particular

$$|f^{-1}(g(b))| \leq |b|.$$

Since $g(b) = (1 - t)f(a) + tf(b) = v$ and $|b| < r$, this says that $f^{-1}(v) \in B_r(0)$, so $v \in f(B_r(0)) = G_r$, as desired.

(b) Argue as in (a) but only allow p to be the point $f(0)$. □

Historical note: The theorem just presented is due to Eduard STUDY (German mathematician, 1862-1930; Professor at Marburg, Greifswald and, after 1903, at Bonn; author of important works on coordinate-free and projective geometry; with contributions to algebra and philosophy); it is a special case of a more general theorem of his about the convexity of image sets under biholomorphic mappings. Cf. pp. 110 ff. of E. STUDY: *Konforme Abbildungung einfach-zusammenhängender Bereiche, Vorlesungen über ausgewählte Gegenstände der Geometrie*; 2. Heft, published in collaboration with W. BLASCHKE, Teubner (1914), Berlin & Leipzig; see also in this connection G. PÓLYA and G. SZEGÖ, vol. I, part 3, problems 317, 318 and vol. II, part 4, problem 163.

The "sehr elementarer Beweis dieses schönen Satzes (very elementary proof of this beautiful theorem)" reproduced above was given in 1929 by T. RADÓ: "Bemerkung über die konformen Abbildungen konvexer Gebiete," *Math. Annalen* **102**(1930), 428-429.

Exercises

Exercise 1. Prove the following sharper version of Schwarz' lemma: If $f : \mathbb{E} \to \mathbb{E}$ is holomorphic with $o_0(f) = n \in \mathbb{N}$, $n \geq 1$, then

$$|f(z)| \leq |z|^n \quad \text{for all } z \in \mathbb{E} \quad \text{and} \quad |f^{(n)}(0)| \leq n!$$

Moreover, $f(z) \equiv az^n$ for some $a \in \partial\mathbb{E}$ if (and only if) either $|f^{(n)}(0)| = n!$ or $|f(c)| = |c|^n$ for some $c \in \mathbb{E} \setminus \{0\}$.

Exercise 2. (SCHWARZ-PICK lemma for $\mathbb{E}$) For $z, w \in \mathbb{E}$ set

$$\Delta(w, z) := \frac{|z - w|}{|\overline{w}z - 1|}.$$

Let $f : \mathbb{E} \to \mathbb{E}$ be holomorphic. Prove that

a) For all $w, z \in \mathbb{E}$, $\Delta(f(w), f(z)) \leq \Delta(w, z)$.

b) The following assertions are equivalent:

   i) $f \in \text{Aut}\,\mathbb{E}$.
   ii) For all $w, z \in \mathbb{E}$, $\Delta(f(w), f(z)) = \Delta(w, z)$.
   iii) There exist two distinct points $a, b \in \mathbb{E}$ such that $\Delta(f(a), f(b)) = \Delta(a, b)$.

Hint. For each $w \in \mathbb{E}$ let g_w denote the involutory automorphism $z \mapsto \frac{z-w}{\overline{w}z-1}$ of $\mathbb{E}$ and apply Schwarz' lemma to the mapping $h_w = g_{f(w)} \circ f \circ g_w$ of $\mathbb{E}$ into $\mathbb{E}$.

Exercise 3. (SCHWARZ-PICK lemma for $\mathbb{H}$) Set $\delta(w, z) := |\frac{z-w}{z-\overline{w}}|$ for $z, w \in \mathbb{H}$ and show that all the conclusions formulated in the preceding exercise, with

δ in the role of Δ, are valid for holomorphic mappings $f : \mathbb{H} \to \mathbb{H}$. *Hint.*
Use the Cayley transformation $h_C : \mathbb{H} \xrightarrow{\sim} \mathbb{E}$ and apply the SCHWARZ-PICK
lemma for $\mathbb{E}$ to the composite $g := h_C \circ g \circ h_C^{-1}$.

Exercise 4. a) Show that for every holomorphic mapping $f : \mathbb{E} \to \mathbb{E}$

(∗) $$\frac{|f'(z)|}{1 - |f(z)|^2} \leq \frac{1}{1 - |z|^2} \qquad \text{for all } z \in \mathbb{E}$$

and if equality holds here for a single $z \in \mathbb{E}$, then in fact equality holds for
all z and $f \in \operatorname{Aut} \mathbb{E}$.
  b) Prove for holomorphic $f : \mathbb{H} \to \mathbb{H}$ assertions like those in a) with the
inequality

$$\frac{|f'(z)|}{\Im f(z)} \leq \frac{1}{\Im z}$$

in the role of (∗).

Exercise 5. Let $\mathcal{G}$ be a subgroup of $\operatorname{Aut} \mathbb{E}$ which contains $\operatorname{Aut}_0 \mathbb{E}$. Show that
either $\mathcal{G} = \operatorname{Aut}_0 \mathbb{E}$ or $\mathcal{G} = \operatorname{Aut} \mathbb{E}$. *Hint.* Let $h_{w,\psi}$ denote the automorphism
$z \mapsto e^{i\psi} \frac{z-w}{\bar{w}z-1}$ of $\mathbb{E}$. Whenever $h_{w,\psi} \in \mathcal{G}$ so does $h_{a,\alpha}$ for each $\alpha \in \mathbb{R}$ and
each $a \in \mathbb{C}$ with $|a| = |w|$. Now consider $h_{|w|,0} \circ h_{|w|,\alpha} \in \mathcal{G}$ for arbitrary
$\alpha \in \mathbb{R}$.

Exercise 6. Let $f : \mathbb{E} \to \mathbb{H}$ be holomorphic, with $f(0) = i$. Show that

  a) $\frac{1-|z|}{1+|z|} \leq |f(z)| \leq \frac{1+|z|}{1-|z|}$ for all $z \in \mathbb{E}$;

  b) $|f'(0)| \leq 2$.

Exercise 7. Let $f : \mathbb{E} \to \mathbb{E}$ be holomorphic, with $f(0) = 0$. Let $n \in \mathbb{N}$,
$n \geq 1$, $\zeta := e^{2\pi i/n}$. Show that

(∗) $|f(\zeta z) + f(\zeta^2 z) + \cdots + f(\zeta^n z)| \leq n|z|^n$ for all $z \in \mathbb{E}$.

Moreover, if there is at least one $c \in \mathbb{E} \setminus \{0\}$ such that equality prevails
in (∗) at $z = c$, then there exists an $a \in \partial \mathbb{E}$ such that $f(z) = az^n$ for all
$z \in \mathbb{E}$. *Hint.* Consider the function $h(z) := \frac{1}{nz^{n-1}} \sum_{j=1}^{n} f(\zeta^j z)$. For the
proof of the implication $f(\zeta z) + \cdots + f(\zeta^n z) = naz^n \Rightarrow f(z) = az^n$, verify
that the function $k(z) := f(z) - az^n$ satisfies

$$k(\zeta z) + \cdots + k(\zeta^n z) = 0 \quad \text{and} \quad |az^n|^2 + 2\Re(az^n \overline{k(\zeta^j z)}) + |k(\zeta^j z)|^2 < 1$$

for every $j \in \{0, 1, \ldots, n-1\}$, and consequently $|k(z)|^2 < n(1 - |z|^{2n})$.

§3   Holomorphic logarithms and holomorphic roots

In 5.4.1 holomorphic logarithm functions $\ell(z)$ were introduced via the requirements that they satisfy $\exp(\ell(z)) = z$. In 7.1.2 we saw that in the slit plane $\mathbb{C}^-$ the principal branch $\log z$ of the logarithm possesses the integral representation $\int_{[1,z]} \frac{d\zeta}{\zeta}$.

If f is any holomorphic function in a domain D, then any holomorphic function g in D which satisfies

$$\exp(g(z)) = f(z)$$

will be called a *(holomorphic) logarithm of f* in D. Of course for f to possess a logarithm in D it must be *zero-free* in D. In the following we will prove existence assertions about holomorphic logarithms, using contour integrals. To be able to formulate things conveniently, we work in subsection 2 with *homologically simply-connected* domains.

From the existence of holomorphic logarithms follows at once the existence of holomorphic roots – cf. subsection 3; a converse of the theorem on roots will be found in subsection 5. In subsection 4 among other things we will derive the integer-valuedness of all integrals $\frac{1}{2\pi i} \int_\gamma \frac{f'(\zeta)}{f(\zeta)} d\zeta$ over closed paths γ.

1. Logarithmic derivative. Existence lemma. If g is a logarithm of f in D, which means that $f = e^g$, then

$$(1) \qquad\qquad\qquad g' = f'/f.$$

Generally, for any zero-free holomorphic function f in D, the quotient f'/f is called the *logarithmic derivative of f*. (This terminology is suggested by the dangerous notation $g = \log f$, which one is inclined to use when a logarithm of f exists, even though it obscures the non-uniqueness issue.) For the logarithmic derivative the product rule $(f_1 f_2)' = f_1' f_2 + f_1 f_2'$ becomes a

Sum formula:

$$(f_1 f_2)'/f_1 f_2 = f_1'/f_1 + f_2'/f_2.$$

Existence lemma. *The following assertions about a zero-free holomorphic function in a domain D are equivalent:*

  i) *There exists a holomorphic logarithm of f in D.*

  ii) *The logarithmic derivative f'/f is integrable in D.*

Proof. i) ⇒ ii) Any holomorphic logarithm of f is, according to (1), a primitive of f'/f.

ii) ⇒ i) We may assume that D is a region. Let $F \in \mathcal{O}(D)$ be a primitive of f'/f in D. Then $h := f \cdot \exp(-F)$ satisfies $h' = 0$ throughout D, and so for some constant a, necessarily non-zero since f is, we have $f = a \exp(F)$. Being non-zero, a has the form $a = e^b$ for some $b \in \mathbb{C}$. Then the function $g := F + b$ satisfies $\exp g = f$. □

2. Homologically simply-connected domains. Existence of holomorphic logarithm functions. A domain D in $\mathbb{C}$ in which *every* holomorphic function $g \in \mathcal{O}(D)$ is integrable is called *homologically simply-connected*.[1] According to the integrability criterion 6.3.2 this property is enjoyed by a domain D exactly when:

$$\int_\gamma g(\zeta)d\zeta = 0 \quad \text{for all } g \in \mathcal{O}(D) \text{ and all closed paths } \gamma \text{ in } D.$$

On the basis of the Cauchy integral theorem 7.1.2 all star-shaped regions in $\mathbb{C}$ are homologically simply-connected. There are however many other kinds of examples.

As an important immediate consequence of the existence lemma we obtain

Existence theorem for holomorphic logarithms. *In a homologically simply-connected domain every zero-free holomorphic function has a holomorphic logarithm.*

Thus in a star-shaped region G, like $\mathbb{C}$ or the slit plane $\mathbb{C}^-$, every zero-free holomorphic function f has the form $f = e^g$, $g \in \mathcal{O}(G)$. Here the functions g can even be written down explicitly: Fix $c \in G$, choose any b with $e^b = f(c)$ and for each $z \in G$ let γ_z be any path in G from c to z. Then one such g is

$$(2) \qquad g(z) := \int_{\gamma_z} \frac{f'(\zeta)}{f(\zeta)}d\zeta + b.$$

It follows in particular that

$$(3) \qquad f(z) = f(c) \exp \int_{\gamma_z} \frac{f'(\zeta)}{f(\zeta)}d\zeta.$$

---

[1]This concept, which describes a *function-theoretic* property of domains, is convenient and useful in many considerations. But in reality it is a superfluous concept: Because sufficiently detailed knowledge of plane topology would reveal that the homologically simply-connected domains in $\mathbb{C}$ are precisely the *topologically* simply-connected ones. This equivalence of a function-theoretic and a topological condition is however not particularly relevant in what follows and we won't go into it any further until the second volume.

These are the natural generalizations of the equations

$$\log z = \int_{[1,z]} \frac{d\zeta}{\zeta} \qquad \text{and} \quad z = \exp(\log z)$$

(where $f(z) := z$).

3. Holomorphic root functions. Let $n \geq 1$ be an integer, D a domain in $\mathbb{C}$. A function $q \in \mathcal{O}(D)$ is called a *(holomorphic)nth root of* $f \in \mathcal{O}(D)$ if $q^n = f$ throughout D.

If D is a region and q is an nth root of $f \neq 0$, then, for $\zeta := \exp(2\pi i/n)$, the functions $q, \zeta q, \zeta^2 q, \ldots, \zeta^{n-1} q$ are all of the nth roots of f.

Proof. Since $f \neq 0$, there is an open disc $B \subset D$ in which f, hence also q, is zero-free; so $1/q \in \mathcal{O}(B)$. If now $\tilde{q} \in \mathcal{O}(D)$ is any nth root of f, then $(\tilde{q}/q)^n = 1$ in B. That is, in B the continuous function $\tilde{q}/q$ takes values in $\{1, \zeta, \ldots, \zeta^{n-1}\}$. By virtue of the connectedness of B, $\tilde{q}/q$ is therefore constant in B: $\tilde{q} = \zeta^k q$ in B for some $0 \leq k < n$. From the identity theorem it follows that $\tilde{q} = \zeta^k q$ throughout D. $\square$

Theorem on roots. *If $g \in \mathcal{O}(D)$ is a logarithm of f in D, then for each $n = 1, 2, 3, \ldots$ the function $\exp(\frac{1}{n}g)$ is an nth root of f.*

Proof. This is immediate from the addition theorem for the exponential, which shows that

$$\left[\exp\left(\frac{1}{n}g\right)\right]^n = \exp g = f.   \square$$

From the existence theorem 1 there now follows immediately an

Existence theorem for holomorphic roots . *If D is homologically simply-connected and $f \in \mathcal{O}(D)$ is zero-free, then for every $n \geq 1$, f has an nth root.*

This is not true of an arbitrary domain: The reader should show that the function $f(z) := z$ has no square-root in the annulus $\{z \in \mathbb{C} : 1 < |z| < 2\}$.

4. The equation $f(z) = f(c) \exp \int_\gamma \frac{f'(\zeta)}{f(\zeta)} d\zeta$. In homologically simply-connected domains $\int_\gamma \frac{f'(\zeta)}{f(\zeta)} d\zeta$ is path-independent. For general domains this integral itself is not path-independent but its exponential is.

Theorem. *Let D be an arbitrary domain, and let f be holomorphic and zero-free in D. For any path γ in D with initial point c and terminal point z, we have*

$$f(z) = f(c) \exp \int_\gamma \frac{f'(\zeta)}{f(\zeta)} d\zeta.$$

Proof. Let $[a, b]$ be the parameter interval of γ and choose finitely many points $a =: t_0 < t_1 < \cdots < t_n := b$ and discs $U_1, \ldots, U_n$ in D in such a way that the path $\gamma_\nu := \gamma|[t_{\nu-1}, t_\nu]$ has its trajectory wholly in U_ν ($\nu = 1, 2, \ldots, n$). Compactness of γ clearly ensures that this can be done. Then (cf. 1.(3))

$$\exp \int_{\gamma_\nu} \frac{f'(\zeta)}{f(\zeta)} d\zeta = \frac{f(\gamma(t_\nu))}{f(\gamma(t_{\nu-1}))}, \qquad 1 \le \nu \le n.$$

Since $\gamma = \gamma_1 + \gamma_2 + \cdots + \gamma_n$, it follows (addition theorem) that

$$\exp \int_\gamma \frac{f'(\zeta)}{f(\zeta)} d\zeta = \prod_{\nu=1}^{n} \exp \int_{\gamma_\nu} \frac{f'(\zeta)}{f(\zeta)} d\zeta = \frac{f(\gamma(b))}{f(\gamma(a))} = \frac{f(z)}{f(c)}.$$

Corollary. *Let f be holomorphic and zero-free in the domain D, and let γ be an arbitrary closed path in D. Then*

$$\int_\gamma \frac{f'(\zeta)}{f(\zeta)} d\zeta \in 2\pi i \mathbb{Z}.$$

Proof. Here $c = z$, so the theorem shows that $\exp \int_\gamma \frac{f'(\zeta)}{f(\zeta)} d\zeta = 1$. Since $\ker(\exp) = 2\pi i \mathbb{Z}$, the stated result follows. $\square$

This corollary will be used in 5.1 to show that the index function is integer-valued. In 13.3.2 the integral $\int_\gamma \frac{f'(\zeta)}{f(\zeta)} d\zeta$ will be used to count the number of zeros and poles of f.

5. The power of square-roots. With the help of corollary 4 a converse to the theorem on roots in subsection 3 can be proved:

(1) *Let M be an infinite subset of $\mathbb{N}$, $f \in \mathcal{O}(D)$ a function which is nowhere locally identically 0. If for each $m \in M$, f has a holomorphic mth root function in D, then f has a holomorphic logarithm in D.*

Proof. First we note that f must be zero-free in D. If namely, $q_m \in \mathcal{O}(D)$ is an mth root of f, then $o_z(f) = m o_z(q_m)$ at each $z \in D$. From which follows $o_z(f) = 0$, for all $z \in D$, since otherwise the right side would be arbitrarily large, whereas the hypothesis says that f has no zero of order ∞.

According to the existence lemma 1 it therefore suffices to show that f'/f is integrable in D, i.e., that $\int_\gamma (f'/f) d\zeta = 0$ for every closed path γ

in D (recall the integrability criterion 6.3.2). Now from $q_m^m = f$ follows $mq_m^{m-1}q_m' = f'$ and so

$$\int_\gamma \frac{f'(\zeta)}{f(\zeta)} d\zeta = m \int_\gamma \frac{q_m'(\zeta)}{q_m(\zeta)} d\zeta, \qquad m \in M.$$

According to corollary 4 both these integrals lie in $2\pi i \mathbb{Z}$. The right side must therefore vanish for some m as m runs through M, for otherwise its modulus is at least $2\pi m$, i.e., it is arbitrarily large. It follows that $\int_\gamma (f'/f) d\zeta = 0$, as desired.   □

We may immediately obtain

Theorem. *Suppose that every zero-free holomorphic function in the domain D has a holomorphic square-root in D. Then every zero-free function in $\mathcal{O}(D)$ also has a holomorphic logarithm and holomorphic nth roots in D, for every $n \in \mathbb{N}$.*

Proof. Choose $M := \{2^k : k \in \mathbb{N}\}$ in (1).   □

Remark. The power of (iterated) square-root extraction was impressively demonstrated by the technique which A. HURWITZ suggested in 1911 for introducing the real logarithm function into elementary analysis; cf. "Über die Einführung der elementaren Funktionen in der algebraischen Analysis," *Math. Annalen* **70**(1911), 33-47 [*Werke* **1**, 706-721].

Exercises

Exercise 1. Determine all pairs of entire functions f_1, f_2 which satisfy $f_1^2 + f_2^2 = 1$.

Exercise 2. Let D be an open neighborhood of 0, $f : D \to \mathbb{C}$ a holomorphic function with $f(0) = 0$. Show that for any $m \in \mathbb{N}$ there is an open neighborhood $U \subset D$ of 0 and a holomorphic function $g : U \to \mathbb{C}$ satisfying $g(z)^m = f(z^m)$ for all $z \in U$.

Exercise 3. Let D_1, D_2 be homologically simply-connected domains. Show that if $D_1 \cap D_2$ is connected, then $D_1 \cup D_2$ is also homologically simply-connected.

Exercise 4. Let D be a domain in $\mathbb{C}$, $a \in D$, $f : D \to \mathbb{C}$ holomorphic with $o_a(f) \in \mathbb{N}$. Prove the equivalence of the following statements:

   i) There is a neighborhood U of a in D and a holomorphic square-root for $f|U$.

ii) $o_a(f)$ is an even integer.

§4 Biholomorphic mappings. Local normal forms

The real function $\mathbb{R} \to \mathbb{R}$, $x \mapsto x^3$ is infinitely differentiable and has a continuous inverse function $\mathbb{R} \to \mathbb{R}$, $y \mapsto \sqrt[3]{y}$, but this inverse function is not differentiable at 0. This phenomenon does not occur in $\mathbb{C}$: holomorphic injections are necessarily biholomorphic (subsection 1). As with functions on the line, functions f in the plane having $f'(c) \neq 0$ are injective in a neighborhood of c. We will give two proofs of this (subsection 2): one via integral calculus (which is also valid for functions on $\mathbb{R}$) and one via power series. Taken together this means that a holomorphic mapping f is locally biholomorphic around every point c where $f'(c) \neq 0$. It is further deducible from this that in the small (i.e., locally) each non-constant holomorphic function f has a unique normal form

$$f(z) = f(c) + h(z)^n$$

provided $f'(c) \neq 0$ (subsection 3). Mapping-theoretically this means that near c, f is a covering map which branches nowhere except possibly at c (subsection 4).

1. Biholomorphy criterion. *Let $f : D \to \mathbb{C}$ be a holomorphic injection. Then $D' := f(D)$ is a domain in $\mathbb{C}$ and $f'(z) \neq 0$ for all $z \in D$.*

The mapping $f : D \to D'$ is biholomorphic and the inverse function f^{-1} satisfies

$$(f^{-1})'(w) = 1/f'(f^{-1}(w)) \qquad \text{for all } w \in D'.$$

Proof. a) Since f is nowhere locally constant, the open mapping theorem affirms that f is open, so D' is a domain; but it follows too that the inverse map $f^{-1} : D' \to D$ is *continuous*, because for every open subset U of D, its f^{-1}-preimage $(f^{-1})^{-1}(U)$, being $f(U)$, is open in D'.

Because f is injective the derivative cannot vanish identically in any disc lying in D, so according to the identity theorem its zero-set $N(f')$ is discrete and closed in D. Since f is an open mapping , the image $M := f(N(f'))$ is *discrete and closed in D'*.

b) Consider $d \in D' \setminus M$ and set $c := f^{-1}(d)$. We have $f(z) = f(c) + (z - c)f_1(z)$, where $f_1 : D \to \mathbb{C}$ is continuous at c and $f_1(c) = f'(c) \neq 0$. For $z = f^{-1}(w)$, $w \in D'$, it follows that $w = d + (f^{-1}(w) - c)f_1(f^{-1}(w))$. The function $q := f_1 \circ f^{-1}$ is continuous at d and $q(d) = f'(c) \neq 0$. Therefore we can transform the last equation into

$$f^{-1}(w) = f^{-1}(d) + (w - d)1/q(w) \qquad \text{for all } w \in D' \text{ near } d.$$

From which we infer that f^{-1} is complex-differentiable at d and that

$$(f^{-1})'(d) = 1/f'(c) = 1/f'(f^{-1}(d)) \qquad \text{for all } d \in D' \setminus M.$$

c) By now we know that f^{-1} is holomorphic in $D' \setminus M$ and continuous throughout D'. It follows from Riemann's continuation theorem 7.3.4 that $f^{-1} \in \mathcal{O}(D')$. The equation $(f^{-1})'(w) \cdot f'(f^{-1}(w)) = 1$, which holds in $D' \setminus M$ by b) therefore remains valid throughout D' by continuity. In particular, $f'(z) \neq 0$ for all $z \in D$.   □

In the proof just given the open mapping theorem, the identity theorem and Riemann's continuation theorem were all used; in this sense the proof is "expensive." But part b) of the proof (a variant of the chain rule) is completely elementary.

2. Local injectivity and locally biholomorphic mappings. In order to be able to apply the biholomorphy criterion one needs conditions which insure the injectivity of holomorphic mappings. Analogous to the situation in $\mathbb{R}$ we have the

Injectivity lemma. *Let $f : D \to \mathbb{C}$ be holomorphic, $c \in D$ a point at which $f'(c) \neq 0$. Then there is a neighborhood U of c with the property that the restriction $f|U : U \to \mathbb{C}$ is injective.*

First proof. We will use the fact that derivatives are uniformly approximable by difference quotients; more precisely (cf. lemma 5.3):

Approximation lemma. *If B is a disc centered at c and lying in D, and f is holomorphic in D, then*

$$(*) \qquad \left| \frac{f(w) - f(z)}{w - z} - f'(c) \right| \leq |f' - f'(c)|_B \qquad \text{for all } w, z \in B , w \neq z.$$

For the proof of this, note that $f(\zeta) - f'(c)\zeta$ is a primitive of $f'(\zeta) - f'(c)$ in D, and so

$$f(w) - f(z) - f'(c)(w - z) = \int_z^w (f'(\zeta) - f'(c))d\zeta \qquad \text{for all } w, z \in B,$$

the integration being along the segment $[z, w] \subset B$. The standard estimate for integrals gives

$$\left| \int_z^w (f'(\zeta) - f'(c)) d\zeta \right| \le |f' - f'(c)|_B |w - z|$$

and therewith the inequality $(*)$.   □

The proof of local injectivity is now trivial: If namely $f'(c) \ne 0$, then continuity of f' means that for appropriate $r > 0$, $B := B_r(c)$ lies in D and

$$|f' - f'(c)|_B < |f'(c)|.$$

Then for $w, z \in B$ with $w \ne z$ we necessarily have $f(w) \ne f(c)$, since otherwise $(*)$ furnishes the contradiction $|f'(c)| < |f'(c)|$. Consequently $f|B : B \to \mathbb{C}$ is injective.   □

Second proof. This one uses the following

Injectivity lemma for power series. *Suppose that $f(z) = \sum a_\nu (z - c)^\nu$ converges in $B := B_r(c)$, $r > 0$ and that $|a_1| > \sum_{\nu \ge 2} \nu |a_\nu| r^{\nu-1}$. Then $f : B \to \mathbb{C}$ is injective.*

This is verified by calculating. Consider $w, z \in B$ with $f(z) = f(w)$. Thus for $p := w - c$ and $q := z - c$ we have $0 = \sum a_\nu (p^\nu - q^\nu)$. Since

$$p^\nu - q^\nu = (w - z)(p^{\nu-1} + p^{\nu-2} q + \cdots + q^{\nu-1}),$$

it follows, in case $w \ne z$, that $-a_1 = \sum_{\nu \ge 2} a_\nu (p^{\nu-1} + \cdots + q^{\nu-1})$ and this entails, bearing in mind that $|p| < r$ and $|q| < r$, the contradiction $|a_1| \le \sum_{\nu \ge 2} |a_\nu| \nu r^{\nu-1}$. Therefore whenever $f(z) = f(w)$ with $z, w \in B$, it must be true that $z = w$.   □

The proof of local injectivity is once again trivial: we consider the Taylor series $\sum a_\nu (z - c)^\nu$ of f around c. Since $a_1 = f'(c) \ne 0$ and $\sum_{\nu \ge 2} \nu |a_\nu| t^{\nu-1}$ is continuous near $t = 0$, there is an $r > 0$ with $\sum_{\nu \ge 2} \nu |a_\nu| r^{\nu-1} < |a_1|$ and consequently $f|B_r(c)$ is injective.   □

A holomorphic mapping $f : D \to \mathbb{C}$ is called *locally biholomorphic around* $c \in D$ if there is an open neighborhood U of c in D such that the restriction $f|U : U \to f(U)$ is biholomorphic. The biholomorphy criterion and the injectivity lemma imply the

Local biholomorphy criterion. *A holomorphic mapping $f : D \to \mathbb{C}$ is locally biholomorphic around $c \in D$ exactly when $f'(c) \ne 0$.*

Proof. In case $f'(c) \neq 0$, the injectivity lemma furnishes an open neighborhood U of c in D such that $f|U : U \to \mathbb{C}$ is injective. According to the biholomorphy criterion, $f|U : U \to f(U)$ is then biholomorphic. The converse follows trivially.

Example. Each function $f_n : \mathbb{C}^\times \to \mathbb{C}$, $z \mapsto z^n$, $n = \pm 1, \pm 2, \ldots$ is everywhere locally biholomorphic but only when $n = \pm 1$ is it biholomorphic.

3. The local normal form. *Suppose $f \in \mathcal{O}(D)$ is not constant in any neighborhood of the point $c \in D$. Then we have:*

1) *Existence assertion: There is a disc $B \subset D$ centered at c and a biholomorphic mapping $h : B \to h(B)$ satisfying*

$$(*) \qquad\qquad f|B = f(c) + h^n,$$

with $n := \nu(f, c)$, the multiplicity of f at c.

2) *Uniqueness assertion: If $\tilde{B} \subset D$ is a disc centered at c and $\tilde{h}$ is holomorphic in $\tilde{B}$, $\tilde{h}'(c) \neq 0$ and for some $m \in \mathbb{N}$*

$$f|\tilde{B} = f(c) + \tilde{h}^m,$$

then $m = n$, and there is an nth root of unity ξ such that $\tilde{h} = \xi h$ in $B \cap \tilde{B}$, where h and B are as in 1).

Proof. ad 1) According to 8.1.4 an equation $f(z) = f(c) + (z - c)^n g(z)$ holds, in which $1 \leq n := \nu(f, c) < \infty$ and g is holomorphic in D with $g(c) \neq 0$. We choose a neighborhood B of c in D so small that g is zero-free therein. The existence theorem 3.3 for holomorphic roots supplies a $q \in \mathcal{O}(B)$ such that $q^n = g|B$. Set $h := (z - c)q \in \mathcal{O}(B)$. Then $(*)$ holds and $h'(c) = q(c) \neq 0$, since $q^n(c) = g(c) \neq 0$. Therefore the biholomorphy criterion insures that B can be shrunk a little if necessary so as to have $h : B \to h(B)$ biholomorphic.

ad 2) In $B \cap \tilde{B}$, $h^n = \tilde{h}^m$, with $h(c) = \tilde{h}(c) = 0$. Since moreover $h'(c) \neq 0$ and $\tilde{h}'(c) \neq 0$, the order of the zero that each of $h, \tilde{h}$ has at c is 1. Therefore $n = o_c(h^n) = o_c(\tilde{h}^m) = m$. Consequently $h^n = \tilde{h}^n$ prevails in $B \cap \tilde{B}$. That is, h and $\tilde{h}$ are both nth roots of h^n. As the latter is not identically the zero function, it follows from 3.3 that $\tilde{h} = \xi h$ in $B \cap \tilde{B}$, for some constant ξ with $\xi^n = 1$. $\qquad\qquad\square$

The representation of $f|B$ in $(*)$ is called the *local normal form* of f near c. The reader should compare the results of this subsection with those of 4.4.3.

4. Geometric interpretation of the local normal form. The geometric significance of the multiplicity $\nu(f,c)$ becomes clear from the following mapping-theoretic interpretation of the existence assertion 3.1:

Theorem. *Let $f \in \mathcal{O}(D)$ be non-constant near $c \in D$. Then there is a neighborhood U of c in D, a biholomorphic mapping $u : U \xrightarrow{\sim} \mathbb{E}$ with $u(c) = 0$ and a linear mapping $v : \mathbb{E} \to V$ onto a disc V with $f(U) = V$ and $v(0) = f(c)$, such that the induced map $f|U : U \to V$ factors as follows:*

$$ U \xrightarrow{u} \mathbb{E} \xrightarrow{z \mapsto z^n} \mathbb{E} \xrightarrow{v} V \qquad \textit{with } n := \nu(f,c). $$

Proof. Write $d := f(c)$. According to theorem 3 there is a disc B about c lying in D and a biholomorphic mapping $h : B \xrightarrow{\sim} h(B)$ such that $f|B = d + h^n$. Since $h(c) = 0$, there is an $r > 0$ such that $B_r(0) \subset h(B)$. We set

$$ U := h^{-1}(B_r(0)) \; , \; u(z) := r^{-1}h(z) \; , \; p(z) := z^n \; , \; v(z) := r^n z + d. $$

Then $u : U \to \mathbb{E}$ is biholomorphic, with $u(c) = 0$, $p : \mathbb{E} \to \mathbb{E}$ is holomorphic, and $f|U = v \circ p \circ u$. Because $V := v(\mathbb{E})$ is a disc centered at $v(0) = d$, the theorem is proved.   □

Non-constant holomorphic mappings thus behave locally as does the mapping $\mathbb{E} \to \mathbb{E}$, $z \mapsto z^n$ near 0. Note that the neighborhood U of c can be chosen "arbitrarily small" and, thanks to the theorem, in each such U there are n distinct points which f maps to a common value different from $f(c)$. It is then natural to declare that the value of f at c is also realized n-fold.

Setting $U^* := U \setminus c$ and $V^* := V \setminus f(c)$, the mapping $f|U^* : U^* \to V^*$ has, in view of the theorem, the following properties: every point in V^* has an open neighborhood $W \subset V^*$ whose preimage $(f|U^*)^{-1}(W)$ consists of n open connected components $U_1, \ldots, U_n$ on which each induced mapping $f|U_\nu : U_\nu \to W$ is biholomorphic $(1 \leq \nu \leq n)$. This state of affairs is expressed in topology as follows:

The mapping $f|U^* : U^* \to V^*$ is a (unlimited, unbranched) holomorphic covering of V^* by U^* having n sheets.

Intuitively these n sheets "branch out" from the point c. In the case $n \geq 2$ (that is, when f is not locally biholomorphic in a neighborhood of c), c is a *branch point* and $f|U : U \to V$ is a *branched* (at c) holomorphic covering. Thus *locally* everything is just exactly as it is for the mapping $z \mapsto z^n$ of $\mathbb{E}$ onto $\mathbb{E}$: the point 0 is a branch point and the covering $z \mapsto z^n$ of $\mathbb{E} \setminus \{0\}$ onto $\mathbb{E} \setminus \{0\}$ is unlimited and unbranched.

5. Compositional factorization of holomorphic functions. If $g : G \to G'$ is a holomorphic mapping then, by virtue of the chain rule, for every holomorphic function h in G' the composite $h(g(z))$ is holomorphic in G. Thus a mapping

$$g^* : \mathcal{O}(G') \to \mathcal{O}(G) \, , \, h \mapsto h \circ g$$

is defined which "lifts" holomorphic functions in G' to holomorphic functions in G. It is the work of an instant to verify that

$g^* : \mathcal{O}(G') \to \mathcal{O}(G)$ *is a* $\mathbb{C}$-*algebra homomorphism and is injective if and only if* g *is not constant.*

Every lifted function $f = g^*(h)$ is constant on the fibers of g. This necessary condition for f to lie in the image set of g^* is also sufficient in certain important cases.

Factorization theorem. *Let* $g : G \to G'$ *be a holomorphic mapping of a region* G *onto a region* G'. *Let* f *be a holomorphic function on* G *which is constant on each* g-*fiber* $g^{-1}(w)$, $w \in G'$. *Then there is a (unique) holomorphic function* h *in* G' *such that* $g^*(h) = f$, *i.e.,* $h(g(z)) = f(z)$ *for all* $z \in G$.

Proof. For each $w \in G'$, $g^{-1}(w)$ is non-empty and $f(g^{-1}(w))$ is a single point. Denoting this point by $h(w)$, a function $h : G' \to G$ is well defined and it satisfies $f = h \circ g$. We have to show that $h \in \mathcal{O}(G')$.

Being non-constant, g is an open mapping (cf. 8.5.1). Consequently $h^{-1}(V) = g(f^{-1}(V))$ is open in G' for every open $V \subset \mathbb{C}$, proving that h is continuous in G'.

The zero-set $N(g')$ of the derivative of g is discrete and closed in G. The fact that g is an open mapping then entails that the set

$$M := \{b \in G' : g^{-1}(b) \subset N(g')\}$$

is discrete and closed in G'. (The reader should check the simple topological argument confirming this.) Therefore to prove h is holomorphic in G', it suffices, by Riemann's continuation theorem 7.3.4, to prove h is holomorphic in $G' \setminus M$. Every point $v \in G' \setminus M$ has at least one g-preimage $c \in G$ at which $g'(c) \neq 0$. According to the local biholomorphy criterion 2, g is therefore locally biholomorphic near c, so there is an open neighborhood V of v in G' and a holomorphic function $\tilde{g} : V \to G$ such that $g \circ \tilde{g} = $ id on V. It follows that $h|V = h \circ g \circ \tilde{g} = f \circ \tilde{g}$, that is, $h \in \mathcal{O}(V)$. This shows that h is holomorphic in $G' \setminus M$ and, as noted earlier, completes the proof of the theorem.

Exercises

Exercise 1. Let $g : D \to \mathbb{C}$, $h : D' \to \mathbb{C}$ be continuous, with $g(D) \subset D'$. Suppose h and $h \circ g$ are holomorphic and h is nowhere locally constant in D. Without using exercise 4 of chapter 1, §3, show that g is also holomorphic.

Exercise 2. The tangent function $\tan z = \frac{\sin z}{\cos z}$ is holomorphic in the disc of radius $\frac{\pi}{2}$ centered at 0. Show that it is locally biholomorphic in this disc and write down the power series of the inverse function at 0.

Exercise 3. Show by means of an example that the condition $|a_1| > \sum_{\nu \geq 2} \nu |a_\nu| r^{\nu-1}$ featuring in the injectivity lemma for power series is not necessary for the injectivity of $f(z) := \sum_{\nu=1}^{\infty} a_\nu z^\nu$.

Exercise 4. For regions G, $\tilde{G}$ in $\mathbb{C}$ let $h : G \to \tilde{G}$ be holomorphic and denote by $\mathcal{O}'(G)$ the $\mathbb{C}$-vector space $\{f' : f \in \mathcal{O}(G)\}$. Show that:

a) The $\mathbb{C}$-*algebra* homomorphism $f \mapsto f \circ h$ of $\mathcal{O}(\tilde{G})$ into $\mathcal{O}(G)$ generally does not map the subspace $\mathcal{O}'(\tilde{G})$ into the subspace $\mathcal{O}'(G)$.

b) But the $\mathbb{C}$-*vector space* homomorphism $\varphi : \mathcal{O}(\tilde{G}) \to \mathcal{O}(G)$ given by $f \mapsto (f \circ h) \cdot h'$ does map $\mathcal{O}'(\tilde{G})$ into $\mathcal{O}'(G)$.

c) If h is biholomorphic and surjective, then φ is an isomorphism and maps $\mathcal{O}'(\tilde{G})$ bijectively onto $\mathcal{O}'(G)$.

§5 General Cauchy theory

The integral theorem and the integral formula were proved in 7.1 and 7.2 only for star-shaped and circular regions, respectively. That's adequate for deriving many important results of function theory. But mathematical curiosity impels us to find the limits of validity of these two key results. Two questions suggest themselves:

For a domain D given in advance, how can we describe the closed paths γ in D for which the Cauchy theorems are valid?

For what kind of domains D are the integral theorem and/or the integral formula valid for all closed paths in D?

In this section both questions will be treated. The first is satisfactorily answered by the so-called principal theorem of the Cauchy theory (subsections 3 and 4): the necessary and sufficient condition is that *the inside of*

γ *lies in* D. Our answer to the second question is only a formal one: the theorem saying that such D are just the (homotopically) simply-connected domains must await the second volume for its full elucidation.

We remind the reader of the agreement made at the end of 6.1.1:

All paths to be considered are piecewise continuously differentiable.

1. The index function $\mathrm{ind}_\gamma(z)$. If γ is a closed path in $\mathbb{C}$ and z a point in $\mathbb{C}$ not lying on γ, we are looking for a measure of how often the path γ winds around the point z. We will show that

$$\mathrm{ind}_\gamma(z) := \frac{1}{2\pi i} \int_\gamma \frac{d\zeta}{\zeta - z} \in \mathbb{C}$$

is an integer and that it measures this "winding" very well. We already know, from theorem 6.2.4, that for every disc B

$$\mathrm{ind}_{\partial B}(z) = \begin{cases} 1 & \text{for} \quad z \in B \\ 0 & \text{for} \quad z \in \mathbb{C} \setminus \overline{B}, \end{cases}$$

which corresponds to the following intuitive state of affairs: all points on the "inside of a circle" are wound around exactly once in the course of a (counterclockwise) traversal of the circle, while no point on the "outside of a circle" is wound around at all during such a circuit.

The number $\mathrm{ind}_\gamma(z)$ defined by (1) will be called the *index* (or also the *winding number*) of γ *with respect to* $z \in \mathbb{C} \setminus \gamma$. Exercise 2 below explains the latter terminology. The considerations of this section are based on corollary 3.4.

Properties of the index function. *Let γ be a given, fixed closed path in $\mathbb{C}$. Then the following hold:*

1) *For every* $z \in \mathbb{C} \setminus \gamma$, $\mathrm{ind}_\gamma(z) \in \mathbb{Z}$.

2) *The function* $\mathrm{ind}_\gamma(z)$, $z \in \mathbb{C} \setminus \gamma$, *is locally constant in* $\mathbb{C} \setminus \gamma$.

3) *For any closed path* γ^* *in* $\mathbb{C}$ *having the same initial point as* γ

$$\mathrm{ind}_{\gamma+\gamma^*}(z) = \mathrm{ind}_\gamma(z) + \mathrm{ind}_{\gamma^*}(z) , \qquad z \in \mathbb{C} \setminus (\gamma \cup \gamma^*);$$

in particular, we have for every $z \in \mathbb{C} \setminus \gamma$ *the rule*

$$\mathrm{ind}_{-\gamma}(z) = -\mathrm{ind}_\gamma(z).$$

Proof. The first claim follows from corollary 3.4, applied to $f(\zeta) := \zeta - z$. The second claim follows from the integer-valuedness and the continuity (proof!) of the index function in $\mathbb{C} \setminus \gamma$. The claims in 3) are trivial to verify. $\qquad\square$

If γ is a closed path in $\mathbb{C}$, the sets $\operatorname{Int} \gamma := \{z \in \mathbb{C} \setminus \gamma : \operatorname{ind}_\gamma(z) \neq 0\}$ and $\operatorname{Ext} \gamma := \{z \in \mathbb{C} \setminus \gamma : \operatorname{ind}_\gamma(z) = 0\}$ are called the *inside* (interior) and the *outside* (exterior) *of* γ, respectively. Because of the different meanings the parenthetic terms have in topology we will not use them in the present context.

We have

(*) $$\mathbb{C} = \operatorname{Int} \gamma \cup \gamma \cup \operatorname{Ext} \gamma,$$

a (disjoint) decomposition of $\mathbb{C}$. Since γ is locally constant it follows that

The sets $\operatorname{Int} \gamma$ *and* $\operatorname{Ext} \gamma$ *are open in* $\mathbb{C}$ *and their topological boundaries satisfy:* $\partial \operatorname{Int} \gamma \subset \gamma$, $\partial \operatorname{Ext} \gamma \subset \gamma$.

For every open disc B, $\operatorname{Int} \partial B = B$, $\operatorname{Ext} \partial B = \mathbb{C} \setminus \overline{B}$ and $\partial \operatorname{Ext} \partial B = \partial \operatorname{Int} \partial B = \partial B$. Analogous equalities hold for triangles, rectangles, etc. We show more generally that

The set $\operatorname{Int} \gamma$ *is bounded; the set* $\operatorname{Ext} \gamma$ *is never empty and always unbounded; more precisely,*

$$\operatorname{Int} \gamma \subset B_r(c) \text{ and } \mathbb{C} \setminus B_r(c) \subset \operatorname{Ext} \gamma$$

whenever $\gamma \subset B_r(c)$.

Proof. The set $V := \mathbb{C} \setminus B_r(c)$ is non-empty, connected and disjoint from the trace of γ, so the index function for γ is constant in V. Since $\lim_{z \to \infty} \int_\gamma \frac{d\zeta}{\zeta - z} = 0$, it follows that this constant value is 0, that is, $V \subset \operatorname{Ext} \gamma$. The inclusion $\operatorname{Int} \gamma \subset B_r(c)$ now follows from (*). $\qquad\square$

For constant paths the inside is empty.

2. The principal theorem of the Cauchy theory. *The following assertions about a closed path γ in a domain D are equivalent:*

i) *For all $f \in \mathcal{O}(D)$ the integral theorem is valid, that is,*

$$\int_\gamma f(\zeta) d\zeta = 0.$$

ii) *For all $f \in \mathcal{O}(D)$ the integral formula*

$$(1) \qquad \mathrm{ind}_\gamma(z)f(z) = \frac{1}{2\pi i} \int_\gamma \frac{f(z)}{\zeta - z} d\zeta , \qquad z \in D \setminus \gamma$$

is valid.

iii) *The inside* Int γ *of* γ *lies wholly in* D.

The equivalence of i) and ii) is quickly seen: i) $\Rightarrow$ ii) Introduce, as was done in 7.2.2, for fixed $z \in D$ the holomorphic difference quotient $g(\zeta) := [f(\zeta) - f(z)](\zeta - z)^{-1}$, $\zeta \in D \setminus \{z\}$, whose value at z is defined to be $g(z) := f'(z)$. By i), $\int_\gamma g(\zeta)d\zeta = 0$ and this yields

$$\frac{1}{2\pi i} \int_\gamma \frac{f(\zeta)}{\zeta - z} d\zeta = \frac{f(z)}{2\pi i} \int_\gamma \frac{d\zeta}{\zeta - z} = \mathrm{ind}_\gamma(z)f(z) \qquad \text{if } z \in D \setminus \gamma.$$

ii) $\Rightarrow$ i) Apply the integral formula, for any fixed $z \in D \setminus \gamma$ whatsoever, to the function $h(\zeta) := (\zeta - z)f(\zeta) \in \mathcal{O}(D)$. Since $h(z) = 0$, it follows that $\int_\gamma f(\zeta)d\zeta = 0$.

Also the proof of i) $\Rightarrow$ iii) is very easy: Since $(\zeta - w)^{-1} \in \mathcal{O}(D)$ as a function of ζ for each point $w \in \mathbb{C} \setminus D$, it follows that

$$\mathrm{ind}_\gamma(w) = \frac{1}{2\pi i} \int_\gamma \frac{d\zeta}{\zeta - w} = 0 \qquad \text{for all } w \in \mathbb{C} \setminus D,$$

that is, Int $\gamma \subset D$.

But the implication iii) $\Rightarrow$ ii) is not so simple to verify. Here lies the real "burden of proof". For a long time people had been searching for a simple argument that the inclusion Int $\gamma \subset D$ has as a consequence the validity of Cauchy's integral formula for γ. In 1971 J. D. DIXON in the paper "A brief proof of Cauchy's integral theorem," *Proc. Amer. Math. Soc.* **29**(1971), 625-626 gave a surprisingly simple such proof, which illustrates once again the power of Liouville's theorem. Dixon writes: "[We] present a very short and transparent proof of Cauchy's theorem. ... The proof is based on simple 'local properties' of analytic functions that can be derived from Cauchy's theorem for analytic functions on a disc. ... We ... emphasize the elementary nature of the proof." It is this proof which we now give.

3. Proof of iii) $\Rightarrow$ ii) after DIXON. Let $f \in \mathcal{O}(D)$ and consider the difference quotients

$$(1) \qquad g(w, z) := \frac{f(w) - f(z)}{w - z} , \; w \neq z ; g(z, z) := f'(z) , \; w, z \in D.$$

Because of the (definitional) equation $\text{ind}_\gamma(z) = \frac{1}{2\pi i}\int_\gamma \frac{d\zeta}{\zeta - z}$, the integral formula 2.(1) is equivalent to the formula

$$(2) \qquad \int_\gamma g(\zeta, z)d\zeta = 0 , \qquad z \in D \setminus \gamma.$$

Lemma. *The difference quotient g formed from any fixed $f \in \mathcal{O}(D)$ is continuous on $D \times D$. The integrated function $h(z) := \int_\gamma g(\zeta, z)d\zeta$, $z \in D$, is holomorphic in D.*

Proof. (1) shows that for each fixed $w \in D$, $z \mapsto g(w, z)$ is holomorphic in D. Therefore only the continuity of g in $D \times D$ needs to be proved and theorem 8.2.2 does the rest. This continuity is immediate and trivial, from the quotient recipe in (1), at each point $(a, b) \in D \times D$ with $a \neq b$. So consider a point $(c, c) \in D \times D$. Let B be an open disc centered at c and lying wholly in D, $\sum a_\nu(z - c)^\nu$ the Taylor series of f in B. One checks easily that for all $w, z \in B$

$$g(w, z) = g(c, c) + \sum_{\nu \geq 2} a_\nu q_\nu(w, z), \quad \text{where}$$

$$q_\nu(w, z) := \sum_{j=1}^{\nu}(w - c)^{\nu - j}(z - c)^{j-1}.$$

Since $|q_\nu(w, z)| \leq \nu t^{\nu-1}$ whenever $|w - c| < t$ and $|z - c| < t$, it follows that for sufficiently small t

$$|g(w, z) - g(c, c)| \leq t(2|a_2| + 3|a_3|t + \cdots + n|a_n|t^{n-2} + \cdots)$$

whenever $w, z \in B_t(c) \subset B$. Since the power series on the right side of this inequality is continuous at $t = 0$, we see that $\lim_{(w,z)\to(c,c)} g(w, z) = g(c, c)$.
□

Equation (2) will now be proved from Liouville's theorem by demonstrating that the function $h \in \mathcal{O}(D)$ admits an extension to an entire function $\tilde{h} \in \mathcal{O}(\mathbb{C})$ which satisfies $\lim_{z\to\infty} \tilde{h}(z) = 0$. In the outside $\text{Ext}\,\gamma$ of γ we consider the function (so-called *Cauchy-transform*)

$$h^*(z) := \int_\gamma \frac{f(\zeta)}{\zeta - z}d\zeta , \qquad z \in \text{Ext}\,\gamma.$$

According to 8.2.2, $h^* \in \mathcal{O}(\text{Ext}\,\gamma)$ and $\lim_{z\to\infty} h^*(z) = 0$. Since (by definition of $\text{Ext}\,\gamma$) $\int_\gamma \frac{d\zeta}{\zeta - z} = 0$ for $z \in \text{Ext}\,\gamma$, it follows that

$$h(z) = h^*(z) \qquad \text{for all } z \in D \cap \text{Ext}\,\gamma.$$

Now the hypothesis Int $\gamma \subset D$ and the identity $\mathbb{C} = \text{Int}\,\gamma \cup \gamma \cup \text{Ext}\,\gamma$ have as a consequence the identity $\mathbb{C} = D \cup \text{Ext}\,\gamma$. Consequently the decree

$$\tilde{h}(z) := \begin{cases} h(z) & \text{for} \quad z \in D \\ h^*(z) & \text{for} \quad z \in \text{Ext}\,\gamma \end{cases}$$

well defines an *entire* function $\tilde{h}$ which is an extension of h. Since the inclusion Int $\gamma \subset B_r(0)$ prevails for all sufficiently large $r > 0$, $\tilde{h}$ and h^* coincide outside $B_r(0)$, for such r, and so like h^*, the function $\tilde{h}$ satisfies $\lim_{z \to \infty} \tilde{h}(z) = 0$. $\qquad \square$

Remark. In the second volume we will give a quite short proof of the implication iii) $\Rightarrow$ i) using Runge's approximation theorem and the residue calculus.

4. Nullhomology. Characterization of homologically simply-connected domains.

Closed paths in D enjoying the equivalent properties i), ii) and iii) of theorem 2, play a key role in function theory. They are said to be *nullhomologous* in D. This terminology comes from algebraic topology and signifies that the path "bounds a piece of area inside D". In $\mathbb{C}^\times$ the circular paths $\gamma := \partial B_r(0)$, $r > 0$, are not nullhomologous, because $\int_\gamma f(\zeta)d\zeta \neq 0$ for $f(z) := 1/z \in \mathcal{O}(\mathbb{C}^\times)$.

The Cauchy function theory of a domain D is simplest when every closed path in D is nullhomologous in D. Theorem 2 and the results of §3 combine to give the following characterization of such domains.

Theorem. *The following assertions about a domain D are equivalent:*

   i) *D is homologically simply-connected.*

   ii) *Every holomorphic function in D is integrable in D.*

   iii) *For every $f \in \mathcal{O}(D)$ and every closed path γ in D*

$$\text{ind}_\gamma(z)f(z) = \frac{1}{2\pi i} \int_\gamma \frac{f(\zeta)}{\zeta - z}d\zeta , \qquad z \in D \setminus |\gamma|.$$

   iv) *The inside Int γ of each closed path γ in D lies wholly in D.*

   v) *Every unit in $\mathcal{O}(D)$ has a holomorphic logarithm in D.*

   vi) *Every unit in $\mathcal{O}(D)$ has a holomorphic square-root in D.*

Proof. The assertions i) through iv) are equivalent by theorem 2 and theorem 3.2. The equivalence of v) and vi) follows from the theorem on roots

3.3 and theorem 3.5. One gets the implication ii) ⇒ v) from the existence lemma 3.1.

For the proof of v) ⇒ iv), consider $a \in \mathbb{C} \setminus D$ and a closed path γ in D. The unit $f(z) := z - a \in \mathcal{O}(D)$ has a holomorphic logarithm in D. Consequently, $f'(z)/f(z) = 1/(z-a)$ is integrable in D (existence lemma 3.1). It follows that $\mathrm{ind}_\gamma(a) = 0$, and this for every $a \in \mathbb{C} \setminus D$. And that says that the inside of γ lies in D. □

In the second volume we will see, among other things, that D is homologically simply-connected just when D has no "holes", and that every homologically simply-connected proper subset of $\mathbb{C}$ is biholomorphic with $\mathbb{E}$ (the Riemann mapping theorem).

Exercises

Exercise 1. Show with appropriate examples that in general neither Int γ nor Ext γ need be connected.

Exercise 2. Let $a, b \in \mathbb{R}$, $a < b$, $\omega : [a, b] \to \mathbb{R}$ and $r : [a, b] \to \mathbb{R}^+$ be continuously differentiable functions. Set $\gamma := re^{2\pi i \omega}$. Show that

$$\frac{1}{2\pi i} \int_\gamma \frac{d\zeta}{\zeta} = \frac{1}{2\pi i} \log \frac{r(b)}{r(a)} + \omega(b) - \omega(a)$$

and use this observation to give a geometric interpretation of the index function.

Exercise 3. a) Let γ be a closed path in $\mathbb{C} \setminus \{0\}$, $n \in \mathbb{N}$ and $g(z) := z^n$. Show that $\mathrm{ind}_{g \circ \gamma}(0) = n \cdot \mathrm{ind}_\gamma(0)$.

b) Let D be a domain in $\mathbb{C}$, $f : D \to f(D)$ a biholomorphic map, $c \in D$, γ a closed path in $D \setminus \{c\}$ such that Int $\gamma \subset D$. Show that $\mathrm{ind}_\gamma(c) = \mathrm{ind}_{f \circ \gamma}(f(c))$. *Hint.* For b): The function

$$h(z) := \begin{cases} 1 & , \quad z = c \\ f'(z) \cdot \dfrac{z - c}{f(z) - f(c)} & , \quad z \in D \setminus \{c\} \end{cases}$$

is holomorphic in D.

§6* Asymptotic power series developments

In this section G will always denote a region which *contains 0 as a boundary point*. We will show that certain holomorphic functions in G, even though in general not even defined at $0 \in \partial G$, can nevertheless be "developed into power series" at 0. A special role is played by *circular sectors at 0*. These are regions of the form

$$S = S(r, \alpha, \beta) := \{z = |z|e^{i\varphi} : 0 < |z| < r \,,\, \alpha < \varphi < \beta\} \,,\, 0 < r \leq \infty.$$

Circular sectors of radius ∞ are also called *angular sectors*. Our principal result is a theorem of RITT about the asymptotic behavior of holomorphic functions at the apex 0 of such circular sectors (subsection 4). This theorem of RITT contains as a special case (subsection 5) the theorem of É. BOREL stated in 7.4.1. For the proof of Ritt's theorem the Weierstrass convergence theorem is the primary tool; for the derivation of Borel's theorem we will need in addition to that the Cauchy estimates for derivatives (cf. subsection 3).

As follow-up literature for the themes of this section we mention W. WASOW, *Asymptotic Expansions for Ordinary Differential Equations*, Interscience Publishers, New York (1965), esp. chapter III, (reprinted by R. E. Krieger, Huntington, New York, 1976).

As in Chapter 4.4 we use $\bar{A}$ to designate the $\mathbb{C}$-algebra of formal power series centered at 0.

1. Definition and elementary properties. A formal power series $\sum a_\nu z^\nu$ is called an *asymptotic development* or *representation of* $f \in \mathcal{O}(G)$ *at* $0 \in \partial G$ if

$$(1) \qquad \lim_{z \to 0} z^{-n} \left[f(z) - \sum_0^n a_\nu z^\nu \right] = 0 \qquad \text{for every } n \in \mathbb{N}.$$

A function $f \in \mathcal{O}(G)$ has *at most one* asymptotic development at 0, since from (1) we get at once the recursion formulas

$$a_0 = \lim_{z \to 0} f(z) \,,\, a_n = \lim_{z \to 0} z^{-n} \left[f(z) - \sum_0^{n-1} a_\nu z^\nu \right] \qquad \text{for } n > 0.$$

We will write $f \sim_G \sum a_\nu z^\nu$ when the series is the asymptotic development of f at 0. Condition (1) for this to happen can be rephrased as

$$(1') \qquad \begin{cases} \text{For each } n \in \mathbb{N} \text{ there exists } f_n \in \mathcal{O}(G), \text{ such that} \\ f(z) = \sum_0^n a_\nu z^\nu + f_n(z) z^n \quad \text{and} \quad \lim_{z \to 0} f_n(z) = 0. \end{cases}$$

The existence of asymptotic developments depends essentially on the region G. Thus, e.g., the function $\exp(1/z) \in \mathcal{O}(\mathbb{C}^\times)$ has no asymptotic development at 0; but by contrast in *every* angular sector $W := S(\infty, \frac{1}{2}\pi + \eta, \frac{3}{2}\pi - \eta)$, $0 < \eta < \frac{1}{2}\pi$, lying in the left half-plane (with aperture less than π) we have

$$(*) \qquad\qquad \exp(1/z) \sim_W \sum a_\nu z^\nu, \text{ in which all } a_\nu = 0.$$

The reader should confirm all this, and also show that $(*)$ is false if $\eta = 0$ is allowed in the definition of W.

If $f \in \mathcal{O}(G)$ has a holomorphic extension $\tilde{f}$ in a region $G \subset \tilde{G}$ with $0 \in \tilde{G}$, then the Taylor development of $\tilde{f}$ in a neighborhood of 0 is the asymptotic development of f at 0. By means of the Riemann continuation theorem and the equality $a_0 = \lim_{z \to 0} f(z)$ it follows immediately that:

If 0 is an isolated boundary point of G, then $f \in \mathcal{O}(G)$ has an asymptotic development $\sum a_\nu z^\nu$ at 0 if and only if f is bounded near 0; in that case, $\sum a_\nu z^\nu$ is the Taylor series of f around 0.

We will denote by $\mathcal{B}$ the set of functions in $\mathcal{O}(G)$ which possess asymptotic developments at 0.

Theorem. $\mathcal{B}$ *is a $\mathbb{C}$-subalgebra of $\mathcal{O}(G)$ and the mapping*

$$\varphi : \mathcal{B} \to \bar{A}, \, f \mapsto \sum a_\nu z^\nu,$$

where $f \sim_G \sum a_\nu z^\nu$, is a $\mathbb{C}$-algebra homomorphism.

The proof is canonical. To see, e.g., that $\varphi(fg) = \varphi(f)\varphi(g)$, write $f(z) = \sum_0^n a_\nu z^\nu + f_n(z) z^n$, $g(z) = \sum_0^n b_\nu z^\nu + g_n(z) z^n$, where for each $n \in \mathbb{N}$, $\lim_{z \to 0} f_n(z) = \lim_{z \to 0} g_n(z) = 0$. For the numbers $c_\nu := \sum_{\kappa + \lambda = \nu} a_\kappa b_\lambda$ then, we have

$$f(z)g(z) = \sum_0^n c_\nu z^\nu + h_n(z) z^n, \quad \text{with}$$

$h_n(z) := \sum_{n+1}^{2n} c_\nu z^{\nu-n} + f_n(z) \sum_0^n a_\nu z^\nu + g_n(z) \sum_0^n b_\nu z^\nu + z^n f_n(z) g_n(z) \in \mathcal{O}(G)$; whence clearly $\lim_{z \to 0} h_n(z) = 0$ for each $n \in \mathbb{N}$. It follows that $fg \in \mathcal{B}$ and $fg \sim_G \sum c_\nu z^\nu$. Since the series $\sum c_\nu z^\nu$ with c_ν thus defined is by definition the product of the series $\sum a_\nu z^\nu$ and $\sum b_\nu z^\nu$ in the ring $\bar{A}$, it further follows that $\varphi(fg) = \varphi(f)\varphi(g)$. $\qquad\square$

The homomorphism φ is generally *not injective*, as the example $\exp(1/z) \sim_L \sum 0 z^\nu$ above demonstrates. In case $G := \mathbb{C}^\times$, φ is injective but *not surjective*. Cf. also subsection 4.

2. A sufficient condition for the existence of asymptotic developments. *Let G be a region with $0 \in \partial G$ and the following property: to every point $z \in G$ is associated a null sequence c_k such that each segment $[c_k, z]$ lies wholly in G. If then $f \in \mathcal{O}(G)$ is a function for which all the limits $f^{(\nu)}(0) := \lim_{z \to 0} f^{(\nu)}(z)$, $\nu \in \mathbb{N}$, exist, the asymptotic development $\sum \frac{f^{(\nu)}(0)}{\nu!} z^\nu$ is valid for f at 0.*

Proof. Choose and fix n (arbitrary) in $\mathbb{N}$. Since $\lim_{z \to 0} f^{(n+1)}(z)$ exists, there is a disc B centered at 0 such that $|f^{(n+1)}|_{B \cap G} \leq M$ for some finite M. Consider any pair c, z with $[c, z] \subset B \cap G$. As in the case of functions on intervals in $\mathbb{R}$, there is for each $m \in \mathbb{N}$ a Taylor formula

$$f(z) = \sum_0^m \frac{f^{(\nu)}(c)}{\nu!}(z - c)^\nu + r_{m+1}(z)$$

with remainder $r_{m+1}(z)$ given by

$$r_{m+1}(z) := \frac{1}{m!} \int_{[c,z]} f^{(m+1)}(\zeta)(z - \zeta)^m d\zeta.$$

(Integration by parts here will drive an inductive proof on m.) For r_{m+1} the standard estimate gives

$$|r_{m+1}(z)| \leq \frac{1}{m!} \sup_{\zeta \in [c,z]} |f^{(m+1)}(\zeta)(z-\zeta)^m||z-c| \leq \frac{1}{m!}|f^{(m+1)}|_{[c,z]}|z-c|^{m+1}.$$

If we take $m = n$ and a null sequence of c_k with $[c_k, z] \subset G$, these considerations yield in the limit

$$\left| f(z) - \sum_0^n \frac{f^{(\nu)}(0)}{\nu!} z^n \right| \leq \frac{M}{n!}|z|^{n+1}, \text{ valid for all } z \in B \cap G.$$

Since M depends on n but not on z, it follows that $\lim_{z \to 0} z^{-n}[f(z) - \sum_0^n \frac{f^{(\nu)}(0)}{\nu!} z^\nu] = 0$, for each $n \in \mathbb{N}$.   □

The hypotheses concerning the limits of the derivatives of f are suggested by the form of Taylor's formula — which leads us to expect that, in case $f(z)$ does have the asymptotic development $\sum a_\nu z^\nu$, then $\nu! a_\nu = \lim_{z \to 0} f^{(\nu)}(z)$. The notation $f^{(\nu)}(0)$ which we chose for $\lim_{z \to 0} f^{(\nu)}(z)$ is (merely) suggestive — naturally $f^{(\nu)}(0)$ is not a derivative.

The hypothesis imposed on G is fulfilled by every circular sector at 0.

3. Asymptotic developments and differentiation. We consider two circular sectors at 0 which have the same radius, $S = S(r, \alpha, \beta)$ and $T = S(r, \gamma, \delta)$. We suppose that $S \neq B_r(0) \setminus \{0\}$, i.e., that $\beta - \alpha \leq 2\pi$. We will say that T is *enveloped by* S, and note this by $T \subset\subset S$, if the inclusion $\{z = |z|e^{i\varphi} : 0 < |z| < r, \gamma \leq \varphi \leq \delta\} \subset S$ holds (cf. the figure below).

Lemma. *Let S, T be circular sectors at 0 with $T \subset\subset S$. Suppose that $g \in \mathcal{O}(S)$ satisfies $\lim_{z \in S, z \to 0} g(z) = 0$. Then it follows that $\lim_{z \in T, z \to 0} zg'(z) = 0$.*

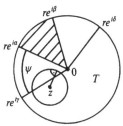

Proof. There is evidently an $a > 0$ such that for every point $z \in T$ with $|z| < \frac{1}{2}r$ the compact disc $\overline{B_{a|z|}(z)}$ lies wholly in S (e.g., $a := \sin \psi$ in the figure). The Cauchy estimates 8.3.1 yield

$$|g'(z)| \leq \frac{1}{a|z|}|g|_{\overline{B_{a|z|}(z)}}, \quad \text{i.e.,} \quad a|zg'(z)| \leq |g|_{\overline{B_{a|z|}(z)}},$$

for all $z \in T$ with $|z| < \frac{1}{2}r$. Since a is constant and by hypothesis the right side of the last inequality converges to 0 as z does, the claim follows. $\quad\square$

Theorem. *Let f be holomorphic in the circular sector $S \neq B_r(0) \setminus \{0\}$ at 0. Suppose that $f \sim_S \sum a_\nu z^\nu$. Then $f' \sim_T \sum_{\nu \geq 1} \nu a_\nu z^{\nu-1}$ for every circular sector $T \subset\subset S$.*

Proof. For each $n \in \mathbb{N}$ there is an $f_n \in \mathcal{O}(S)$ satisfying $\lim_{z \in S, z \to 0} f_n(z) = 0$ and $f(z) = \sum_0^n a_\nu z^\nu + f_n(z)z^n$. It follows that

$$f'(z) = \sum_1^n \nu a_\nu z^{\nu-1} + g_n(z)z^{n-1}, \quad \text{with } g_n(z) := zf'_n(z) + nf_n(z) \in \mathcal{O}(S).$$

According to the lemma each g_n has the requisite limiting behavior, i.e., $\lim_{z \in T, z \to 0} g_n(z) = 0$. $\quad\square$

It now follows quickly that for circular sectors the limit conditions on the $f^{(n)}$ in subsection 2 are also necessary. More precisely, we have

Corollary. *If f is holomorphic in the circular sector S at 0 and $f \sim_S \sum a_\nu z^\nu$, then $\lim_{z \in T, z \to 0} f^{(n)}(z) = n!a_n$ for every $n \in \mathbb{N}$ and every circular sector $T \subset\subset S$.*

Proof. From n successive applications of the theorem we get

$$f^{(n)}(z) \sim_T \sum_{\nu \geq n} \nu(\nu - 1) \cdots (\nu - n + 1) a_\nu z^{\nu - n}.$$

According to definition 1.(1) this implies that $\lim_{z \in T, z \to 0} f^{(n)}(z) = n! a_n$.

□

The corollary just proved is essential to the arguments presented in subsection 5.

4. The theorem of RITT. The question of what conditions a power series must satisfy in order to occur as an asymptotic development has a surprisingly simple answer for circular sectors S at 0: there are no such conditions. For *every* formal power series $\sum a_\nu z^\nu$ (thus even for such monsters as $\sum \nu^\nu z^\nu$) we will construct a holomorphic function f in S which satisfies $f \sim_S \sum a_\nu z^\nu$. The idea of the construction is simple: Replace the given series by a function series of the type

$$f(z) := \sum_0^\infty a_\nu f_\nu(z) z^\nu$$

in which the "convergence factors" $f_\nu(z)$ are to be chosen as follows:

1) The series should converge normally in S; this requires that $f_\nu(z)$ become small very quickly as ν grows.

2) $f \sim_S \sum a_\nu z^\nu$ should hold; this requires that for each fixed ν, $f_\nu(z)$ converge rapidly to 1 as z approaches 0.

We will see that functions of the form

$$f_\nu(z) := 1 - \exp(-b_\nu / \sqrt{z}) , \text{ with } \sqrt{z} := e^{\frac{1}{2} \log z} \in \mathcal{O}(\mathbb{C}^-),$$

have the desired properties if the $b_\nu > 0$ are properly chosen. To this end we need the following

Lemma. *Let $S := S(r, -\pi + \psi, \pi - \psi)$, $0 < \psi < \pi$ be a circular sector at 0 in the slit plane $\mathbb{C}^-$. Then the function $h(z) := 1 - \exp(-b/\sqrt{z})$, $b \in \mathbb{R}$, $b > 0$ is holomorphic in $\mathbb{C}^-$ and has the following properties:*

1) $|h(z)| \leq b/|\sqrt{z}|$ *for $z \in S$.*

2) $\lim_{z \in S, z \to 0} z^{-m}(1 - h(z)) = 0$ *for every $m \in \mathbb{N}$.*

Proof. 1) Every $z \in S$ has the form $z = |z| e^{i\varphi} \in \mathbb{C}^\times$ with $|\varphi| < \pi - \psi$. Since $|\frac{1}{2}\varphi| < \frac{1}{2}\pi$ and $\cos x$ is positive in the interval $(-\frac{1}{2}\pi, \frac{1}{2}\pi)$, and $b > 0$, it follows that $w := b/\sqrt{z}$ satisfies $\Re w = b e^{-\frac{1}{2} \log |z|} \cos \frac{1}{2}\varphi > 0$. Therefore

$|h(z)| = |1 - \exp(-b/\sqrt{z})| \le b/|\sqrt{z}|$ for $z \in S$, upon recalling (cf. 6.2.2)
that generally $|1 - e^{-w}| \le |w|$ whenever $\Re w > 0$.

2) We have $z^{-m}(1 - h(z)) = z^{-m} \exp(-b/\sqrt{z})$ and so

$$|z^{-m}(1 - h(z))| = |z^{-m}||\exp(-b/\sqrt{z})| = |z|^{-m} \exp(-b|z|^{-\frac{1}{2}} \cos(\varphi/2)).$$

For $z \in S$, $|\varphi| < \pi - \psi$, so $\cos \frac{1}{2}\varphi > \cos \frac{1}{2}(\pi - \psi) = \sin \frac{1}{2}\psi$. Since $b > 0$, it
follows that

$$|z^{-m}(1 - h(z))| < |z|^{-m} \exp(-b|z|^{-\frac{1}{2}} \sin(\psi/2)) \qquad \text{for } z \in S.$$

Set $t := b/\sqrt{|z|}$ and note that $\sin \frac{1}{2}\psi > 0$. We then obtain

$$\lim_{z \in S, z \to 0} |z^{-m}(1 - h(z))| \le b^{-2m} \lim_{t \to +\infty} t^{2m} e^{-t \sin \frac{1}{2}\psi} = 0 \text{ , for every } m \in \mathbb{N},$$

since for $q > 0$, e^{-tq} decays more rapidly than any power of t as t approaches
0.   □

A circular sector $S = S(r, \alpha, \beta)$ is called *proper* if $B_r(0) \setminus S$ has interior
points, i.e., if $\beta - \alpha < 2\pi$. [Generally a point x in a subset A of a topological
space X is called an *interior point* of A if there is a neighborhood of x in
X which is wholly contained in A.] We now maintain that

Theorem of RITT. *If S is a proper circular sector at 0, then to every
formal power series $\sum a_\nu z^\nu$ there corresponds a holomorphic function f in
S such that $f \sim_S \sum a_\nu z^\nu$.*

Proof. If $z \mapsto e^{i\gamma}z$ rotates S into the circular sector S^* and if $f^* \in \mathcal{O}(S^*)$
satisfies $f^*(z) \sim_{S^*} \sum a_\nu e^{i\gamma\nu} z^\nu$, then $f(z) := f^*(e^{-i\gamma}z) \in \mathcal{O}(S)$ satisfies
$f(z) \sim_S \sum a_\nu z^\nu$. Since S is a proper sector, after such a rotation we may
assume that S has the form $S(r, -\pi + \psi, \pi - \psi)$, with $0 < \psi < \pi$. Obviously
we need only consider angular sectors, that is, the case $r = \infty$. We set

$$b_\nu := (|a_\nu|\nu!)^{-1} \text{ if } a_\nu \ne 0 , \qquad b_\nu := 0 \text{ otherwise,} \qquad \nu \in \mathbb{N}.$$

Then define $f_\nu(z) := 1 - \exp(-b_\nu/\sqrt{z})$ and $f(z) := \sum_0^\infty a_\nu f_\nu(z) z^\nu$. According to assertion 1) of the lemma

$$|a_\nu f_\nu(z) z^\nu| \le |b_\nu a_\nu z^{\nu - 1/2}| \le |\frac{1}{\nu!} z^\nu||z|^{-1/2} \qquad \text{for } z \in S.$$

Since $\sum_0^\infty \frac{1}{\nu!} z^\nu$ converges normally in $\mathbb{C}$, the series defining f converges
normally in S. It follows from Weierstrass' convergence theorem (cf. 8.4.2)
that $f \in \mathcal{O}(S)$. The first sum on the right side of the equation $z^{-n}(f(z) -$
$\sum_0^n a_\nu z^\nu) = -\sum_0^n a_\nu (1 - f_\nu(z)) z^{-(n-\nu)} + \sum_{n+1}^\infty a_\nu f_\nu(z) z^{\nu - n}$ converges to
0 as z converges to 0 through S, because each summand has this property,

according to statement 2) of the lemma. For the second sum on the right
with $z \in S$ and $|z| < 1$ we have the estimate

$$\left| \sum_{n+1}^{\infty} a_\nu f_\nu(z) z^{\nu-n} \right| \leq \sum_{n+1}^{\infty} |a_\nu f_\nu(z) z^{\nu-n}| \leq \sum_{n+1}^{\infty} |z|^{\nu-\frac{1}{2}-n} = \frac{\sqrt{|z|}}{1-|z|},$$

and therefore this sum also converges to 0 when z converges to 0 through
S.   □

Since every *convex* region G in $\mathbb{C}$ with $0 \in \partial G$ lies in some proper angular
sector with vertex 0 (proof!), it follows in particular that RITT's theorem
remains valid if the S there is replaced by any convex region G having 0 as
a boundary point.

In the terminology of subsection 1 we have shown that

*For every proper circular sector at 0 the homomorphism $\varphi : B \to \bar{A}$ is
surjective.*

Special cases of the theorem proved here were proved by the American mathematician J. F. RITT in "On the derivatives of a function at a
point," *Annals of Math.* (2) **18**(1916), 18-23. He used somewhat different
convergence factors $f_\nu(z)$ but indicated (p. 21) that the factors we have
used, involving $\sqrt{z}$ in the denominator of the argument of the exponential
function, were probably the best suited for the construction. On this point
compare also pp. 41, 42 of WASOW's book.

From this theorem of RITT we immediately get

5. Theorem of É. BOREL .

Let $q_0, q_1, q_2 \ldots$ be any sequence of real
numbers, $I := (-r, r)$, $0 < r < \infty$ an interval in $\mathbb{R}$. Then there exists a
function $g : I \to \mathbb{R}$ with the following properties:

1) g is real-analytic in $I \setminus \{0\}$, that is, g is representable by a convergent
power series in a neighborhood of each point of $I \setminus \{0\}$.

2) g is infinitely often real-differentiable in I and $g^{(n)}(0) = q_n$ for every
$n \in \mathbb{N}$.

Proof. Choose a proper circular sector S at 0 of radius r which contains
$I \setminus \{0\}$. Theorem 4 furnishes an $f \in \mathcal{O}(S)$ with $f \sim_S \sum \frac{q_\nu}{\nu!} z^\nu$. If we set
$g(x) := \Re f(x)$ for $x \in I \setminus \{0\}$, then $g : I \setminus \{0\} \to \mathbb{R}$ is real-analytic and so,
in particular all the derivatives $g^{(n)} : I \setminus \{0\} \to \mathbb{R}$ exist. Since q_n is real,

$$\lim_{x \to 0} g^{(n)}(x) = \lim_{x \to 0} f^{(n)}(x) = n! \frac{q_n}{n!} = q_n \qquad \text{for each } n \in \mathbb{N}$$

(cf. subsection 3). This shows that $g^{(n)}$ can be extended to a continuous
function $I \to \mathbb{R}$ by assigning it the value q_n at 0. Now if $u, v : I \to \mathbb{R}$ are

continuous, u is differentiable in $I \setminus \{0\}$ with $u' = v$ there, then u is also differentiable at 0 and $u'(0) = v(0)$. This is an easy consequence of the mean value theorem of the differential calculus. Since $g^{(n)}$ is the derivative of $g^{(n-1)}$ in $I \setminus \{0\}$, it follows from the fact just described that this equality prevails at 0 too. Consequently, $g^{(n)} : I \to \mathbb{R}$ is the nth derivative of $g = g^{(0)}$ throughout I. Since $g^{(n)}(0) = q_n$, the theorem is proven.   □

RITT actually re-discovered the Borel theorem; not until after writing his paper (cf. the introduction thereof) did he learn of Borel's dissertation, in which in fact only the existence of an infinitely differentiable function g in I with prescribed derivatives at 0 was proven.

It is hard to understand why, but textbooks on real analysis have scarcely picked up on Borel's theorem. You find it in, say, the book of R. NARASIM-HAN: *Analysis on Real and Complex Manifolds*, North-Holland (1968), Amsterdam, on pp. 28-31 for the case $\mathbb{R}^n$; and set as a problem for infinitely differentiable mappings between Banach spaces on p. 192 of J. DIEUDONNÉ: *Foundations of Modern Analysis*, vol. I, Academic Press (1969), New York & London.

There are as well some quite short real proofs (which however do not show that g can even be chosen to be real-analytic in $I \setminus \{0\}$). One can, e.g., proceed as follows [after H. MIRKIL: "Differentiable functions, formal power series and moments," *Proc. Amer. Math. Soc.* **7**(1956), 650-652]: First manufacture – say, via Cauchy's famous $\exp(-1/x^2)$ example – an infinitely differentiable function $\varphi : \mathbb{R} \to \mathbb{R}$ which satisfies $\varphi = 1$ in $[-1, 1]$, $\varphi = 0$ in $\mathbb{R} \setminus (-2, 2)$. Then set

$$g_\nu(x) := \frac{q_\nu}{\nu!} x^\nu \varphi(r_\nu x) , \qquad \nu \in \mathbb{N}$$

with positive numbers $r_0, r_1, r_2, \ldots$ so chosen as to make

$$|g_\nu^{(n)}|_{\mathbb{R}} < 2^{-\nu} \qquad \text{for } n = 0, 1, \ldots, \nu - 1$$

and for all $\nu \in \mathbb{N}$. Because φ has compact support, such choices are possible. Standard elementary convergence theorems of real analysis show that $g(x) := \sum g_\nu(x)$ is infinitely differentiable in $\mathbb{R}$. Since φ is constant in $[-1, 1]$, it follows that for $x \in [-r_\nu^{-1}, r_\nu^{-1}]$

$$g_\nu^{(n)}(x) = \frac{q_\nu}{(\nu - n)!} x^{\nu - n} \varphi(r_\nu x) \qquad \text{for } n = 0, 1, \ldots, \nu ; \text{ and}$$

$$g_\nu^{(n)}(x) = 0 \qquad \text{for } n > \nu.$$

We see therefore that $g_\nu^{(n)}(0) = 0$ for $\nu \neq n$ and $g_n^{(n)}(0) = q_n$; whence $g^{(n)}(0) = q_n$, for all $n \in \mathbb{N}$.

Chapter 10

Isolated Singularities. Meromorphic Functions

Functions with singularities are well known from calculus; e.g., the functions

$$\frac{1}{x} \ , \ x \sin \frac{1}{x} \ , \ \exp\left(-\frac{1}{x^2}\right) \ , \ x \in \mathbb{R} \setminus \{0\}$$

are singular at the origin. Although the problem of classifying isolated singularities cannot be satisfactorily solved for functions defined only on $\mathbb{R}$, the situation is quite different in the complex domain. In section 1 we show that isolated singularities of holomorphic functions can be described in a simple way. In section 2, as an application of the classification we study the automorphisms of punctured domains, showing among other things that every automorphism of $\mathbb{C}$ is linear.

In section 3 the concept of holomorphic function will be considerably broadened – meromorphic functions will be introduced. In this larger function algebra it is also possible to perform division. Just as for holomorphic functions, there is an identity theorem.

§1 Isolated singularities

If f is holomorphic in a domain D with the exception of a point $c \in D$, i.e., holomorphic in $D \setminus \{c\}$, then c is called an *isolated singularity of f*. Our goal in this section is to show that for holomorphic functions there are just three kinds of isolated singularities:

1) *removable* singularities, which upon closer examination turn out not to be singularities at all;

2) *poles*, which arise from reciprocals of holomorphic functions with zeros. In every neighborhood of a pole the function *uniformly exceeds every
bound*;

3) *essential* singularities, in every neighborhood of which the function
behaves so erratically that its values come arbitrarily close to every complex
number.

Singularities such as the real functions $|x|$ or $x \sin \frac{1}{x}$ have at the origin
are not to be found in complex function theory.

In the sequel we will always write simply $D \setminus c$ instead of $D \setminus \{c\}$.

1.   Removable singularities.   Poles. An isolated singularity c of a
function $f \in \mathcal{O}(D \setminus c)$ is called *removable* if f is holomorphically extendable
over c (cf. 7.3.4).

Examples. The functions $\dfrac{z^2 - 1}{z - 1}$, $\dfrac{z}{e^z - 1}$ have removable singularities at 1
and 0, respectively.

From Riemann's continuation theorem 7.3.4 follows directly the

Removability theorem. *The point c is a removable singularity of $f \in
\mathcal{O}(D \setminus c)$ if f is bounded in $U \setminus c$ for some neighborhood $U \subset D$ of c.*

Thus if c is not a removable singularity of $f \in \mathcal{O}(D \setminus c)$, then f is *not*
bounded near c. We might then ask whether $(z - c)^n f$ is bounded near c
for a sufficiently large power $n \in \mathbb{N}$. If this occurs, then c is called a *pole
of f* and the natural number

$$m := \min\{\nu \in \mathbb{N} : (z - c)^\nu f \text{ is bounded near } c\} \geq 1$$

is called the *order of the pole c of f*. The order of a pole is thus always
positive. Poles of the first order are called *simple*. For $m \geq 1$, the function
$(z - c)^{-m}$ has a pole of order m at c.

Theorem. *For $m \in \mathbb{N}$, $m \geq 1$ the following assertions concerning $f \in
\mathcal{O}(D \setminus c)$ are equivalent:*

i) *f has a pole of order m at c.*

ii) *There is a function $g \in \mathcal{O}(D)$ with $g(c) \neq 0$ such that*

$$f(z) = \frac{g(z)}{(z - c)^m} \qquad \text{for } z \in D \setminus c.$$

iii) *There is an open neighborhood U of c lying in D and an $h \in \mathcal{O}(U)$ which is zero-free in $U \setminus c$ and has a zero of order m at c, such that $f = 1/h$ in $U \setminus c$.*

iv) *There is a neighborhood U of c lying in D and positive finite constants M_*, M^* such that for all $z \in U \setminus c$*

$$M_* |z - c|^{-m} \leq |f(z)| \leq M^* |z - c|^{-m}.$$

Proof. i) $\Rightarrow$ ii) Since $(z - c)^m f \in \mathcal{O}(D \setminus c)$ is bounded near c, the removability theorem furnishes a $g \in \mathcal{O}(D)$ with $g = (z - c)^m f$ in $D \setminus c$. If $g(c)$ were 0, then it would follow that g has the form $(z - c)\tilde{g}$, with $\tilde{g} \in \mathcal{O}(D)$ and consequently $\tilde{g} = (z - c)^{m-1} f$ in $D \setminus c$. This would imply that $(z - c)^{m-1} f$ is bounded near c and since $m - 1 \in \mathbb{N}$, that would violate the minimality of m.

ii) $\Rightarrow$ iii) Since $g(c) \neq 0$, g is zero-free in some open neighborhood $U \subset D$ of c. Then $(z - c)^m / g(z) \in \mathcal{O}(U)$ furnishes the desired function h.

iii) $\Rightarrow$ iv) If U is chosen small enough, then h has the form $(z-c)^m \tilde{h}$ for an $\tilde{h} \in \mathcal{O}(U)$ with $M_* := \inf_{z \in U}\{|\tilde{h}(z)|^{-1}\} > 0$ and $M^* := \sup_{z \in U}\{|\tilde{h}(z)|^{-1}\} < \infty$. Since $|f(z)| = |z - c|^{-m} |\tilde{h}(z)|^{-1}$, the claim follows.

iv) $\Rightarrow$ i) The inequality $|(z - c)^m f(z)| \leq M^*$ for $z \in U \setminus c$ shows that $(z - c)^m f$ is bounded near c, whereas the inequality $|(z - c)^{m-1} f(z)| \geq M_* |z - c|^{-1}$ shows that $(z - c)^{m-1} f$ is not bounded near c. Consequently, c is a pole of f of order m. $\square$

Because of the equivalence between i) and iii) poles arise basically via the *formation of reciprocals*. The equivalence of i) and iv) characterizes poles via the *behavior of the values of f near c*. We say that f *increases uniformly to* ∞ *around* c, written $\lim_{z \to c} f(z) = \infty$, if for every finite M there is a neighborhood U of c in D such that $\inf_{z \in U \setminus c} |f(z)| \geq M$. (The reader should satisfy himself that $\lim_{z \to c} f(z) = \infty$ obtains precisely when $\lim_{n \to \infty} |f(z_n)| = \infty$ for every sequence $z_n \in D \setminus c$ with $\lim_{n \to \infty} z_n = c$.) Another equivalent statement is: $\lim_{z \to c} 1/f(z) = 0$. Therefore the following less precise version of iv) is a direct consequence of the equivalence i) $\Leftrightarrow$ iii):

Corollary. *The function $f \in \mathcal{O}(D \setminus c)$ has a pole at c if and only if $\lim_{z \to c} f(z) = \infty$.*

2. Development of functions about poles. *Let f be holomorphic in $D \setminus c$ and let c be a pole of f of order m. Then there exist complex numbers $b_1, \ldots, b_m$, with $b_m \neq 0$, and a holomorphic function $\tilde{f}$ in D such that*

$$(1) \quad f(z) = \frac{b_m}{(z - c)^m} + \frac{b_{m-1}}{(z - c)^{m-1}} + \cdots + \frac{b_1}{z - c} + \tilde{f}(z), \quad z \in D \setminus c.$$

The numbers $b_1, \ldots, b_m$ and the function $\tilde{f}$ are uniquely determined by f.
   Conversely, every function $f \in \mathcal{O}(D \setminus c)$ which satisfies equation (1) has a pole of order m at c.

Proof. According to theorem 1 there is a holomorphic function g in D with $g(c) \neq 0$ and $f(z) = (z - c)^{-m} g(z)$ for all $z \in D \setminus c$. g is uniquely determined in $D \setminus c$ by f and this equation; hence by continuity it is uniquely determined throughout D by f. The Taylor series of g at c can be written in the form

$$g(z) = b_m + b_{m-1}(z - c) + \cdots + b_1(z - c)^{m-1} + (z - c)^m \tilde{f}(z)$$

with $b_m = g(c) \neq 0$ and $\tilde{f}$ holomorphic in a disc $B \subset D$ centered at c. Inserting this representation of g into the equation $f(z) = (z - c)^{-m} g(z)$, gives (1) in the punctured disc $B \setminus c$. We simply use (1) to *define* $\tilde{f}$ in $D \setminus B$. The uniqueness claims are clear from the uniqueness of g. Just as clear is the converse assertion in the theorem.   □

   The series (1) is a "Laurent series with finite principal part"; such series and generalizations of them will be intensively studied in chapter 12. From (1) follows

(2) $$f'(z) = \frac{-mb_m}{(z - c)^{m+1}} + \cdots + \frac{-b_1}{(z - c)^2} + \tilde{f}'(z).$$

Since $mb_m \neq 0$, it is therefore clear that

If c is a pole of order $m \geq 1$ of $f \in \mathcal{O}(D \setminus c)$, then $f' \in \mathcal{O}(D \setminus c)$ has a pole of order $m + 1$ at c; in the development of f' about c no term $a/(z - c)$ occurs.

   The number 1 is thus never the order of a pole of the derivative of a holomorphic function whose only isolated singularities are poles. We can show a little more: there is no holomorphic function with isolated singularities of any kind whose derivative has a pole anywhere of first order:

If f is holomorphic in $D \setminus c$ and f' has a pole of order k at c, then $k \geq 2$ and f has a pole of order $k - 1$ at c.

Proof. We may assume that $c = 0$. From the development theorem we have for appropriate $h \in \mathcal{O}(D)$ and $d_1, \ldots, d_k \in \mathbb{C}$ with $d_k \neq 0$

$$f'(z) = d_k z^{-k} + \cdots + d_1 z^{-1} + h(z) , \; z \in D \setminus 0.$$

For every disc B centered at 0 with $\overline{B} \subset D$, the fact that f is a primitive of f' in $D \setminus 0$ yields

$$0 = \int_{\partial B} f' d\zeta = 2\pi d_1 + \int_{\partial B} h d\zeta$$

since $\int_{\partial B} \zeta^{-\nu} d\zeta = 0$ for $\nu > 1$. But by the Cauchy integral theorem $\int_{\partial B} h d\zeta = 0$, and so it follows that $d_1 = 0$. Since $d_k \neq 0$, it must be that $k > 1$. Let $H \in \mathcal{O}(B)$ be a primitive of $h|B$ and define

$$F(z) := -\frac{1}{k-1} d_k z^{-(k-1)} - \cdots - d_2 z^{-1} + H(z)$$

for $z \in B \setminus 0$. Then $f' = F'$, and $f = F +$ const., in $B \setminus 0$. Therefore, along with F, the function f has a pole at 0, of order $k - 1$.

3. Essential singularities. Theorem of CASORATI and WEIERSTRASS. An isolated singularity c of $f \in \mathcal{O}(D \setminus c)$ is called *essential* if c is neither a removable singularity nor a pole of f. For example, the origin is an essential singularity of $\exp(z^{-1})$ (cf. exercise 2).

If f has an essential singularity at c then, on the one hand each product $(z - c)^n f(z)$, $n \in \mathbb{N}$, is unbounded near c and on the other hand there exist sequences $z_n \in D \setminus c$ with $\lim z_n = c$, such that $\lim f(z_n)$ exists and is *finite*. Using the idea of a dense set introduced in 0.2.3, we can show more:

Theorem (CASORATI, WEIERSTRASS). *The following assertions about a function f which is holomorphic in $D \setminus c$ are equivalent:*

i) *The point c is an essential singularity of f.*

ii) *For every neighborhood $U \subset D$ of c, the image set $f(U \setminus c)$ is dense in $\mathbb{C}$.*

iii) *There exists a sequence z_n in $D \setminus c$ with $\lim z_n = c$, such that the image sequence $f(z_n)$ has no limit in $\mathbb{C} \cup \{\infty\}$.*

Proof. i) $\Rightarrow$ ii) by *reductio ad absurdum*. Assume that there is a neighborhood $U \subset D$ of c such that $f(U \setminus c)$ is not dense in $\mathbb{C}$. This means that there is some $B_r(a)$ with $r > 0$ and $B_r(a) \cap f(U \setminus c) = \emptyset$, that is, $|f(z) - a| \geq r$ for all $z \in U \setminus c$. The function $g(z) := 1/(f(z) - a)$ is thus holomorphic in $U \setminus c$ and bounded there by r^{-1}, and so has c as a removable singularity. It follows that $f(z) = a + 1/g(z)$ has a removable singularity at c in case $\lim_{z \to c} g(z) \neq 0$ and a pole at c in case $\lim_{z \to c} g(z) = 0$. Thus in either case there is no essential singularity at c, contrary to the hypothesis i).

The implications ii) $\Rightarrow$ iii) $\Rightarrow$ i) are trivial. □

Since non-constant holomorphic functions are open mappings, in the situation of the CASORATI-WEIERSTRASS theorem every set $f(U \setminus c)$ is (in case U is open) even *open and dense* in $\mathbb{C}$. Far more can actually be shown: $f(U \setminus c)$ *is always either the whole of $\mathbb{C}$* (as in the case $f(z) = \sin(z^{-1})$) *or $\mathbb{C}$ with just one point deleted* (as in the case $f(z) = \exp(z^{-1})$, in which 0 is the one point never taken as an f-value). This is the famous *great theorem* of PICARD, which we can't derive here.

As a simple consequence of the Casorati-Weierstrass theorem we record the

Theorem of CASORATI-WEIERSTRASS for entire functions. *If f is a transcendental entire function, then for every $a \in \mathbb{C}$ there exists a sequence z_n in $\mathbb{C}$ with $\lim |z_n| = \infty$ and $\lim f(z_n) = a$.*

This is obviously a consequence of the general theorem and the following

Lemma. *The entire function $f \in \mathcal{O}(\mathbb{C})$ is transcendental if and only if the function $f^\times \in \mathcal{O}(\mathbb{C}^\times)$ defined by $f^\times(z) := f(z^{-1})$ has an essential singularity at 0.*

Proof. Let $f(z) = \sum a_\nu z^\nu$ be entire and suppose that 0 is not an essential singularity of $f^\times$. Then for all sufficiently large $n \in \mathbb{N}$, $z^n f^\times(z) = \sum_0 a_\nu z^{n-\nu} \in \mathcal{O}(\mathbb{C}^\times)$ is holomorphically continuable over 0. For these n the Cauchy integral theorem yields

$$0 = \int_{\partial \mathbb{E}} \zeta^n f^\times(\zeta) d\zeta = \sum_0 a_\nu \int_{\partial \mathbb{E}} \zeta^{n-\nu} d\zeta = 2\pi i a_{n+1},$$

which means that f is a polynomial – of degree at most n.

Conversely, suppose f is not transcendental, that is, f is a polynomial $a_0 + a_1 z + \cdots + a_n z^n$. Then

$$f^\times(z) = f(z^{-1}) = a_n z^{-n} + \cdots + a_1 z^{-1} + a_0$$

and according to theorem 2 the origin is either a pole (if $n > 0$) of $f^\times$ or a removable singularity of $f^\times$ (if $n = 0$); so in either case 0 is not an essential singularity of $f^\times$.   □

The lemma is also a direct consequence of theorem 12.2.3, which we will prove later.

4. Historical remarks on the characterization of isolated singularities. The description of poles in terms of growth behavior as well as the series development theorem 2 are to be found as early as 1851 in RIE-MANN ([R], Art. 13). The word "pole" was introduced in 1875 by BRIOT and BOUQUET ([BB], 2nd ed., p. 15). WEIERSTRASS used the phrase "außerwesentliche singuläre Stelle (inessential singular point)" for the opposite of the "wesentlich singulären Stellen (essential singular points)" ([W₃], p.78).

It is customary to designate the implication i) ⇒ ii) in theorem 3 as the CASORATI-WEIERSTRASS theorem. It was discovered in 1868 by the Italian mathematician Felice CASORATI (1835-1890, Professor at Padua). The proof reproduced here goes back to him ("Un teorema fondamentale nella teorica delle discontinuità delle funzioni," *Opere* 1, 279-281). WEIERSTRASS presented the result in 1876, independently of CASORATI. He formulated it thus ([W₃], p. 124):

"Hiernach ändert sich die Function $f(x)$ in einer unendlich kleinen Umgebung der Stelle c in der Art discontinuirlich, dass sie jedem willkürlich angenommenen Werthe beliebig nahe kommen kann, für $x = c$ also einen bestimmten Werthe nicht besitzt. (Accordingly the function $f(x)$ varies so discontinuously in an infinitely small neighborhood of the point c that it can come as close as desired to any prescribed value. So it cannot possess a determinate value at $x = c$.)"

The CASORATI-WEIERSTRASS theorem for entire functions was known to BRIOT and BOUQUET by 1859, although their formulation of it ([BB], 1st ed., §38) is incorrect. The state of the theory around 1882 is beautifully reviewed in O. HÖLDER's article "Beweis des Satzes, dass eine eindeutige analytische Function in unendlicher Nähe einer wesentlich singulären Stelle jedem Werth beliebig nahe kommt," *Math. Annalen* **20**(1882), 138-143. For a detailed history of the Casorati-Weierstrass theorem and a discussion of priorities, see E. NEUENSCHWANDER, "The Casorati-Weierstrass theorem (studies in the history of complex function theory I)," *Historia Math.* **5**(1978), 139-166.

Exercises

Exercise 1. Classify the isolated singularities of each of the following functions and in case of poles specify the order:

a) $\dfrac{z^4}{(z^4 + 16)^2}$,

b) $\dfrac{1 - \cos z}{\sin z}$,

c) $\dfrac{z}{e^z - z + 1}$,

d) $\dfrac{z^2 - \pi^2}{\sin^2 z}$,

e) $\dfrac{1}{e^z - 1} - \dfrac{1}{z - 2\pi i}$,

f) $\dfrac{1}{\cos(1/z)}$.

Exercise 2. Show that the function $\exp(1/z) \in \mathcal{O}(\mathbb{C}^\times)$ has neither a removable singularity nor a pole at 0.

Exercise 3. Show that a non-removable singularity c of $f \in \mathcal{O}(D \setminus c)$ is always an essential singularity of $\exp \circ f$.

Exercise 4. Let $c \in D$ open $\subset \mathbb{C}$, $f \in \mathcal{O}(D \setminus c)$, P a non-constant polynomial. Show that c is a removable singularity or a pole or an essential singularity of f if and only if it is a removable singularity or a pole or an essential singularity, respectively, of $P \circ f$.

§2* Automorphisms of punctured domains

The results of section 1 permit us to extend automorphisms of $D \setminus c$ to automorphisms of D. It thereby becomes possible to determine the groups $\operatorname{Aut}\mathbb{C}$ and $\operatorname{Aut}\mathbb{C}^\times$ explicitly. Furthermore, we can exhibit bounded regions which have no automorphisms at all except the identity map (conformal rigidity).

1. Isolated singularities of holomorphic injections. *Let A be a discrete and relatively closed subset of D and let $f : D \setminus A \to \mathbb{C}$ be holomorphic and injective. Then:*

   a) *no point $c \in A$ is an essential singularity of f;*

   b) *if $c \in A$ is a pole of f, then c has order 1 ;*

   c) *if every point of A is a removable singularity of f, then the holomorphic continuation $\tilde{f} : D \to \mathbb{C}$ is injective.*

Proof. a) Let B be an open disc containing c and satisfying $B \cap A = \{c\}$ and $D' := D \setminus (A \cup \overline{B}) \neq \emptyset$. Then $f(D')$ is non-empty and open (Open Mapping Theorem). Because of injectivity $f(B \setminus c)$ does not meet the set $f(D')$, consequently is not dense in $\mathbb{C}$. By the CASORATI-WEIERSTRASS theorem c is therefore not an essential singularity of f.

b) Consider a pole $c \in A$ of order $m \geq 1$. There is a neighborhood $U \subset D$ of c with $U \cap A = \{c\}$ such that $g := (1/f)|U$ is holomorphic and has a zero of order m at c (cf. Theorem 1.1). Because f is injective so is the function $g : U \setminus \{c\} \to \mathbb{C} \setminus \{0\}$. Consequently, $g : U \to \mathbb{C}$ is injective. According to theorem 9.4.1 then $g'(c) \neq 0$; that is, $m = 1$.

c) Suppose there are two different points $a, a' \in D$ with $p := \tilde{f}(a) = \tilde{f}(a')$. Choose disjoint open discs B, B' containing a, a', respectively, and satisfying $B \setminus a \subset D \setminus A$, $B' \setminus a' \subset D \setminus A$. Then $\tilde{f}(B) \cap \tilde{f}(B')$ is a neighborhood of p and accordingly there exist points $b \in B \setminus a$, $b' \in B' \setminus a'$ with $f(b) = f(b')$. Since b, b' both lie in $D \setminus A$ and are unequal, the injectivity of f is compromised. This contradiction proves that no such a, a' exist.   □

2. The groups $\operatorname{Aut} \mathbb{C}$ and $\operatorname{Aut} \mathbb{C}^{\times}$. Every mapping $\mathbb{C} \to \mathbb{C}$, $z \mapsto az + b$, $a \in \mathbb{C}^{\times}$, $b \in \mathbb{C}$ is biholomorphic, and in particular a holomorphic injection. We show that conversely

Theorem. *Every injective holomorphic mapping $f : \mathbb{C} \to \mathbb{C}$ is linear, that is, of the form*

$$f(z) = az + b, \quad a \in \mathbb{C}^{\times}, b \in \mathbb{C}.$$

Proof. Along with f the function $f^{\times} : \mathbb{C}^{\times} \to \mathbb{C}$ defined by $f^{\times}(z) := f(z^{-1})$ is also injective. Taking $D := \mathbb{C}$, $A := \{0\}$ in theorem 1a), we learn that 0 is an inessential singularity of $f^{\times} \in \mathcal{O}(\mathbb{C}^{\times})$. According to lemma 1.3 f is then a polynomial, and so f' is too. But the injectivity of f forces f' to be zero-free, which by the Fundamental Theorem of Algebra means that f' is constant, f therefore linear.   □

Holomorphic injections $\mathbb{C} \to \mathbb{C}$ thus always map $\mathbb{C}$ biholomorphically *onto* $\mathbb{C}$. It follows in particular that

$$\operatorname{Aut} \mathbb{C} = \{f : \mathbb{C} \to \mathbb{C}, z \mapsto az + b : a \in \mathbb{C}^{\times}, b \in \mathbb{C}\}.$$

This so-called *affine* group of $\mathbb{C}$ is non-abelian. The set

$$T := \{f \in \operatorname{Aut} \mathbb{C} : f(z) = z + b, b \in \mathbb{C}\}$$

of translations is an abelian normal subgroup of $\operatorname{Aut} \mathbb{C}$. The plane $\mathbb{C}$ is homogeneous with respect to T.

The group $\operatorname{Aut} \mathbb{C}$ being considered here is not to be confused with the group of *field* automorphisms of $\mathbb{C}$.   □

The mappings $z \mapsto az$ and $z \mapsto az^{-1}$, $a \in \mathbb{C}^{\times}$, are automorphisms of $\mathbb{C}^{\times}$. The converse of this observation is contained in the following

Theorem. *Every injective holomorphic mapping $f : \mathbb{C}^{\times} \to \mathbb{C}^{\times}$ has either the form*

$$f(z) = az \quad or \quad f(z) = az^{-1}, \quad a \in \mathbb{C}^{\times}.$$

Proof. According to theorem 1, with $D := \mathbb{C}$, $A := \{0\}$ there are two possible cases:

a) The origin is a removable singularity of f. The holomorphic continuation $f : \mathbb{C} \to \mathbb{C}$ is then injective. It follows from the preceding theorem that f has the form $f(z) = az + b$ for appropriate $a \in \mathbb{C}^\times$, $b \in \mathbb{C}$. Since $f(\mathbb{C}^\times) \subset \mathbb{C}^\times$ while $f(-ba^{-1}) = 0$, it follows that $b = 0$.

b) The origin is a pole (of order 1) of f. Since $w \mapsto w^{-1}$ is an automorphism of $\mathbb{C}^\times$, $z \mapsto g(z) := 1/f(z)$ is another injective holomorphic mapping of $\mathbb{C}^\times$ into itself. Since 0 is a zero of g by theorem 1.1, it follows from a) that g has the form $g(z) = dz$, $d \in \mathbb{C}^\times$, and so $f(z) = az^{-1}$ for $a := d^{-1}$.

$\square$

Holomorphic injections $\mathbb{C}^\times \to \mathbb{C}^\times$ thus always map $\mathbb{C}^\times$ biholomorphically *onto* $\mathbb{C}^\times$. It follows in particular that

$$\operatorname{Aut} \mathbb{C}^\times = \{ f : \mathbb{C}^\times \to \mathbb{C}^\times, z \mapsto az; a \in \mathbb{C}^\times \} \cup$$
$$\{ f : \mathbb{C}^\times \to \mathbb{C}^\times, z \mapsto az^{-1}; a \in \mathbb{C}^\times \}.$$

This group is non-abelian. It decomposes into two "connected components each isomorphic to $\mathbb{C}^\times$." The component $L := \{ f : \mathbb{C}^\times \to \mathbb{C}^\times, z \mapsto az; a \in \mathbb{C}^\times \}$ is an abelian normal subgroup of $\operatorname{Aut} \mathbb{C}^\times$ and the punctured plane $\mathbb{C}^\times$ is homogeneous with respect to L.

3. Automorphisms of punctured bounded domains. For every subset M of D the set

$$\operatorname{Aut}_M D := \{ f \in \operatorname{Aut} D : f(M) = M \}$$

of all automorphisms of D which map M (bijectively) onto itself, constitutes a subgroup of $\operatorname{Aut} D$. If M consists of a single point c, then this is none other than the isotropy group of c with respect to $\operatorname{Aut} D$ introduced in 9.2.2. If $D \setminus M$ is again a domain, then via restriction to $D \setminus M$ every $f \in \operatorname{Aut}_M D$ determines an automorphism of $D \setminus M$. A group homomorphism from $\operatorname{Aut}_M D$ into $\operatorname{Aut}(D \setminus M)$ is thereby defined. If $D \setminus M$ has interior points in each connected component of D, then this (restriction) mapping is injective – for in this case any $g \in \operatorname{Aut}_M D$ which is the identity map on $D \setminus M$ is necessarily the identity map on D, on account of the identity theorem. In particular we have

If M is relatively closed in D and has no interior, then $\operatorname{Aut}_M D$ is isomorphic in a natural way to a subgroup of $\operatorname{Aut}(D \setminus M)$.

In interesting cases $\operatorname{Aut}_M D$ is in fact the whole group $\operatorname{Aut}(D \setminus M)$.

Theorem. *If D is bounded and has no isolated boundary points, then for every discrete and relatively closed subset A of D, the homomorphism $\operatorname{Aut}_A D \to \operatorname{Aut}(D \setminus A)$ is bijective.*

Proof. All that needs to be shown is that for each $f \in \operatorname{Aut}(D \setminus A)$ there exists an $\tilde{f} \in \operatorname{Aut}_A D$ with $f = \tilde{f}|(D \setminus A)$. Since f and $g := f^{-1}$ map $D \setminus A$

into $D \setminus A$, a subset of the bounded set D, these functions are bounded. Since A is discrete and relatively closed in D, f and g are bounded and holomorphic in a punctured neighborhood of each point of A. By Riemann's continuation theorem they extend to holomorphic functions $\tilde{f} : D \to \mathbb{C}$, $\tilde{g} : D \to \mathbb{C}$. According to theorem 1.c) both $\tilde{f}$ and $\tilde{g}$ are injective.

Next we show that $\tilde{f}(D) \subset D$. Since $\tilde{f}$ is continuous, $\tilde{f}(D)$ at least lies in the closure $\overline{D}$ of D. Suppose there were a point $p \in D$ with $\tilde{f}(p) \in \partial D$. Then p would necessarily lie in A and there would be a disc B around p with $(B \setminus p) \subset (D \setminus A)$ because A is discrete. Since $\tilde{f}$ is an open mapping, $\tilde{f}(B)$ would be a neighborhood of $\tilde{f}(p)$. Then since $\tilde{f}$ is injective, it would follow that

$$\tilde{f}(B) \setminus \tilde{f}(p) = \tilde{f}(B \setminus p) = f(B \setminus p) \subset D,$$

which says that $\tilde{f}(p)$ is an isolated boundary point of D, contrary to the hypothesis that D has no such points. Therefore no such p exists and $\tilde{f}(D) \subset D$ is confirmed. In exactly the same way we show that $\tilde{g}(D) \subset D$. This established, the composites $\tilde{f} \circ \tilde{g} : D \to \mathbb{C}$ and $\tilde{g} \circ \tilde{f} : D \to \mathbb{C}$ are well defined. Since these maps agree with $f \circ g = g \circ f = \mathrm{id}$ on the sense subset $D \setminus A$ of D, we have $\tilde{f} \circ \tilde{g} = \tilde{g} \circ \tilde{f} = \mathrm{id}$ on D, that is, $\tilde{f} \in \mathrm{Aut}\, D$. Finally, because $\tilde{f}(D \setminus A) = D \setminus A$, it follows that $\tilde{f}(A) = A$, meaning that $\tilde{f} \in \mathrm{Aut}_A D$. □

Example. The group $\mathrm{Aut}\, \mathbb{E}^\times$ of the punctured open unit disc $\mathbb{E}^\times := \mathbb{E} \setminus 0$ is isomorphic to the circle group S_1:

$$\mathrm{Aut}\, \mathbb{E}^\times = \{f : \mathbb{E}^\times \to \mathbb{E}^\times, z \mapsto az \,; a \in S_1\}.$$

Proof. As a result of our theorem, $\mathrm{Aut}\, \mathbb{E}^\times = \mathrm{Aut}_0 \mathbb{E}$. Therefore the claim follows from theorem 9.2.2. □

The theorem proved in this subsection is a continuation theorem for automorphisms of $D \setminus A$ to automorphisms of D, and the boundedness of D is essential, as the example $D := \mathbb{C}$, $A := \{0\}$, $f(z) := 1/z$ shows. The theorem is likewise false for bounded domains which have isolated boundary points: Take, for example, $D = \mathbb{E}^\times$ (of which 0 is an isolated boundary point) and $A := \{c\}$, where $c \in \mathbb{E}^\times$. We then have $\mathrm{Aut}_c \mathbb{E}^\times = \{\mathrm{id}\}$, by virtue of the preceding example; while the automorphism

$$z \mapsto \frac{z - c}{\bar{c}z - 1}$$

of $\mathbb{E}$ which interchanges 0 and c restricts to an automorphism of $\mathbb{E}^\times \setminus \{c\}$ different from the identity. Cf. also corollary 1 in the next subsection.

4. Conformally rigid regions. A domain D is called (conformally) *rigid* if its only automorphism is the identity map. We want to construct some bounded rigid regions and by way of preparation prove

Theorem. *Let A be a finite non-empty subset of $\mathbb{E}^\times$. Then there is a natural group monomorphism $\pi : \mathrm{Aut}\,(\mathbb{E}^\times \setminus A) \longrightarrow \mathrm{Perm}\,(A \cup \{0\})$ into the permutation group of the set $A \cup \{0\}$. (This permutation group is of course isomorphic to a symmetric group $\mathfrak{S}_n$.)*

Proof. Since $\mathbb{E}^\times \setminus A = \mathbb{E} \setminus (A \cup \{0\})$, we have $\mathrm{Aut}\,(\mathbb{E}^\times \setminus A) = \mathrm{Aut}\,_{A\cup\{0\}}\mathbb{E}$ from theorem 3. Every automorphism f of $\mathbb{E}^\times \setminus A$ thus maps $A \cup \{0\}$ bijectively onto itself, that is, induces a permutation $\pi(f)$ of $A \cup \{0\}$. It is clear that the correspondence $f \mapsto \pi(f)$ is a group homomorphism $\pi : \mathrm{Aut}\,(\mathbb{E}^\times \setminus A) \to \mathrm{Perm}\,(A \cup \{0\})$. Because a non-identity automorphism of $\mathbb{E}$ can fix at most one point (theorem 9.2.3) and $A \neq \emptyset$, π is injective.

Corollary 1. *Each group $\mathrm{Aut}\,(\mathbb{E}^\times \setminus c)$, $c \in \mathbb{E}^\times$, is isomorphic to the cyclic group $\mathfrak{S}_2$; the mapping $g(z) := \frac{z-c}{cz-1}$ is the only non-identity automorphism of $\mathbb{E}^\times \setminus c$.*

Proof. According to the theorem $\mathrm{Aut}\,(\mathbb{E}^\times \setminus c)$ is isomorphic to a subgroup of $\mathrm{Perm}\,\{0, c\} \cong \mathfrak{S}_2$; on the other hand, g does belong to $\mathrm{Aut}\,(\mathbb{E}^\times \setminus c)$, by theorem 2.3.3.

Corollary 2. *Suppose $a, b \in \mathbb{E}^\times$, $a \neq b$. Then $\mathrm{Aut}\,(\mathbb{E}^\times \setminus \{a,b\}) \neq \{\mathrm{id}\}$ if and only if at least one of the following four relations obtains between a and b:*

$$a = -b \quad or \quad 2b = a + \bar{a}b^2 \quad or \quad 2a = b + \bar{b}a^2 \quad or$$

$$|a| = |b| \quad and \quad a^2 + b^2 = ab(1 + |b|^2).$$

Proof. Because $\mathrm{Aut}\,(\mathbb{E}^\times \setminus \{a,b\}) = \mathrm{Aut}\,_{\{0,a,b\}}\mathbb{E}$ (theorem 3), theorem 9.2.2 insures that every $f \in \mathrm{Aut}\,(\mathbb{E}^\times \setminus \{a,b\})$ has the form $f(z) = e^{i\varphi}\frac{z-w}{\bar{w}z-1}$, for appropriate $\varphi \in \mathbb{R}$, $w \in \mathbb{E}$. Now $f \neq \mathrm{id}$ is the case precisely when $f : \{0,a,b\} \to \{0,a,b\}$ is not the identity permutation. *Five* cases are possible, of which we will discuss two:

$$f(0) = 0\,,\ f(a) = b\,,\ f(b) = a \Leftrightarrow f(z) = e^{i\varphi}z\,,\ \text{with both } e^{i\varphi}a = b\,,\ e^{i\varphi}b = a.$$

That occurs exactly when $e^{i2\varphi} = 1$, that is, when $e^{i\varphi} = \pm 1$, or equivalently, since $a \neq b$ by hypothesis, when $a = -b$. The second case we consider is:

$$f(0) = a\,,\ f(a) = b\,,\ f(b) = 0 \Leftrightarrow f(z) = e^{i\varphi}\frac{z-b}{\bar{b}z-1},$$

with both $a = e^{i\varphi}b$ and $b(\bar{b}a - 1) = e^{i\varphi}(a - b)$.

This leads to the case $|a| = |b|$ and $a^2 + b^2 = ab(1 + |b|^2)$. The remaining three cases are treated analogously.

Consequence. *The region $\mathbb{E} \setminus \{0, \frac{1}{2}, \frac{3}{4}\}$ is rigid.*

Exercises

Exercise 1. Show that if $f : \mathbb{C}^\times \to \mathbb{C}$ is holomorphic and injective, then for some $c \in \mathbb{C}$ $f(\mathbb{C}^\times) = \mathbb{C} \setminus c$.

Exercise 2. Show that $\text{Aut}\,(\mathbb{C} \setminus \{0, 1\})$ is comprised of exactly the following six functions: $z \mapsto z$, $z \mapsto z^{-1}$, $z \mapsto 1 - z$, $z \mapsto (1 - z)^{-1}$, $z \mapsto z(z - 1)^{-1}$ and $z \mapsto (z - 1)z^{-1}$.

Exercise 3. Investigate for which $z \in \mathbb{H}$ the region $\mathbb{H} \setminus \{i, 2i, z\}$ is rigid.

§3 Meromorphic functions

Holomorphic functions with poles have played such a prominent role in function theory from the beginning that very early a special name was introduced for them. As early as 1875 BRIOT and BOUQUET called such functions *meromorphic* ([BB], 2nd ed., p.15): "Lorsqu'une fonction est holomorphe dans une partie du plan, excepté en certains pôles, nous dirons qu'elle est *méromorphe* dans cette partie du plan, c'est-à-dire semblable aux fractions rationnelles. (When a function is holomorphic in part of the plane except for certain poles, we say that it is *meromorphic* in that part of the plane; that is to say, it resembles the rational fractions.)"

Meromorphic functions may not only be added, subtracted and multiplies but even – and therein lies their great advantage over holomorphic functions – divided by one another. This makes their algebraic structure simpler in comparison to that of the holomorphic functions. In particular, the meromorphic functions in a region form a field.

In subsections 1 through 3 the algebraic foundations of the theory of meromorphic functions will be discussed; in subsection 4 the order function o_c will be extended to meromorphic functions.

1. Definition of meromorphy. A function f is called *meromorphic* in D, if there is a discrete subset $P(f)$ of D (dependent of course on f) such that f is holomorphic in $D \setminus P(f)$ and has a pole at each point of $P(f)$. The set $P(f)$ is called the *pole-set* of f; obviously this set is always *relatively closed in D*.

We remark explicitly that the case of an empty pole-set is allowed:

The holomorphic functions in D are also meromorphic in D.

Since $P(f)$ is discrete and relatively closed in D it follows, just as for a-places (cf. 8.1.3), that

The pole-set of each function meromorphic in D is either empty, finite, or countably infinite.

A meromorphic function f in D having a non-empty pole-set can't map the whole of D into $\mathbb{C}$. In view of corollary 1.1 it is natural and convenient to choose the element ∞ as the function value at each pole:

$$f(z) := \infty \qquad \text{for } z \in P(f).$$

Meromorphic functions in D are thus special mappings $D \to \mathbb{C} \cup \{\infty\}$.

Examples. 1) Every *rational* function

$$h(z) := \frac{a_0 + a_1 z + \cdots + a_m z^m}{b_0 + b_1 z + \cdots + b_n z^n}, \qquad b_n \neq 0 \,, \, m, n \in \mathbb{N},$$

is meromorphic in $\mathbb{C}$, the pole-set is *finite* and is contained in the zero-set of the denominator polynomial.

2) The cotangent function $\cot \pi z = \cos \pi z / \sin \pi z$ is meromorphic, but not rational; its pole-set is countably infinite:

$$P(\cot \pi z) = Z(\sin \pi z) = \mathbb{Z}.$$

A function is called *meromorphic at c* if it is meromorphic in a neighborhood of c. According to the development theorem 1.2 every such function f which is non-zero has a representation

$$f(z) = \sum_{\nu = m}^{\infty} a_\nu (z - c)^\nu$$

around c, with *uniquely determined* numbers $a_\nu \in \mathbb{C}$ and $m \in \mathbb{Z}$ such that $a_m \neq 0$. If $m < 0$, $\sum_m^{-1} a_\nu (z - c)^\nu$ is called the *principal part* of f at c. In case $m \leq 0$ the principal part of f is defined to be 0.

From the expansions $\sin \pi z = (-1)^n \pi (z - n) +$ higher powers of $(z - n)$ and $\cos \pi z = (-1)^n + (-1)^{n+1} \pi (z - n)^2 / 2 +$ higher powers of $(z - n)$ it follows that

$$(1) \qquad \pi \cot \pi z = \frac{1}{z - n} + \text{power series in } (z - n), \qquad \text{for every } n \in \mathbb{Z}.$$

This equation will be used to obtain the partial fraction series of the cotangent in 11.2.1.

2. The $\mathbb{C}$-algebra $\mathcal{M}(D)$ of the meromorphic functions in D.

For the totality of meromorphic functions in D there is no generally accepted symbol. But recently, especially in the theory of functions of several complex variables, the notation, which we will use,

$$\mathcal{M}(D) := \{h : h \text{ meromorphic in } D\}$$

has gained the ascendancy. Clearly $\mathcal{O}(D) \subsetneq \mathcal{M}(D)$.

Meromorphic functions may be added, subtracted and multiplied. If, say, $f, g \in \mathcal{M}(D)$ with pole-sets $P(f)$, $P(g)$ are given, then $P(f) \cup P(g)$ is also discrete and relatively closed in D and in $D \setminus (P(f) \cup P(g))$ each of f and g, hence also $f \pm g$ and $f \cdot g$, are holomorphic. For each $c \in P(f) \cup P(g)$ there are natural numbers m, n and a neighborhood U of c lying in D with $U \cap (P(f) \cup P(g)) = \{c\}$, such that $(z - c)^m f(z)$ and $(z - c)^n g(z)$ are each bounded in $U \setminus c$. (Cf. theorem 1.1; $m = 0$ in case $c \notin P(f)$ and $n = 0$ in case $c \notin P(g)$.) Then each of the three functions

$$(z - c)^{m+n} \cdot [f(z) \dot{\pm} g(z)]$$

is bounded in $U \setminus c$. The point c is then either a removable singularity or a pole of the various functions $f \dot{\pm} g$. Thus the pole-sets of these functions are subsets of $P(f) \cup P(g)$ and as such are discrete and relatively closed in D. From this it follows that $f \dot{\pm} g \in \mathcal{M}(D)$. The rules of calculating with holomorphic functions imply that

$\mathcal{M}(D)$ *is a $\mathbb{C}$-algebra (with respect to pointwise addition, subtraction and multiplication). The $\mathbb{C}$-algebra $\mathcal{O}(D)$ is a $\mathbb{C}$-subalgebra of $\mathcal{M}(D)$. For all $f, g \in \mathcal{M}(D)$ the pole-sets satisfy*

$$P(-f) = P(f) \, , \, P(f \dot{\pm} g) \subset P(f) \cup P(g).$$

$P(f \dot{\pm} g)$ is generally a proper subset of $P(f) \cup P(g)$. For example, with $D := \mathbb{C}$, $f(z) := 1/z$, $g(z) := z - 1/z$ we have $P(f) = P(g) = \{0\}$, but $P(f + g) = \emptyset \neq P(f) \cup P(g)$; while for $f(z) := 1/z$ and $g(z) := z$ we have $P(f) = \{0\}$, $P(g) = \emptyset$ and $P(fg) = \emptyset \neq P(f) \cup P(g)$. □

Like $\mathcal{O}(D)$, the $\mathbb{C}$-algebra $\mathcal{M}(D)$ is closed under differentiation; more precisely (on the basis of results from 1.2):

Along with f, its derivative f' is also meromorphic in D. These two functions have the same pole-set: $P(f) = P(f')$; and if q is the principal part of f at a pole, then q' is the principal part of f' there.

3. Division of meromorphic functions. In the ring $\mathcal{O}(D)$ of holomorphic functions in D, division by an element g is possible just when g is zero-free in D. But in the ring $\mathcal{M}(D)$ we can – and this is of great advantage – also divide by functions which have zeros. By the *zero-set $Z(f)$ of a meromorphic function $f \in \mathcal{M}(D)$* we understand the zero-set of the holomorphic function $f|(D \setminus P(f)) \in \mathcal{O}(D \setminus P(f))$. Clearly $Z(f)$ is relatively closed in D and $Z(f) \cap P(f) = \emptyset$.

Theorem on units. *The following assertions about a meromorphic function* $u \in \mathcal{M}(D)$ *are equivalent:*

i) u *is a unit in* $\mathcal{M}(D)$*, that is,* $u\tilde{u} = 1$ *for some* $\tilde{u} \in \mathcal{M}(D)$*.*

ii) *The zero-set* $Z(u)$ *is discrete in* D*.*

When i) *holds,* $P(\tilde{u}) = Z(u)$ *and* $Z(\tilde{u}) = P(u)$*.*

Proof. i) $\Rightarrow$ ii) The equation $u\tilde{u} = 1$ immediately implies for $c \in D$

$$u(c) = 0 \Leftrightarrow \tilde{u}(c) = \infty \qquad \text{and} \qquad u(c) = \infty \Leftrightarrow \tilde{u}(c) = 0,$$

which means that $Z(u) = P(\tilde{u})$ and $P(u) = Z(\tilde{u})$. In particular, being the pole-set of a meromorphic function in D, $Z(u)$ is discrete in D.

ii) $\Rightarrow$ i) The set $A := Z(u) \cup P(u)$ is discrete and relatively closed in D. In $D \setminus A$, $\tilde{u} := 1/u$ is holomorphic. Every point of $Z(u)$ is a pole of $\tilde{u}$ (cf. theorem 1.1) and every point $c \in P(u)$ is a removable singularity (and a zero) of $\tilde{u}$ because $\lim_{z \to c} 1/u(z) = 0$. This means that $\tilde{u} \in \mathcal{M}(D)$. □

On the basis of this theorem the quotient of two elements $f, g \in \mathcal{M}(D)$ exists in the ring $\mathcal{M}(D)$ exactly when $Z(g)$ is discrete in D. In particular, $f/g \in \mathcal{M}(D)$ for any $f, g \in \mathcal{O}(D)$ when $Z(g)$ is discrete in D.

An important consequence of the theorem on units is the

Corollary. *The* $\mathbb{C}$*-algebra of all meromorphic functions in a region is a field.*

Proof. If $f \in \mathcal{M}(G)$ is not the zero element and G is a region, then $G \setminus P(f)$ is a region (proof!) and $f|(G \setminus P(f))$ is a holomorphic function which is not the zero element of $\mathcal{O}(G \setminus P(f))$. Therefore $Z(f)$ is discrete in G (cf. 8.1.3) and so by the theorem on units, f is a unit in $\mathcal{M}(G)$. That is, every element of $\mathcal{M}(G) \setminus \{0\}$ is a unit. □

The field $\mathcal{M}(\mathbb{C})$ contains the field $\mathbb{C}(z)$ of rational functions as a *proper* subfield, since, e.g., $\exp(z)$, $\cot(z) \notin \mathbb{C}(z)$.

Every integral domain lies in a smallest field, its so-called *quotient field*. The quotient field of $\mathcal{O}(G)$, which consists of all quotients f/g with $f, g \in \mathcal{O}(G)$ and $g \neq 0$, consequently lies in the field $\mathcal{M}(G)$. A fact which even for $G = \mathbb{C}$ is not trivial and which we will only be able to prove in the second volume (via Weierstrass' product theorem) is:

The field $\mathcal{M}(G)$ *is the quotient field of* $\mathcal{O}(G)$*.*

In 9.4.5 we associated to every non-constant holomorphic mapping $g : G \to G'$ a monomorphic lifting $g^* : \mathcal{O}(G') \to \mathcal{O}(G)$, $h \mapsto h \circ g$. We now show that

The mapping $g^* : \mathcal{O}(G') \to \mathcal{O}(G)$ extends to a $\mathbb{C}$-algebra monomorphism $g^* : \mathcal{M}(G') \to \mathcal{M}(G)$ of the field of meromorphic functions in G' into the field of meromorphic functions in G. For every $h \in \mathcal{M}(G')$, we have $P(g^*(h)) = g^{-1}(P(h))$.

Proof. Since g is not constant and each set $P(h)$, $h \in \mathcal{M}(G')$, is discrete and relatively closed in G', the set $g^{-1}(P(h))$ is always discrete and relatively closed in G (see 8.1.3). $h \circ g$ is holomorphic in $G \setminus g^{-1}(P(h))$. Since in addition

$$\lim_{z \to c} (h \circ g)(z) = \lim_{w \to g(c)} h(w) = \infty \qquad \text{for every } c \in g^{-1}(P(h)),$$

we infer that $g^*(h) := h \circ g$ is a meromorphic function in G with pole-set $g^{-1}(P(h))$. Evidently the mapping g^* so-defined is a $\mathbb{C}$-algebra monomorphism of $\mathcal{M}(G')$ into $\mathcal{M}(G)$. □

Next the Identity Theorem 8.1.1 will be generalized to

The Identity Theorem for Meromorphic Functions. *The following statements about a pair of meromorphic functions f, g in a region G are equivalent:*

i) *$f = g$.*

ii) *The set $\{w \in G \setminus P(f) \cup P(g)) : f(w) = g(w)\}$ has a cluster point in $G \setminus (P(f) \cup P(g))$.*

iii) *There is a point $c \in G \setminus (P(f) \cup P(g))$ such that $f^{(n)}(c) = g^{(n)}(c)$ for all $n \in \mathbb{N}$.*

Proof. If G is a region, $G \setminus (P(f) \cup P(g))$ is also a region. Also f and g are each holomorphic in the latter region. Therefore the asserted equivalences follow from 8.1.1. □

4. The order function o_c. If $f \not\equiv 0$ is meromorphic at c, then f has a unique development

$$f(z) = \sum_{m}^{\infty} a_\nu (z - c)^\nu \qquad \text{with } a_\nu \in \mathbb{C} \text{ , } m \in \mathbb{Z} \text{ and } a_m \neq 0$$

(cf. subsection 1). The integer m which is *uniquely determined* by this equation is called *the order of f at c* and denoted $o_c(f)$. If f is in fact holomorphic at c, then this is the order already introduced in 8.1.4. From the definition it is immediate that

For an f which is meromorphic at c:

1) *f is holomorphic at c $\Leftrightarrow$ $o_c(f) \geq 0$.*

2) *In case $m = o_c(f) < 0$, c is a pole of f of order $-m$.*

The poles of f are therefore just those points where the order of f is negative. It is unfortunate that the word "order" has acquired a double meaning in connection with meromorphic functions: at a point c, f *always* has an order, possibly negative, and it *may* have a pole, the latter being necessarily of positive order. Thus forewarned, the reader will not be confused about this in the future.

As in 8.1.4 we now again have the

Rules of computation for the order function. *For all functions f, g which are meromorphic at c*

1) $o_c(fg) = o_c(f) + o_c(g)$ *(product rule)*;

2) $o_c(f + g) \geq \min\{o_c(f), o_c(g)\}$, *with equality whenever $o_c(f) \neq o_c(g)$.*

The proof is remanded to the reader.

Let $\mathcal{M}_c$, $\mathcal{O}_c$ denote the set of all functions which are meromorphic or holomorphic, respectively, at c. Consider two such functions equal if they coincide in some (perhaps smaller than the domain of either) neighborhood of c. $\mathcal{O}_c$ is an integral domain and $\mathcal{M}_c$ is in a natural way its quotient field. The order function introduced above is nothing but the natural extension of the order function of $\mathcal{O}_c$ to a non-archimedean valuation of $\mathcal{M}_c$; on this point cf. 4.4.3.

Exercises

Exercise 1. Show that if f is meromorphic in D and has a finite set of poles, then there is a rational function h with $P(h) = P(f)$ and $(f-h)|D \in \mathcal{O}(D)$.

Exercise 2. a) Prove the equivalence of the following statements about a pair of functions f and g which are meromorphic in a region G:

i) $f = g$.

ii) The set $\{w \in G \setminus (P(f) \cup P(g)) : f(w) = g(w)\}$ has a cluster point in G.

b) Find an example of an $f \in \mathcal{O}(\mathbb{C}^\times)$ which is not the function 0 but satisfies $f(\frac{1}{n}) = 0$ for all non-zero integers n.

Exercise 3. Let $f \in \mathcal{M}(\mathbb{C})$ satisfy $|f(z)| \leq M|z|^n$ for all $z \in \mathbb{C} \setminus P(f)$ with $|z| > r$, for some finite constants M, r and some $n \in \mathbb{N}$. Show that f is a rational function.

Chapter 11

Convergent Series of Meromorphic Functions

In 1847 the Berlin mathematician Gotthold EISENSTEIN (known to students of algebra from his irreducibility criterion) introduced into the theory of the trigonometric functions the series

$$\sum_{\nu=-\infty}^{\infty} \frac{1}{(z+\nu)^k}, \qquad k = 1, 2, \ldots$$

which nowadays are frequently named after him. These Eisenstein series are the simplest examples of normally convergent series of meromorphic functions in $\mathbb{C}$. In this chapter we will first introduce in section 1 the general concepts of compact and normally convergent series of meromorphic functions. In section 2 the partial fraction decomposition

$$\pi \cot \pi z = \frac{1}{2} + \sum_{1}^{\infty} \frac{2z}{z^2 - \nu^2} = \frac{1}{z} + \sum_{1}^{\infty} \left(\frac{1}{z+\nu} + \frac{1}{z-\nu} \right)$$

of the cotangent function will be studied; it is one of the most fruitful series developments in classical analysis. In section 3 by comparing coefficients from the Taylor series of $\sum_{1}^{\infty} \frac{2z}{z^2 - \nu^2}$ and $\pi \cot \pi z - \frac{1}{z}$ around 0 we secure the famous Euler identities

$$\sum_{1}^{\infty} \frac{1}{\nu^{2n}} = (-1)^{n-1} \frac{(2\pi)^{2n}}{2(2n)!} B_{2n}, \qquad n = 1, 2, \ldots.$$

In section 4 we sketch Eisenstein's approach to the trigonometric functions.

§1   General convergence theory

In the definition of convergence of series of meromorphic functions the poles of the summands not unexpectedly cause difficulties. Since we want the limit function in any case to be itself meromorphic in D, it is not unreasonable to demand that in every compact subset of D only finitely many of the summands really have any poles. This "dispersion of poles," which will be found to prevail in all later applications, is really the only new feature here; everything else proceeds just as in the convergence theory of holomorphic functions.

1. Compact and normal convergence. A series $\sum f_\nu$ of functions f_ν each of which is meromorphic in D is called *compactly convergent in D*, if to each compact $K \subset D$ there corresponds an index $m = m(K) \in \mathbb{N}$ such that:

 1) For each $\nu \geq m$ the pole-set $P(f_\nu)$ is disjoint from K, and

 2) The series $\sum_{\nu \geq m} f_\nu | K$ converges uniformly on K.

The series $\sum f_\nu$ is called *normally convergent in D* if 1) holds but in place of 2) the stronger

2') $\sum_{\nu \geq m} |f_\nu|_K < \infty$

prevails.

Conditions 2) and 2') make sense because, thanks to the *"pole-dispersion condition"* 1), the functions f_ν with $\nu \geq m$ are all pole-free, hence continuous, in K. Another consequence of this condition is that the set $\bigcup_0^\infty P(f_\nu)$ is discrete and relatively closed in D. It is clear that 1) and 2), or 1) and 2'), hold for all compact subsets of D if they hold for all closed discs lying in D.

As before, normal convergence implies compact convergence. If all the functions f_ν are actually holomorphic in D, then requirement 1) is vacuous and we are back to talking about compact or normal convergence of series of holomorphic functions.

Compactly convergent series of meromorphic functions have meromorphic limit functions. More precisely,

Convergence theorem. *Let $f_\nu \in \mathcal{M}(D)$ and $\sum f_\nu$ be compactly (respectively, normally) convergent in D. Then there is precisely one meromorphic function f in D with the following property:*

If U is an open subset of D and for some $m \in \mathbb{N}$ none of the functions f_ν with $\nu \geq m$ has any poles in U, then the series $\sum_{\nu \geq m} f_\nu | U$ of holomorphic functions converges compactly (respectively, normally) in U to an $F \in \mathcal{O}(U)$ such that

(1) $f | U = f_0 | U + f_1 | U + \cdots + f_{m-1} | U + F.$

In particular, f is holomorphic in $D \setminus \bigcup_0^\infty P(f_\nu)$, i.e., $P(f) \subset \bigcup_0^\infty P(f_\nu)$.

The proof is a simple exercise. Naturally we call the function f the sum of the series $\sum f_\nu$ and write $f = \sum f_\nu$. It should be noted that due to the pole-dispersion condition, for every *relatively compact* subdomain $U \subset D$ equation (1) holds for appropriate m and $F \in \mathcal{O}(U)$. [Recall that a subset M of a metric space X is called *relatively compact* in X if its closure $\overline{M}$ in X is compact.] You can quickly develop a sound sense for calculating with series of meromorphic functions by just keeping in mind the following simplifying rule of thumb:

In every relatively compact subdomain U of D, after subtraction of finitely many initial terms what remains is a series of functions which are holomorphic in U and this series converges compactly (respectively, normally) to a holomorphic function in U.

2. Rules of calculation. One confirms in a microsecond that:

If $f = \sum f_\nu$, $g = \sum g_\nu$ are compactly (respectively, normally) convergent series of meromorphic functions in D, then for every $a, b \in \mathbb{C}$ the series $\sum(af_\nu + bg_\nu)$ converges compactly (respectively, normally) in D to $af + bg$.

If $f_\nu \in \mathcal{M}(D)$, and the series $\sum f_\nu$ converges normally in D, then so does every one of its subseries; likewise we have (cf. 3.3.1):

Rearrangement theorem. *If $f_\nu \in \mathcal{M}(D)$ and $\sum_0^\infty f_\nu$ converges normally in D to f, then for every bijection $\tau : \mathbb{N} \to \mathbb{N}$ the rearranged series $\sum_0^\infty f_{\tau(\nu)}$ converges normally in D to f.*

Also valid is the

Differentiation theorem. *If $f_\nu \in \mathcal{M}(D)$ and $\sum f_\nu = f$ converges compactly (respectively, normally) in D, then for every $k \geq 1$ the k-times term-wise differentiated series $\sum f_\nu^{(k)}$ converges compactly (respectively, normally) in D to $f^{(k)}$.*

Proof. It suffices to consider the case $k = 1$. Given an open and relatively compact set $U \subset D$, choose m so large that f_ν is holomorphic in U for every $\nu \geq m$. Then $\sum_{\nu \geq m} f_\nu | U$ converges compactly (respectively, normally) in U to a function $F \in \mathcal{O}(U)$ for which 1.(1) holds. We have $f_\nu' | U = (f_\nu | U)' \in \mathcal{O}(U)$ and by 8.4.2 the series $\sum_{\nu \geq m} f_\nu' | U$ converges compactly (respectively, normally) in U to $F' \in \mathcal{O}(U)$. This establishes that $\sum f_\nu'$ converges compactly (respectively, normally) throughout D. Due to 1.(1), its sum $g \in \mathcal{M}(D)$ satisfies

$$g|U = f_0'|U + \cdots + f_{m-1}'|U + F' = (f_0|U + \cdots + f_{m-1}|U + F)' = (f|U)'.$$

This proves that $g = f'$.   □

There is no direct analog of the theorem in 8.4.2 on products of series. To see this, assume $f_\mu, g_\nu \in \mathcal{M}(D)$, $f = \sum f_\mu$ and $g = \sum g_\nu$ are normally convergent in D and form a product series $\sum h_\lambda$, in which the h_λ run through all the products $f_\mu g_\nu$ exactly once. In general there is no guarantee that the sequence h_λ satisfies the "pole-dispersion condition." However $\sum h_\lambda$ does converge normally in $D \setminus \bigcup_{\mu,\nu} (P(f_\mu) \cup P(g_\nu))$.

3. Examples. For any $r > 0$ the inequalities

$$|z \pm n|^k \geq (n - r)^k \qquad \text{for } k \geq 1 , n \in \mathbb{N} , |z| \leq r < n$$

hold. From them we infer for $K := B_r(0)$ the estimates

$$\left| \frac{1}{z+n} - \frac{1}{n} \right|_K \leq \frac{r}{n(n-r)} \qquad \text{for } |n| > r ;$$

$$\left| \frac{1}{(z \pm n)^k} \right|_K \leq \frac{1}{(n-r)^k} \qquad \text{for } k \geq 1 , n > r.$$

Since the series $\sum n^{-k}$ ($k > 1$) and $\sum (n(n-r))^{-1}$ converge, and since every compactum in $\mathbb{C}$ lies in some disc $B_r(0)$, we see that (cf. 3.3.2):

The four series

$$\sum_1^\infty \left(\frac{1}{z+\nu} - \frac{1}{\nu} \right) , \ \sum_1^\infty \left(\frac{1}{z-\nu} + \frac{1}{\nu} \right) , \ \sum_0^\infty \frac{1}{(z+\nu)^k} , \ \sum_0^\infty \frac{1}{(z-\nu)^k}$$

(where $k \geq 2$) are normally convergent in $\mathbb{C}$ to meromorphic functions.

Addition of the first two of these series shows that $\sum_1^\infty \frac{2z}{z^2 - \nu^2}$ is also normally convergent in $\mathbb{C}$.

Besides the series $\sum_0^\infty f_\nu$, one has to consider more general series of the form

$$\sum_{-\infty}^\infty f_\nu := \sum_{-\infty}^{-1} f_\nu + \sum_0^\infty f_\nu, \quad \text{where } \sum_{-\infty}^{-1} f_\nu \text{ means } \lim_{n\to\infty} \sum_{-n}^{-1} f_\nu.$$

Such a series of functions is said to converge (absolutely) at $c \in \mathbb{C}$ if both the series $\sum_{-\infty}^{-1} f_\nu(c)$ and $\sum_0^\infty f_\nu(c)$ converge (absolutely). Compact or normal convergence of $\sum_{-\infty}^\infty f_\nu$ means compact or normal convergence of both of $\sum_{-\infty}^{-1} f_\nu$ and $\sum_0^\infty f_\nu$. Such generalized series will play a significant role later (cf. 12.3.1) in the theory of Laurent series.

From the preceding it is now clear that

The series of meromorphic functions in $\mathbb{C}$ given by

$$\sum_{-\infty}^{\infty}{}' \left(\frac{1}{z+\nu} + \frac{1}{\nu} \right) = \sum_{1}^{\infty} \frac{2z}{z^2 - \nu^2} \quad \text{and} \quad \sum_{-\infty}^{\infty} \frac{1}{(z+\nu)^k} \ , \ k \geq 2$$

are each normally convergent in $\mathbb{C}$. *(The standard abbreviation* $\sum_{-\infty}^{\infty}{}' :=$

$\sum_{-\infty}^{-1} + \sum_{0}^{\infty}$ *is being used here.)*

Exercises

Exercise 1. Show that the series $\sum_{\nu \geq 1} \frac{(-1)^{\nu-1}}{z+\nu}$ converges compactly in $\mathbb{C}$ but not normally.

Exercise 2. Show that the following series are normally convergent in $\mathbb{C}$:
  a) $\sum_{n=0}^{\infty} \left(\frac{2^n}{z-2^n} + 1 \right)$
  b) $\sum_{n=1}^{\infty} \left(\frac{n^2}{(z-n)^2} - 1 - \frac{2z}{n} \right)$.

Exercise 3. Set $a_n := \frac{n-1}{n}$ for $n \geq 2$. Show that:
  a) $\sum_{n=2}^{\infty} \left\{ \frac{1}{z-a_n} + \frac{1}{a_n} \sum_{k=0}^{n} \left(\frac{z}{a_n} \right)^k \right\}$
is normally convergent in $\mathbb{E}$ and diverges at every point of $\mathbb{C} \setminus \mathbb{E}$;
  b) $\sum_{n=2}^{\infty} \left\{ \frac{1}{z-a_n} + \sum_{k=0}^{n} (\frac{1}{n})^k (\frac{1}{1-z})^{k+1} \right\}$
is normally convergent in $\mathbb{C} \setminus \{1\}$.

§2   The partial fraction development of $\pi \cot \pi z$

On the basis of 1.3 the equations

$$\varepsilon_1(z) := \lim_{n \to \infty} \sum_{-n}^{n} \frac{1}{z+\nu} = \frac{1}{z} + \sum_{1}^{\infty} \left(\frac{1}{z+\nu} + \frac{1}{z-\nu} \right) = \frac{1}{z} + \sum_{1}^{\infty} \frac{2z}{z^2 - \nu^2}$$

involve series of meromorphic functions which converge normally in $\mathbb{C}$ and so by the convergence theorem 1.1 the function ε_1 thus defined is meromorphic in $\mathbb{C}$. For its esthetic and suggestive value we write

$$\sum_{-\infty}^{\infty} e := \lim_{n \to \infty} \sum_{-n}^{n};$$

this is the so-called "Eisenstein summation." Thus

$$\varepsilon_1(z) = \sum_{-\infty}^{\infty} e \frac{1}{z + \nu}.$$

(One should note that $\sum_{-\infty}^{\infty} \frac{1}{z+\nu}$ does not exist according to the conventions established in 1.2.) Also we have

$$\varepsilon_1(z) = \frac{1}{z} + \sum_{-\infty}^{\infty}{}' \left(\frac{1}{z+\nu} - \frac{1}{\nu} \right)$$

The study of this function is the central concern of this section. We begin by characterizing the cotangent function.

1.   The cotangent and its double-angle formula.   The identity $\pi\cot \pi z = \varepsilon_1(z)$. The function $\pi\cot\pi z$ is holomorphic in $\mathbb{C} \setminus \mathbb{Z}$ and every point $m \in \mathbb{Z}$ is a first-order pole at which the principal part is $(z - m)^{-1}$; see 10.3.1(1). Furthermore this is an odd function and it satisfies (see 5.2.5) the

Double-angle formula

$$2\pi \cot 2\pi z = \pi \cot \pi z + \pi \cot \pi(z + \tfrac{1}{2}),$$

We will show that these properties characterize the cotangent.

Lemma. *Let the function g be holomorphic in $\mathbb{C} \setminus \mathbb{Z}$ and have principal part $(z - m)^{-1}$ at each $m \in \mathbb{Z}$. Suppose further that g is an odd function and satisfies the duplication formula*

$$2g(2z) = g(z) + g(z + \tfrac{1}{2}).$$

Then $g(z) = \pi \cot \pi z$ for all $z \in \mathbb{C} \setminus \mathbb{Z}$.

Proof. The function $h(z) := g(z) - \pi \cot \pi z$ is entire and odd and satisfies

$(*)$ $$2h(2z) = h(z) + h(z + \tfrac{1}{2}), \quad h(0) = 0.$$

Were h not identically 0, the maximum principle 8.5.2 would furnish a $c \in \overline{B_2(0)}$ such that $|h(z)| < |h(c)|$ for all $z \in B_2(0)$. Since both $\frac{1}{2}c$ and $\frac{1}{2}(c+1)$ lie in $B_2(0)$, it would follow that

$$|h(\tfrac{1}{2} c) + h(\tfrac{1}{2} c + \tfrac{1}{2})| \leq |h(\tfrac{1}{2} c)| + |h(\tfrac{1}{2}(c+1)))| < 2|h(c)|,$$

in contradiction with (∗). Therefore h must indeed be the function 0. □

Now follows quickly the

Theorem. *The cotangent function has in* $\mathbb{C} \setminus \mathbb{Z}$ *the series representations*

$$(1) \qquad \pi \cot \pi z = \varepsilon_1(z) = \sum_{-\infty}^{\infty} e \frac{1}{z + \nu} \quad = \quad \frac{1}{z} + {\sum_{-\infty}^{\infty}}' \left(\frac{1}{z + \nu} - \frac{1}{\nu} \right)$$

$$= \quad \frac{1}{z} + \sum_{1}^{\infty} \frac{2z}{z^2 - \nu^2}.$$

Proof. From the definition of ε_1 we immediately infer that it is holomorphic in $\mathbb{C} \setminus \mathbb{Z}$ and has principal part $(z - m)^{-1}$ at each $m \in \mathbb{Z}$. It also follows directly from that definition that $\varepsilon_1(-z) = -\varepsilon_1(z)$. And we verify by routine algebra that the partial sums $s_n(z) = \frac{1}{z} + \sum_1^n \left\{ \frac{1}{z+\nu} + \frac{1}{z-\nu} \right\}$ satisfy

$$s_n(z) + s_n(z + \tfrac{1}{2}) = 2 s_{2n}(z) + \frac{1}{2z + 2n + 1},$$

from which, after passage to the limit on n, we acquire the duplication formula $2\varepsilon_1(2z) = \varepsilon_1(z) + \varepsilon_1(z + \tfrac{1}{2})$. The preceding lemma thus guarantees that $\varepsilon_1(z) = \pi \cot \pi z$. □

The equation (1) is called the *partial fraction representation* of $\pi \cot \pi z$. A second, quite different proof of it, which goes back to EISENSTEIN, will be given in 4.2.

2. Historical remarks on the cotangent series and its proof. The partial fraction series for $\pi \cot \pi z$ was quite familiar to EULER; by 1740 he knew the more general formula ("De seriebus quibusdam considerationes," *Opera Omnia* (1) **14**, 407-462)

$$\frac{\pi}{n} \frac{\cos[\pi(w - z)/2n]}{\sin[\pi(w + z)/2n] - \sin[\pi(w - z)/2n]} =$$

$$\frac{1}{z} + \sum_{1}^{\infty} \left[\frac{2w}{(2\nu - 1)^2 n^2 - w^2} - \frac{2z}{(2\nu)^2 n^2 - z^2} \right],$$

which for $n := 1$, $w := -z$ becomes the cotangent series. In 1748 he incorporated the cotangent series into his *Introductio* (cf. [E], §178 bottom).

The double angle formula for the cotangent was used as early as 1868 by H. SCHRÖTER to get the partial fraction series "in the most elementary way"; cf. "Ableitung der Partialbruch- und Produkt-Entwickelungen für die trigonometrischen Funktionen," *Zeitschr. Math. u. Physik* **13**, 254-259. The elegant proof of the equality $\pi \cot \pi z = \varepsilon_1(z)$ reproduced in subsection 1 was published in 1892 by Friedrich Hermann SCHOTTKY in a now forgotten paper "Über das Additionstheorem der Cotangente ⋯," *Jour. für Reine u. Angew. Math.* **110**, 324-337; cf. in particular p. 325. Gustav HERGLOTZ (German mathematician, 1881-1953; from 1909-1925 Professor at Leipzig, thereafter at Göttingen; teacher of Emil ARTIN) observed that in Schottky's proof one doesn't need the maximum principle at all. Because it is elementary to prove the

Lemma (HERGLOTZ). *Every function h which is holomorphic in a disc $B_r(0)$, $r > 1$, and satisfies the duplication formula*

$$(*) \quad 2h(2z) = h(z) + h(z + \tfrac{1}{2}), \quad \text{whenever } z, z + \tfrac{1}{2}, 2z \text{ all lie in } B_r(0),$$

is constant.

Proof. From $(*)$ follows $4h'(2z) = h'(z) + h'(z + \tfrac{1}{2})$. Choose $1 < t < r$ and let M denote the maximum of $|h'|$ in the compact disc $\overline{B_t(0)}$. We notice that $\tfrac{1}{2}z$ and $\tfrac{1}{2}z + \tfrac{1}{2}$ lie in $B_t(0)$ whenever z does and we apply the above identity involving h' with $\tfrac{1}{2}z$ in the role of z; it yields, for all $z \in B_t(0)$

$$4|h'(z)| \le |h'(\tfrac{1}{2}z)| + |h'(\tfrac{1}{2}z + \tfrac{1}{2})| \le M + M.$$

Therefore $4M \le M + M$, $M = 0$. That is, $h' = 0$; so h is constant, in $B_t(0)$ hence throughout $B_r(0)$. □

This proof, which makes a factor of 4 out of the 2, is called the HERGLOTZ trick; it furnishes particularly easy access to the partial fraction representation of the cotangent in $\mathbb{C}$. HERGLOTZ used this trick in his lectures but never published it. The first explicit appearance of it in print was in the 1950 original German edition of CARATHÉODORY [5], pp. 268-271; besides this it has occurred in some 1936 mimeographed lecture notes of S. BOCHNER at Princeton on functions of several complex variables. And ARTIN had used the Herglotz trick in connection with the gamma function in his little monograph *The Gamma Function*, Holt, Rinehart and Winston (1964), New York (see p. 26), whose German original appeared in 1931.

It may be noted that equation 1(1) can be secured for *real* z somewhat more immediately: The function h satisfying the duplication formula $(*)$ is *real-valued* and *continuous* on $\mathbb{R}$ and, as the difference of two odd functions, it is *odd*. Given

$r \geq 1$, let M denote the maximum of $|h|$ over the interval $[-r, r]$. If M were positive, then $h(0) = 0$ and $h(-x) = -h(x)$ for all x would insure the existence of a smallest *positive* real number $2t \leq r$ with $|h(2t)| = M$. The identity (*) would yield $2M \leq |h(t)| + |h(t + \frac{1}{2})|$ and the fact that $t, t + \frac{1}{2} \in [-r, r]$ would give $|h(t)| \leq M$ and $|h(t + \frac{1}{2})| \leq M$ as well; from which $|f(t)| = M$ would follow, in violation of the minimality of t. Therefore it must be that $M = 0$, and since $r \geq 1$ is arbitrary, this says that $h = 0$ throughout $\mathbb{R}$. This verifies 1(1) for real z. But to get the formula for all complex z from this we would have to invoke the Identity Theorem.

3. Partial fraction series for $\dfrac{\pi^2}{\sin^2 \pi z}$ **and** $\dfrac{\pi}{\sin \pi z}$. Noting the identity $(\cot z)' = -(\sin z)^{-2}$ and applying the differentiation theorem 1.2 to the normally convergent series $\dfrac{1}{z} + \displaystyle\sum_{-\infty}^{\infty}{}' \left(\dfrac{1}{z+\nu} - \dfrac{1}{\nu} \right)$, we deduce from the equation $\varepsilon_1(z) = \pi \cot \pi z$ the classical partial fraction development

$$(1) \qquad \frac{\pi^2}{\sin^2 \pi z} = \sum_{-\infty}^{\infty} \frac{1}{(z + \nu)^2}.$$

Another differentiation yields

$$(2) \qquad \pi^3 \frac{\cot \pi z}{\sin^2 \pi z} = \sum_{-\infty}^{\infty} \frac{1}{(z + \nu)^3}.$$

From the identities $\pi \tan \frac{1}{2}\pi z = \pi \cot \frac{1}{2}\pi z - 2\pi \cot \pi z$ (cf. 5.2.5), and $\pi \cot \pi z = \varepsilon_1(z)$ it follows that

$$(3) \qquad \pi \tan \tfrac{1}{2}\pi z = \sum_{0}^{\infty} \frac{4z}{(2\nu + 1)^2 - z^2}.$$

The formula $\frac{\pi}{\sin \pi z} = \pi \cot \pi z + \pi \tan \frac{1}{2}\pi z$ (cf. 5.2.5) further supplies us with the equation

$$(4) \qquad \frac{\pi}{\sin \pi z} = \frac{1}{z} + \sum_{1}^{\infty} (-1)^\nu \frac{2z}{z^2 - \nu^2}.$$

From (4) and the obvious relation $\frac{2z}{z^2 - \nu^2} = \frac{1}{z+\nu} + \frac{1}{z-\nu}$ we obtain the classical partial fraction development

$$(5) \qquad \frac{\pi}{\sin \pi z} = \sum_{-\infty}^{\infty} \frac{(-1)^\nu}{z + \nu}.$$

In (5) we may group the summands corresponding to indices ν and $-(\nu + 1)$, for $\nu \in \mathbb{N}$, and apply the identity $\cos \pi z = \sin \pi(z + \frac{1}{2})$ to verify that

(6)
$$\frac{\pi}{\cos \pi z} = 2 \sum_{0}^{\infty} (-1)^{\nu} \frac{(\nu + \frac{1}{2})}{(\nu + \frac{1}{2})^2 - z^2}.$$

When $z = 0$ we have the Leibniz series $\frac{\pi}{4} = 1 - \frac{1}{3} + \frac{1}{5} - + \cdots$. Amusing series for $\pi/\sqrt{2}$ arise from (4) and (6) with the choice $z = 1/4$.

4*. Characterizations of the cotangent by its addition theorem and by its differential equation. According to 5.2.5

$$\cot z = i \frac{e^{2iz} + 1}{e^{2iz} - 1} , \quad \cot(w + z) = \frac{\cot w \cot z - 1}{\cot w + \cot z} , \quad (\cot z)' + (\cot z)^2 + 1 = 0.$$

We will show that the second of these identities, the addition theorem, and the third, the differential equation, each characterize cot z. To this end we need the following

Lemma. *Let g be meromorphic in the region G. Then the differential equation $g' + g^2 + 1 = 0$ holds only for the family of functions*

$$g(z) \equiv i, \quad g(z) = i \frac{ae^{2iz} + 1}{ae^{2iz} - 1} , \quad a \text{ arbitrary in } \mathbb{C}.$$

Proof. That the functions listed do satisfy the differential equation is a direct and routine calculation. Conversely, consider $g \in \mathcal{M}(G)$ which satisfies the differential equation but is not the constant function i. Then the "Cayley transform" $f := (g + i)/(g - i)$ also belongs to $\mathcal{M}(G)$ and it satisfies the differential equation $f' = 2if$. From theorem 5.1.1 it follows that $f(z) = a \exp(2iz)$. At first this is valid only in the region $G \setminus P(f)$ but, after appeal to the Identity Theorem 10.3.4, it holds throughout G. Since $g = i(f + 1)/(f - 1)$, the claim about the form of g follows.

Remark. The trick in the foregoing proof is the passage to the "Cayley transform" of g. This is the device that linearizes the "Riccati" differential equation $y' + y^2 = 1 = 0$.

Theorem. *The following statements concerning a function g which is meromorphic in a neighborhood U of 0 are equivalent:*

i) *The principal part of g at 0 is $\frac{1}{z}$ and for all $w, z \in U \setminus P(g)$ such that $w + z \in U \setminus P(g)$*

$$g(w + z) = \frac{g(w)g(z) - 1}{g(w) + g(z)} \qquad (\text{Addition Theorem}) .$$

ii) *g has a pole at 0 and satisfies $g' + g^2 + 1 = 0$.*

iii) $g(z) = \cot z$.

Proof. i) $\Rightarrow$ ii) From the addition theorem it follows that

$$g'(z) = \lim_{h \to 0} \frac{g(z+h) - g(z)}{h} = -\lim_{h \to 0} \frac{g(z)^2 + 1}{hg(z) + hg(h)} = -g(z)^2 - 1,$$

because $\lim_{h \to 0} hg(h) = 1$ and $\lim_{h \to 0} hg(z) = 0$ for all $z \in U \setminus P(g)$.

ii) $\Rightarrow$ iii) Since $g(z) \not\equiv i$, it has the form $i\frac{ae^{2iz} + 1}{ae^{2iz} - 1}$ for some $a \in \mathbb{C}$, according to the preceding lemma. Since g has a pole at 0, the denominator must vanish at 0, which means that $a = 1$ and consequently $g(z) = \cot z$.

iii) $\Rightarrow$ i) Clear.

Exercises

Exercise 1. From the formulas of subsection 3 derive, by means of differentiation or of simple identities between trigonometric functions, the following partial fraction developments:

$$\pi^2 \frac{\sin \pi z}{\cos^2 \pi z} = \sum_{-\infty}^{\infty} \frac{(-1)^\nu}{(z + \nu - \frac{1}{2})^2} \quad,$$

$$\pi \frac{\cos \frac{\pi}{2}(w - z)}{\sin \frac{\pi}{2}(w + z) - \sin \frac{\pi}{2}(w - z)} = \frac{1}{z} + \sum_{1}^{\infty} \left[\frac{2w}{(2\nu - 1)^2 - w^2} - \frac{2z}{(2\nu)^2 - z^2} \right],$$

due to EULER, 1740.

Exercise 2. Give the partial fraction developments for the following functions:

a) $f(z) = (e^z - 1)^{-1}$,

b) $f(z) = \pi(\cos \pi z - \sin \pi z)^{-1}$.

Hints. For a) use the partial fraction development of $\cot \pi z$. For b) remember that $\cos \frac{\pi}{4} = \sin \frac{\pi}{4} = \frac{1}{\sqrt{2}}$.

§3   The Euler formulas for $\sum_{\nu \geq 1} \nu^{-2n}$

The first order of business in this section is to determine the numbers $\zeta(2n)$ for all $n \geq 1$. We also derive an interesting identity between Bernoulli numbers, and finally we briefly discuss the Eisenstein series $\varepsilon_k(z)$, $k \geq 2$.

1. Development of $\varepsilon_1(z)$ around 0 and Euler's formula for $\zeta(2n)$.

The function $\varepsilon_1(z) - z^{-1}$ is holomorphic in the unit disc $\mathbb{E}$ and has poles at ± 1. Its Taylor series at 0, which thus has radius of convergence exactly 1, can be explicitly written:

$$(1) \quad \varepsilon_1(z) = \frac{1}{z} - \sum_1^\infty q_{2n} z^{2n-1}, \ z \in \mathbb{E}^\times, \ \text{where } q_{2n} := 2\zeta(2n) = 2\sum_{\nu \geq 1} \frac{1}{\nu^{2n}}.$$

Proof. $-\frac{2}{\nu^{2n}}$ is the $(2n-2)$th Taylor coefficient of $2(z^2 - \nu^2)^{-1}$, as one sees from the geometric series expansion $\frac{2}{z^2-\nu^2} = \frac{-2}{\nu^2}\sum_{\mu=0}^\infty \left(\frac{z}{\nu}\right)^{2\mu}$. Since the series $\sum_{\nu \geq 1} 2(z^2 - \nu^2)^{-1}$ converges compactly in $\mathbb{E}$, it follows from Weierstrass's double series theorem 8.4.2 that its $(2n-2)$th Taylor coefficient is $\sum_{\nu \geq 1} \frac{-2}{\nu^{2n}}$, that is, $-q_{2n}$. But $\sum_{\nu \geq 1} 2(z^2 - \nu^2)^{-1}$ is an even function, so all Taylor coefficients of odd index vanish, and so this series has $\sum_1^\infty -q_{2n} z^{2n-2}$ as Taylor series about 0. Since $\varepsilon_1(z) = z^{-1} + z\sum_1^\infty 2(z^2 - \nu^2)^{-1}$ for all $z \in \mathbb{E}^\times$, the claim (1) follows. $\square$

The Bernoulli numbers B_{2n} were introduced in 7.5.1. In a few lines we can now acquire the famous

Formulas of EULER:

$$\zeta(2n) = (-1)^{n-1}\frac{(2\pi)^{2n}}{2(2n)!} B_{2n}, \qquad n = 1, 2, \ldots$$

Proof. From (1) and 7.5.2(1) we have in a neighborhood of 0

$$z^{-1} - \sum_1^\infty q_{2n} z^{2n-1} = \varepsilon_1(z) = \pi\cot\pi z = z^{-1} + \sum_1^\infty (-1)^n \frac{2^{2n}}{(2n)!} B_{2n}\pi^{2n} z^{2n-1}.$$

Since $q_{2n} = 2\zeta(2n)$, comparison of coefficients here leads to the desired conclusion. $\square$

From the formulas of EULER we can infer incidentally that the Bernoulli numbers $B_2, B_4, \ldots, B_{2n}, \ldots$ have alternating signs (as was already hinted at in the equations (2) in 7.5.1). Moreover, the unboundedness assertion concerning the sequence B_{2n} in 7.5.1 can now be made more precise: since $1 < \sum \nu^{-2n} < 2$ for every $n \geq 1$, it follows that

$$2\frac{(2n)!}{(2\pi)^{2n}} < |B_{2n}| < 4\frac{(2n)!}{(2\pi)^{2n}}; \qquad \text{in particular, } \lim\left|\frac{B_{2n+2}}{B_{2n}}\right| = \infty.$$

It further follows from $1 < \zeta(2n) < 2$, the identity $\frac{|B_{2n}|}{(2n)!} = \frac{2\zeta(2n)}{(2\pi)^{2n}}$ and the Cauchy-Hadamard formula that the Taylor series of $z/(e^z - 1)$ about 0 has radius of convergence 2π; this is the "rather tedious" determination of this radius of convergence without examining the zeros of the denominator, which was mentioned in 7.4.5.

The Euler formulas will be generalized in 14.3.4.

2. Historical remarks on the Euler $\zeta(2n)$-formulas. As early as 1673 during LEIBNIZ's first visit to London J. PELL, an expert in series summation, had posed to him the problem of summing the reciprocals of the squares. LEIBNIZ had maintained in youthful exuberance that he could sum *any* series, but Pell's query made clear to him his limitations. The brothers Jakob and Johann BERNOULLI (the latter was EULER's teacher) also expended quite a lot of effort in vain trying to find the value of the sum $1 + \frac{1}{4} + \frac{1}{9} + \frac{1}{16} + \cdots$.

Finally, in the year 1734, using the product formula he had discovered for the sine, EULER in his work "De summis serierum reciprocarum" (*Opera Omnia* (1) **14**, 73-86) proved his famous identities

$$\sum_{1}^{\infty} \frac{1}{\nu^2} = \frac{\pi^2}{6}, \ \sum_{1}^{\infty} \frac{1}{\nu^4} = \frac{\pi^4}{90}, \ \sum_{1}^{\infty} \frac{1}{\nu^6} = \frac{\pi^6}{945}, \ \sum_{1}^{\infty} \frac{1}{\nu^8} = \frac{\pi^8}{9450}, \ \cdots$$

It is often said of the first of these identities that it is among the most beautiful of all Euler's formulas.

Euler's problems in summing the series $\sum \nu^{-2n}$ were described in detail by P. STÄCKEL in a note entitled "Eine vergessene Abhandlung Leonhard Eulers über die Summe der reziproken Quadrate der natürlichen Zahlen," *Biblio. Math.*(3) **8**(1907/08), 37-54 (also included in Euler's *Opera Omnia* (1) **14**, 156-176). Also interesting is the article "Die Summe der reziproken Quadratzahlen," by O. SPIESS in the *Festschrift* to the 60th birthday of Prof. Andreas Speiser, Orell Füssli Verlag, Zürich 1945, pp. 66-86; and the paper by R. AYOUB entitled "Euler and the Zeta function," *Amer. Math. Monthly* **81**(1974), 1067-1086.

3. The differential equation for ε_1 and an identity for the Bernoulli numbers. Because $(\cot z)' = -1 - (\cot z)^2$ (cf. 5.2.5), it follows that

$$(1) \qquad\qquad \varepsilon_1' = -\varepsilon_1^2 - \pi^2;$$

the function ε_1 thus solves the differential equation $y' = -y^2 - \pi^2$. With the help of (1) we get an elegant but not very well known recursion formula for the numbers $\zeta(2n)$, namely

$$(2) \qquad (n + \tfrac{1}{2})\zeta(2n) = \sum_{\substack{k+\ell=n \\ k \geq 1, \ell \geq 1}} \zeta(2k)\zeta(2\ell) \qquad \text{for } n > 1, \qquad \zeta(2) = \frac{\pi^2}{6}.$$

Proof. Using the differentiation theorem 1.2 and 1(1), we get

$$\varepsilon_1'(z) = -\frac{1}{z^2} - 2\sum_1^\infty (2n-1)\zeta(2n)z^{2n-2}.$$

From 1(1) it follows, on the basis of the differentiation theorem 1.2 and the product theorem 8.4.2, that

$$\varepsilon_1^2(z) = \frac{1}{z^2} - 4\sum_1^\infty \zeta(2n)z^{2n-2} + 4\sum_{n=2}^\infty \sum_{\substack{k+\ell=n \\ k\geq 1, \ell\geq 1}} \zeta(2k)\zeta(2\ell)z^{2n-2} \ , \ z\in \mathbb{E}^\times.$$

Substituting these into (1) and comparing coefficients of like powers of z yields (2).   □

If we use the Euler formulas for $\zeta(2n)$, then we get from (2)

$$(3) \qquad (2n+1)B_{2n} + (2n)! \sum_{\substack{k+\ell=n \\ k\geq 1, \ell\geq 1}} \frac{1}{(2k)!(2\ell)!} B_{2k}B_{2\ell} = 0 \ , \qquad n\geq 2.$$

Equations (2) and (3) and their derivation from the differential equation of the cotangent function were brought to my attention by Professor M. KOECHER.

4. The Eisenstein series $\varepsilon_k(z) := \sum_{-\infty}^\infty \frac{1}{(z+\nu)^k}$ are, according to 1.3, normally convergent in $\mathbb{C}$ for all integers $k\geq 2$ and consequently, by the convergence theorem 1.1, they represent meromorphic functions in $\mathbb{C}$. It is immediate from the definition that ε_k is holomorphic in $\mathbb{C}\setminus\mathbb{Z}$ and that at each $n\in\mathbb{Z}$ it has a pole of order k and principal part $1/(z-n)^k$. The functions $\varepsilon_{2\ell}$ are *even* and the functions $\varepsilon_{2\ell+1}$ are *odd*. The series for $\varepsilon_1(z)$ is unexceptional and has this same general form if we agree to use the Eisenstein summation convention $\sum_e$ for it. In 2.3 we saw that

$$\varepsilon_2(z) = \frac{\pi^2}{\sin^2 \pi z} \ , \qquad \varepsilon_3(z) = \pi^3 \frac{\cot \pi z}{\sin^2 \pi z}.$$

Thus we have $\varepsilon_3 = \varepsilon_2\varepsilon_1$, an identity that certainly cannot be perceived directly from the series representations (on this point see also 4.3).

The periodicity theorem. *Let $k\geq 1$ be an integer, $\omega\in\mathbb{C}$. Then*

$$\varepsilon_k(z+\omega) = \varepsilon_k(z) \text{ for all } z\in\mathbb{C} \Leftrightarrow \omega\in\mathbb{Z}.$$

Proof. If $\varepsilon_k(z+\omega) = \varepsilon_k(z)$, then along with 0, ω is also a pole of ε_k, so that $\omega\in\mathbb{Z}$. Since the series may be rearranged at will due to the fact of their normal convergence, we get for every k, $\varepsilon_k(z+1) = \varepsilon_k(z)$. From this it follows that $\varepsilon_k(z+n) = \varepsilon_k(z)$ for all $n \in \mathbb{Z}$.   □

From the differentiation theorem 1.2 it follows that

$$(1) \qquad \varepsilon_k' = -k\varepsilon_{k+1} \qquad \text{for } k \geq 1.$$

(For ε_1 use the normally convergent series $\dfrac{1}{z} + \displaystyle\sum_{-\infty}^{\infty}{}' \left(\dfrac{1}{z+\nu} - \dfrac{1}{\nu} \right)$.) From this via induction on k we get

$$(2) \qquad \varepsilon_k = \frac{(-1)^{k-1}}{(k-1)!} \varepsilon_1^{(k-1)} \qquad \text{for } k \geq 2.$$

From the development 1.(1) of ε_1 it follows (again inductively) that

$$(3) \qquad \varepsilon_k(z) = \frac{1}{z^k} + (-1)^k \sum_{2n \geq k} \binom{2n-1}{k-1} q_{2n} z^{2n-k} \qquad \text{for } k \geq 2$$

and in particular

$$(4) \qquad \varepsilon_2(z) = \frac{1}{z^2} + q_2 + 3q_4 z^2 + \cdots, \quad \varepsilon_3(z) = \frac{1}{z^3} - 3q_4 z - 10q_6 z^3 - \cdots$$

§4*. The EISENSTEIN theory of the trigonometric functions

The theory of the trigonometric functions, which nowadays is almost always based on that of the complex exponential function, can also be developed *ab ovo* from the Eisenstein functions ε_k and simple non-linear relations between them. This construction of the theory of the circular functions was sketched in passing by EISENSTEIN in 1847 in a work [Ei] that today is famous and in which, for example, Weierstrass' $\wp$-function and its differential equation also feature. EISENSTEIN writes (p.396):

"Die Fundamental-Eigenschaften dieser einfach-periodischen Functionen ergeben sich aus der Betrachtung einer einzigen identischen Gleichung, nämlich der folgenden (The fundamental properties of these simply-periodic functions reveal themselves through consideration of a single identity, namely the following):

$$(a) \qquad \frac{1}{p^2 q^2} = \frac{1}{(p+q)^2} \left(\frac{1}{p^2} + \frac{1}{q^2} \right) + \frac{2}{(p+q)^3} \left(\frac{1}{p} + \frac{1}{q} \right)."$$

The p and q here are indeterminates and one confirms (a) by direct calculation or (more simply) by differentiating the obvious identity $p^{-1}q^{-1} =$

$(p+q)^{-1}(p^{-1}+q^{-1})$ with respect to p and with respect to q. EISENSTEIN gets all the important propositions about his series by virtuoso manipulations with the identity (a).

EISENSTEIN was a pupil of Karl Heinrich SCHELLBACH (German mathematician, 1805-1892, professor of mathematics and physics at the Friedrich-Wilhelm Gymnasium in Berlin and from 1843 onward concurrently teacher of mathematics at the general military school in Berlin). In 1845 SCHELLBACH published, in the school-program of his Gymnasium, a treatise entitled *Die einfachsten periodischen Functionen*, in which for the first time functions like

$$\sum_{-\infty}^{\infty} f(x+s) \qquad \text{and} \qquad \prod_{-\infty}^{\infty} f(x+\lambda)$$

were employed in the construction of periodic functions. This treatise of Schellbach's had a big influence on EISENSTEIN (cf. [Ei], p.401).

In 1976 in the second chapter of his *Ergebnisse* monograph [We] André WEIL gave a concise presentation of Eisenstein's theory, at the same time expanding the calculations involved. "Man wird bei diesen Ausführungen an ein musikalisches Analogon, die Diabelli-Variationen von Beethoven erinnert (This presentation brings to mind a musical analog, the Diabelli variations of Beethoven)." – E. HLAWKA in the *Monatshefte für Math.* **83**(1977), p. 225. WEIL chose the notation ε_k in honor of EISENSTEIN, who himself wrote (k,z) instead of $\varepsilon_k(z)$ ([Ei], p. 395).

In what follows we present the beginnings of Eisenstein's theory, after [We]. We will only work with the first four functions $\varepsilon_1, \varepsilon_2, \varepsilon_3, \varepsilon_4$. The identity $\varepsilon_1(z) = \pi \cot \pi z$ will be proven anew, independently of the considerations of the preceding sections, save for using theorem 2.4 on the solutions of the differential equation $g' + g^2 + 1 = 0$.

1. The addition theorem.

$$\varepsilon_2(w)\varepsilon_2(z) - \varepsilon_2(w)\varepsilon_2(w+z) - \varepsilon_2(z)\varepsilon_2(w+z) = 2\varepsilon_3(w+z)[\varepsilon_1(w) + \varepsilon_1(z)].$$

Proof (following [We], p.8). We set $p := z + \mu$, $q := w + \nu - \mu$ in (a) and get

$$\frac{1}{(z+\mu)^2(w+\nu-\mu)^2} - \frac{1}{(w+z+\nu)^2}\left(\frac{1}{(z+\mu)^2} + \frac{1}{(w+\nu-\mu)^2}\right)$$

$$= \frac{2}{(w+z+\nu)^3}\left(\frac{1}{z+\mu} + \frac{1}{w+\nu-\mu}\right).$$

Eisenstein summation over μ with ν fixed gives

$$\sum_{\mu=-\infty}^{\infty}{}_e \frac{1}{(z+\mu)^2(w+\nu-\mu)^2} - \frac{1}{(w+z+\nu)^2}[\varepsilon_2(z) + \varepsilon_2(w+\nu)]$$

$$= \frac{2}{(w+z+\nu)^3}[\varepsilon_1(z) + \varepsilon_1(w+\nu)].$$

Since ν is a period of ε_k (cf. the periodicity theorem 3.4), we may write $\varepsilon_2(w)$ instead of $\varepsilon_2(w+\nu)$ and $\varepsilon_1(w)$ instead of $\varepsilon_1(w+\nu)$. Moreover, because it converges normally, the first sum on the left coincides with the ordinary sum $\sum_{-\infty}^{\infty}$. After we make all these simplifications and sum over ν, we obtain

$$\sum_{\nu=-\infty}^{\infty} \sum_{\mu=-\infty}^{\infty} \frac{1}{(z+\mu)^2(w+\nu-\mu)^2} - \varepsilon_2(w+z)[\varepsilon_2(z) + \varepsilon_2(w)]$$

$$= 2\varepsilon_3(w+z)[\varepsilon_1(z) + \varepsilon_1(w)].$$

Because of normal convergence it is legitimate to interchange the two summation processes on the left. After doing so, we recall that $\varepsilon_2(w - \mu) = \varepsilon_2(w)$. The double sum then becomes

$$\sum_{\mu=-\infty}^{\infty} \frac{1}{(z+\mu)^2} \sum_{\nu=-\infty}^{\infty} \frac{1}{((w-\mu)+\nu)^2} = \sum_{\mu=-\infty}^{\infty} \frac{\varepsilon_2(w-\mu)}{(z+\mu)^2} = \varepsilon_2(w)\varepsilon_2(z),$$

which proves the addition theorem.

2. Eisenstein's basic formulas. The addition theorem 1 was not explicitly formulated by EISENSTEIN. Rather he derives the identities

(1) $$3\varepsilon_4(z) = \varepsilon_2^2(z) + 2\varepsilon_1(z)\varepsilon_3(z)$$

(2) $$\varepsilon_2^2(z) = \varepsilon_4(z) + 2q_2\varepsilon_2(z)$$

directly from (a) ([Ei], 396-398). We now get these basic formulas of Eisenstein from the addition theorem ([We], p.8). To this end we need the following easily verified statements (to derive (+) use 3.4(1)):

For every $z \in \mathbb{C} \setminus \mathbb{Z}$ and every integer $k \geq 1$ there is a neighborhood of $w = 0$ in which

$$\varepsilon_k(w+z) = \sum_{\nu \geq 0} \frac{1}{\nu!} \varepsilon_k^{(\nu)}(z)w^{\nu};$$

in particular,

(+) $$\begin{cases} \varepsilon_1(w+z) &= \varepsilon_1(z) - \varepsilon_2(z)w + \varepsilon_3(z)w^2 - \varepsilon_4(z)w^3 + - \cdots \\ \varepsilon_2(w+z) &= \varepsilon_2(z) - 2\varepsilon_3(z)w + 3\varepsilon_4(z)w^2 - + \cdots \\ \varepsilon_3(w+z) &= \varepsilon_3(z) - 3\varepsilon_4(z)w + 6\varepsilon_5(z)w^2 - + \cdots \end{cases} \qquad \square$$

We now prove equation (1): For fixed $z \in \mathbb{C} \setminus \mathbb{Z}$ the functions appearing in the addition theorem are meromorphic functions of w. Develop each around $w = 0$ and compare the constant (as far as w is concerned) terms. The development 3.4(4) for ε_2 together with the equation (+) above for $\varepsilon_2(w+z)$ yield for the function on the left side of the identity in the addition theorem, bearing in mind that $\varepsilon_2(w)\varepsilon_2(z)$ cancels and $\varepsilon_2(z)\varepsilon_2(w + z)$ has "constant" term $\varepsilon_2^2(z)$,

$$-\left(\frac{1}{w^2} + q_2 + \cdots\right)\left(-2\varepsilon_3(z)w + 3\varepsilon_4(z)w^2 + \cdots\right) - \varepsilon_2^2(z) + \cdots$$

$$= -3\varepsilon_4(z) - \varepsilon_2^2(z) + \cdots;$$

here the latter ellipsis indicates terms in $w^{-1}, w, w^2, \ldots$ For the function on the right side of the identity in the addition theorem we use the development 3.1(1) for ε_1 and the equation (+) for $\varepsilon_3(w + z)$ to obtain

$$2(\varepsilon_3(z) - 3\varepsilon_4(z)w + \cdots)\left(\frac{1}{w} - q_2 w + \cdots\right) + 2\varepsilon_3(z)\varepsilon_1(z) + \cdots$$

$$= -6\varepsilon_4(z) + 2\varepsilon_3(z)\varepsilon_1(z) + \cdots.$$

From which follows $-3\varepsilon_4(z) - \varepsilon_2^2(z) = -6\varepsilon_4(z) + 2\varepsilon_3(z)\varepsilon_1(z)$, a re-statement of (1).

The proof of (2) is carried out similarly. We again fix $z \in \mathbb{C} \setminus \mathbb{Z}$, but this time consider $\zeta := w + z$ as the variable in the addition theorem. We carry out the development around $\zeta = 0$, set $w = \zeta - z$ and compare the constant (as far as ζ is concerned) terms. Using the development 3.4(4) for $\varepsilon_2(z)$ as well as the equation (+) for $\varepsilon_2(\zeta - z)$ [realize that $\varepsilon_{2\ell}$ is an even and $\varepsilon_{2\ell+1}$ is an odd function], we see that the "constant" term on the left in the addition theorem is $\varepsilon_2^2(z) - 2q_2\varepsilon_2(z) - 3\varepsilon_4(z)$. On the basis of 3.4(4) and (+), $\varepsilon_3(\zeta)\varepsilon_1(\zeta - z)$ has constant term $-\varepsilon_4(z)$. Also $\varepsilon_3(\zeta)\varepsilon_1(z)$ is an odd function of ζ and consequently has no constant term. Therefore the constant term on the right in the addition theorem is $-2\varepsilon_4(z)$. (2) then follows immediately.

3. More Eisenstein formulas and the identity $\varepsilon_1(z) = \pi\cot\pi z$.

Eliminating $\varepsilon_4(z)$ between 2(1) and (2) yields

$$(1) \qquad \varepsilon_1(z)\varepsilon_3(z) = \varepsilon_2^2(z) - 3q_2\varepsilon_2(z).$$

If we differentiate (1), and take into account 3.4(1), we obtain $\varepsilon_2\varepsilon_3 = \varepsilon_1\varepsilon_4 + 2q_2\varepsilon_3$. Use 2(2) to eliminate ε_4 here, and, after division by $\varepsilon_2 - 2q_2$, get

$$(2) \qquad \varepsilon_3(z) = \varepsilon_1(z)\varepsilon_2(z).$$

Insert this into (1) and divide by ε_2, and it follows that

$$(3) \qquad\qquad \varepsilon_1^2(z) = \varepsilon_2(z) - 3q_2.$$

From the relations already garnered the reader can draw the

Conclusion ([Ei], p.400) *Each function ε_k is a real polynomial in ε_1.*

On account of $\varepsilon_2 = -\varepsilon_1'$, equation (3) may also be viewed as the differential equation

$$(4) \qquad\qquad \varepsilon_1'(z) = -\varepsilon_1^2(z) - 3q_2$$

for the function ε_1. From (4) alone we now obtain (anew)

$$(*) \quad \varepsilon_1(z) = \pi \cot \pi z \quad \text{and} \quad \frac{1}{6}\pi^2 = \sum_{\nu \geq 1} \frac{1}{\nu^2} \quad \left(\text{and therefore } q_2 = \tfrac{1}{3}\pi^2\right).$$

Proof. Let a be the positive square-root of $3q_2 = 6\sum_{\nu \geq 1} \nu^{-2}$. For $g(z) := a^{-1}\varepsilon_1(a^{-1}z) \in \mathcal{M}(\mathbb{C})$, the differential equation $g' + g^2 + 1 = 0$ holds. Since g has a pole at the origin, it follows from theorem 2.4 that $g(z) = \cot z$, and so $\varepsilon_1(z) = a \cot az$. Since $\mathbb{Z}$ is the set of periods of $\varepsilon_1(z)$, while $\pi a^{-1}\mathbb{Z}$ is the set of periods of $\cot az$ (note that $\mathrm{per}(\cot) = \pi\mathbb{Z}$ by 5.2.5), it follows that $\mathbb{Z} = \pi a^{-1}\mathbb{Z}$, and so $a = \pi$ since a is positive. $\qquad\square$

The addition formula for the cotangent says (cf. 5.2.5)

$$\varepsilon_1(w + z) = \frac{\varepsilon_1(w)\varepsilon_1(z) - \pi^2}{\varepsilon_1(w) + \varepsilon_1(z)}.$$

EISENSTEIN also proves this formula by direct manipulation of series ([Ei], pp. 408, 409); interested readers are referred to [We], pp. 8, 9.

4. Sketch of the theory of the circular functions according to EISENSTEIN. The foregoing considerations show that basically the theory of the trigonometric functions can be developed from Eisenstein's function ε_1 alone. One first defines π as $\sqrt{3q_2}$ and makes the equation $\pi \cot \pi z = \varepsilon_1(z)$ the *definition of the cotangent*. All the other circular functions can now be reduced to ε_1. If we recall the formula

$$\frac{1}{\sin z} = \frac{1}{2}\left(\cot \frac{z}{2} - \cot \frac{z + \pi}{2}\right)$$

(which is mentioned in passing on p.409 of [Ei]), then it is clear that in the putative Eisenstein theory the equation

$$\frac{\pi}{\sin \pi z} = \frac{1}{2}\left[\varepsilon_1\left(\frac{z}{2}\right) - \varepsilon_1\left(\frac{z+1}{2}\right)\right]$$

should be elevated to the status of definition of the sine. The partial fraction development

$$\frac{\pi}{\sin \pi z} = \frac{1}{2}\sum_{-\infty}^{\infty}{}_e\frac{2}{z+2\nu} - \frac{1}{2}\sum_{-\infty}^{\infty}{}_e\frac{2}{z+1+2\nu} = \sum_{-\infty}^{\infty}\frac{(-1)^\nu}{z+\nu}$$

is an incidental bonus. Because $\cos \pi z = \sin \pi(z + \frac{1}{2})$, we can regard

$$\frac{\pi}{\cos \pi z} = \frac{1}{2}\left[\varepsilon_1\left(\frac{2z+1}{4}\right) - \varepsilon_1\left(\frac{2z+3}{4}\right)\right]$$

as the definition of the cosine.

Also the exponential function can be defined by means of ε_1 alone: For the function

$$e(z) := \frac{\varepsilon_1(z) + \pi i}{\varepsilon_1(z) - \pi i} = \frac{1 + \pi i z + \cdots}{1 - \pi i z + \cdots} \in \mathcal{M}(\mathbb{C})$$

it follows at once, recalling $-\varepsilon_1' = \varepsilon_1^2 + \pi^2$, that

$$e'(z) = -2\pi i\frac{\varepsilon_1'(z)}{(\varepsilon_1(z) - \pi i)^2} = 2\pi i\frac{\varepsilon_1^2(z) + \pi^2}{(\varepsilon_1(z) - \pi i)^2} = 2\pi i e(z).$$

Since $e(0) = 1$, theorem 5.1.1 and the Identity Theorem tell us that the function $e(z)$ just introduced is in fact $\exp(2\pi i z)$.

It seems that the construction of the theory of the circular functions sketched here has never been consistently carried out in all detail this way. Even so, due to lack of space, we shall have to forego doing it here. One advantage of the Eisenstein approach is that the periodicity of the circular functions is evident on the basis of the explicit form of the series for ε_1.

Exercise

Exercise. Using the duplication formula $2\varepsilon_1(2z) = \varepsilon_1(z) + \varepsilon_1(z + \frac{1}{2})$, show that

$$\varepsilon_1(z)\varepsilon_1(z + \tfrac{1}{2}) + \pi^2 = 0.$$

What does this formula say about the classical trigonometric functions?

N.H. ABEL 1802–1829

F.G.M. EISENSTEIN 1823–1852

J. LIOUVILLE 1809–1882

H.A. SCHWARZ 1843–1921

Line drawings by Martina Koecher

Chapter 12

Laurent Series and Fourier Series

At quantopere doctrina de seriebus infinitis Analysin sublim-
iorem amplificaveret, nemo est, qui ignoret (There is nobody
who does not know the extent to which the theory of infinite
series has enriched higher analysis). – L. EULER 1748, *Introductio.*

In this chapter we discuss two types of series which, after power series,
are among the most important series in function theory: *Laurent series*
$\sum_{-\infty}^{\infty} a_\nu(z-c)^\nu$ and *Fourier series* $\sum_{-\infty}^{\infty} c_\nu e^{2\pi i \nu z}$. The theory of Laurent
series is a theory of power series in annuli; WEIERSTRASS éven called Lau-
rent series power series too (cf. [W₂], p.67). Fourier series are Laurent
series around $c = 0$ with $e^{2\pi i z}$ taking over the role of z; their great importance lies in the fact that periodic holomorphic functions can be developed
in such series. A particularly important Fourier series is the *theta series*
$\sum_{-\infty}^{\infty} e^{-\nu^2 \pi \tau} e^{2\pi i \nu z}$, which gave quite a decisive impulse to 19th-century
mathematics.

§1 Holomorphic functions in annuli and Laurent series

Let $r, s \in \mathbb{R} \cup \{\infty\}$ with $0 \leq r < s$. The open subset

$$A_{r,s}(c) := \{z \in \mathbb{C} : r < |z - c| < s\}$$

of $\mathbb{C}$ is called the *annulus* or *circular ring around* c with *inner radius* r
and *outer radius* s. When $s < \infty$, $A_{0,s}(c) = B_s(c) \setminus c$, a punctured disc,

and $A_{0,\infty}(0)$ is the punctured plane $\mathbb{C}^\times$. In contexts where there is no possibility of misunderstanding the notation $A_{r,s}(c)$ is shortened to just A.

The annulus A with radii s and r is naturally the intersection

$$A = A^+ \cap A^- \quad \text{with } A^+ := B_s(c) \text{ and } A^- := \{z \in \mathbb{C} : |z - c| > r\}.$$

This notation will be used extensively in the sequel. As in earlier chapters, the boundary $\partial B_\rho(c)$ of the disc will be denoted by S_ρ.

1. Cauchy theory for annuli. The point of departure for the theory of holomorphic functions in circular rings is the

Cauchy integral theorem for annuli. *Let f be holomorphic in the annulus A around c with radii r and s. Then*

$$(1) \qquad \int_{S_\rho} f d\zeta = \int_{S_\sigma} f d\zeta \quad \text{for all } \rho, \sigma \in \mathbb{R} \text{ with } r < \rho \leq \sigma < s.$$

We intend to give three proofs of this basic theorem. In all of them we may take $c = 0$.

First proof (by reduction to theorem 7.1.2 via decomposition into convex regions). Let ρ be given. We choose a ρ' with $r < \rho' < \rho$ and determine on $S_{\rho'}$ the vertices of a regular n-gon which lies wholly in the annulus with radii r and ρ'. This inclusion occurs for all sufficiently large n. In the figure $n = 6$.

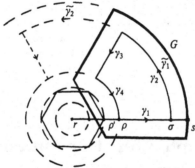

As this figure shows, $\tilde\gamma_1 := \gamma_1 + \gamma_2 + \gamma_3 + \gamma_4$ is a closed path in the truncated circular sector G, a convex region lying in the domain of holomorphy of f, for any σ with $\rho \leq \sigma < s$. Consequently $\int_{\tilde\gamma_1} f d\zeta = 0$. Analogously, $\int_{\tilde\gamma_2} f d\zeta = 0$ for the path $\tilde\gamma_2$ (partly shown in the figure) which begins with the piece $-\gamma_3$. Continuing, we define closed paths $\tilde\gamma_\nu$, $\nu = 1, 2, \ldots, n$, with $-\gamma_1$ featuring as a piece of $\tilde\gamma_n$. It then follows that

$$0 = \sum_{\nu=1}^n \int_{\tilde\gamma_\nu} f d\zeta = \int_{S_\sigma} f d\zeta - \int_{S_\rho} f d\zeta,$$

since the integrals over the various radial components of the paths $\tilde{\gamma}_\nu$ cancel out.

Second proof (by reduction to theorem 7.1.2 via the exponential mapping). We assume that $0 < r < s < \infty$ and pick $a, \alpha, \beta, b \in \mathbb{R}$ with $e^a = r$, $e^\alpha = \rho$, $e^\beta = \sigma$, $e^b = s$. By means of $z \mapsto \exp z$ the boundary ∂R of the rectangle $R := \{z \in \mathbb{C} : \alpha < \Re z < \beta , |\Im z| < \pi\}$ is mapped onto the (closed) path $\Gamma := \sum_1^4 \exp(\gamma_\nu) = S_\sigma + \gamma + S_\rho - \gamma$ (cf. the figure below and 5.2.3).

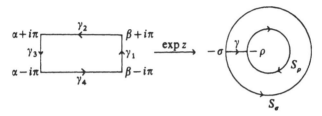

From the transformation rule 6.2.1 it follows that

$$\int_\Gamma f(\zeta)d\zeta = \int_{\partial R} f(\exp z) \exp z \, dz.$$

Since the convex region $G := \{z \in \mathbb{C} : a < \Re z < b\}$ which contains ∂R is mapped into A by $\exp z$, the integrand on the right is holomorphic in G and the integral is consequently 0 by theorem 7.1.2. The claim follows from this.

Third proof (by interchanging integration and differentiation). Since each $f \in \mathcal{O}(A)$ can be written in the form $f(z) = z^{-1}g(z)$ with $g \in \mathcal{O}(A)$, it suffices to show that for each $g \in \mathcal{O}(A)$ the function

$$J(t) := \int_{S_t} \frac{g(\zeta)}{\zeta} d\zeta = i \int_0^{2\pi} g(te^{i\varphi})d\varphi , \qquad t \in (r, s)$$

is constant. According to well-known theorems of real analysis, $J(t)$ is differentiable and

$$\begin{aligned} J'(t) &= i \int_0^{2\pi} \frac{d}{dt} g(te^{i\varphi})d\varphi = i \int_0^{2\pi} g'(te^{i\varphi})e^{i\varphi}d\varphi \\ &= t^{-1} \int_{S_r} g'(\zeta)d\zeta \qquad \text{for } t \in (r, s). \end{aligned}$$

The last integral is 0 because g' has a primitive. Thus $J'(t) \equiv 0$, so J is constant. □

Remark. From a "higher point of view" the integrals over S_σ and S_ρ are equal because these paths can be deformed into one another while staying in A. But we won't go into the general theory of integration for such "homotopic" paths until the second volume.   □

From the integral theorem follows (as did the analogous fact for discs in 7.2.2)

The Cauchy integral formula for annuli. *Let f be holomorphic in the domain D. Suppose that the annulus $A = A^+ \cap A^-$ about $c \in D$ lies, along with its boundary, in D. Then*

$$
\begin{aligned}
f(z) &= \frac{1}{2\pi i} \int_{\partial A} \frac{f(\zeta)}{\zeta - z} d\zeta \\
&= \frac{1}{2\pi i} \int_{\partial A^+} \frac{f(\zeta)}{\zeta - z} d\zeta - \frac{1}{2\pi i} \int_{\partial A^-} \frac{f(\zeta)}{\zeta - z} d\zeta \qquad \text{for all } z \in A.
\end{aligned}
$$

Proof. For fixed $z \in A$ the function

$$
g(\zeta) := \begin{cases} \dfrac{f(\zeta) - f(z)}{\zeta - z} & \text{for} \quad \zeta \in D \setminus z \\ f'(z) & \text{for} \quad \zeta = z \end{cases}
$$

is continuous in D and holomorphic in $D \setminus z$. From theorem 7.3.4 it follows that in fact $g \in \mathcal{O}(D)$. Therefore the integral theorem for annuli insures that $\int_{\partial A^+} g d\zeta = \int_{\partial A^-} g d\zeta$; that is,

$$
\int_{\partial A^-} \frac{f(\zeta)}{\zeta - z} d\zeta - f(z) \int_{\partial A^-} \frac{d\zeta}{\zeta - z} = \int_{\partial A^+} \frac{f(\zeta)}{\zeta - z} d\zeta - f(z) \int_{\partial A^+} \frac{d\zeta}{\zeta - z}.
$$

The second integral in this equation vanishes because $|z| > r$ and the fourth integral has the value $2\pi i$ because $|z| < s$.   □

The integral formula is the key to the

2. Laurent representation in annuli. First we introduce a convenient locution: if h is a complex-valued function in an *unbounded* domain W, then we write $\lim_{z \to \infty} h(z) = b \in \mathbb{C}$, if for every neighborhood V of b there is a finite R such that $h(z) \in V$ for all $z \in W$ with $|z| \geq R$. Of course it should be noted that this definition depends on W. In what follows W will be the exterior of a disc, that is, a set A^-.

Theorem. *Let f be holomorphic in the annulus $A = A^+ \cap A^-$ with center c and radii r, s. Then there are two functions, $f^+ \in \mathcal{O}(A^+)$ and $f^- \in \mathcal{O}(A^-)$ such that*

$$
f = f^+ + f^- \qquad \text{in} \quad A \qquad \text{and} \qquad \lim_{z \to \infty} f^-(z) = 0.
$$

These conditions uniquely determine the functions f^+ and f^- as follows: For every $\rho \in (r, s)$

$$f^+(z) = \frac{1}{2\pi i} \int_{S_\rho} \frac{f(\zeta)}{\zeta - z} d\zeta , \quad z \in B_\rho(c);$$

$$f^-(z) = \frac{-1}{2\pi i} \int_{S_\rho} \frac{f(\zeta)}{\zeta - z} d\zeta , \quad z \in \mathbb{C} \setminus \overline{B_\rho(c)}.$$

Proof. a) Existence: the function

$$f_\rho^+(z) := \frac{1}{2\pi i} \int_{S_\rho} \frac{f(\zeta)}{\zeta - z} d\zeta , \quad z \in B_\rho(c),$$

is holomorphic in $B_\rho(c)$. For $\sigma \in (\rho, s)$ we have $f_\rho^+ = f_\sigma^+ | B_\rho(c)$, thanks to the integral theorem. Hence there is a function $f^+ \in \mathcal{O}(A^+)$ which in each $B_\rho(c)$ coincides with f_ρ^+. In the same way a function $f^- \in \mathcal{O}(A^-)$ is defined by the prescription

$$f^-(z) := f_\sigma^-(z) := \frac{-1}{2\pi i} \int_{S_\sigma} \frac{f(\zeta)}{\zeta - z} d\zeta , \text{ whenever } r < \sigma < \min\{s, |z - c|\}.$$

Application of the integral formula to all the different annuli A' centered at c with $\overline{A'} \subset A$ confirms that the representation $f = f^+ + f^-$ holds throughout A. The standard estimate for integrals gives, for $z \in A^-$,

$$|f^-(z)| \leq \sigma \max_{\zeta \in S_\sigma} |f(\zeta)(\zeta - z)^{-1}| \leq \frac{\sigma}{|z - c| - \sigma} |f|_{S_\sigma},$$

whence $\lim_{z \to \infty} f^-(z) = 0$.

b) Uniqueness: Let $g^+ \in \mathcal{O}(A^+)$, $g^- \in \mathcal{O}(A^-)$ be other functions satisfying $f = g^+ + g^-$ in A and $\lim_{z \to \infty} g^-(z) = 0$. Then $f^+ - g^+ = g^- - f^-$ in A and consequently the recipe

$$h := \left\{ \begin{array}{ll} f^+ - g^+ & \text{in} \quad A^+ \\ g^- - f^- & \text{in} \quad A^- \end{array} \right.$$

well defines an entire function h which satisfies $\lim_{z \to \infty} h(z) = 0$. It follows from Liouville's theorem that $h \equiv 0$, and so $f^+ = g^+$ and $f^- = g^-$.   □

The representation of f as the sum $f^+ + f^-$ is known as the *Laurent representation* (or the *Laurent separation*) of f in A. The function f^- is called the *principal part*, the function f^+ the *regular part* of f.

If f is meromorphic in $D \setminus c$, then the representation of f described in the development theorem 10.1.2 is nothing but the Laurent representation of f in $B \setminus c$ (where $B \subset D$ and $r = 0$). In particular, the notion of principal part introduced here for Laurent developments generalizes the notion of the principal part of a meromorphic function at a pole.

3. Laurent expansions. Expressions of the form $\sum_{-\infty}^{\infty} a_\nu(z-c)^\nu$ are called *Laurent series around c*. The series

$$\sum_{-\infty}^{1} a_\nu(z-c)^\nu = \sum_{1}^{\infty} a_{-\nu}(z-c)^{-\nu} \text{ , respectively, } \sum_{0}^{\infty} a_\nu(z-c)^\nu$$

are called its *principal part*, respectively, its *regular part*. Laurent series are thus special examples of the series of functions $\sum_{-\infty}^{\infty} f_\nu(z)$ introduced in 11.1.3. Therefore in particular, the concepts of absolute, compact and normal convergence are available for Laurent series in annuli.

Laurent series are generalized power series. The corresponding generalization of the CAUCHY-TAYLOR representation theorem is the

LAURENT expansion theorem. *Every function f which is holomorphic in the annulus A of center c and radii r, s is developable in A into a unique Laurent series,*

$$(1) \qquad f(z) = \sum_{-\infty}^{\infty} a_\nu(z-c)^\nu,$$

which converges normally in A to f. Furthermore, we have

$$(2) \qquad a_\nu = \frac{1}{2\pi i} \int_{S_\rho} \frac{f(\zeta)}{(\zeta-c)^{\nu+1}} d\zeta \qquad \text{for all } r < \rho < s \text{ , } \nu \in \mathbb{Z}.$$

Proof. Let $f = f^+ + f^-$ be the Laurent separation of f in $A = A^+ \cap A^-$, as provided by theorem 2. The regular part $f^+ \in \mathcal{O}(A^+)$ of f has a Taylor development $\sum a_\nu(z-c)^\nu$ in $A^+ = B_s(c)$, by the CAUCHY-TAYLOR theorem of 7.3.1. Observe that the principal part $f^- \in \mathcal{O}(A^-)$ of f also admits a simple series development in $A^- = \{z \in \mathbb{C} : |z-c| > r\}$. We see this as follows. The mapping

$$w \mapsto z := c + w^{-1} \text{ of } B_{r^{-1}}(0) \setminus 0 \text{ onto } A^-$$

is biholomorphic with inverse $z \mapsto w = (z-c)^{-1}$ and

$$g(w) := f^-(c + w^{-1}) \text{ defines an element of } \mathcal{O}(B_r(0) \setminus 0).$$

We have $\lim_{w \to 0} g(w) = 0$ on account of $\lim_{z \to \infty} f^-(z) = 0$, and so, by the Riemann continuation theorem, $g(0) := 0$ extends g holomorphically over 0. This extended g therefore has a Taylor development $g(w) = \sum_{\nu \geq 1} b_\nu w^\nu \in \mathcal{O}(B_{r^{-1}}(0))$, which in fact converges normally in $B_{r^{-1}}(0)$. Since $f^-(z) = g((z-c)^{-1})$ for $z \in A^-$, we obtain from this the representation $f^-(z) = \sum_{\nu \geq 1} b_\nu(z-c)^{-\nu}$, which converges normally to f^- in A^-. With the notation $a_{-\nu} := b_\nu$, $\nu \geq 1$, this series can be written as $f^-(z) = \sum_{-\infty}^{-1} a_\nu(z-c)^\nu$. In summary, we have thus found a Laurent series $\sum_{-\infty}^{\infty} a_\nu(z-c)^\nu$ which converges normally in A to f. The uniqueness follows as soon as the equations (2) are verified. To this end, consider *any* series satisfying (1) and for each $n \in \mathbb{Z}$ the equation

$$(z-c)^{-n-1}f(z) = \sum_{\nu=-\infty}^{-1} a_{\nu+n+1}(z-c)^\nu + \sum_{\nu=0}^{\infty} a_{\nu+n+1}(z-c)^\nu$$

which, thanks to the convergence being normal, may be integrated termwise. After this is done over the circle S_ρ, only the summand corresponding to $\nu = -1$ survives:

$$\int_{S_\rho} (z-c)^{-n-1}f(z)dz = a_n \int_{S_\rho} (z-c)^{-1}dz = 2\pi i a_n , \quad n \in \mathbb{Z}. \qquad \square$$

We call (1) the *Laurent expansion* (or *development*) of f about c in A.

4. Examples. The determination of Laurent coefficients by means of the integral formulas 3(2) is only possible in rare cases. More often, known Taylor series are exploited to develop f into a Laurent series. One that proves adequate for most examination questions is the geometric series.

1) The function $f(z) = 1/(1+z^2)$ is holomorphic in $\mathbb{C} \setminus \{i, -i\}$. Let c be any point in the upper half-plane $\mathbb{H}$. Then (see the figure below) $|c-i| < |c+i|$, since $\Im c > 0$, and in the annulus A centered at c with inner radius $r := |c-i|$, outer radius $s := |c+i|$, f has a Laurent expansion, which can be quickly found with the aid of the partial fraction decomposition

$$\frac{1}{1+z^2} = \frac{1}{2i}\frac{1}{z-i} + \frac{(-1)}{2i}\frac{1}{z+i} , \quad z \in \mathbb{C} \setminus \{i, -i\}.$$

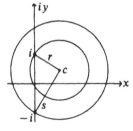

If we set

$$f^+(z) := \frac{-1}{2i}\frac{1}{z+i} , \qquad f^-(z) := \frac{1}{2i}\frac{1}{z-i},$$

then evidently $f = f^+|A + f^-|A$ is the Laurent separation of f in A. The associated series are

$$f^+(z) = \frac{-1}{2i(i+c)}\frac{1}{1+\frac{z-c}{i+c}} = \sum_{0}^{\infty} \frac{1}{2i}\frac{(-1)^{\nu+1}}{(i+c)^{\nu+1}}(z-c)^\nu , \quad |z-c| < s,$$

$$f^-(z) = \frac{1}{2i(z-c)} \frac{1}{1 - \frac{i-c}{z-c}} = \sum_{-\infty}^{-1} \frac{1}{2i} \frac{1}{(i-c)^{\nu+1}} (z-c)^\nu , \quad |z-c| > r.$$

It is to be noted that the case $c = i$ is allowed. What form do $f^+(z)$ and $f^-(z)$ then assume? We should note further that $(1 + z^2)^{-1}$ also has a Laurent expansion in the exterior $\{z \in \mathbb{C} : |z - c| > s\}$ of the larger disc. What does it look like?

2) The function $f(z) = 6/[z(z+1)(z-2)]$ is holomorphic in $\mathbb{C}\setminus\{0,-1,2\}$ and consequently has *three* Laurent developments around 0: one in the punctured unit disc $\mathbb{E}^\times$, one in the annulus $\{z \in \mathbb{C} : 1 < |z| < 2\}$ and one in the exterior $\{z \in \mathbb{C} : |z| > 2\}$ of the disc $B_2(0)$. Using the partial fraction decomposition of f, determine the corresponding Laurent developments.

3) The function $\exp(z^{-k}) \in \mathcal{O}(\mathbb{C}^\times)$ has the Laurent development around 0 given by

$$\exp(z^{-k}) = 1 + \frac{1}{1!}\frac{1}{z^k} + \frac{1}{2!}\frac{1}{z^{2k}} + \cdots + \frac{1}{n!}\frac{1}{z^{nk}} + \cdots , \qquad k = 1, 2, \ldots$$

5. Historical remarks on the theorem of LAURENT. In the year 1843 CAUCHY reported to the French Academy (*C. R. Acad. Sci. Paris* **17**, p. 938; also in his *Œuvres* (1) **8**, 115-117) about a work of P. A. LAURENT (1813-1854, engineer in the army and active in the construction of the port of Le Havre) entitled "Extension du théorème de M. Cauchy relatif à la convergence du développement d'une fonction suivant les puissances ascendantes de la variable." Here LAURENT shows that Cauchy's theorem on the representability via power series of holomorphic functions in discs is even valid in annuli, if series in which negative powers of $z - c$ occur, are allowed. The original work of LAURENT was never published. Only in 1863, thanks to the dedication of his widow, did his "Mémoire sur la théorie des imaginaires, sur l'équilibre des températures et sur l'équilibre d'élasticité" appear in *Jour. de l'École Polytech.* **23**, 75-204; it contains his proof, the exposition of which is unfortunately very cumbersome (esp. pp. 106, 145).

In his 1843 report CAUCHY talks more about himself than about LAURENT's result. He emphasizes that LAURENT arrived at his theorem by a meticulous analysis of his own proof of power series developability. Nevertheless he does declare that "L'extension donnée par M. Laurent $\cdots$ nous paraît digne de remarque (the extension given by M. Laurent $\cdots$ seems to us worthy of note)." LAURENT proves his theorem by using, as we did in the text, the Cauchy integral method, which he generalized. To this day there is no proof which does not, however cryptically, use complex integrals. (Cf. in this connection also the next section.)

The integral theorem for annuli is to be found in CAUCHY's 1840 *Exercices D'Analyse* (*Œuvres* (2) **11**, p.337); he formulated it however without integrals and in terms of mean values. In his report on Laurent's work he says that the latter's theorem follows immediately from this ("Le théorème de M. Laurent peut se déduire immédiatement ···", p.116).

The theorem under discussion had already been proved by WEIER-STRASS [W₁] in 1841, a work which was not published until 1894. Many authors consequently call the theorem the LAURENT-WEIERSTRASS theorem. On this issue of the name, KRONECKER bitingly says (cf. [Kr], p.177): "Diese Entwicklung wird manchmals Laurent'scher Satz bezeichnet; aber da sie eine unmittelbare Folge des Cauchy'schen Integrals ist, so ist es unnütz, einen besonderen Urheber zu nennen (This development is sometimes designated as Laurent's theorem, but since it is an immediate consequence of Cauchy's integral, it is useless to name one particular author)." KRONECKER doesn't vouchsafe a word about his colleague WEIERSTRASS.

The independence of the integral 1 (1) from the radius is the heart of Weierstrass' work [W₁]; it says there (p.57): "..., d.h. der Werth des Integrals ist für alle Werthe von x_0, deren absoluter Betrag zwischen den Grenzen A, B enthalten ist, derselbe (..., i.e., the value of the integral is the same for all values of x_0 whose modulus is contained between the limits $A, B[= r, s]$)."

Throughout his life WEIERSTRASS never gave much prominence to this result, possibly due to the integrals in his proof (on which point cf. also 6.1.3 and 8.2.4). For example, in 1896 PRINGSHEIM in his paper "Ueber Vereinfachungen in der elementaren Theorie der analytischen Functionen," *Math. Annalen* **47**, 121-154 expressed surprise that in his lectures WEIER-STRASS "weder explicite bewiesen noch direct angewendet (neither explicitly proved nor directly applied)" the theorem.[1] PRINGSHEIM laments that this theorem had not yet "den ihm eigentlich zukommenden Platz erhalten hat (secured the place it really deserves)" in *elementary function theory* – by which he understood that part of the theory of holomorphic functions which is based solely on power series without any use of integrals. He rightly points out that it seems that "die elementare Functionentheorie *ohne* den Laurent'schen Satz keinerlei Hülfsmittel zu besitzen (elementary function theory without Laurent's theorem possesses no means whatsoever)" of inferring, e.g., Riemann's continuation theorem, even when the function has already been continuously extended over the isolated singularity c and moreover all the power series which represent f in a punctured neighborhood of c converge absolutely at c. (Cf. also 2.2.)

In 1896 PRINGSHEIM considered it "dringend wünschenswerth (urgently desirable)" to ground the theorem of LAURENT in the "möglichst elementaren Weg (most elementary possible way)." He believed this goal could be achieved through the "Einführung gewisser Mittelwerthe an Stelle der sonst benutzten Integrale (introduction of certain mean values in place of the otherwise-used integrals)." His elementary direct proof of Laurent's

---

[1]PRINGSHEIM seems to have learned about this youthful work of Weierstrass only as his own paper was in press (cf. footnote p. 123).

theorem works in a very contrived way however. As he himself says (p.125) – his mean values may "stets als Specialfälle bestimmter Integrale ansehen (always be viewed as special cases of definite integrals)". For this reason his proof, which is only integral-free insofar as its inner workings aren't examined, has not caught on. PRINGSHEIM set out his "pure methodology" in the 1223-page long work [P]. To which treatise might be applied the very words (p.124) with which PRINGSHEIM appraised a so-called elementary proof of Laurent's theorem by MITTAG-LEFFLER: "Die Consequenz der Methode [wird] auf Kosten der Einfachheit allzu theuer erkauft (the consequences of the method are bought all too dearly at the expense of simplicity)."

6*. Deviation of LAURENT'S theorem from the CAUCHY–TAYLOR theorem.

The existence proof carried out in section 3 rests on the Laurent representation of theorem 2 and thereby on the Cauchy integral formula for annuli. The view is sometimes maintained that in fact Cauchy's theory for annuli is essential to proving Laurent's theorem. But this is not so. As early as 1884 L. SCHEEFFER (1859-1885) had reduced the theorem to the Cauchy-Taylor theorem in a short paper "Beweis des Laurent'schen Satzes", in *Acta Math.* **4**, 375-380. In what follows we are going to reproduce this forgotten proof. We use the notation of section 3 and suppose $c = 0$; thus $A = A_{r,s}(0)$. First we show that

Lemma. *Let $f \in \mathcal{O}(A)$ and suppose there is an annulus $A' \subset A$ centered at 0 such that f has a Laurent development around 0 in A':*

$$f(z) = \sum_{-\infty}^{\infty} a_\nu z^\nu , \qquad z \in A'.$$

Then this Laurent series is in fact normally convergent throughout A to f.

Proof. Because of the identity theorem the only issue is the normal convergence throughout A of the given Laurent series. Let $A' = A_{\rho,\sigma}(0)$. It suffices to consider each of the two special cases $\rho = r$ and $\sigma = s$. By considering $f(z^{-1}) \in \mathcal{O}(A_{s^{-1},r^{-1}})$ instead of $f(z) \in \mathcal{O}(A_{r,s})$, the second of these cases is reduced to the first, which we now treat.

The series $\sum_{-\infty}^{-1} a_\nu z^\nu$ and $\sum_0^\infty a_\nu z^\nu$ converge normally in $A_{r,\infty}(0)$ and $B_\sigma(0)$, respectively. (Recall the definitions in 11.1.3.) Therefore the function

$$f_1(z) := \begin{cases} f(z) - \sum_{-\infty}^{-1} a_\nu z^\nu , & z \in A \\ \\ \sum_0^\infty a_\nu z^\nu , & z \in B_\sigma(0) \end{cases}$$

is well defined and holomorphic in $B_s(0)$. Since $\sum_0^\infty a_\nu z^\nu$ is the Taylor series of f_1 at 0, this series converges normally throughout $B_s(0)$ according to the CAUCHY-TAYLOR representation theorem 7.3.2. Therefore $\sum_{-\infty}^\infty a_\nu z^\nu$ converges normally in A. □

Let us further remark that it suffices to prove the existence of Laurent developments for *odd* functions in $\mathcal{O}(A)$. This is because an arbitrary function $f \in \mathcal{O}(A)$

can always be realized as a sum $f_1(z) + zf_2(z)$ with odd functions

$$f_1(z) := \tfrac{1}{2}\left(f(z) - f(-z)\right), \; f_2(z) := \tfrac{1}{2}z^{-1}(f(z) + f(-z)), \qquad z \in A.$$

From Laurent developments of f_1 and f_2 we immediately get one for f at 0 in A.

We now begin the proof proper. Thus we are given an odd function f holomorphic in the annulus $A = A_{r,s}(0)$. Because of what was proved in the lemma, we can shrink s and expand r if necessary and thereby assume without loss of generality that $0 < r < s < \infty$. By introducing the new variable $v := (\sqrt{rs})^{-1}z$ and working with it, we find we may further assume without loss of generality that $rs = 1$. Then necessarily $s > 1$. We conscript the function $q : \mathbb{C}^\times \to \mathbb{C}$, $z \mapsto \tfrac{1}{2}(z + z^{-1})$ and use the following property of it, established by the reader in Exercises 3 and 4, §1 of Chapter 2:

Every q-fiber $q^{-1}(b)$, $b \neq \pm 1$, consists of two distinct points $a, a^{-1} \in \mathbb{C}^\times$. Suppose that $s > 1 + \sqrt{2}$, and set $R := \tfrac{1}{2}(s - s^{-1}) > 1$, $\sigma := R + \sqrt{R^2 - 1}$, $\rho := \sigma^{-1}$. Then on the one hand the q-preimage of the disc $B := B_R(0)$ is contained in A and on the other hand $q^{-1}(B)$ completely contains the annulus $A' := A_{\rho,\sigma} \subset A$.

We first suppose that $s > 1 + \sqrt{2}$, so that the properties just recorded become available. Since $q^{-1}(B) \subset A$, $f(z) + f(z^{-1})$ is holomorphic in $q^{-1}(B)$ and evidently constant on each q-fiber. Since moreover q maps $q^{-1}(B)$ onto B, the factorization theorem 9.4.5 is applicable. It furnishes a $g \in \mathcal{O}(B)$ for which

$$f(z) + f(z^{-1}) = g(q(z)) \qquad \text{for all } z \in q^{-1}(B).$$

Let $\sum_0^\infty a_\mu w^\mu = g(w)$ be the Taylor series of g in B. It follows that for $z \in A' \subset q^{-1}(B)$

$$f(z) + f(z^{-1}) = \sum_{\mu=0}^\infty a_\mu 2^{-\mu}(z + z^{-1})^\mu = \sum_{\mu=0}^\infty \sum_{\nu=0}^\mu a_\mu 2^{-\mu}\binom{\mu}{\nu}z^{\mu-2\nu}.$$

We will show that for every compact $K \subset A'$

(∗)
$$\sum_{\mu=0}^\infty \sum_{\nu=0}^\mu \left| a_\mu 2^{-\mu}\binom{\mu}{\nu}z^{\mu-2\nu}\right|_K < \infty.$$

With this result in hand, the summands in the preceding double sum can be ordered according to powers of z and we get, from the sharpened form of the rearrangement theorem 3.3.1, a Laurent development

$$f(z) + f(z^{-1}) = \sum_{-\infty}^\infty b_\nu z^\nu$$

which is normally convergent throughout A'. Thanks to the foregoing lemma, this Laurent development is actually valid throughout A. Analogous considerations involving the mapping $\tilde{q} : \mathbb{C}^\times \to \mathbb{C}$ given by $\tilde{q}(z) := -iq(iz) = \tfrac{1}{2}(z - z^{-1})$ furnish a Laurent development

$$f(z) + f(-z^{-1}) = \sum_{-\infty}^{\infty} \tilde{b}_\nu z^\nu$$

which converges normally throughout A. Since f is *odd*, addition gives us

$$f(z) = \sum_{-\infty}^{\infty} \frac{1}{2}(b_\nu + \tilde{b}_\nu)z^\nu, \qquad z \in A.$$

In proving (*) it suffices to look at K of the form $K = \overline{A}_{u,v}$, where $\rho < u < v < \sigma$ and $uv = 1$. Now for all $\mu, \nu \in \mathbb{N}$, $|z^{\mu-2\nu}|_K = v^{|\mu-2\nu|}$. Using the trivial identity $\binom{\mu}{\mu-\nu} = \binom{\mu}{\nu}$, it follows easily that

$$\sum_{\nu=0}^{\mu} \binom{\mu}{\nu} |z^{\mu-2\nu}|_K \leq 2(v + 1/v)^\mu = 2^{\mu+1} q(v)^\mu.$$

Since $q(v)$ lies in the disc B of convergence of $\sum a_\mu w^\mu$, we get

$$\sum_{\mu=0}^{\infty} \sum_{\nu=0}^{\mu} \left| a_\mu 2^{-\mu} \binom{\mu}{\nu} z^{\mu-2\nu} \right|_K \leq 2 \sum_{\mu=0}^{\infty} |a_\mu q(v)^\mu| < \infty$$

and therewith (*).

Now let us take up the case of arbitrary $s > 1$. There is a (smallest) natural number n for which $s^{2^n} > 1 + \sqrt{2}$. By what was proved above, every function which is holomorphic in $A_{r^{2^n}, s^{2^n}}$ has a Laurent development in that annulus. Consequently our work is completed by n applications of the following fact:

If every function in $\mathcal{O}(A_{r^2, s^2})$ has a Laurent development around 0 in A_{r^2, s^2}, then every function in $\mathcal{O}(A)$ has a Laurent development around 0 in $A := A_{r,s}$.

We now prove the preceding statement. As we have noted, it suffices to prove the conclusion for every *odd* function $f \in \mathcal{O}(A)$. The image of A under the square map $z \mapsto z^2$ is A_{r^2, s^2} and every point in A_{r^2, s^2} has exactly two distinct pre-images, which are negatives of one another, in A. Since $zf(z)$ is an even function, 9.4.5 furnishes an $h \in \mathcal{O}(A_{r^2, s^2})$ such that $zf(z) = h(z^2)$ for all $z \in A$. By assumption h has a Laurent development $h(w) = \sum_{-\infty}^{\infty} a_\nu w^\nu$ in A_{r^2, s^2}. Then $\sum_{-\infty}^{\infty} a_\nu z^{2\nu-1}$ is evidently a Laurent development of f around 0 in A.   □

Remarks on the SCHEEFFER proof. In the proof of the lemma as well as in the proof proper the *global* aspect of the power series development of a holomorphic function (namely, that the Taylor series at a point represents the function in as large a disc about the point as lies in the domain) was used decisively. [By contrast, it was only for convenience that the factorization lemma 9.4.5 was invoked; at the appropriate points *ab initio* arguments could have been given, as e.g., SCHEEFFER did.] The representation theorem 7.3.2 which affirms this is derived from the Cauchy integral formula 7.2.2. In this sense SCHEEFFER's proof is therefore not integration-free. But it does yield the Cauchy theory in annuli as a corollary: Because for every $f \in \mathcal{O}(A)$, as soon as we have a Laurent development $\sum_{-\infty}^{\infty} a_\nu z^\nu$ for it, there follows at once the fundamental equation 1.1(1)

$$\int_{S_\rho} f d\zeta = 2\pi i a_{-1} = \int_{S_\sigma} f d\zeta \qquad \text{for all } \rho, \sigma \in \mathbb{R} \text{ with } r < \rho \leq \sigma < s.$$

Nevertheless, Cauchy's integral formula for an annulus constitutes the natural approach to the Laurent expansion theorem. SCHEEFFER's baroque route is not recommended for lectures.

Exercises

Exercise 1. Develop the following functions into their Laurent series in the indicated annuli:

a) $f(z) = \frac{4z - z^2}{(z^2 - 4)(z+1)}$ in $A_{1,2}(0)$, $A_{2,\infty}(0)$, $A_{0,1}(-1)$,

b) $f(z) = \frac{1}{(z-c)^n}$, $(n \in \mathbb{N}, n \geq 1, c \in \mathbb{C}^\times)$ in $A_{|c|,\infty}(0)$, $A_{0,\infty}(c)$,

c) $f(z) = \frac{1}{(z+2)^2(z^2-9)}$ in $A_{2,3}(0)$, $A_{3,\infty}(0)$, $A_{0,1}(-2)$,

d) $f(z) = \sin\left(\frac{z}{z-1}\right)$ in $A_{0,\infty}(1)$,

e) $f(z) = \sin\left(\frac{z-1}{z}\right)$ in $\mathbb{C}^\times$,

f) $f(z) = (\exp(z^{-1}))^{-1}$ in $\mathbb{C}^\times$.

Exercise 2. Let f, g be holomorphic in $A = A_{r,s}(c)$, $0 \leq r < s \leq \infty$ and have in A Laurent representations

$$\sum_{\nu=-\infty}^{\infty} a_\nu(z - c)^\nu \quad \text{and} \quad \sum_{\mu=-\infty}^{\infty} b_\mu(z - c)^\mu,$$

respectively. Show that the series $c_k := \sum_{\nu=-\infty}^{\infty} a_\nu b_{k-\nu}$ converges in $\mathbb{C}$ for each $k \in \mathbb{Z}$ and that $\sum_{k=-\infty}^{\infty} c_k(z-c)^k$ is the Laurent representation in A of fg. *Hint.* The kth Laurent coefficient of fg is given by $\frac{1}{2\pi i} \int_{S_\rho(c)} \frac{f(\zeta)g(\zeta)}{(\zeta-c)^{k+1}} d\zeta$ for any $r < \rho < s$.

Exercise 3. Let A be a non-empty open annulus centered at 0, $f \in \mathcal{O}(A)$ and $\sum_{\nu=-\infty}^{\infty} a_\nu z^\nu$ the Laurent representation of $f(z)$ in A. Prove that f is even (that is, $f(z) = f(-z)$ for all $z \in A$) if and only if $a_\nu = 0$ for all odd ν and f is odd (that is, $f(z) = -f(z)$ for all $z \in A$) if and only if $a_\nu = 0$ for all even ν.

Exercise 4. Let A be a non-empty open annulus centered at 0, $f \in \mathcal{O}(A)$ a unit.

a) Show that if $\sum_{\nu=-\infty}^{\infty} a_\nu z^\nu$ is the Laurent development of f'/f in A, then $n := a_{-1}$ is an integer and f has a holomorphic logarithm in A exactly when this integer is 0.

b) Prove the *lemma on units* for $\mathcal{O}(A)$: In the notation of a), there exists $g \in \mathcal{O}(A)$ such that $f(z) = z^n e^{g(z)}$ for all $z \in A$; and if $f(z) = z^m e^{h(z)}$ for all $z \in A$ and some $m \in \mathbb{Z}$, $h \in \mathcal{O}(A)$, then $m = n$.

Hint. For a) utilize corollary 9.3.4.

Exercise 5. (Bessel functions) For $\nu \in \mathbb{Z}$, $w \in \mathbb{C}$ let $J_\nu(w)$ designate the coefficient of z^ν in the Laurent series of the function $\exp[\frac{1}{2}(z - z^{-1})w] \in \mathcal{O}(\mathbb{C}^\times)$; thus $\exp[\frac{1}{2}(z - z^{-1})w] = \sum_{\nu=-\infty}^{\infty} J_\nu(w)z^\nu$. Show that

a) $J_{-\nu}(w) = J_\nu(w)$ for all $\nu \in \mathbb{Z}$, $w \in \mathbb{C}$.

b) $J_\nu(w) = \frac{1}{2\pi} \int_0^{2\pi} \cos(\nu\varphi - w\sin\varphi)d\varphi$.

c) Each function $J_\nu : \mathbb{C} \to \mathbb{C}$ is holomorphic. Its power series at 0 is (for each $\nu \geq 0$)

$$J_\nu(w) = \sum_{k=0}^{\infty} \frac{(-1)^k}{k!} \frac{(\frac{1}{2}w)^{2k+\nu}}{(\nu + k)!}.$$

The functions J_ν, $\nu \geq 0$, are called *Bessel functions (of the first kind)*. J_ν satisfies the *Bessel differential equation* $z^2 f''(z) + z f'(z) + (z^2 - \nu^2)f(z) = 0$.

§2   Properties of Laurent series

In this section many elementary assertions about power series are extended to cover Laurent series. In addition, we show how the Laurent development of a holomorphic function at an isolated singularity leads to a simple characterization of the singularity type in terms of the Laurent coefficients.

1. Convergence and identity theorems. On the basis of theorem 1.3, every function f which is holomorphic in an annulus A centered at c is developable into a Laurent series which converges normally in A to f. In order to get a converse to this statement, we associate to every Laurent series $\sum_{-\infty}^{\infty} a_\nu(z-c)^\nu$ the radius of convergence s of its regular part and the radius of convergence $\hat{r}$ of the power series $\sum_{\nu\geq1} a_{-\nu}w^\nu$. We set $r := \hat{r}^{-1}$, meaning $r = 0$ if $\hat{r} = \infty$ and $r = \infty$ if $\hat{r} = 0$, and we demonstrate the

Convergence theorem for Laurent series. *If $r < s$, the Laurent series $\sum_{-\infty}^{\infty} a_\nu(z - c)^\nu$ converges normally in the open annulus $A := A_{r,s}(c)$ to a function which is holomorphic in A; the Laurent series converges at no point of $\mathbb{C} \setminus \bar{A}$.*

If $r \geq s$, the Laurent series converges in no open subset of $\mathbb{C}$.

Proof. We set

$$f^+(z) := \sum_{0}^{\infty} a_\nu(z - c)^\nu \in \mathcal{O}(B_s(c)) \ , \ g(w) := \sum_{1}^{\infty} a_{-\nu} w^\nu \in \mathcal{O}(B_{\hat{r}}(0)).$$

Then $\sum_{-\infty}^{-1} a_\nu(z - c)^\nu$ converges normally in $\mathbb{C} \setminus \overline{B_r}(c)$ to the function $f^-(z) := g((z - c)^{-1}) \in \mathcal{O}(\mathbb{C} \setminus \overline{B_r}(c))$. Therefore if $r < s$, the Laurent series converges normally in $B_s(c) \cap (\mathbb{C} \setminus \overline{B_r}(c)) = A$ to $f^+ + f^- \in \mathcal{O}(A)$.

The remaining statements of the theorem follow from the convergence behavior of the power series f^+, g in their respective discs of convergence. (Use theorem 4.1.2.)   □

In function theory we are only concerned with Laurent series having $r < s$. Laurent series with $r \geq s$ are uninteresting, since there is *no meaningful calculus for them*: For $L := \sum_{-\infty}^{\infty} z^\nu$ with $r = s = 1$, formal calculation leads to $z \cdot L = \sum_{-\infty}^{\infty} z^{\nu+1}$, that is, back to L, so that one gets $(z - 1)L = 0$ which can't happen in function theory.

For Laurent series we have a simple

Identity theorem. *If $\sum_{-\infty}^{\infty} a_\nu(z - c)^\nu$ and $\sum_{-\infty}^{\infty} b_\nu(z - c)^\nu$ are Laurent series which each converge uniformly on a circle S_ρ, $\rho > 0$, to the same limit function f, then*

$$(1) \qquad a_\nu = b_\nu = \frac{1}{2\pi\rho^\nu} \int_0^{2\pi} f(c + \rho e^{i\varphi}) e^{-i\nu\varphi} d\varphi \ , \qquad \nu \in \mathbb{Z}.$$

Proof. First note that these integrals exist, because f is necessarily continuous on S_ρ. The uniformity of the convergence is the reason for this continuity and it also justifies interchanging the integration with the limit process which determines f. Then equation (1) emerges from the "orthogonality relations" among the exponentials (on which point cf. also 8.3.2).   □

Naturally the equations (1) are nothing other than the formulas (2) from 1.3. The assumption that a Laurent series about c converge compactly on a circle S_ρ centered at c is fulfilled in all cases where the series actually converges in an annulus around c which contains S_ρ. If we let $L(A)$ be the set of Laurent series which (normally) converge in the annulus A, then the identity theorem and the theorem of Laurent together prove that

The mapping $\mathcal{O}(A) \mapsto L(A)$ which assigns to every holomorphic function in A its Laurent series around c is bijective.

The holomorphic functions in A and the convergent Laurent series in A thus correspond *biuniquely*.

Historical remark. CAUCHY proved the above identity theorem for Laurent series in 1841 (*Œuvres* (1) **6**, p.361). He hypothesized merely the pointwise convergence of the two series to the same limit function on S_ρ and then blithely integrated term-wise (which is inadmissible). Thereupon LAURENT communicated his investigations to the Paris Academy and in an accompanying letter remarked (*C. R. Acad. Sci. Paris* **17**(1843), p. 348) that he was in possession of convergence conditions "for all the series developments heretofore used by mathematicians."

2. The Gutzmer formula and Cauchy inequalities. *If the Laurent series $\sum_{-\infty}^{\infty} a_\nu (z - c)^\nu$ converges uniformly on the circle S_ρ centered at c to $f : S_\rho \to \mathbb{C}$, then the Gutzmer formula holds:*

$$(1) \quad \sum_{-\infty}^{\infty} |a_\nu|^2 \rho^{2\nu} = \frac{1}{2\pi} \int_0^{2\pi} |f(c + \rho e^{i\varphi})|^2 d\varphi \leq M(\rho)^2 \, , \; M(\rho) := |f|_{S_\rho}.$$

In particular, the Cauchy inequalities

$$(2) \qquad\qquad |a_\nu| \leq \frac{M(\rho)}{\rho^\nu} \qquad \text{prevail for all } \nu \in \mathbb{Z}.$$

The proof is analogous to that of 8.3.2. $\qquad\qquad\qquad\qquad\qquad$ □

If the Laurent series is holomorphic in some annulus $A_{r,s}(c)$ with $r < \rho < s$, then one can naturally get the inequalities (2) directly and immediately from 1.3(2). WEIERSTRASS' proof in [W$_2$], pp. 68, 69, which was reproduced in 8.3.5, was actually carried out by him for Laurent series.

With the help of the inequalities (2) we can understand at a glance and better than before why the Riemann continuation theorem is valid. Namely, if $\sum_{-\infty}^{\infty} a_\nu (z - c)^\nu$ is the Laurent development of f in a neighborhood of the isolated singularity c and if $M < \infty$ is a bound for f near c, then for all sufficiently small radii ρ the estimates $\rho^\nu |a_\nu| \leq M$ hold for all $\nu \in \mathbb{Z}$. Since $\lim_{\rho \to 0} \rho^\nu = \infty$ for every $\nu \leq -1$, the only way out is for a_ν to be 0 for every such ν. That is, the Laurent series is really a power series and consequently via $f(c) := a_0$, f is holomorphically extended over c. It was in just this way that WEIERSTRASS proved the continuation theorem in 1841 ([W$_1$], p.63).

3. Characterization of isolated singularities. The theorem of Laurent makes possible a new approach to the classification of isolated singularities of holomorphic functions. If f is holomorphic in $D \setminus c$, where $c \in D$, then there is a uniquely determined Laurent series $\sum_{-\infty}^{\infty} a_\nu (z - c)^\nu$ which represents f in every punctured disc $B_r(c) \setminus c$ lying in $D \setminus c$. We call this series *the Laurent development of f* at (or *around*) c and show

Theorem (Classification of isolated singularities). *Suppose $c \in D$ is an isolated singularity of $f \in \mathcal{O}(D \setminus c)$ and that*

$$f(z) = \sum_{-\infty}^{\infty} a_\nu (z - c)^\nu$$

is the Laurent development of f at c. Then c is

1) *a removable singularity $\Leftrightarrow a_\nu = 0$ for all $\nu < 0$,*

2) *a pole of order $m \geq 1 \Leftrightarrow a_\nu = 0$ for all $\nu < -m$ and $a_{-m} \neq 0$,*

3) *an essential singularity $\Leftrightarrow a_\nu \neq 0$ for infinitely many $\nu < 0$.*

Proof. ad 1) The singularity c is removable exactly when there is a Taylor series at c which represents f near c. Because of the uniqueness of the Laurent development, this occurs just when the Laurent series is already a Taylor series, i.e., $a_\nu = 0$ for all $\nu < 0$.

ad 2) On the basis of theorem 10.1.2 we know that c is a pole of order m exactly when an equation

$$f(z) = \frac{b_m}{(z - c)^m} + \cdots + \frac{b_1}{z - c} + \tilde{f}(z) \qquad \text{with } b_m \neq 0$$

holds in a punctured neighborhood of c, $\tilde{f}$ being given by a power series convergent in that whole neighborhood. Again due to the uniqueness of the Laurent development, this occurs just if $a_\nu = 0$ for all $\nu < -m$ and $a_{-m} = b_m \neq 0$.

ad 3) An essential singularity occurs at c exactly when neither case 1) nor case 2) prevails, that is, when a_ν is non-zero for infinitely many $\nu < 0$. □

It now follows trivially that $\exp z^{-1}$ and $\cos z^{-1}$ have essential singularities at the origin, since their Laurent series

$$\sum_0^\infty \frac{1}{\nu!} \frac{1}{z^\nu} \qquad \text{and} \qquad \sum_0^\infty \frac{(-1)^\nu}{(2\nu)!} \frac{1}{z^{2\nu}}$$

each have principal parts containing infinitely many non-zero terms. Furthermore, lemma 10.1.3 also follows directly. □

We want to emphasize again that

If c is an isolated singularity of f, the principal part u_c the Laurent development of f at c is holomorphic in $\mathbb{C} \setminus c$.

This follows from theorem 1.2 because in this case $A^- = \mathbb{C} \setminus c$.   □

A Laurent series at the point c in the punctured disc $B \setminus c$, when regarded as the sum of the series of meromorphic functions $f_\nu(z) := a_\nu(z - c)^\nu$ in B, converges normally in B in the sense of 11.1.1 exactly when its principal part is finite (for only then is the pole-dispersion condition met).

Exercises

Exercise 1. Determine the region of convergence of each of the following Laurent series:

a) $\sum_{\nu=-\infty}^{\infty} \frac{z^\nu}{|\nu|!}$,

b) $\sum_{\nu=-\infty}^{\infty} \frac{(z - 1)^{2\nu}}{\nu^2 + 1}$,

c) $\sum_{\nu=-\infty}^{\infty} c^\nu (z - d)^\nu$, $c \in \mathbb{C}^\times, d \in \mathbb{C}$,

d) $\sum_{\nu=-\infty}^{\infty} \frac{(z - 3)^{2\nu}}{(\nu^2 + 1)^\nu}$

Exercise 2. Let $f(z) = \sum_{\nu=0}^{\infty} a_\nu z^\nu$ be a power series with positive radius of convergence. Determine the region of convergence of the Laurent series $\sum_{\nu=-\infty}^{\infty} a_{|\nu|} z^\nu$ and identify the function which it represents there.

Exercise 3. Let $0 < r < s < \infty$, $A = A_{r,s}(0)$ and $f \in \mathcal{O}(A)$. Suppose that $\lim_{n \to \infty} f(z_n) = 0$ either for every sequence $\{z_n\} \subset A$ with $|z_n| \to r$ or for every sequence $\{z_n\} \subset A$ with $|z_n| \to s$. Show that then $f(z) = 0$ for all $z \in A$. Investigate whether this conclusion remains valid in either limiting case $r = 0$ or $s = \infty$.

Exercise 4. For each of the following functions classify the isolated singularity at 0 and specify the principal part of the Laurent development there:

a) $\frac{\sin z}{z^n}$, $n \in \mathbb{N}$;

b) $\frac{z}{(z + 1) \sin(z^n)}$, $n \in \mathbb{N}$;

c) $\cos(z^{-1}) \sin(z^{-1})$;

d) $(1 - z^{-n})^{-k}, \quad n, k \in \mathbb{N} \setminus \{0\}.$

§3   Periodic holomorphic functions and Fourier series

The simplest holomorphic functions with complex period $\omega \neq 0$ are the entire functions $\cos(\frac{2\pi}{\omega} z)$, $\sin(\frac{2\pi}{\omega} z)$ and $\exp(\frac{2\pi i}{\omega} z)$. Series of the form

$$(1) \qquad \sum_{-\infty}^{\infty} c_\nu \exp\left(\frac{2\pi i}{\omega} \nu z\right), \qquad c_\nu \in \mathbb{C}$$

are called Fourier series of period ω. The goal of this section is to show that every holomorphic function with period ω admits development into a normally convergent Fourier series (1). The proof is accomplished by representing every such function f in the form $f(z) = F(\exp(\frac{2\pi i}{\omega} z))$, where F is holomorphic in an annulus (cf. subsection 2); the Laurent development of F then automatically furnishes the Fourier development of f.

1. Strips and annuli. In what follows ω will always designate a non-zero complex number. A region G is called ω-*invariant* if $z \pm \omega \in G$ for every $z \in G$. This evidently occurs just when every translation $z \mapsto z + n\omega$ ($n \in \mathbb{Z}$) induces an automorphism of G. For every pair $a, b \in \mathbb{R}$ with $a < b$ the set

$$T_\omega := T_\omega(a, b) := \{z \in \mathbb{C} : a < \Im\left(\frac{2\pi}{\omega} z\right) < b\}$$

will be called the *strip* determined by ω, a and b. The argument of ω determines the "direction" of the strip, as the example in the figure illustrates.

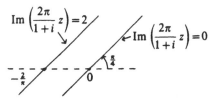

Strips T_ω are ω-invariant and in fact convex, so if $d \in T_\omega$ then the whole interval $[d, d+\omega]$ and indeed the whole line $d + \omega\mathbb{R}$ lies in T_ω. We also want to admit $a = -\infty$ and $b = \infty$. Evidently $T_\omega(-\infty, b)$, with $b \in \mathbb{R}$, is an open half-plane, while $T_\omega(-\infty, \infty)$ is all of $\mathbb{C}$. The figure shows $T_{1+i}(0, 2)$; $\pi/4$ is the argument of $\omega = 1 + i$.

By means of $z \mapsto z/\omega$ the strip $T_\omega(a, b)$ is mapped biholomorphically onto $T_1(a, b)$. Therefore most problems reduce to the case $\omega = 1$. On the other hand, $z \mapsto w := \exp(2\pi i z)$ maps the straight line $L_s := \{z \in \mathbb{C} : \Im(2\pi z) = s\}$ onto the circle $\{w \in \mathbb{C} : |w| = e^{-s}\}$, for each $s \in \mathbb{R}$. From this observation it follows immediately that

The strip $T_1(a, b)$ is mapped by

$$h : T_1(a, b) \to A_{e^{-b}, e^{-a}}(0) \ , \ z \mapsto w := \exp(2\pi i z)$$

holomorphically onto the annulus $A_{e^{-b}, e^{-a}}(0)$ centered at 0 and having inner radius e^{-b}, outer radius e^{-a}.

As special cases of this, $h(T_1(a, \infty))$ is the *punctured disc* $\{w \in \mathbb{C} : 0 < |w| < e^{-a}\}$ and $h(\mathbb{C}) = \mathbb{C}^\times$.

In the sequel we will write just A for $A_{e^{-b}, e^{-a}}(0)$.

2. Periodic holomorphic functions in strips.

If f is holomorphic in an ω-invariant region G, then for every $z \in G$ and $n \in \mathbb{Z}$, the number $f(z + n\omega)$ is well defined. The function f is called *periodic* in G with *period* ω, or simply *ω-periodic* in G, if

$$f(z + \omega) = f(z) \qquad \text{holds for all } z \in G.$$

The ostensibly more general identities $f(z + n\omega) = f(z)$ for all $z \in G$, $n \in \mathbb{Z}$, follow trivially.

The set $\mathcal{O}_\omega(G)$ of all ω-periodic holomorphic functions in the ω-invariant region G is evidently *a $\mathbb{C}$-subalgebra of $\mathcal{O}(G)$ which is closed in $\mathcal{O}(G)$ with respect to compact convergence.*

Now again let $\omega = 1$ and G be the strip T_1. The holomorphic mapping $h : G \to A$ of G onto the annulus A, considered in subsection 1, induces an algebra monomorphism $h^* : \mathcal{O}(A) \to \mathcal{O}(G)$ via $F \mapsto f := F \circ h$ which "lifts" every holomorphic function F on A to a holomorphic function $f(z) := F(\exp(2\pi i z))$ on G with period 1. The image algebra $h^*(\mathcal{O}(A))$ is thus contained in the algebra $\mathcal{O}_1(G)$. In fact, $\mathcal{O}_1(G)$ is precisely the range of h^*. We have namely, as an immediate consequence of the factorization theorem 9.4.5,

Theorem. *For every 1-periodic holomorphic function f on G there is exactly one holomorphic function F on A such that $f(z) = F(\exp(2\pi i z))$ for all $z \in G$.*

So, in summary, the mapping $h^* : \mathcal{O}(A) \to \mathcal{O}_1(G)$ is a $\mathbb{C}$-algebra isomorphism.

3. The Fourier development in strips. In any strip T_ω all the (complex) Fourier series $\sum_{-\infty}^{\infty} c_\nu \exp(\frac{2\pi i}{\omega}\nu z)$, $c_\nu \in \mathbb{C}$, which are normally convergent there define ω-periodic holomorphic functions. The fundamental insight here is that all ω-periodic $f \in \mathcal{O}(T_\omega)$ are thereby accounted for.

Theorem. *Let f be holomorphic and ω-periodic in the strip $G = T_\omega$. Then f can be expanded into a unique Fourier series*

$$(1) \qquad f(z) = \sum_{-\infty}^{\infty} c_\nu \exp\left(\frac{2\pi i}{\omega}\nu z\right)$$

which is normally convergent to f in G. (The convergence is uniform in every substrip $T_\omega(a', b')$ of T_ω, where $a < a' < b' < b$.)
For every point $d \in G$ we have

$$(2) \qquad c_\nu = \frac{1}{\omega}\int_{[d,d+\omega]} f(\zeta)\exp\left(-\frac{2\pi i}{\omega}\nu\zeta\right)d\zeta, \qquad \nu \in \mathbb{Z}.$$

Proof. We again restrict ourselves to the case $\omega = 1$. According to theorem 2 there is a unique holomorphic function F in the annulus $A := \{w \in \mathbb{C} : e^{-b} < |w| < e^{-a}\}$ such that $f(z) = F(\exp(2\pi i z))$. The function F has a unique Laurent development in A:

$$F(w) = \sum_{-\infty}^{\infty} c_\nu w^\nu \qquad \text{with} \qquad c_\nu = \frac{1}{2\pi i}\int_S F(\xi)\xi^{-\nu-1}d\xi$$

and S any circle with center 0 which lies in A. This settles the question of the existence of the representation (1). The uniqueness and convergence assertions here follow from the corresponding ones for Laurent series.

The interval $[d, d+1]$ is parametrized by $\zeta(t) := d + \frac{1}{2\pi}t$ and the circle S by $\xi(t) := qe^{it}$, $t \in [0, 2\pi]$. Here $q := \exp(2\pi i d) \in A$. It follows that

$$\frac{1}{2\pi i}\int_S F(\xi)\xi^{-\nu-1}d\xi \;=\; \frac{1}{2\pi}\int_0^{2\pi} f(\zeta(t))(qe^{it})^{-\nu}dt$$

$$=\; \int_{[d,d+1]} f(\zeta)\exp(-2\pi i\nu\zeta)d\zeta,$$

that is, (2) is valid for the coefficients c_ν, in view of 1.3(2).   □

In simple cases, as with Laurent series, the Fourier series of a function can be exhibited directly without recourse to the integral formulas (2) for the coefficients. We discuss some.

4. Examples. 1) The Euler formulas

$$\cos z = \frac{1}{2}e^{-iz} + \frac{1}{2}e^{iz}, \qquad \sin z = -\frac{1}{2i}e^{-iz} + \frac{1}{2i}e^{iz}$$

are the complex Fourier series of $\cos z$ and $\sin z$ in $\mathbb{C}$.

2) The function $\frac{1}{\cos z}$ is holomorphic in the open upper half-plane and in the open lower half-plane and it has period $\omega := 2\pi$. Since $\frac{1}{\cos z} = 2e^{iz}\frac{1}{1+e^{2iz}}$, we have

$$\frac{1}{\cos z} = \begin{cases} -i - \sum_1^\infty 2ie^{2i\nu z} & \text{if } \Im z > 0, \\[2mm] i + \sum_{-\infty}^{-1} 2ie^{2i\nu z} & \text{if } \Im z < 0. \end{cases}$$

as the corresponding Fourier developments.

3) The function $\cot z$ is holomorphic in the open upper half-plane and in the open lower-half plane and it has period π. From $\cot z = i\left(1 - \frac{2}{1-e^{2iz}}\right)$ we get the respective Fourier developments

$$\cot z = \begin{cases} \frac{(-2\pi i)^k}{(k-1)!}\sum_1^\infty \nu^{k-1}e^{2\pi i\nu} & \text{if } \Im z > 0, \\[2mm] -\frac{(-2\pi i)^k}{(k-1)!}\sum_{-\infty}^{-1} \nu^{k-1}e^{2\pi i\nu z} & \text{if } \Im z < 0. \end{cases}$$

4) Since $\varepsilon_1(z) = \pi \cot \pi z$ and $(k-1)!\varepsilon_k = (-1)^{k-1}\varepsilon_1^{(k-1)}$ (see 11.3.4(2)), from differentiation of (3) we get the Fourier developments of all the Eisenstein functions for $k \geq 2$, namely

$$\varepsilon_k(z) = \begin{cases} \frac{(-2\pi i)^k}{(k-1)!}\sum_1^\infty \nu^{k-1}e^{2\pi i\nu} & \text{if } \Im z > 0, \\[2mm] -\frac{(-2\pi i)^k}{(k-1)!}\sum_{-\infty}^{-1} \nu^{k-1}e^{2\pi i\nu z} & \text{if } \Im z < 0. \end{cases}$$

The reader should derive the Fourier series of $\tan z$, $(\sin z)^{-1}$ and $(\cos z)^{-2}$.

5. Historical remarks on Fourier series.

As early as 1753 D. BERNOULLI and L. EULER used trigonometric series

$$\frac{1}{2}a_0 + \sum_1^\infty (a_\nu \cos \nu x + b_\nu \sin \nu x), \qquad a_\nu, b_\nu \in \mathbb{R}$$

in solving the differential equation $\frac{\partial^2 y}{\partial t^2} = \alpha^2 \frac{\partial^2 y}{\partial x^2}$ of a vibrating string. The real creator of the theory of trigonometric series is however Jean Baptiste Joseph de FOURIER (born 1768, took part in Napoleon's Egyptian campaign; later a politician by profession, close collaborator of Napoleon's for years, among other capacities as prefect of the Isère département; did physics and mathematics in his scant spare time, died in Paris 1830). It is to honor him that such series are called Fourier series. FOURIER developed the theory of his series starting in 1807. The point of departure was the problem of heat conduction in a solid body, which leads to the "heat equation" (cf. 4.1). Although the physical applications were more important for FOURIER than the new mathematical insights being uncovered (cf. the famous statement of JACOBI in 4.5 contrasting his and FOURIER's attitudes), he nevertheless perceived at once the great significance of trigonometric series within the so-called pure mathematics and occupied himself intensively with it. He published his investigations in Paris in 1822; this fundamental book, which is exciting to read even today, was entitled *La Théorie Analytique de la Chaleur* (see also vol. I of his *Œuvres*, or the 1878 English translation by A. FREEMAN, published by Cambridge University Press and reprinted by Dover Publ. Co., Inc. (1955), New York). A very good historical presentation of the development of the theory in the first half of the 19th century was given in 1854 by RIEMANN in his Göttingen Habilitationsschrift entitled *Ueber die Darstellbarkeit einer Function durch eine trigonometrische Reihe* (*Werke*, 227-264).

§4 The theta function

The focus of this section is the theta function

$$\theta(z, \tau) := \sum_{-\infty}^{\infty} e^{-\nu^2 \pi \tau} e^{2\pi i \nu z},$$

which by its very form is a Fourier series (in z). "Die Eigenschaften dieser Transcendenten lassen sich durch Rechnung leicht erhalten, weil sie durch unendliche Reihen mit einem Bildungsgesetz von elementarer Einfachheit dargestellt werden können (The properties of these transcendental functions are easily obtained via calculation because they can be represented by infinite series whose formation law is of elemental simplicity)." Thus spoke FROBENIUS in 1893 in his induction speech before the Berlin Academy (*Gesammelte Abhand.* **2**, p. 575).

After the necessary convergence proof (subsection 1) our first business will be the construction (subsection 2) of *doubly periodic* meromorphic functions by means of the theta function. In subsection 4 from the Fourier development of $e^{-z^2 \pi \tau} \theta(i \tau z, \tau)$ we will get the classical *Transformation formula*

$$\theta(z, \frac{1}{\tau}) = \sqrt{\tau} e^{-z^2 \pi \tau} \theta(i \tau z, \tau),$$

in the course of which the famous equation

$$\int_{-\infty}^{\infty} e^{-x^2} dx = \sqrt{\pi}$$

for the error integral drops out as a by-product, though to be sure we have first to derive (in subsection 3 via Cauchy's integral formula) a certain "translation-invariance" of this integral.

Historical remarks on the theta function and the error integral will be found in subsections 5 and 6.

1. The convergence theorem. *The theta series*

$$\theta(z, \tau) = \sum_{-\infty}^{\infty} e^{-\nu^2 \pi \tau} e^{2\pi i \nu z}$$

is normally convergent in the region $\{(z, \tau) \in \mathbb{C}^2 : \Re \tau > 0\}$.

Proof. In the product region $\mathbb{C}^\times \times \mathbb{E}$ the Laurent series

$$I(w, q) := \sum_{-\infty}^{\infty} q^{\nu^2} w^\nu$$

is normally convergent (by the ratio criterion 4.1.4). Since $I(e^{2\pi i z}, e^{-\pi \tau}) = \theta(z, \tau)$ and since $|e^{-\pi \tau}| < 1$ for all τ with $\Re \tau > 0$, the claim follows. $\quad\square$

We designate by $\mathbb{T}$ the open right half-plane $\{\tau \in \mathbb{C} : \Re \tau > 0\}$. (The capital "t" should remind us that this set involves the variable "τ".) From the preceding theorem and general theory, $\theta(z, \tau)$ is continuous in $\mathbb{C} \times \mathbb{T}$ and for each fixed value of one of the variables it is a holomorphic function of the other. We think of τ principally as a parameter; then $\theta(z, \tau)$ is a non-constant entire function of z. We also have

$$\theta(z, \tau) = 1 + 2 \sum_{1}^{\infty} e^{-\nu^2 \pi \tau} \cos(2\pi \nu z), \qquad (z, \tau) \in \mathbb{C} \times \mathbb{T}.$$

Because of the normal convergence, the theta series may be differentiated term-wise arbitrarily often with respect to either variable. Calculating in this way one easily confirms that

$$\frac{\partial^2 \theta}{\partial z^2} = 4\pi \frac{\partial \theta}{\partial \tau} \qquad \text{in } \mathbb{C} \times \mathbb{T}.$$

The theta function therefore satisfies the partial differential equation $\frac{\partial^2 y}{\partial x^2} = 4\pi \frac{\partial y}{\partial \tau}$, which is central in the theory of heat conduction (τ being regarded as a time parameter).

2. Construction of doubly periodic functions. We first notice the trivial identity

$$(1) \qquad\qquad \theta(z+1,\tau) = \theta(z,\tau).$$

Furthermore

$$\theta(z+i\tau,\tau) = \sum_{-\infty}^{\infty} e^{-\nu^2\pi\tau - 2\pi\nu\tau} e^{2\pi i\nu z} = e^{\pi\tau - 2\pi iz} \sum_{-\infty}^{\infty} e^{-(\nu+1)^2\pi\tau} e^{2\pi i(\nu+1)z},$$

that is, upon rephrasing the latter sum in terms of ν instead of $\nu + 1$,

$$(2) \qquad\qquad \theta(z+i\tau,\tau) = e^{\pi\tau} e^{-2\pi iz} \theta(z,\tau).$$

The theta function consequently has period 1 in z and the "quasi-period $i\tau$ with periodicity factor $e^{\pi\tau} e^{-2\pi iz}$". This behavior enables us to construct doubly periodic functions.

Theorem. *For each $\tau \in \mathbb{T}$ the function*

$$E_\tau(z) := \frac{\theta(z+\frac{1}{2},\tau)}{\theta(z,\tau)}$$

is meromorphic and non-constant in $\mathbb{C}$. It satisfies

$$(3) \qquad E_\tau(z+1) = E_\tau(z) \qquad and \qquad E_\tau(z+i\tau) = -E_\tau(z).$$

Proof. Obviously $E_\tau(z)$ is meromorphic in $\mathbb{C}$ and on account of (1) and (2), equations (3) hold. If there were a $\sigma \in \mathbb{T}$ for which $E_\sigma(z)$ were constant, equal to $a \in \mathbb{C}$, say, then $\theta(z+\frac{1}{2},\sigma) = a\theta(z,\sigma)$ would hold for all $z \in \mathbb{C}$. Now

$$\theta(z+\tfrac{1}{2},\sigma) = \sum_{-\infty}^{\infty}(-1)^\nu e^{-\nu^2\pi\sigma} e^{2\pi i\nu z}$$

is the Fourier development of $\theta(z+\frac{1}{2},\sigma)$. The preceding identity and uniqueness of Fourier developments would then lead to the contradiction that $a = (-1)^\nu$ for all $\nu \in \mathbb{Z}$. $\square$

A meromorphic function in $\mathbb{C}$ is called *doubly periodic* or *elliptic* provided it has two real-linearly independent periods. On the basis of the foregoing theorem it is clear, since $\Re\tau \neq 0$, that

The functions $E_\tau(z)$ and $E_\tau(z)^2$, for each $\tau \in \mathbb{T}$, are non-constant doubly periodic functions having the two periods 1 and $2i\tau$, and 1 and $i\tau$, respectively.

3. The Fourier series of $e^{-z^2\pi\tau}\theta(i\tau z, \tau)$. For our discussion of this function we will need the following "translation-invariance" of the error integral:

$$(1) \qquad \int_{-\infty}^{\infty} e^{-b(x+a)^2}\, dx = \sqrt{b^{-1}} \int_{-\infty}^{\infty} e^{-x^2}\, dx \qquad \text{for all } a \in \mathbb{C},\, b \in \mathbb{R}^+.$$

(The meaning and criteria for the existence of improper integrals are briefly reviewed in 14.1.0 and in what follows we will use $\mathbb{R}^+$, as was frequently done in 5.4.4, to designate the set of positive real numbers.)

Proof. From the power series representation we obtain $e^{bx^2} > \frac{(bx^2)^2}{2!}$ for all $x \in \mathbb{R}$ and all $b > 0$. Therefore $x^2 e^{-bx^2} < 2b^{-2}x^{-2} \to 0$ as $|x| \to \infty$ and consequently the integral $\int_{-\infty}^{\infty} e^{-bx^2}\, dx$ exists. We consider the entire function $g(z) := e^{-bz^2}$. According to the Cauchy integral theorem we have for all finite $r, s > 0$ (see the above figure)

$$(*) \qquad \int_{-r}^{s} g\, dx + \int_{\gamma_1 + \gamma_3} g\, d\zeta = \int_{\gamma_2} g\, d\zeta.$$

Using $\gamma_1(t) := s + it$, $0 \le t \le q$ and the standard estimate 6.2.2, it follows that

$$\left| \int_{\gamma} g\, d\zeta \right| \le q \max_{0 \le t \le q} \left| e^{-b(s+it)^2} \right| = M e^{-bs^2},$$

where the constant $M := q e^{bq^2}$ is independent of s. Since $b > 0$, it follows that

$$\lim_{s \to \infty} \int_{\gamma_1} g\, d\zeta = 0, \quad \text{and analogously} \quad \lim_{r \to \infty} \int_{\gamma_3} g\, d\zeta = 0.$$

Because $\gamma_2(t) = t + a$, $t \in [-r - p, s - p]$, we have $\int_{\gamma_2} g\, d\zeta = \int_{-r-p}^{s-p} e^{-b(t+a)^2}\, dt$. From what has been noted so far, the left side of $(*)$ has a limit in $\mathbb{C}$ as $r, s \to \infty$. Therefore the existence of the improper integral of $e^{-b(t+a)^2}$ as well as the equality

$$\int_{-\infty}^{\infty} e^{-b(t+a)^2}\, dt = \int_{-\infty}^{\infty} e^{-bt^2}\, dt$$

follows from (∗). Substitution of $x := \sqrt{b}t$ on the right side of this gives us (1).   □

Anyone who has no anxiety about permuting the order of differentiation and improper integration, can derive (1) as follows: The function $h(a) := \int_{-\infty}^{\infty} e^{-b(x+a)^2} dx$ has throughout $\mathbb{C}$ the (complex) derivative

$$h'(a) = -\int_{-\infty}^{\infty} 2b(x+a)e^{-b(x+a)^2}\, dx = e^{-b(x+a)^2}\Big|_{-\infty}^{\infty} = 0.$$

Therefore h is constant: $h(a) = h(0)$. To adequately justify this argument requires however a bit of work.   □

Now denote by $\sqrt{\tau}$ the holomorphic square-root function in $\mathbb{T}$ which is uniquely determined according to 9.3.2 by the specification $\sqrt{1} = 1$. Then with the help of equation (1) we can show

Theorem. *For each* $\tau \in \mathbb{T}$, $e^{-z^2\pi\tau}\theta(i\tau z, z)$ *is an entire function of* z *having period 1 and Fourier expansion*

$$(2) \qquad e^{-z^2\pi\tau}\theta(i\tau z, z) = (\sqrt{\tau})^{-1}\sum_{-\infty}^{\infty} e^{-n^2\pi/\tau} e^{2\pi i n z}$$

Proof. The series definition of $\theta(z, \tau)$ gives

$$(3) \qquad e^{-z^2\pi\tau}\theta(i\tau z, z) = \sum_{-\infty}^{\infty} e^{-(z+\nu)^2\pi\tau}, \qquad (z, \tau) \in \mathbb{C} \times \mathbb{T}.$$

For each $\tau \in \mathbb{T}$ this is an entire function of z which visibly has period 1 (shift of summation index). Consequently according to theorem 3.4 there is valid for it the equation

$$e^{-z^2\pi\tau}\theta(i\tau z, z) = \sum_{n=-\infty}^{\infty} c_n(\tau)e^{2\pi i n z} \quad \text{for all} \quad (z, \tau) \in \mathbb{C} \times \mathbb{T},$$

where

$$c_n(\tau) := \int_0^1 e^{-t^2\pi\tau}\theta(i\tau t, \tau)e^{-2\pi i n t}\, dt.$$

Due to the identity $(t + \nu)^2 + 2itn/\tau = (t + \nu + in/\tau)^2 - 2\nu in/\tau + n^2/\tau^2$ and the normal convergence of the theta function, it follows from (3) that

$$c_n(\tau) = \sum_{\nu=-\infty}^{\infty}\int_0^1 e^{-(t+\nu)^2\pi\tau}e^{-2\pi i n t}\, dt = \sum_{\nu=-\infty}^{\infty}\int_0^1 e^{-\pi\tau(t+\nu+in/\tau)^2 - \pi n^2/\tau}\, dt.$$

Since $\int_0^1 e^{-\pi\tau(t+\nu+in/\tau)^2}\, dt = \int_\nu^{\nu+1} e^{-\pi\tau(t+in/\tau)^2}\, dt$, equation (1) converts the last expression for $c_n(\tau)$ into

$$c_n(\tau) = e^{-n^2\pi/\tau} \int_{-\infty}^{\infty} e^{-\pi\tau(t+in/\tau)^2}\, dt = \frac{1}{\sqrt{\pi\tau}} e^{-n^2\pi/\tau} \int_{-\infty}^{\infty} e^{-t^2}\, dt,$$

provided τ is not just in $\mathbb{T}$ but $\tau \in \mathbb{R}^+$. Assembling the pieces,

$$(*) \qquad e^{-z^2\pi\tau}\theta(i\tau z, \tau) = \frac{1}{\sqrt{\pi\tau}} \int_{-\infty}^{\infty} e^{-t^2}\, dt \sum_{-\infty}^{\infty} e^{-n^2\pi/\tau} e^{2\pi i n z}$$

for all $(z, \tau) \in \mathbb{C} \times \mathbb{R}^+$. Taking $z := 0$, $\tau := 1$, this yields

$$\theta(0, 1) = \left(\frac{1}{\sqrt{\pi}} \int_{-\infty}^{\infty} e^{-t^2}\, dt \right) \theta(0, 1).$$

Since $\theta(0, 1) = \sum_{-\infty}^{\infty} e^{-\nu^2\pi} > 0$, it follows that $\int_{-\infty}^{\infty} e^{-t^2}\, dt = \sqrt{\pi}$. $(*)$ then becomes the equation (2), for the points $(z, \tau) \in \mathbb{C} \times \mathbb{R}^+$. For fixed $z \in \mathbb{C}$, both sides of (2) are holomorphic functions of $\tau \in \mathbb{T}$, so we can now cite the identity theorem 8.1.1 to affirm the validity of (2) for all $(z, \tau) \in \mathbb{C} \times \mathbb{T}$. $\square$

4. Transformation formulas of the theta function. From the Fourier expansion 3.(2) we immediately deduce the

Transformation formula:

$$\theta\left(z, \frac{1}{\tau}\right) = \sqrt{\tau}\, e^{-z^2\pi\tau}\theta(i\tau z, \tau)\,, \qquad (z, \tau) \in \mathbb{C} \times \mathbb{T};$$

an equation which can also be written in the "real" form

$$\sum_{-\infty}^{\infty} e^{-n^2\pi\tau - 2n\pi\tau z} = \frac{e^{z^2\pi\tau}}{\sqrt{\tau}} \left(1 + 2\sum_{1}^{\infty} e^{-n^2\pi/\tau}\cos(2n\pi z) \right).$$

The function

$$\theta(\tau) := \theta(0, \tau) = \sum_{-\infty}^{\infty} e^{-\nu^2\pi\tau}\,, \qquad \tau \in \mathbb{T},$$

is the classical theta function (the "theta-null-value"); for it there is the

Transformation formula:

$$\theta\left(\frac{1}{\tau}\right) = \sqrt{\tau}\,\theta(\tau).$$

Considerable numerical power is concealed in this identity. For example, if we set $q := e^{-\pi\tau}$ and $r := e^{-\pi/\tau}$, then it says that

$$1 + 2q + 2q^4 + 2q^9 + \cdots = \sqrt{1/\tau}(1 + 2r + 2r^4 + 2r^9 + \cdots).$$

If q is only a little smaller than 1 (that is, if the positive number τ is very small), then the series on the left converges very slowly; but in this case r is very small and a few summands from the right side suffice to calculate the value of the sum with high precision.

The transformation formula for $\theta(z, \tau)$ is only the tip of an iceberg of interesting equations involving the theta function. In 1893 FROBENIUS (*loc. cit.*, pp. 575,6) said: "In der Theorie der Thetafunctionen ist es leicht, eine beliebig grosse Menge von Relationen aufzustellen, aber die Schwierigkeit beginnt da, wo es sich darum handelt, aus diesem Labyrinth von Formeln einen Ausweg zu finden. Die Beschäftigung mit jenen Formelmassen scheint auf die mathematische Phantasie eine verdorrende Wirkung auszuüben (In the theory of the theta functions it is easy to assemble arbitrarily large collections of relations; but the difficulty begins when it becomes a question of finding one's way out of this Labyrinth. Preoccupation with such masses of formulas seems to have a dessicating effect on the mathematical imagination)."

5. Historical remarks on the theta function. In the year 1823 POISSON considered the theta function $\theta(\tau)$ for positive real arguments τ and derived the transformation formula $\theta(\tau^{-1}) = \sqrt{\tau}\theta(\tau)$ (*Jour. de l'École Polytechn.* 12 Cahier 19, p. 420). RIEMANN used this formula in 1859 in his revolutionary, short paper "Über die Anzahl der Primzahlen unter einer gegebenen Grösse" (*Werke*, 145-153) when studying the function $\psi(\tau) := \sum_1^{\infty} e^{-\nu^2 \pi \tau} = \frac{1}{2}(\theta(\tau) - 1)$, in order to get "einen sehr bequemen Ausdruck der Function $\zeta(s)$ (a very convenient expression of the function $\zeta(s)$)" – p.147.

Carl Gustav Jacob JACOBI (born 1804 in Potsdam, 1826-1844 professor at Königsberg and founder of the Königsberg school, from 1844 academician at the Prussian Academy of Science in Berlin; died 1851 of smallpox; one of the most important mathematicians of the 19th century; the JACOBI biography by L. KOENIGSBERGER, Teubner-Verlag (1904), Leipzig is very informative) studied θ-series systematically from 1825 on and founded his theory of elliptic functions with them. This was in his *Fundamenta Nova Theoriae Functionum Ellipticarum* published in Königsberg in 1829 (*Gesammelte Werke* 1, 49-239), a work of extraordinary richness. It closes with an analytic proof of LAGRANGE's theorem that every natural number is a sum of four squares. Our transformation formula is but a special case of more general transformation equations in JACOBI (cf., say, p. 235 *loc. cit.*).

JACOBI secured the properties of the θ-functions purely algebraically. Since the time of LIOUVILLE's lectures, methods from Cauchy's function theory have predominated, as is, e.g., already the case in the book [BB].

We speak of the theta function because JACOBI himself happened to designate the function $\theta(z, \tau)$ that way (using Θ instead of θ). In his memorial address on JACOBI, DIRICHLET (*Werke* **2**, p.239) says: "... die Mathematiker würden nur eine Pflicht der Dankbarkeit erfüllen, wenn sie sich vereinigten, [dieser Funktion] JACOBI's Namen beizulegen, um das Andenken des Mannes zu ehren, zu dessen schönsten Entdeckungen es gehört, die innere Natur und die hohe Bedeutung dieser Transcendente zuerst erkannt zu haben (... mathematicians would simply be fulfilling an obligation of gratitude if they were to unite in bestowing JACOBI's name on this function, thus honoring the memory of the man among whose most beautiful discoveries was to have first discerned the inner nature and enormous significance of this transcendent)."

As shown in subsection 1, the function $\theta(z, \tau)$ satisfies the heat equation. So it is not surprising that as early as 1822 – seven years before the appearance of Jacobi's *Fundamenta Nova* – theta functions show up in Fourier's *La Théorie Analytique de la Chaleur* (cf., e.g., *Œuvres* **1**, p. 295 and 298); however FOURIER did not perceive the great mathematical significance of these functions[2] To him the value of mathematics generally lay in its applications, but JACOBI did not recognize such criteria. His point of view is wonderfully expressed in his letter of July 2, 1830 to LEGENDRE (*Gesammelte Werke* **1**, p.454/5): "Il est vrai que M. Fourier avait l'opinion que le but principal des mathématiques était l'utilité publique et l'explication des phénomènes naturels; mais un philosophe comme lui aurait dû savoir que le but unique de la science, c'est l'honneur de l'esprit humain ... (It is true that M. Fourier had the opinion that the principal aim of mathematics was public usefulness and the explanation of natural phenomena; but a philosopher like him should have known that the unique aim of science is the honor of the human mind ...)."

6. Concerning the error integral. In the proof of theorem 3 the equality

$$(1) \qquad \int_{-\infty}^{\infty} e^{-x^2} dx = \sqrt{\pi}$$

was an incidental by-product. This integral is frequently called the Gauss error integral. It already occurs implicitly in the famous work on the calculus of probability by DE MOIVRE entitled *The Doctrine of Chances* (first edition 1718); the reprint of the third edition 1967 by Chelsea Publ. Co. contains a biography of DE MOIVRE on pp. 243-259. (Abraham DE MOIVRE, 1667-1754, Huguenot; emigrated to London after the repeal of the Edict of Nantes in 1685; 1697 member of the Royal Society and later

---

[2]In 1857 in his first lectures on the theory of elliptic functions WEIERSTRASS said of the heat equation (cf. L. KOENIGSBERGER, *Jahresber. DMV* **25**(1917), 394-424, esp. p. 400): "..., die schon Fourier für die Temperatur eines Drahtes aufgestellt, in der er jedoch diese wichtige Transzendente nicht erkannt hat (..., which Fourier had already proposed for the temperature of a wire, but in which he did not perceive these important transcendents)."

of the academies in Paris and Berlin; discovered the "Stirling formula" $n! \approx \sqrt{2\pi n}(n/e)^n$ before STIRLING; served on a committee of the Royal Society adjudicating the priority dispute between NEWTON and LEIBNIZ over the discovery of the infinitesimal calculus; in his old age NEWTON is said to have replied when someone asked him a mathematical question: "Go to Mr. DE MOIVRE; he knows these things better than I do.") GAUSS never laid any claim to the integral and in fact in 1809 in his "Theoria Motus Corporum Coelestium" (*Werke* **7**, p.244; English translation by Charles H. Davis, Dover Publ. Co., Inc. (1963), New York) he indicated LAPLACE as the inventor; later he corrected himself and said (*Werke* **7**, p. 302) that (1) in the form

$$\int_0^1 \sqrt{\ln(1/x)}dx = \frac{1}{2}\sqrt{\pi}$$

was already in a 1771 work of EULER ("Evolutio formulae integralis $\int x^{f-1}$ $dx(\ell x)^{\frac{m}{n}}$ integratione a valore $x = 0$ ad $x = 1$ extensa," *Opera Omnia* (1) **17**, 316-357). In Euler (p.333) even the more general formula

$$(2) \quad \int_0^1 (\ln(1/x))^{\frac{2n-1}{2}} dx = \frac{1}{2} \cdot \frac{3}{2} \cdot \frac{5}{2} \cdots \frac{2n-1}{2}\sqrt{\pi} , \qquad n = 1, 2, \ldots$$

is to be found; via the substitution $x := e^{-t^2}$, that is, $t = \sqrt{\ln(1/x)}$, this goes over directly into the formula

$$(3) \quad \int_{-\infty}^{\infty} x^{2n}e^{-x^2}dx = \frac{(2n)!}{n!4^n}\sqrt{\pi} , \qquad n \in \mathbb{N},$$

which may also be found on p. 269 of LAPLACE's "Mémoire sur les approximations des formules qui sont fonctions de très grands nombres" (*Œuvres* **10**, 209-291). However, the equations (3) follow immediately by induction from the more basic (1), using integration by parts ($f'(x) := x^{2n}$, $g(x) := e^{-x^2}$) and the fact that $\lim_{|x|\to\infty} x^{2n+1}e^{-x^2} = 0$.

EULER knew the special cases $n = 1, 2$ of formula (2) as early as 1729 (*Opera Omnia* (1) **14**, 1-24, esp. pp. 10,11) but the equality (1) seems not to have come up explicitly with EULER.

In the theory of the gamma function $\Gamma(z) = \int_0^\infty t^{z-1}e^{-t}dt$ (which will be taken up in the second volume) equation (1) is just a trivial case of EULER's functional equation $\Gamma(z)\Gamma(1 - z) = \frac{\pi}{\sin \pi z}$, since $\Gamma(\frac{1}{2}) = \int_0^\infty \frac{e^{-t}}{\sqrt{t}}dt = 2\int_0^\infty e^{-x^2}dx$ ($x := \sqrt{t}$). EULER knew the equality $\Gamma(\frac{1}{2}) = \sqrt{\pi}$, but nowhere does he mention the simple proof that via substitution as above one thereby gets the equality $\Gamma(\frac{1}{2}) = 2\int_0^\infty e^{-x^2}dx$ (cf. *Opera Omnia* (1) **19**, p. LXI). In chapter 14 using the residue calculus we will give further proofs of the formula $\int_{-\infty}^\infty e^{-x^2}dx = \sqrt{\pi}$.

One gets the value I of the error integral most quickly by reduction to a certain double integral which arises from iterated integrals (EULER had evaluated other integrals this way; cf. *Opera Omnia* (1) **18**, pp. 70,71). Using the polar coordinates $x = r\cos\varphi$, $y = r\sin\varphi$, the double integral is evaluated thus:

$$I^2 = \int_{-\infty}^{\infty} e^{-x^2}dx \int_{-\infty}^{\infty} e^{-y^2}dy = \iint_{\mathbb{R}^2} e^{-x^2-y^2}dxdy$$

$$= \int_0^{2\pi}\int_0^{\infty} e^{-r^2}rdrd\varphi = \pi \int_0^{\infty} e^{-t}dt = \pi.$$

This proof is due to POISSON and was used in 1891 by E. PICARD in his *Traité d'analyse*, vol. I, pp. 102-104.

There is yet another quite elementary way to determine the value of I. Following J. VAN YZEREN, "Moivre's and Fresnel's integrals by simple integration," *Amer. Math. Monthly* **86**(1979), 691-693, we set

$$e(t) := \int_{-\infty}^{\infty} \frac{e^{-t(1+x^2)}}{1+x^2}dx, \quad t \geq 0.$$

Then

$$e(t) = e^{-t}\int \frac{e^{-tx^2}}{1+x^2}dx \leq e^{-t}\int \frac{1}{1+x^2}dx = \pi e^{-t},$$

whence $\lim_{t\to\infty} e(t) = 0$. Differentiating under the integral sign gives

$$(*) \quad e'(t) = -\int_{-\infty}^{\infty} e^{-t(1+x^2)}dx = -e^{-t}\int_{-\infty}^{\infty} e^{-tx^2}dx = -I(\sqrt{t})^{-1}e^{-t};$$

which together with $\lim_{t\to\infty} e(t) = 0$ yields

$$e(t) = -\int_t^{\infty} e'(u)du = I\int_t^{\infty} \frac{e^{-u}}{\sqrt{u}}du = 2I\int_{\sqrt{t}}^{\infty} e^{-x^2}dx.$$

Set $t = 0$ to obtain $I^2 = \pi = e(0)$. The perilous appearing interchange of differentiation and integration which is involved in $(*)$ is readily and rigorously justified: For $p > 0$, $h \in \mathbb{R}\setminus 0$ we have $ph < e^{ph} - 1 < phe^{ph}$, from which follows

$$he^{-p(t+h)} < \frac{e^{-pt} - e^{-p(t+h)}}{p} < he^{-pt} \quad \text{for all } t \in \mathbb{R}.$$

Suppose now that $t > 0$, $t+h > 0$ and $p := 1+x^2$. Integration with respect to x from $-\infty$ to ∞ gives $h(\sqrt{t+h})^{-1}e^{-(t+h)}I \leq e(t) - e(t+h) \leq h(\sqrt{t})^{-1}e^{-t}I$, and so upon dividing by h and letting $h \to 0$, we see that $e'(t) = -I(\sqrt{t})^{-1}e^{-t}$.

In closing let us note that the translation-invariance formula 3.(1) holds also in the more general form

$$\int_{-\infty}^{\infty} e^{-u(x+a)^2} dx = \int_{-\infty}^{\infty} e^{-ux^2} dx \qquad \text{for all } a \in \mathbb{C} \text{ , } u \in \mathbb{T}.$$

The proof from subsection 3 for $u = b > 0$ carries over almost verbatim to the present context where only $\Re u > 0$. From this and either theorem 7.1.6* or the Exercise in §1 of Chapter 7 follows:

For all $(u, v, w) \in \mathbb{T} \times \mathbb{C} \times \mathbb{C}$

$$\int_{-\infty}^{\infty} e^{-(ux^2 + 2vx + w)} dx = \frac{\sqrt{\pi}}{\sqrt{u}} e^{(v^2/u) - w},$$

where $\sqrt{u} = \sqrt{|u|} e^{i\varphi}$ with $|\varphi| < \frac{1}{4}\pi$.

In turn we immediately get from this

(#) $$\int_0^{\infty} e^{-ux^2} \cos(vx) dx = \frac{\sqrt{\pi}}{2\sqrt{u}} e^{-v^2/4u}.$$

For real values of the parameters u and v this last formula is to be found in the book *Théorie analytique des probabilités* of LAPLACE, on page 96 of the 3rd edition, 1820 (1st ed. was 1812, Paris). By decomposing into real and imaginary parts the reader can derive two real integral formulas from (#).

Chapter 13

The Residue Calculus

As early as the 18th century many real integrals were evaluated by passing up from the real domain to the complex (passage du réel á l'imaginaire). Especially EULER (*Calcul intégral*), LEGENDRE (*Exercices de Calcul Intégral*) and LAPLACE made use of this method at a time when the theory of the complex numbers had not yet been rigorously grounded and "all convergence questions still lay under a thick fog." The attempt to put this procedure on a secure foundation lead CAUCHY to the residue calculus.

In this chapter we will develop the theoretical foundations of this calculus and in the next chapter use it to evaluate classical real integrals. In order to be able to formulate things with ease and in the requisite degree of generality, we work with winding numbers and nullhomologous paths. The Residue Theorem 1.4 is a natural generalization of the Cauchy integral formula. The classical literature on the residue calculus is quite extensive. Especially deserving of mention is the 1904 booklet [Lin] by the Finnish mathematician Ernst LINDELÖF (1870-1946); it is still quite readable today and contains many historical remarks. The monograph by MITRINOVIĆ and KEČKIĆ is a modern descendant of it which is somewhat broader in scope and contains a short biography of Cauchy.

§1   The residue theorem

At the center of this section stands the concept of the residue, which is discussed in considerable detail and illustrated with examples (subsections 2 and 3). The residue theorem itself is the natural generalization of Cauchy's integral theorem to holomorphic functions which have isolated singularities. Its proof is reduced to the integral theorem by means of Laurent's expansion theorem.

In order to be able to formulate the residue theorem in sufficient generality we call on the index function $\text{ind}_\gamma(z)$ which was introduced in 9.5.1. The special situation in which this number equals 1 will be considered first, in subsection 1.

1. Simply closed paths. A closed path γ is called *simply closed* if

$$\text{Int}\,\gamma \neq \emptyset \quad \text{and} \quad \text{ind}_\gamma(z) = 1 \quad \text{for all } z \in \text{Int}\,\gamma.$$

Here we are using the notation introduced in 9.5.1:

$$\text{ind}_\gamma(z) = \frac{1}{2\pi i} \int_\gamma \frac{d\zeta}{\zeta - z} \in \mathbb{Z}, \quad z \in \mathbb{C} \setminus \gamma$$

and

$$\text{Int}\,\gamma = \{z \in \mathbb{C} \setminus \gamma : \text{ind}_\gamma(z) \neq 0\}.$$

Simply closed paths are particularly amenable. According to theorem 6.2.4 every circle is simply closed. In what follows we offer further examples, sufficient for the applications of the residue calculus to be made in the next chapter.

0) The *boundaries* of *circular segments* and of *triangles* and *rectangles* are *simply closed*.

Proof. Let A be an open circular segment; this expression is used *faute de mieux* to designate a chord-arc region of the kind shown shaded in the figure below, viz., the intersection of an open disc and an open half-plane. We have $\partial A = \partial B - \partial A'$, with $\partial B = \gamma + \gamma''$ a circle and $\partial A' = \gamma'' - \gamma'$ the boundary of the circular segment A' complementary to A (cf. the figure).

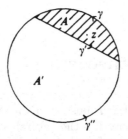

It follows that $\text{ind}_{\partial A}(z) = 1 - \text{ind}_{\partial A'}(z)$ for $z \in A$. Points z in A lie in the outside of the curve $\partial A'$, as we see by joining z to ∞ by a ray which is disjoint from A'. Constancy of the index on this line and its vanishing at ∞ then give $\text{ind}_{\partial A'}(z) = 0$. Thus $\text{ind}_{\partial A}(z) = 1$ and ∂A is simply closed.

This proof can be repeated if more of A is excised by chords of the containing circle. Since every triangle and every rectangle arises from such excisions on a containing circle, each has a simply closed boundary. (Cf.

the assertion that $\mathrm{ind}_{\partial R}(z) = 1$ for $z \in R$, about the rectangular boundary ∂R, with Exercise 1 in §1 of Chapter 6.)

1) The *boundary* of every open *circular sector* A (left-hand figure below) is *simply closed*.

Proof. We have $\partial A = \gamma + \gamma_1 + \gamma_2$ and $\partial A = \partial B - \partial A'$, with A' denoting the sector complementary to A in the whole disc B. It follows that $\mathrm{ind}_{\partial A}(z) = 1 - \mathrm{ind}_{\partial A'}(z)$ for $z \in A$. As before, $\mathrm{ind}_{\partial A'}$ is constant on a radial path from z to ∞ which lies outside A', so $\mathrm{ind}_{\partial A'}(z) = 0$.

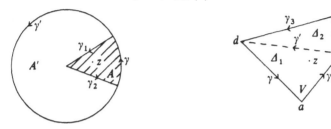

2) The *boundary* of every open *convex n-gon* V is *simply closed*.

Proof. Convex n-gons are decomposable into triangles. For example, in the right-hand figure above $\partial V = \partial\Delta_1 + \partial\Delta_2$. For $z \in \Delta_1$, $\mathrm{ind}_{\partial\Delta_1}(z) = 1$ by 0), and $\mathrm{ind}_{\partial\Delta_2}(z) = 0$ because z lies outside Δ_2. Thus $\mathrm{ind}_{\partial V}(z) = 1$. Since V is a region and the index is locally constant, it follows that $\mathrm{ind}_{\partial V}(z) = 1$ for all $z \in V$.

3) The *boundary* of every *circularly indented quadrilateral* $\widehat{V}$ of the kind shown in the left-hand figure below, in which γ' is a circular arc, is *simply closed*.

Proof. The quadrilateral $V = abcd$ is convex and $\partial\widehat{V} = \partial V - \partial A$, where A is the circular segment being excised. Since each $z \in \widehat{V}$ lies outside A the claim $\mathrm{ind}_{\partial\widehat{V}}(z) = 1$ follows from the preceding example.

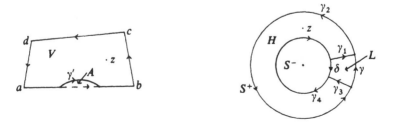

4) The *boundary* of every open *circular horseshoe* H is *simply closed*.

Proof. We have (cf. the right-hand figure above)

$$\partial H = \gamma_1 + \gamma_2 + \gamma_3 + \gamma_4 = \gamma_1 + S^+ - \gamma + \gamma_3 - S^- - \delta,$$

where S^-, S^+ are the inner and outer circles, respectively. Now $\mathrm{ind}_{S^+}(z) = 1$ for $z \in H$. And every $z \in H$ lies outside of S^- and outside of the circular quadrangle L bounded by $\gamma - \gamma_1 + \delta - \gamma_3$. It follows that $\mathrm{ind}_{\partial H}(z) = 1$ for $z \in H$.

The preceding examples can be multiplied at will. They are all special instances of the following theorem, the usual proof of which rests on the integral theorem of STOKES:

A closed path $\gamma : [a, b] \to \mathbb{C}$ is simply closed under the following circumstances:

1) *γ is homeomorphic to the circle S_1 (that is, $\gamma : [a, b) \to \mathbb{C}$ is injective).*

2) *Int $\gamma \neq \emptyset$ and lies "to the left of γ" (that is, if the tangent line $\gamma(u) + \gamma'(u)t$, $t \in \mathbb{R}$, to γ at any point of γ where $\gamma'(u) \neq 0$, is "rotated by $\frac{1}{2}\pi$," the resulting line $\gamma(u) + i\gamma'(u)t$ has points lying in Int γ for which t is positive and arbitrarily small.)*

According to 9.5.2–3 for every path γ which is nullhomologous in a domain D the general Cauchy integral formula

$$\mathrm{ind}_\gamma(z)f(z) = \frac{1}{2\pi i} \int_\gamma \frac{f(\zeta)}{\zeta - z}d\zeta \qquad \text{for all } f \in \mathcal{O}(D) , z \in D \setminus \gamma$$

is valid. In particular then, for a *simply closed* path γ which is *nullhomologous in D* we always have

$$f(z) = \frac{1}{2\pi i} \int_\gamma \frac{f(\zeta)}{\zeta - z}d\zeta \qquad \text{for all } f \in \mathcal{O}(D) , z \in \text{Int }\gamma.$$

We want to explicitly emphasize

Cauchy's integral formula for convex n-gons. *Let V be an open, convex n-gon with $\overline{V} \subset D$ and $n \geq 3$. Then every function $f \in \mathcal{O}(D)$ satisfies the equation*

$$\frac{1}{2\pi i} \int_{\partial V} \frac{f(\zeta)}{\zeta - z}d\zeta = \left\{ \begin{array}{ll} f(z), & \text{for} \quad z \in V \\ 0, & \text{for} \quad z \in D \setminus \overline{V}. \end{array} \right.$$

For the simply closed path ∂V is nullhomologous in D: There is an open n-gon V' (obtained by dilating V by a similarity transformation) with $\overline{V} \subset V' \subset D$, and so by Cauchy's integral theorem for star-shaped regions we even have

$$\int_{\partial V} g d\zeta = 0 \qquad \text{for all } g \in \mathcal{O}(V').$$

2. The residue. If f is holomorphic in $D \setminus c$ and $\sum_{-\infty}^{\infty} a_\nu(z-c)^\nu$ is its Laurent development in a punctured disc $B^\times$ centered at c, then according to 12.1.3(2)

$$a_{-1} = \frac{1}{2\pi i} \int_S f(\zeta)d\zeta$$

for every circle S centered at c and lying in B. Thus of all the Laurent coefficients only a_{-1} remains after integration of f. For this reason it is called the *residue of f at c* and we write

$$\operatorname{res}_c f := a_{-1}.$$

The residue of f is defined at every isolated singularity of f.

Theorem. *The residue at c of $f \in \mathcal{O}(D \setminus c)$ is the unique complex number a such that the function $f(z) - a(z-c)^{-1}$ has a primitive in a punctured neighborhood of c.*

Proof. If $\sum_{-\infty}^{\infty} a_\nu(z-c)^\nu$ is the Laurent series of f in $B^\times := B_r(c) \setminus c$, then the function $F(z) := \sum_{\nu \neq -1} \frac{1}{\nu+1} a_\nu(z-c)^{\nu+1} \in \mathcal{O}(B^\times)$ has derivative $F' = f - a_{-1}(z-c)^{-1}$. Therefore a function H is a primitive of $f - a(z-c)^{-1}$ in some punctured neighborhood of c if and only if $(F-H)' = (a - a_{-1}) \cdot (z-c)^{-1}$ there. Since according to 6.3.2 $(z-c)^{-1}$ has no primitive around c, an H with $H = F+$ const. exists if and only if $a = a_{-1}$. $\square$

If f is holomorphic at c, then $\operatorname{res}_c f = 0$. However, this last equation can hold for other reasons too. For example,

$$\operatorname{res}_c \left[\frac{1}{(z-c)^n} \right] = 0 \qquad \text{for } n \geq 2.$$

For every $f \in \mathcal{O}(D \setminus c)$ we always have $\operatorname{res}_c f' = 0$, because the Laurent development of f' around c has the form $\sum_0^\infty \nu a_\nu(z-c)^{\nu-1}$ and consequently contains no term $a(z-c)^{-1}$. $\square$

We discuss next some rules for calculating with residues. Immediately clear is the $\mathbb{C}$-*linearity*

$$\operatorname{res}_c(af + bg) = a \operatorname{res}_c f + b \operatorname{res}_c g \qquad \text{for } f, g \in \mathcal{O}(D \setminus c) \,, \, a, b \in \mathbb{C}.$$

Decisive for many applications is the fact that residues, though defined as integrals, can often be calculated algebraically. The matter is especially simple at a first order pole:

Rule 1). *If c is a simple pole of f, then*

$$\mathrm{res}_c f = \lim_{z \to c} (z - c) f(z).$$

Proof. This is trivial, because $f = a_{-1}(z - c)^{-1} + h$, with h holomorphic in a neighborhood of c. $\square$

From this rule we obtain a result which is extremely useful in practice:

Lemma. *Suppose g and h are holomorphic in a neighborhood of c, $g(c) \neq 0$, $h(c) = 0$ and $h'(c) \neq 0$. Then $f := g/h$ has a simple pole at c and*

$$\mathrm{res}_c f = \frac{g(c)}{h'(c)}.$$

Proof. Since h has the Taylor development $h(z) = h'(c)(z - c) + \cdots$ about c, it follows that

$$\lim_{z \to c}(z - c)f(z) = \lim_{z \to c}(z - c)\frac{g(z)}{h(z)} = \frac{g(c)}{h'(c)} \neq 0.$$

Consequently, c is a pole of f of the first order with residue $g(c)/h'(c)$. $\square$

There is no comparably handy criterion for the determination of residues at poles of higher order:

Rule 2). *If $f \in \mathcal{M}(D)$ has a pole at c of order at most m and if g is the holomorphic continuation of $(z - c)^m f(z)$ over c, then*

$$\mathrm{res}_c f = \frac{1}{(m - 1)!} g^{(m-1)}(c).$$

Proof. Near c we have $f = \frac{b_m}{(z - c)^m} + \cdots + \frac{b_1}{z - c} + h$, with h holomorphic at c as well. Then $g(z) = b_m + b_{m-1}(z - c) + \cdots + b_1(z - c)^{m-1} + \cdots$ is the Taylor series of g at c, so that $\mathrm{res}_c f = b_1 = \frac{1}{(m-1)!} g^{(m-1)}(c)$ follows. $\square$

There is no simple procedure for calculating residues at essential singularities.

3. Examples. 1) For the function $f(z) = \frac{z^2}{1 + z^4}$, the point $c := \exp(\frac{\pi i}{4}) = \frac{1}{\sqrt{2}}(1 + i)$ is a simple pole. Since $c^{-1} = \bar{c}$, it follows from lemma 1 that

$$\mathrm{res}_c f = \frac{c^2}{4c^3} = \frac{1}{4}\bar{c} = \frac{1}{4\sqrt{2}}(1 - i).$$

The points ic, $-c$ and $-ic$ are likewise first order poles of f; for them one finds

$$\operatorname{res}_{ic} f = -\frac{i}{4}\bar{c}\,, \qquad \operatorname{res}_{-c} f = -\frac{1}{4}\bar{c}\,, \qquad \operatorname{res}_{-ic} f = \frac{i}{4}\bar{c}.$$

2) Let $n \in \mathbb{N} \setminus \{0\}$, $g \in \mathcal{O}(\mathbb{C})$ be such that $g(c) \neq 0$ for any $c \in \mathbb{C}$ satisfying $c^n = -1$. Then the function $f(z) := \frac{g(z)}{1+z^n}$ has a simple pole at each such c, with

$$\operatorname{res}_c f = \frac{g(c)}{nc^{n-1}} = -\frac{c}{n}g(c).$$

3) Suppose $p \in \mathbb{R}$ and $p > 1$. Then the rational function $\tilde{R}(z) = \frac{4z}{(z^2 + 2pz + 1)^2}$ has poles of order two at each of the points $c := -p + \sqrt{p^2 - 1} \in \mathbb{E}$ and $d := -p - \sqrt{p^2 - 1} \in \mathbb{C} \setminus \mathbb{E}$. Since $z^2 + 2pz + 1 = (z - c)(z - d)$, $g(z) := 4z(z - d)^{-2}$ is the holomorphic continuation of $(z - c)^2 \tilde{R}(z)$ over c. We have $g'(c) = -4(c+d)(c-d)^{-3}$, and so from Rule 2,2)

$$\operatorname{res}_c \tilde{R} = \frac{p}{(\sqrt{p^2 - 1})^3}.$$

4) *Let g, h be holomorphic near c and c be an a-point of g of multiplicity $\nu(g, c)$. Then*

$$\operatorname{res}_c \left(h(z) \frac{g'(z)}{g(z) - a} \right) = h(c)\nu(g, c).$$

Proof. Setting $n := \nu(g, c)$, we have $g(z) = a + (z - c)^n \tilde{g}(z)$ near c, with $\tilde{g}$ holomorphic near c and $\tilde{g}'(c) \neq 0$ (cf. 8.1.4). It follows that near c

$$\frac{g'(z)}{g(z) - a} = \frac{n(z - c)^{n-1}\tilde{g}(z) + (z - c)^n \tilde{g}'(z)}{(z - c)^n \tilde{g}(z)}$$

$$= \frac{n}{z - c} + \text{holomorphic function} .$$

From this the claim follows. $\qquad\square$

We emphasize the special case:

If g has a zero at c of order $o_c(g) < \infty$, then

$$\operatorname{res}_c \left(\frac{g'}{g} \right) = o_c(g).$$

A proof analogous to that of 4) establishes:

5) *If g has a pole at c and h is holomorphic at c, then*

$$\operatorname{res}_c\left(h(z)\frac{g'(z)}{g(z)-a}\right) = h(c)o_c(g) \qquad \text{for all } a \in \mathbb{C}.$$

Application of 4) yields a

Transformation rule for residues. *Let $g : \tilde{D} \to D$, $\tau \mapsto z := g(\tau)$ be holomorphic, with $g(\tilde{c}) = c$, $g'(\tilde{c}) \neq 0$. Then*

$$\operatorname{res}_c f = \operatorname{res}_{\tilde{c}}((f \circ g)g') \qquad \text{for all } f \in \mathcal{M}(D).$$

Proof. Set $a := \operatorname{res}_c f$. According to theorem 1 there is a punctured neighborhood $V^\times$ of c and an $F \in \mathcal{O}(V^\times)$ such that $F'(z) = f(z) - \frac{a}{z - c}$. Then $g^{-1}(V^\times)$ is a punctured neighborhood of $\tilde{c}$ and for τ in that neighborhood

$$(F \circ g)'(\tau) = F'(g(\tau))g'(\tau) = f(g(\tau))g'(\tau) - a\frac{g'(\tau)}{g(\tau) - c}.$$

Since derivatives always have residue 0 (subsection 1), it follows that

$$\operatorname{res}_{\tilde{c}}((f \circ g)g') = a \operatorname{res}_{\tilde{c}}\left(\frac{g'(\tau)}{g(\tau) - c}\right).$$

Since, due to $g'(\tilde{c}) \neq 0$, g has at $\tilde{c}$ a c-point of multiplicity 1, the present claim follows from 4). □

The transformation rule says that the residue concept becomes invariant if it is applied to differential forms instead of functions.

4. The residue theorem. *Let γ be a nullhomologous path in the domain D, A a finite subset of $D \setminus \gamma$. Then*

(1)
$$\boxed{\frac{1}{2\pi i} \int_\gamma h d\zeta = \sum_{c \in \operatorname{Int}\gamma} \operatorname{ind}_\gamma(c) \cdot \operatorname{res}_c h}$$

for every holomorphic function h in $D \setminus A$.

Remark. Since $\operatorname{res}_z h = 0$ in case $z \notin A$, the sum on the right side of (1) is really only extended over $c \in A \cap \operatorname{Int}\gamma$, and so there is no convergence question.

Proof. Let $A = \{c_1, \ldots, c_n\}$. We consider the principal part $h_\nu = b_\nu(z - c_\nu)^{-1} + \tilde{h}_\nu$ of the Laurent development of h at c_ν, where $\tilde{h}_\nu$ contains all the summands featuring powers $(z - c_\nu)^k$ with $k \leq -2$. According to 12.2.3, h_ν is holomorphic in $\mathbb{C} \setminus c_\nu$. Since $\tilde{h}_\nu$ has a primitive in $\mathbb{C} \setminus c_\nu$ and $c_\nu \notin \gamma$, it follows from the very definition of the index that

$$(*) \qquad \int_\gamma h_\nu d\zeta = b_\nu \int_\gamma \frac{d\zeta}{\zeta - c_\nu} = 2\pi i b_\nu \mathrm{ind}_\gamma(c_\nu) , \qquad 1 \leq \nu \leq n.$$

Since $h - (h_1 + \cdots + h_n)$ is holomorphic in D and γ is nullhomologous in D, $\int_\gamma (h - h_1 - \cdots - h_n)d\zeta = 0$ by the integral theorem. Because of $(*)$ and the fact that $b_\nu = \mathrm{res}_{c_\nu} h$, it follows that

$$\frac{1}{2\pi i} \int_\gamma h d\zeta = \frac{1}{2\pi i} \sum_1^n \int_\gamma h_\nu d\zeta = \sum_1^n \mathrm{ind}_\gamma(c_\nu) \cdot \mathrm{res}_{c_\nu} h.$$

Since $\mathrm{ind}_\gamma(c_\nu) = 0$ whenever $c_\nu \in \mathrm{Ext}\, \gamma$, this is just equation (1).   □

On the right side of equation (1) we find residues, which depend on the function $h \in \mathcal{O}(D \backslash A)$ and are analytically determinable, as well as winding numbers, which depend on the path and are generally not susceptible to direct calculation. Thus the residue theorem becomes especially elegant for *simply closed paths*:

If $\gamma \subset D$ is simply closed and nullhomologous in D, then under the hypotheses of the residue theorem we have the

Residue formula: $\dfrac{1}{2\pi i} \displaystyle\int_\gamma h d\zeta = \sum_{c \in \mathrm{Int}\, \gamma} \mathrm{res}_c h.$ □

In later applications D will always be a star-shaped region; the hypothesis that γ be nullhomologous in D is then automatically fulfilled by every $\gamma \subset D$.

The general Cauchy integral formula is a special case of the residue theorem. For if f is holomorphic in D and z is a point in D, then $\zeta \mapsto f(\zeta)(\zeta - z)^{-1}$ is holomorphic in $D \setminus z$ and has the residue $f(z)$ at z. Therefore for every path γ which is nullhomologous in D the equation

$$\frac{1}{2\pi i} \int_\gamma \frac{f(\zeta)}{\zeta - z} d\zeta = \mathrm{ind}_\gamma(z) f(z) \qquad \text{for } z \in D \setminus \gamma$$

holds.

5. Historical remarks on the residue theorem. Cauchy's first investigations in function theory were the beginnings of the residue calculus as well. The memoir $[C_1]$ from the year 1814, which we have repeatedly cited, had as its primary goal the development of general methods for calculating definite integrals by passing from the reals to the complexes. The "singular integrals" introduced by CAUCHY at that time ($[C_1]$, p.394) are in the final analysis the first residue integrals. "Der Sache nach kommt das Residuum bereits in der Jacobi'schen Doctor-Dissertation [aus dem Jahre

1825] vor (In essence the residue already occurs in Jacobi's doctoral dissertation [from the year 1825])" – quoted from p.170 of [Kr]. The word "residue" was first used by CAUCHY in 1826 (*Œuvres* (2) **6**, p. 23), but to be sure the definition there is quite complicated. For details of these matters we refer to LINDELÖF's book [Lin], especially pp. 12 ff.

As applications of his theory CAUCHY derived anew virtually all the then known integral formulas; for example, "la belle formule d'Euler, relative à l'intégrale"

$$\int_0^\infty \frac{x^{a-1}dx}{1+x^b} \qquad ([C_1], \text{ p.432});$$

and in addition he discovered many new integral formulas (also see the next chapter).

Nevertheless POISSON was not particularly impressed with the treatise [C_1], for he wrote (cf. CAUCHY's *Œuvres* (2) **2**, 194-198): " $\cdots$ je n'ai remarqué aucune intégrale qui ne fût pas déjà connue $\cdots$ (I have not noticed any integral which was not already known $\cdots$)".

Naturally nullhomologous paths aren't to be found in CAUCHY's writing. Moreover, because he was unaware of essential singularities, he worked exclusively with functions having at worst polar singularities.

Exercises

Exercise 1. Determine $\text{res}_c f$ for all isolated singularities $c \in \mathbb{C}$ of the following:

a) $f(z) = \dfrac{z^2 + z + 5}{z(z^2 + 1)^2}$

b) $f(z) = (z^4 + a^4)^{-2}$

c) $f(z) = \dfrac{(z-1)^2}{(e^z - 1)^3}$

d) $f(z) = \dfrac{z^2}{(z-2)^2(\cos z - 1)^3}$

e) $f(z) = \cos\left(\dfrac{1-z}{z}\right)$

f) $f(z) = \tan^3 z$

g) $f(z) = \cos\left(\dfrac{z}{1-z}\right)$

h) $f(z) = \sin(1 + z^{-1})\cos(1 + z^{-2})$.

Exercise 2. Show that $\mathrm{res}_0 \exp(z + z^{-1}) = \sum_{n=0}^{\infty} \frac{1}{n!(n+1)!}$.

Exercise 3. Let f, g be holomorphic around c with $g(c) = g'(c) = 0$ and $g''(c) \neq 0$. Show that

$$\mathrm{res}_c(f/g) = [6f'(c)g''(c) - 2f(c)g'''(c)]/(3g''(c)^2).$$

Exercise 4. With the help of the residue theorem show that

a) $\displaystyle\int_{\partial B_2(0)} \frac{1}{\sin^2 z \cos z} dz = 0$

b) $\displaystyle\int_{\partial E} \frac{\sin z}{z^4(z^2 + 2)} dz = -\frac{2}{3}\pi i$

c) $\displaystyle\int_{\gamma} \frac{e^{\pi z}}{(z^2 + 1)} dz = -\pi$, where γ is the boundary of $B_2(0) \cap \mathbb{H}$.

Exercise 5. Let $q(z)$ be a rational function having denominator of degree at least 2 more than the degree of its numerator. Show that

$$\sum_{c \in \mathbb{C}} \mathrm{res}_c q = 0.$$

Exercise 6. a) Let f be holomorphic in $\mathbb{C}$ with the exception of *finitely* many (isolated) singularities. Then $z^{-2}f(z^{-1})$ is holomorphic in a punctured neighborhood of 0. Show that

$$\mathrm{res}_0(z^{-2}f(z^{-1})) = \sum_{c \in \mathbb{C}} \mathrm{res}_c f.$$

b) Show that $\displaystyle\int_{\partial E} \frac{5z^6 + 4}{2z^7 + 1} dz = 5\pi i$.

c) With the help of a) give new solutions to exercise 1 g) and h).

§2 Consequences of the residue theorem

Probably the most famous application of the residue theorem is its use in generating a formula for the number of zeros and poles of a meromorphic function. We will derive this formula from a more general one. As a special application we discuss the theorem of ROUCHÉ.

1. The integral $\frac{1}{2\pi i} \int_\gamma F(\zeta) \frac{f'(\zeta)}{f(\zeta) - a} d\zeta$. *Let f be meromorphic in D and have only finitely many poles. Suppose γ is a nullhomologous path in D which avoids all the poles of f. For any complex number a such that the fiber $f^{-1}(a)$ is finite and disjoint from γ, and for any $F \in \mathcal{O}(D)$, we then have*

$$\frac{1}{2\pi i}\int_\gamma F(\zeta)\frac{f'(\zeta)}{f(\zeta)-a}d\zeta \;=\; \sum_{c\in f^{-1}(a)} \mathrm{ind}_\gamma(c)\cdot\nu(f,c)\cdot F(c)$$

$$+ \sum_{d\in P(f)} \mathrm{ind}_\gamma(d)\cdot o_d(f)\cdot F(d).$$

Remark. Only finitely many summands are involved here and only the a-points and poles which lie inside of γ play a role.

Proof. According to the residue theorem 1.4

$$\frac{1}{2\pi i}\int_\gamma F(\zeta)\frac{f'(\zeta)}{f(\zeta)-a}d\zeta = \sum_{z\in D\setminus\gamma} \mathrm{ind}_\gamma(z)\mathrm{res}_z\left(F(\zeta)\frac{f'(\zeta)}{f(\zeta)-a}\right)$$

The only possible points at which the function $F(z)\dfrac{f'(z)}{f(z)-a}$ can have a non-zero residue are the points of the pole-set $P(f)$ or the fiber $f^{-1}(a)$. If f is holomorphic at $c\in D$ and has an a-point at c of multiplicity $\nu(f,c)$, then by 1.3,4)

$$\mathrm{res}_c\left(F(z)\frac{f'(z)}{f(z)-a}\right) = F(c)\nu(f,c).$$

If on the other hand c is a pole of f of order $o_c(f)$, then by 1.3,5)

$$\mathrm{res}_c\left(F(z)\frac{f'(z)}{f(z)-a}\right) = F(c)o_c(f).$$

The claim follows from these three equations. $\qquad\square$

In most applications we find γ is simply closed. Then (under the hypotheses of the theorem) we have, for example

$$\frac{1}{2\pi i}\int_\gamma \zeta^n\frac{f'(\zeta)}{f(\zeta)}d\zeta = \sum c^n o_c(f),$$

where the sum extends over all zeros and poles of f inside γ.

The theorems just proved make possible explicit local descriptions via integrals of the inverses of biholomorphic mappings.

Let $f : D \xrightarrow{\sim} D^*$, $z \mapsto w := f(z)$ be *biholomorphic with inverse function* $f^{-1} : D^* \xrightarrow{\sim} D$ *given by* $w \mapsto z := f^{-1}(w)$. *If* B *is any compact disc contained in* D, *then the mapping* $f^{-1}|f(B)$ *is given by the formula*

$$f^{-1}(w) = \frac{1}{2\pi i}\int_{\partial B} \zeta\frac{f'(\zeta)}{f(\zeta)-w}d\zeta\,, \qquad w\in f(B).$$

Proof. Since ∂B is simply closed and nullhomologous in D and the pole-set $P(f) = \emptyset$, the integral on the right has the value

$$\frac{1}{2\pi i} \int_{\partial B} \zeta \frac{f'(\zeta)}{f(\zeta) - w} d\zeta = \sum_{c \in f^{-1}(w)} \nu(f,c)c.$$

Because f is biholomorphic, each fiber $f^{-1}(w)$ consists of exactly one point and $\nu(f, f^{-1}(w)) = 1$. Therefore the sum on the right is just the number $f^{-1}(w)$.

2. A counting formula for the zeros and poles. If f is meromorphic in D and M is a subset of D in which there are only finitely many a-points and poles of f, then the numbers

$$\text{Anz}_f(a, M) := \sum_{c \in f^{-1}(a) \cap M} \nu(f,c) , \qquad a \in \mathbb{C} ;$$

$$\text{Anz}_f(\infty, M) := \sum_{c \in P(f) \cap M} |o_c(f)|$$

are finite; we call them the *number* (= *Anzahl* in German, hence the notation), *counted according to multiplicity*, of a-points, respectively, poles of f in M. Immediately from theorem 1 follows

Theorem. *Let f be meromorphic and have only finitely many poles in D; let $\gamma \subset D \setminus P(f)$ be a simply closed path which is nullhomologous in D. If $a \in \mathbb{C}$ is any number whose fiber $f^{-1}(a)$ is finite and disjoint from γ, then*

$$(1) \qquad \frac{1}{2\pi i} \int_\gamma \frac{f'(\zeta)}{f(\zeta) - a} d\zeta = \text{Anz}_f(a, \text{Int}\,\gamma) - \text{Anz}_f(\infty, \text{Int}\,\gamma).$$

A special case of equation (1) is the famous

A counting formula for zeros and poles. *Let f be meromorphic in D and have only finitely many zeros and poles. Let γ be a simply closed path in D such that no zeros or poles of f lie on γ. Then*

$$(1') \qquad \frac{1}{2\pi i} \int_\gamma \frac{f'(\zeta)}{f(\zeta)} d\zeta = N - P,$$

where $N := \text{Anz}_f(0, \text{Int}\,\gamma)$ *and* $P := \text{Anz}_f(\infty, \text{Int}\,\gamma)$.

Formula $(1')$ yields incidentally yet another proof of the Fundamental Theorem of Algebra. Namely, if $p(z) = z^n + a_1 z^{n-1} + \cdots + a_n \in \mathbb{C}[z]$, $n \geq 1$ and r is chosen large enough that $|p(z)| \geq 1$ for all $|z| \geq r$, then for such z

$$\frac{p'(z)}{p(z)} = \frac{nz^{n-1} + \cdots}{z^n + \cdots} = \frac{n}{z} + \text{ terms in } \frac{1}{z^\nu} , \qquad \nu \geq 2.$$

If we now integrate over $\partial B_r(0)$, it follows from $(1')$ that $N = n$, since p has no poles in $\mathbb{C}$. The claim follows because $n \geq 1$.

3. ROUCHÉ's theorem. *Let f and g be holomorphic in D and let γ be a simply closed path which is nullhomologous in D and which satisfies*

$$(*)\qquad |f(\zeta) - g(\zeta)| < |g(\zeta)| \qquad \text{for all } \zeta \in \gamma.$$

Then f and g have the same number of zeros inside γ:

$$\text{Anz}_f(0, \text{Int}\,\gamma) = \text{Anz}_g(0, \text{Int}\,\gamma).$$

Proof. We can assume without loss of generality that f and g each have only finitely many zeros in D. The function $h := f/g$ is meromorphic in D and $(*)$ insures that there is a neighborhood U of γ, $U \subset D$, in which h is actually holomorphic, with

$$|h(z) - 1| < 1 \quad \text{for } z \in U, \quad \text{i.e., } h(U) \subset B_1(1) \subset \mathbb{C}^-.$$

Consequently, $\log h$ is well defined in U and furnishes a primitive for h'/h in U. Since f and g are both zero-free on γ by $(*)$, and $h'/h = f'/f - g'/g$ there, it follows that

$$0 = \frac{1}{2\pi i} \int_\gamma \frac{f'(\zeta)}{f(\zeta)} d\zeta - \frac{1}{2\pi i} \int_\gamma \frac{g'(\zeta)}{g(\zeta)} d\zeta.$$

The theorem then follows from formula $(1')$ of the preceding subsection. □

Here is a second proof: The functions $h_t := g + t(f - g)$, $0 \leq t \leq 1$, are holomorphic in D. Because of $(*)$ they satisfy $|h_t(\zeta)| \geq |g(\zeta)| - |f(\zeta) - g(\zeta)| > 0$ for all $\zeta \in \gamma$. All the functions h_t are thus zero-free on γ and so by theorem 2

$$\text{Anz}_{h_t}(0, \text{Int}\,\gamma) = \frac{1}{2\pi i} \int_\gamma \frac{h_t'(\zeta)}{h_t(\zeta)} d\zeta, \qquad 0 \leq t \leq 1.$$

The right side of this equation depends continuously on t and according to 9.3.4 it is integer-valued. Hence it must be constant. In particular, $\text{Anz}_{h_0}(0, \text{Int}\,\gamma) = \text{Anz}_{h_1}(0, \text{Int}\,\gamma)$. Since $h_0 = g$ and $h_1 = f$, the claim follows. □

Remark. *The conclusion of ROUCHÉ's theorem remains valid if instead of the inequality $(*)$ we only demand*

$$(**)\qquad |f(\zeta) - g(\zeta)| < |f(\zeta)| + |g(\zeta)| \qquad \text{for all } \zeta \in \gamma.$$

Since once again this condition entails that g is zero-free on γ, we form $h := f/g$. This function assumes no non-positive real value on γ; for if $h(a) = r \leq 0$ with

$a \in \gamma$, then (**) would imply the contradiction $|r - 1| < |r| + 1$. Consequently, we can choose U, as in the first proof, so that $h \in \mathcal{O}(U)$ and $h(U) \subset \mathbb{C}^-$

The second proof can also be carried through, because each h_t is again zero-free on γ. Indeed, $h_0 = g$ and for $1 \geq t > 0$ from $h_t(a) = 0$, $a \in \gamma$, the equation $h(a) = 1 - 1/t \leq 0$ would follow.   □

We will give five typical applications of Rouché's theorem. In each one must find, for the given f, an appropriate comparison function g whose zeros are known and which satisfies the inequality (*).

1) *Yet another proof of the Fundamental Theorem of Algebra*: Given $f(z) = z^n + a_{n-1} z^{n-1} + \cdots + a_0$, with $n \geq 1$, we set $g(z) := z^n$. Then for sufficiently large r and all $|\zeta| = r$, $|f(\zeta) - g(\zeta)| < |g(\zeta)|$ (by the growth lemma), whence follows $\operatorname{Anz}_f(0, B_r(0)) = \operatorname{Anz}_g(0, B_r(0)) = n$.

2) One can wring information about the zeros of a function from knowledge of the zeros of its Taylor polynomials. More precisely:

If g is a polynomial of degree less than n, $f(z) = g(z) + z^n h(z)$ is the Taylor development of f in a neighborhood of $\overline{B}$, where $B := B_r(0)$, and if $r^n |h(\zeta)| < |g(\zeta)|$ for all $\zeta \in \partial B$, then f and g have equally many zeros in B. – This is clear from ROUCHÉ. For example, the polynomial $f(z) = 3 + az + 2z^4$, where $a \in \mathbb{R}$ and $a > 5$, has exactly one zero in $\mathbb{E}$ because $3 + az$ does and $2 < |3 + a\zeta|$ for all $\zeta \in \partial \mathbb{E}$.

3) *If h is holomorphic in a neighborhood of $\overline{\mathbb{E}}$ and $h(\partial \mathbb{E}) \subset \mathbb{E}$, then h has exactly one fixed point in $\mathbb{E}$.* – If $f(z) := h(z) - z$, and $g(z) := -z$ we have

$$|f(\zeta) - g(\zeta)| = |h(\zeta)| < 1 = |g(\zeta)| \text{for all } \zeta \in \partial \mathbb{E};$$

therefore $h(z) - z$ and $-z$ have the same number of zeros in $\mathbb{E}$, that is, there is exactly one $c \in \mathbb{E}$ with $h(c) = c$.

4) *For every real number $\lambda > 1$ the function $f(z) := ze^{\lambda - z} - 1$ has exactly one zero in $\mathbb{E}$ and it is real and positive.* – Setting $g(z) := ze^{\lambda - z}$, $1 = |f(\zeta) - g(\zeta)| < |g(\zeta)|$ prevails for all $\zeta \in \partial \mathbb{E}$, because $\lambda > 1$; therefore f and g have the same number of zeros in $\mathbb{E}$, viz., exactly one. It is real and positive because, thanks to $f(0) = -1$, $f(1) = e^{\lambda - 1} - 1 > 0$ and the intermediate value theorem, the real-valued function $f|_{[0,1]}$ has at least one zero in the interval $(0, 1)$.

5) *Proof of the theorem of* HURWITZ. We use the notation employed in 8.5.5. First consider the case that U is a disc. We have $\varepsilon := \min\{|f(\zeta)| : \zeta \in \partial U\} > 0$. So we can choose n_U large enough that $|f_n - f|_{\partial U} < \varepsilon$ for all $n \geq n_U$. Then $|f_n(\zeta) - f(\zeta)| < |f(\zeta)|$ for all $\zeta \in \partial U$ and all such n. Our assertion therefore follows from Rouché's theorem (with f in the role of g and f_n in the role of f). Now for an arbitrary U, f has only finitely many zeros in the compactum $\overline{U}$ (by the identity theorem). Consequently there exist pairwise disjoint discs $U_1, \ldots, U_k$ ($k \in \mathbb{N}$) such that f is zero-free in the compactum $K := \overline{U} \setminus \bigcup_1^k U_\nu$. Then almost all the f_n are also

zero-free in K. Therefore the conclusion of the theorem is reduced to the case already settled of discs.   □

Historical note.   The French mathematician Eugène ROUCHÉ (1832-1910) proved his theorem in 1862 in the "Mémoire sur la série de Lagrange" (*Jour. l'École Imp. Polytechn.* **22** (no. 39), 193-224). He formulated it as follows (pp. 217/218, but we use our notation)

Let α be a constant such that on the boundary ∂B of $B := B_r(0)$

$$\left| \alpha \frac{f(z)}{g(z)} \right| < 1$$

holds for a pair of functions f and g which are holomorphic in a neighborhood of $\overline{B}$. Then the equations $g(z) - \alpha \cdot f(z) = 0$ and $g(z) = 0$ have an equal number of roots in B.

ROUCHÉ used logarithm functions in his proof. In 1889 HURWITZ formulated ROUCHÉ's theorem as a lemma and proved his theorem (as in 5) above) with it. Cf. the citations in 8.5.5. ROUCHÉ's name is not mentioned by HURWITZ.

The sharper version of ROUCHÉ's theorem based on the inequality (**) is to be found on p.156 of the 1962 textbook of T. ESTERMANN: *Complex Numbers and Functions*, Athlone Press, London.

Exercises

Exercise 1. Determine the number (counted according to multiplicity) of zeros of the following functions in the indicated domains:

  a) $z^5 + \frac{1}{3}z^3 + \frac{1}{4}z^2 + \frac{1}{3}$ in $\mathbb{E}$ and in $B_{1/2}(0)$.

  b) $z^5 + 3z^4 + 9z^3 + 10$ in $\mathbb{E}$ and in $B_2(0)$.

  c) $9z^5 + 5z - 3$ in $\{z \in \mathbb{C} : \frac{1}{2} < |z| < 5\}$.

  d) $z^8 + z^7 + 4z^2 - 1$ in $\mathbb{E}$ and in $B_2(0)$.

Exercise 2. Let $p(z) = z^n + a_{n-1}z^{n-1} + \cdots + a_0$ be a normalized polynomial with coefficients $a_j \in \mathbb{C}$ and $n \geq 1$. Show that there is a point $c \in \partial\mathbb{E}$ where $|p(c)| \geq 1$.

Exercise 3. Show that if $\lambda \in \mathbb{R}$, $\lambda > 1$, then the function $f(z) := \lambda - z - e^{-z}$ has exactly one zero in the closed right half-plane, that it is real and that it lies in $B_1(\lambda)$. *Hint.* In ROUCHÉ's theorem set $g(z) := \lambda - z$ and $\gamma := \partial B_1(\lambda)$.

Exercise 4. Let $\{c_n\}$ be a strictly decreasing sequence of positive real numbers. Then $f(z) := \sum_{n=0}^{\infty} c_n z^n$ defines a holomorphic function in $\mathbb{E}$. Show that f has no zeros in $\mathbb{E}$. *Hint.* Show that the partial sums of f have no zeros in $\mathbb{E}$ and deduce from this that f has no zeros in $B_r(0)$ for every $0 < r < 1$. (Cf. also Exercise 3 in §1, Chapter 0.)

Chapter 14

Definite Integrals and the Residue Calculus

Le calcul des résidus constitue la source naturelle des intégrales définies (E. LINDELÖF)

The residue calculus is eminently suited to evaluating real integrals whose integrands have no known explicit antiderivatives. The basic idea is simple: The real interval of integration is incorporated into a closed path γ in the complex plane and the integrand is then extended into the region bounded by γ. The extension is required to be holomorphic there except for isolated singularities. The integral over γ is then determined from the residue theorem, and the needed residues are computed algebraically. EULER, LAPLACE and POISSON needed considerable analytic inventiveness to find their integrals. But today it would be more a question of proficiency in the use of the Cauchy formulas. Nevertheless there is no canonical method of finding, for a given integrand and interval of integration, the best path γ in $\mathbb{C}$ to use.

We will illustrate the techniques with a selection of typical examples in sections 1 and 2, "but even complete mastery does not guarantee success" (AHLFORS [1], p.154). In each case it is left to the reader to satisfy himself that the path of integration being employed is simply closed. In section 3 the Gauss sums will be evaluated residue-theoretically.

§1 Calculation of integrals

The examples assembled in this section are very simple. But everyone studying the subject should master the techniques of dealing with these types of integrals – this circle of ideas is a popular source of examination questions. First we are going to recall some simple facts from the theory of improper integrals. For details we refer the reader to Edmund LANDAU's book *Differential and Integral Calculus*, Chelsea Publ. Co. (1950), New York (especially Chapter 28).

0. Improper integrals. If $f : [a, \infty) \to \mathbb{C}$ is continuous, then, as all readers know, we set

$$\int_a^\infty f(x)dx := \lim_{s \to \infty} \int_a^s f(x)dx$$

whenever the limit on the right exists; $\int_a^\infty f(x)dx$ is called an *improper* integral. Calculations with such integrals obey some rather obvious rules, e.g.,

$$\int_a^\infty f(x)dx = \int_a^b f(x)dx + \int_b^\infty f(x)dx \qquad \text{for all } b > a.$$

Improper integrals of the form $\int_{-\infty}^a f(x)dx$ are defined in a similar way. Finally we set

$$\int_{-\infty}^\infty f(x)dx := \int_{-\infty}^a f(x)dx + \int_a^\infty f(x)dx = \lim_{r,s \to \infty} \int_{-r}^s f(x)dx$$

whenever $f : \mathbb{R} \to \mathbb{C}$ is continuous and the two limits involved both exist. It is to be emphasized that r and s have to be allowed to run to ∞ *independently of each* other; that is, the existence of $\lim_{r \to \infty} \int_{-r}^r f(x)dx$ does not imply the existence of $\int_{-\infty}^\infty f(x)dx$ as we are defining the latter. The function $f(x) \equiv x$ demonstrates this convincingly.

Basic to the theory of improper integrals is the following

Existence criterion. *If $f : [a, \infty) \to \mathbb{C}$ is continuous and there is a $k > 1$ such that $x^k f(x)$ is bounded then $\int_a^\infty f(x)dx$ exists.*

This follows rather easily from the Cauchy convergence criterion. The hypothesis $k > 1$ is essential, since, for example, $\int_2^\infty \frac{dx}{x \log x}$ does not exist even though $x(x \log x)^{-1} \to 0$ as $x \to \infty$. Also, though the handy word "criterion" was used, the boundedness of $x^k f(x)$ for some $k > 1$ is only a sufficient and certainly not a necessary condition for the existence of $\int_a^\infty f(x)dx$. For example, both improper integrals

$$\int_0^\infty \frac{\sin x}{x}dx \ , \quad \int_0^\infty \sin(x^2)dx$$

exist, although there is no $k > 1$ such that either $x^k \frac{\sin x}{x}$ or $x^k \sin(x^2)$ is bounded. In the second of these two examples the integrand does not even tend to 0 as $x \to \infty$; a phenomenon probably first pointed out by DIRICHLET in 1837 (*Jour. für die Reine und Angew. Math.* **17**, p.60; *Werke* I, p.263).

The existence criterion applies *mutatis mutandis* to integrals of the form $\int_{-\infty}^{a} f(x)dx$ as well. Almost all the improper integrals known in 1825 are to be found in CAUCHY [C$_2$]. From the extensive further classical literature on (improper) integrals we mention Dirichlet's *Vorlesungen über die Lehre von den einfachen und mehrfachen bestimmten Integralen* (held in the summer of 1854; published in 1904 by Vieweg Verlag in Braunschweig) and Kronecker's *Vorlesungen über die Theorie der einfachen und der vielfachen Integrale* (held in the winter 1883/84 and in the summers 1885, 1887, 1889 and 1891 and then finally as a six-hour course; cf. [Kr]).

1. Trigonometric integrals $\int_0^{2\pi} R(\cos\varphi, \sin\varphi)d\varphi$. Let $R(x, y)$ be a *complex-valued rational function* of $(x, y) \in \mathbb{R}^2$ which is *finite on the circle* $\partial \mathbb{E}$. Then for

$$\tilde{R}(z) := z^{-1}R(\tfrac{1}{2}(z + z^{-1}), \tfrac{1}{2i}(z - z^{-1}))$$

we have

(1)
$$\int_0^{2\pi} R(\cos\varphi, \sin\varphi)d\varphi = 2\pi \sum_{w \in \mathbb{E}} \mathrm{res}_w \tilde{R}.$$

Proof. For $0 \le \varphi \le 2\pi$ and $\zeta := e^{i\varphi}$, we have $\cos\varphi = \tfrac{1}{2}(\zeta + \zeta^{-1})$, $\sin\varphi = \tfrac{1}{2i}(\zeta - \zeta^{-1})$ and so

$$\int_0^{2\pi} R(\cos\varphi, \sin\varphi)d\varphi = \frac{1}{i} \int_{\partial \mathbb{E}} R(\tfrac{1}{2}(\zeta + \zeta^{-1}), \tfrac{1}{2i}(\zeta - \zeta^{-1})) \cdot \zeta^{-1}d\zeta$$

The equality (1) follows from this and the residue theorem.   □

Examples. 1) To evaluate $\displaystyle\int_0^{2\pi} \frac{d\varphi}{1 - 2p\cos\varphi + p^2}, p \in \mathbb{C} \setminus \partial\mathbb{E}$, use $R(x, y) = (1 - 2px + p^2)^{-1}$ and

$$\tilde{R}(z) = \frac{1}{z} \frac{1}{1 - pz - pz^{-1} + p^2} = \frac{1}{(z - p)(1 - pz)}.$$

$\tilde{R}$ has exactly one pole in $\mathbb{E}$, and it has order 1; namely at p if $|p| < 1$ and at p^{-1} if $|p| > 1$. Therefore from the preceding theorem it follows that

$$\int_0^{2\pi} \frac{d\varphi}{1 - 2p\cos\varphi + p^2} = \begin{cases} \frac{2\pi}{1 - p^2} & , \text{ if } |p| < 1 \\[2mm] \frac{2\pi}{p^2 - 1} & , \text{ if } |p| > 1. \end{cases}$$

2) To determine $\displaystyle\int_0^{2\pi} \frac{d\varphi}{(p + \cos\varphi)^2}$, $p \in \mathbb{R}$ and $p > 1$, use $R(x, y) = (p + x)^{-2}$ and, accordingly,

$$\tilde{R}(z) = \frac{1}{z}(p + \tfrac{1}{2}(z + z^{-1}))^{-2} = \frac{4z}{(z^2 + 2pz + 1)^2}.$$

According to 13.1.3,3) this function has precisely one pole in $\mathbb{E}$, at $c := -p + \sqrt{p^2 - 1}$; this is a second order pole and the residue there is $p(\sqrt{p^2 - 1})^{-3}$. The theorem consequently gives us

$$\int_0^{2\pi} \frac{d\varphi}{(p + \cos\varphi)^2} = \frac{2\pi p}{(\sqrt{p^2 - 1})^3} \qquad \text{for } p > 1.$$

Remark. The method of the theorem can also be applied to integrals of the form

$$\int_0^{2\pi} R(\cos\varphi, \sin\varphi) \cdot \cos m\varphi \cdot \sin n\varphi \, d\varphi$$

with $m, n \in \mathbb{Z}$, since $\cos m\varphi = \tfrac{1}{2}(\zeta^m + \zeta^{-m})$ and $\sin n\varphi = \tfrac{1}{2i}(\zeta^n - \zeta^{-n})$.

2. Improper integrals $\int_{-\infty}^{\infty} f(x)dx$. D will denote a domain which contains the closed upper half-plane $\overline{\mathbb{H}} = \mathbb{H} \cup \mathbb{R}$ and $\Gamma(r) : [0, \pi] \to \overline{\mathbb{H}}$ the map $\varphi \mapsto re^{i\varphi}$, the part of the perimeter of the disc $B_r(0)$ which lies in $\overline{\mathbb{H}}$ (cf. the figure).

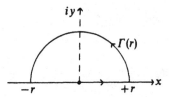

Theorem. *Let f be holomorphic in D, except possibly at finitely many points none of which is real. Suppose that $\int_{-\infty}^{\infty} f(x)dx$ exists and that $\lim_{z\to\infty} zf(z) = 0$. Then*

(1)
$$\int_{-\infty}^{\infty} f(x)dx = 2\pi i \sum_{w \in \mathbb{H}} \mathrm{res}_w f.$$

Proof. All the singularities of f lie in $B_r(0)$ if r is large enough. For such r it follows from the residue theorem that

(*)
$$\int_{-r}^{r} f(x)dx + \int_{\Gamma(r)} f(\zeta)d\zeta = 2\pi i \sum_{w \in \mathbb{H}} \mathrm{res}_w f.$$

The standard estimate for integrals gives $|\int_{\Gamma(r)} f(\zeta)d\zeta| \leq \pi r |f|_{\Gamma(r)}$. Since $\lim_{r \to \infty} r|f|_{\Gamma(r)} = 0$ by the very meaning of the hypothesis $\lim_{z \to \infty} zf(z) = 0$, (1) follows from (*). $\square$

It is easy to generate functions f which fulfill the hypotheses of this theorem. To this end we use the

Growth lemma for rational functions. *Let $p, q \in \mathbb{C}[z]$ be polynomials of degree m and n, respectively. Then positive real numbers K, L, R exist such that*

$$K|z|^{m-n} \leq \left|\frac{p(z)}{q(z)}\right| \leq L|z|^{m-n} \qquad \text{for all } z \in \mathbb{C} \text{ with } |z| \geq R.$$

Proof. According to 9.1.1 there is an $R > 0$ and positive numbers K_1, K_2, L_1, L_2 such that $K_1|z|^m \leq |p(z)| \leq L_1|z|^m$ and $K_2|z|^n \leq |Q(z)| \leq L_2|z|^n$ for all $|z| \geq R$. Therefore the numbers $K := K_1 L_2^{-1}$ and $L := L_1 K_2^{-1}$ do the required job.

Corollary. *If $f(z) = \frac{p(z)}{q(z)} \in \mathbb{C}(z)$ and the degree of the denominator exceeds that of the numerator by ℓ, then $\lim_{z \to \infty} z^k f(z) = 0$ for every $k \in \mathbb{N}$ satisfying $0 \leq k < \ell$.*

In particular, the hypotheses of the theorem are fulfilled by $f = p/q$ if q has no zeros in $\mathbb{R}$ and its degree exceeds that of p by 2 or more.

For the proof use the existence criterion for improper integrals from section 0.

Example. Consider $f(z) := \frac{z^2}{1 + z^4}$. This rational function has exactly two poles in $\mathbb{H}$, each of order 1, namely $c := \exp(\frac{1}{4}i\pi)$ and ic. From 13.1.3,1) we have that $\text{res}_c f = \frac{1}{4}\bar{c}$, $\text{res}_{ic} f = -\frac{i}{4}\bar{c}$. Since $\bar{c} - i\bar{c} = (1-i)\bar{c}$ and $\bar{c} = \frac{1}{\sqrt{2}}(1-i)$, we see that

$$\int_{-\infty}^{\infty} \frac{x^2}{1+x^4}dx = 2\pi i \frac{(1-i)^2}{4\sqrt{2}} = \frac{\pi}{\sqrt{2}}.$$

Innumerable integrals (a few of which are presented in the exercises at the end of this section) can be evaluated with the aid of (1).

3.   The integral $\int_0^{\infty} \frac{x^{m-1}}{1+x^n}dx$ for $m, n \in \mathbb{N}$, $0 < m < n$. The integrand $f(z) := \frac{z^{m-1}}{1+z^n}$ has, according to 13.1.3,2), a first order pole at $c := \exp(\frac{1}{n}i\pi)$ with $\text{res}_c f = -\frac{1}{n}c^m$. To evaluate the integral we will not

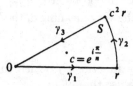

use a semi-circle as auxiliary path but rather integrate along the boundary $\gamma_1 + \gamma_2 + \gamma_3$ of a circular sector S of radius $r > 1$ (cf. the figure above). Since f is holomorphic in $S \setminus c$ (!), it follows (the residue theorem!) that

$$(*) \qquad \int_0^r f(x)dx + \int_{\gamma_2} f(\zeta)d\zeta + \int_{\gamma_3} f(\zeta)d\zeta = -\frac{2\pi i}{n}c^m.$$

The path $-\gamma_3$ is given by $\zeta(t) = tc^2$, $t \in [0, r]$; therefore, taking account of the fact that $c^{2n} = 1$,

$$\int_{\gamma_3} f(\zeta)d\zeta = -\int_0^r \frac{t^{m-1}c^{2m-2}}{1 + t^n c^{2n}}c^2 dt = -c^{2m}\int_0^r \frac{t^{m-1}}{1 + t^n}dt = -c^{2m}\int_0^r f(x)dx.$$

Because $|\int_{\gamma_2} f(\zeta)d\zeta| \leq |f|_{\gamma_2}\frac{2\pi}{n}r$ and $\lim_{r \to \infty} |f|_{\gamma_2}\frac{2\pi}{n}r = 0$ (due to $m < n$), passage to the limit converts $(*)$ into

$$(c^{2m} - 1)\int_0^\infty f(x)dx = \frac{2\pi i}{n}c^m.$$

Now $c^m(c^{2m} - 1)^{-1} = (c^m - c^{-m})^{-1} = (2i\sin\frac{m}{n}\pi)^{-1}$, since $c = e^{i\pi/n}$. It follows that

$$(1) \qquad \int_0^\infty \frac{x^{m-1}}{1 + x^n}dx = \frac{\pi}{n}(\sin\frac{m}{n}\pi)^{-1} \qquad \text{for all } m, n \in \mathbb{N} \text{ with } 0 < m < n.$$

This formula was known to EULER in 1743 (*Opera Omnia* (1) **17**, p.54).

In case m is odd and $n = 2q$ is even, we have $\int_0^\infty \frac{x^{m-1}}{1+x^n}dx = \frac{1}{2}\int_{-\infty}^\infty \frac{x^{m-1}}{1+x^n}dx$. The second integral here can also be evaluated with the help of theorem 2, thus: the points $c_\nu := c^{2\nu+1}$, $0 \leq \nu \leq q - 1$ constitute the totality of the poles of f in $\mathbb{H}$, each being of the first order and giving f respective residues $-n^{-1}c_\nu^m$. Since

$$\sum_0^{q-1} c_\nu^m = c^m \sum_0^{q-1} c^{2m\nu} = c^m\frac{c^{mn} - 1}{c^{2m} - 1} = \frac{(-1)^m - 1}{c^m - c^{-m}}$$

and $c^m - c^{-m} = 2i\sin\frac{m}{n}\pi$, it follows in this case from theorem 2 that

$$\frac{1}{2}\int_{-\infty}^\infty \frac{x^{m-1}}{1 + x^n}dx = \pi i \sum_{w \in \mathbb{H}} \text{res}_w f = -\frac{\pi i}{n}\sum_0^{q-1} c_\nu^m = \frac{\pi}{n}\left(\sin\frac{m}{n}\pi\right)^{-1}$$

Exercises

Exercise 1. For $a > 1$ show that

a) $\displaystyle\int_0^{2\pi} \frac{d\varphi}{a + \sin\varphi} = \frac{2\pi}{\sqrt{a^2 - 1}}$

b) $\displaystyle\int_0^{2\pi} \frac{\sin(2\varphi)d\varphi}{(a + \cos\varphi)(a - \sin\varphi)} = -4\pi\left(1 - \frac{2a\sqrt{a^2 - 1}}{2a^2 - 1}\right).$

Hint. For b) use the abbreviation $w := \sqrt{a^2 - 1}$ and in calculating the residues refrain from multiplying out the products that intervene.

Exercise 2. Verify that for $n \geq 1$

$$\int_0^{2\pi} \frac{(1 - 4\sin^2\varphi)^n \cos(2\varphi)}{2 - \cos\varphi}\,d\varphi = \frac{2\pi(91 - 52\sqrt{3})}{\sqrt{3}}.$$

Exercise 3. Prove the identities

a) $\displaystyle\int_{-\infty}^{\infty} \frac{2x^2 + x + 1}{x^4 + 5x^2 + 4}\,dx = \frac{5}{6}\pi$

b) $\displaystyle\int_{-\infty}^{\infty} \frac{dx}{(1 + x^2)^{n+1}} = \frac{\pi}{2^{2n}}\frac{(2n)!}{(n!)^2},\ n \in \mathbb{N}$

c) $\displaystyle\int_{-\infty}^{\infty} \frac{dx}{(x^4 + a^4)^2} = \frac{3}{8}\frac{\sqrt{2}}{a^7}\pi$ for $a > 0.$

§2 Further evaluations of integrals

In this section we will discuss improper integrals which are more complicated than those considered so far. With the example $\int_0^\infty \frac{\sin x}{x}dx$ we will
illustrate that the method of residues is not always the most advantageous
approach.

1. Improper integrals $\int_{-\infty}^{\infty} g(x)e^{iax}dx$. When $f(z)$ has the form
$g(z)e^{iaz}$ for $a \in \mathbb{R}$ and certain g, the hypotheses of theorem 1.2 can be
weakened. As path of integration we will use, in place of the semi-circle
$\Gamma(r)$, the upper part $\gamma_1 + \gamma_2 + \gamma_3$ of the boundary of a square Q in $\mathbb{H}$ having
vertices $-r, s, s + iq, -r + iq$, where r and s are positive and $q := r + s$. See
the figure below.

Theorem. *Let g be holomorphic in $\mathbb{C}$ except possibly at finitely many
places, none real, and suppose that $\lim_{z \to \infty} g(z) = 0$. Then*

(1) $\displaystyle\int_{-\infty}^{\infty} g(x)e^{iax}dx = \begin{cases} 2\pi i \sum_{w \in \mathbb{H}} \mathrm{res}_w(g(z)e^{iaz}) & \text{in case} \quad a > 0 \\[2ex] -2\pi i \sum_{-w \in \mathbb{H}} \mathrm{res}_w(g(z)e^{iaz}) & \text{in case} \quad a < 0. \end{cases}$

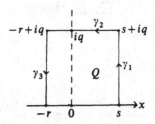

Proof. First consider $a > 0$. Choose r, s large enough that all the singularities of g in $\mathbb{H}$ lie in the square Q. We maintain that

$$(*) \quad I_\nu := \int_{\gamma_\nu} g(\zeta)e^{ia\zeta}d\zeta \quad (\nu = 1, 2, 3) \quad \text{satisfy} \lim I_\nu = 0 \text{ as } r, s \to \infty.$$

The first part of (1) will then follow from the residue theorem.
    Since $(-\gamma_2)(t) = t + iq$, $t \in [-r, s]$, and $|e^{ia\zeta}| = e^{-a\Im\zeta}$, we have

$$|I_2| \le |g(\zeta)e^{ia\zeta}|_{\gamma_2} \cdot (r + s) \le |g|_{\gamma_2} \cdot e^{-aq} \cdot q \le |g|_{\gamma_2} \text{ as soon as } e^{aq} > q.$$

Since $\gamma_1(t) = s + it$, $t \in [0, q]$, it further follows that

$$|I_1| \le \int_0^q |g(s+it)|e^{-at}dt \le |g|_{\gamma_1} \int_0^q e^{-at}dt = |g|_{\gamma_1}a^{-1}(1-e^{-aq}) \le |g|_{\gamma_1}a^{-1}.$$

Similarly one sees that $|I_3| \le |g|_{\gamma_3}a^{-1}$. But because of the hypothesis $\lim_{z\to\infty} g(z) = 0$, we have $\lim_{r,s\to\infty} |g|_{\gamma_\nu} = 0$ for each $\nu = 1, 2, 3$. Therefore $(*)$ follows from the preceding estimates of the $|I_\nu|$.
    As to the case $a < 0$, one simply considers squares in the lower half-plane and estimates the integrals analogously, being careful to note that once again $aq > 0$, since now $q < 0$.   □

Remarks. Integrals of the type (1) are called FOURIER *transforms* (when viewed as functions of a). The reader will have noticed that the inequality for $|I_1|$ involved a trivial but useful sharper version of the standard estimate. Specifically, for non-negative continuous functions u, v on the interval $I = [a, b] \subset \mathbb{R}$, $\int_a^b u(t)v(t)dt \le |u|_I \int_a^b v(t)dt$. Finally, notice that it would not have been expedient in the above proof to have used semi-circular paths of integration as in 1.2: The estimation process would have been more troublesome and, worse, only the existence of $\lim_{r\to\infty} \int_{-r}^r g(x)e^{iax}dx$ would have been secured and this does not imply the existence of $\int_{-\infty}^\infty g(x)e^{iax}dx$.   □

    The limit condition $\lim_{z\to\infty} g(z) = 0$ of the theorem is fulfilled by any rational function g in which *the degree of the denominator exceeds that of the numerator* (cf. Corollary 1.2). Thus, for example, for all positive real numbers a and complex b with $\Re b > 0$

$$(2) \qquad \int_{-\infty}^{\infty} \frac{e^{iax}}{x-ib}\, dx = 2\pi i e^{-ab}\,, \qquad \int_{-\infty}^{\infty} \frac{e^{iax}}{x+ib}\, dx = 0.$$

Since $x\cos ax$ and $\sin ax$ are odd functions and $\lim_{r\to\infty}\int_{-r}^{r} f(x)dx = 0$ for every odd continuous function $f : \mathbb{R} \to \mathbb{C}$, from addition and subtraction in (2) we get the formulas (due to LAPLACE (1810))

$$(3) \qquad \int_0^{\infty} \frac{\beta\cos\alpha x}{x^2+\beta^2}\, dx = \int_0^{\infty} \frac{x\sin\alpha x}{x^2+\beta^2}\, dx = \tfrac{1}{2}\pi e^{-\alpha\beta} \quad \alpha, \beta \text{ real and positive.}$$

By using this formula at the forbidden value $\beta = 0$, CAUCHY uncritically inferred that

$$\int_0^{\infty} \frac{\sin x}{x}\, dx = \frac{1}{2}\pi$$

(see p.60 of the *Ostwald's Klassiker* version of [C_2]); this formula is nevertheless correct and we will derive it in section 3.

In the theorem we may subject g to the additional hypothesis that it be real-valued on $\mathbb{R}$. Then, since $\cos ax = \Re e^{iax}$ and $\sin ax = \Im e^{iax}$ for $x \in \mathbb{R}$, it follows from (1) that

$$(4) \qquad \int_{-\infty}^{\infty} g(x)\cos ax\, dx = -2\pi\Im\left(\sum_{w\in\mathbb{H}} \mathrm{res}_w(g(z)e^{iaz})\right)\,, \; a > 0;$$

$$(5) \qquad \int_{-\infty}^{\infty} g(x)\sin ax\, dx = 2\pi\Re\left(\sum_{w\in\mathbb{H}} \mathrm{res}_w(g(z)e^{iaz})\right)\,, \; a > 0;$$

with corresponding equations for the case $a < 0$.

The reader should try to derive formulas (3) directly from (4) and (5).

2. Improper integrals $\int_0^{\infty} q(x)x^{a-1}dx$. For $a \in \mathbb{C}$ and $z = |z|e^{i\varphi} \in \mathbb{C}^{\times}$, $0 \le \varphi < 2\pi$, we set

$$\ln z := \log|z| + i\varphi\,, \qquad z^a := \exp(a\ln z).$$

These functions are holomorphic in the plane *slit along the positive real axis*; i.e., in the set

$$\tilde{\mathbb{C}} := \mathbb{C} \setminus \{t \in \mathbb{R} : t \ge 0\}.$$

It should be noted that $\ln z$ is not the principal branch of the logarithm and that correspondingly z^a is not the usual power function. Nevertheless, $x^a = e^{a \log x}$ for $z = x$ real and positive. We need the following proposition:

If I is a compact interval on the positive real axis and $\varepsilon > 0$, then

$$(*) \qquad \lim_{\varepsilon \to 0} (x + i\varepsilon)^a = x^a \,, \qquad \lim_{\varepsilon \to 0} (x - i\varepsilon)^a = x^a e^{2\pi i a} \qquad \text{for all } a \in \mathbb{C}$$

and the convergence is uniform for $x \in I$.

This is clear, on account of $\lim_{\varepsilon \to 0} \ln(x + i\varepsilon) = \log x$ and $\lim_{\varepsilon \to 0} \ln(x - i\varepsilon) = \log x + 2\pi i$.

Theorem. *Let $q \in \mathcal{M}(\mathbb{C})$ have only finitely many poles, all lying in $\tilde{\mathbb{C}}$ and suppose an $a \in \mathbb{C} \setminus \mathbb{Z}$ is given which satisfies*

$$(L) \qquad \lim_{z \to 0} q(z)z^a = 0 \qquad \text{and} \qquad \lim_{z \to \infty} q(z)z^a = 0.$$

Then it follows that

$$(1) \qquad \int_0^\infty q(x)x^{a-1}dx = \frac{2\pi i}{1 - e^{2\pi i a}} \sum_{w \in \tilde{\mathbb{C}}} \operatorname{res}_w(q(z)z^{a-1}).$$

Proof. Let ε, r, s be positive. We consider the path $\gamma := \gamma_1 + \gamma_2 + \gamma_3 + \gamma_4$, in which γ_1 and γ_3 are intervals on the lines $\Im z = \varepsilon$ and $\Im z = -\varepsilon$, respectively, and γ_2 and γ_4 are arcs of circles centered at 0 and of radius s and r, respectively. We make ε and r so small and s so large that the region G (in the figure) bounded by γ contains all the poles of q.

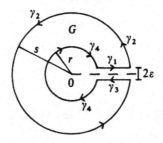

Since γ is simply closed, according to 13.1.1,4), the residue theorem asserts that

$$\int_\gamma q(\zeta)\zeta^{a-1}d\zeta = 2\pi i \sum_{w \in \tilde{\mathbb{C}}} \operatorname{res}_w(q(z)z^{a-1}),$$

independent of ε, r, s. We consider the separate integrals over the component curves $\gamma_1, \ldots, \gamma_4$. Since $s = |\zeta|$ for $\zeta \in \gamma_2$, we have

$$\left| \int_{\gamma_2} q(\zeta)\zeta^{a-1}d\zeta \right| \leq |q(\zeta)\zeta^{a-1}|_{\gamma_2} \cdot 2\pi s = 2\pi |\zeta^a q(\zeta)|_{\gamma_2}.$$

An analogous estimate holds for the path γ_4. Therefore, due to (L)

$$\lim_{r \to 0} \lim_{\varepsilon \to 0} \int_{\gamma_4} q(\zeta)\zeta^{a-1}d\zeta = 0 = \lim_{s \to \infty} \lim_{\varepsilon \to 0} \int_{\gamma_2} q(\zeta)\zeta^{a-1}d\zeta.$$

Since γ_1 and $-\gamma_3$ are given by $x \mapsto x + i\varepsilon$ and $x \mapsto x - i\varepsilon$, respectively, with the same parameter interval $I \subset \mathbb{R}$, it follows from (∗) that

$$\lim_{\varepsilon \to 0} \int_{\gamma_1} q(\zeta)\zeta^{a-1}d\zeta = \int_r^s q(x)x^{a-1}dx,$$

$$\lim_{\varepsilon \to 0} \int_{\gamma_3} q(\zeta)\zeta^{a-1}d\zeta = -e^{2\pi i a} \int_r^s q(x)x^{a-1}dx.$$

Putting everything together, we see finally that $\int_\gamma q(\zeta)\zeta^{a-1}d\zeta$ converges to $(1 - e^{2\pi i a}) \int_0^\infty q(x)x^{a-1}dx$ as $\varepsilon \to 0$, $r \to 0$ and $s \to +\infty$.   □

Integrals of the type (1) are called MELLIN *transforms* (as functions of a). By taking note of the fact that $2\pi i(1 - e^{2\pi i a})^{-1} = -\pi e^{-\pi i a}(\sin \pi a)^{-1}$, we get the

Corollary. *Let q be a rational function which has no poles on the positive real axis (including 0); suppose that the degree of the denominator of q exceeds that of its numerator. Then*

$$(2) \qquad \int_0^\infty q(x)x^{a-1}dx = -\frac{\pi e^{-\pi i a}}{\sin \pi a} \sum_{w \in \tilde{\mathbb{C}}} \mathrm{res}_w(q(z)z^{a-1})$$

for all $a \in \mathbb{C}$ with $0 < \Re a < 1$.

Proof. We have $|z^a q(z)| \leq e^{2\pi |\Im a|}|z|^{\Re a}|q(z)|$. Since q is holomorphic at 0 and $\Re a > 0$, the first of the limit equations (L) evidently holds. Since we have the estimate $|q(z)| \leq M|z|^{-1}$ for some $0 < M < \infty$ and all large z, the second limit equation in (L) follows from the fact that $\Re a < 1$.   □

Example. We will determine $\int_0^\infty \dfrac{x^{a-1}}{x + e^{i\varphi}}dx$, where $a \in \mathbb{R}$, $0 < a < 1$ and $-\pi < \varphi < \pi$. The function $q(z) := \frac{1}{z + e^{i\varphi}}$ has a pole of the first order at $-e^{i\varphi}$ (and no other poles). Since $-e^{i\varphi} = e^{i(\varphi + \pi)}$, we have

$$\text{res}_{-e^{i\varphi}}(q(z)z^{a-1}) = e^{i(\varphi+\pi)(a-1)} = -e^{i(a-1)\varphi}e^{\pi i a}$$

and so from (2)

(3) $$\int_0^\infty \frac{x^{a-1}}{x + e^{i\varphi}}dx = \frac{\pi}{\sin \pi a} \cdot e^{i(a-1)\varphi}, \quad a \in \mathbb{R}, \ 0 < a < 1, \ -\pi < \varphi < \pi.$$

The integral (3) with $\varphi = 0$ plays a role in the theory of the gamma and beta functions. It reflects the equation

$$B(a, 1-a) = \int_0^\infty \frac{x^{a-1}}{1+x}dx = \Gamma(a)\Gamma(1-a) = \frac{\pi}{\sin \pi a}.$$

With the help of (3) the integrals $\int_0^\infty \frac{x^{m-1}}{x^n + e^{i\varphi}}dx$, $m, n \in \mathbb{N}$, $-\pi < \varphi < \pi$, can be elegantly determined. We substitute $t := x^n$ and find

(4) $$\int_0^\infty \frac{x^{m-1}}{x^n + e^{i\varphi}}dx = \frac{1}{n}\int_0^\infty \frac{t^{m/n-1}}{t + e^{i\varphi}}dt = \frac{\pi}{n}\left(\sin \frac{m}{n}\pi\right)^{-1}e^{i(m/n-1)\varphi},$$

$$0 < m < n, \ -\pi < \varphi < \pi,$$

a special case of which (when $\varphi = 0$) is 1.3(1). Multiply numerator and denominator on the left by the conjugate $x^n + e^{-i\varphi}$ of the denominator and then equate the imaginary parts of both sides of (4) to get

(5) $$\int_0^\infty \frac{x^{m-1}}{x^{2n} + 2x^n \cos \varphi + 1}dx = \frac{\pi}{n}\frac{\sin(1-m/n)\varphi}{\sin \frac{m}{n}\pi \cdot \sin \varphi},$$

for $0 < m < n$, $-\pi < \varphi < \pi$. This formula is to be found in EULER, 1785 (*Opera Omnia* (1) **18**, p.202).

3. The integrals $\int_0^\infty \frac{\sin^n x}{x^n}dx$. By no later than 1781 EULER knew the equation $\int_0^\infty \frac{\sin x}{x}dx = \frac{1}{2}\pi$ (cf. *Opera Omnia* (1) **19**, pp. 226, 227). The attempt to derive this from theorem 1 on the basis of the evident relation $\int_0^\infty \frac{\sin x}{x}dx = \Im \int_0^\infty \frac{e^{ix}}{x}dx$ won't succeed without further effort because $z^{-1}e^{iz}$ has a pole at 0, while $z^{-1}\sin z$ is holomorphic throughout $\mathbb{C}$. Here the limitations of the residue calculus become clear. To be sure one can extend theorem 1 to cover such situations and then determine this integral (for that see, e.g., [7], chapter V, example 2.7 or [10], pp. 155-156.) But the following procedure, which is less well known in the literature, is much more convenient. First, with the help of the partial fraction development

$$\frac{\pi^2}{\sin^2 \pi z} = \sum_{-\infty}^\infty \frac{1}{(z+\nu)^2} \quad \text{[cf. 11.2.3(1)]}$$ we can, in an amusing way, get the formula

(1)
$$\int_0^\infty \frac{\sin^2 x}{x^2}\,dx = \frac{1}{2}\pi.$$

Proof. Cauchy's criterion for improper integrals settles the question of the existence of this integral at a glance. In the partial fraction development replace z by $\pi^{-1}z$ and use the fact that $\sin^2(z + \nu\pi) = \sin^2 z$ to re-write that development as

$$\sum_{-\infty}^{\infty} \frac{\sin^2(z + \nu\pi)}{(z + \nu\pi)^2} = 1.$$

Here we may integrate over $[0, \pi]$ term-wise (why?), to get

$$\pi = \sum_{-\infty}^{\infty} \int_0^\pi \frac{\sin^2(x + \nu\pi)}{(x + \nu\pi)^2}\,dx = \sum_{-\infty}^{\infty} \int_{\nu\pi}^{(\nu+1)\pi} \frac{\sin^2 x}{x^2}\,dx = \int_{-\infty}^{\infty} \frac{\sin^2 x}{x^2}\,dx. \quad \Box$$

Now for $0 < s < \infty$ integration by parts yields

$$\int_0^s \frac{\sin^2 x}{x^2}\,dx = -\frac{\sin^2 x}{x}\Big|_0^s + \int_0^s \frac{\sin 2x}{x}\,dx = -\frac{\sin^2 s}{s} + \int_0^{2s} \frac{\sin x}{x}\,dx.$$

From (1) and the triviality $\lim_{s\to\infty} \frac{\sin^2 s}{s} = 0$ follows the existence of $\int_0^\infty \frac{\sin x}{x}\,dx$ as well as the evaluation

(2)
$$\int_0^\infty \frac{\sin x}{x}\,dx = \frac{1}{2}\pi.$$

The derivation of equations (1) and (2) shows that in calculating improper integrals the path through the complexes is not always to be recommended. (In this connection it ought to be noted that the formula $\varepsilon_2(x) = \pi^2(\sin\pi x)^{-2}$ which we employed above can also be gotten by real analysis methods; cf. 11.2.2.) In a similar situation KRONECKER wrote somewhat sarcastically (cf. [Kr], p.84): "Wir brauchen hier übrigens zum Zwecke dieser Beweise das Gebiet der reellen Größen nicht zu verlassen. *Der Glaube an die Unwirksamkeit des Imaginären* trägt auch hier wie anderweitig gute Früchte (here moreover, for the purposes of these proofs, we do not need to leave the realm of real magnitudes. *The belief in the inefficacy of the imaginaries* bears good fruit here too in another way)."

Other proofs of (2) may be found in G. H. HARDY, "The integral $\int_0^\infty \frac{\sin x}{x}\,dx$", *Math. Gazette* **5**(1909-11), 98-103 and "Further remarks on the integral $\int_0^\infty \frac{\sin x}{x}\,dx$," *ibid.* **8**(1915-16), 301-303 (= *Collected Papers* V, 528-533 and 615-618).

The integral (1) occurs at prominent points in the literature. It is needed, e.g., in the proof of the theorem of WIENER and IKEHARA which forms the basis of what is probably the shortest proof of the prime number theorem; see K. CHANDRASEKHARAN, *Introduction to Analytic Number Theory*, Grundlagen der math. Wissenschaften 148, Springer-Verlag (1968), pp. 124 and 126.

Let us write $2x$ instead of x and recall the identity $\sin^2 2x = 4(\sin^2 x - \sin^4 x)$. It then follows from (1) that

$$\int_0^\infty \frac{\sin^2 x - \sin^4 x}{x^2} dx = \frac{1}{4}\pi, \quad \text{whence} \quad \int_0^\infty \frac{\sin^4 x}{x^2} dx = \frac{1}{4}\pi.$$

From this (via integration by parts) you can derive

$$\int_0^\infty \frac{\sin^4 x}{x^4} dx = \frac{1}{3}\pi.$$

For every natural number $n \geq 4$ the integral $I_n := \int_0^\infty \frac{\sin^n x}{x^n} dx$ exists. All the numbers I_n are rational multiples of π; e.g.,

$$I_3 = \frac{3}{8}\pi, \qquad I_5 = \frac{115}{384}\pi, \qquad I_6 = \frac{11}{40}\pi$$

and in general we have

$$I_n = \frac{\pi}{(n-1)!2^n} \sum_{0 \leq \nu < n/2} (-1)^\nu \binom{n}{\nu} (n-2\nu)^{n-1}, \quad n \geq 1,$$

an elegant derivation of which is given by T. M. APOSTOL in *Math. Magazine* **53**(1980), 183.

Exercises

Exercise 1. Show that

a) $\displaystyle\int_{-\infty}^\infty \frac{x}{1+x^2} e^{2ix} dx = i\pi e^{-2}$

b) $\displaystyle\int_0^\infty \frac{\cos x}{(1+x^2)^3} dx = \frac{7\pi}{16e}$

c) $\displaystyle\int_{-\infty}^\infty \frac{e^{2ix}}{x^4 + 10x^2 + 9} dx = \frac{\pi}{8}(e^{-2} - \frac{1}{3}e^{-6})$

d) $\displaystyle\int_{-\infty}^\infty \frac{\cos x}{(x^2 + a^2)^2} dx = \frac{\pi(1+a)}{2a^3 e^a}, \; a > 0$

e) $\displaystyle\int_{-\infty}^\infty \frac{\sin^2 x}{x^2 + a^2} dx = \frac{1}{2}\frac{\pi}{a}(1 - e^{-2a}).$

Exercise 2. Prove that

a) $\displaystyle\int_0^\infty \frac{\sqrt{x}}{x^2 + a^2} dx = \frac{\pi}{\sqrt{2a}}, \; a > 0$

b) $\displaystyle\int_0^\infty \frac{\sqrt{x}}{(x^2+4)^2}\,dx = \frac{1}{32}\pi.$

Exercise 3. Let q be a rational function with no poles in $[0,\infty)$. Suppose that the degree of its denominator exceeds that of its numerator by at least 2. Show by a modification of the proof of theorem 2 that

$$\int_0^\infty q(z)\,dz = -\sum_{w\in\tilde{\mathbb{C}}} \mathrm{res}_w(q(z)\ln z).$$

(Here $\tilde{\mathbb{C}}$ and $\ln z$ are as defined at the beginning of subsection 2.)

Exercise 4. Calculate the following integrals with the aid of exercise 3:

a) $\displaystyle\int_0^\infty \frac{dx}{x^3+x^2+x+1}$

b) $\displaystyle\int_{-1}^\infty \frac{dx}{x^4-x^3-3x^2-x+3}.$

§3 Gauss sums

In mid-May of 1801 GAUSS recorded in his diary (note 118) the formulas which nowadays are written together as the single sum formula

$$(1) \qquad \sum_0^{n-1} e^{\frac{2\pi i}{n}\nu^2} = \frac{1+(-i)^n}{1-i}\sqrt{n}$$

and designated as *Gauss sums*. With the help of his summation formulas GAUSS also derived that year in his *Disquisitiones Arithmeticae* (*Werke* 1, pp. 442, 443; English translation by Arthur A. CLARKE, revised edition (1986), Springer-Verlag) the law of quadratic reciprocity, without however determining exactly the sign of the square-root that intervenes in it. The determination of this sign proved to be extremely difficult and only after several years of effort did GAUSS finally succeed, on August 30, 1805. Quite agitated, in a letter of September 3, 1805 to OLBERS (*Werke* 10, part 1, p.24/25) he sketched how he had wrestled with the problem: "Die Bestimmung des Wurzelzeichens ist es gerade, was mich immer gequält hat. Dieser Mangel hat mir alles Übrige, was ich fand, verleidet; und seit 4 Jahren wird selten eine Woche hingegangen sein, wo ich nicht einen oder den andern vergeblichen Versuch, diesen Knoten zu lösen, gemacht hätte · · · Endlich vor ein paar Tagen ist's gelungen – aber nicht meinem mühsamen Suchen, sondern bloss durch die Gnade Gottes möchte ich sagen. Wie der Blitz einschlägt, hat sich das Räthsel gelöst. (The determination of the sign of the square-root is exactly what had always tormented me. This deficiency ruined for me all the other things I had found. For 4 years a week seldom passed without my making one or another futile attempt to loosen this knot $\cdots$ Finally a few days ago I succeeded – not through my laborious searching but merely by the grace of God I might say. Just like lightning striking, the puzzle resolved itself.)"

In subsection 2 we will give a proof of (1) by means of the residue calculus and as a by-product find once again the value of the error integral. This "diabolic proof" is due to L. J. MORDELL, "On a simple summation of the series $\sum_{s=0}^{n-1} e^{2s^2\pi i/n}$," *Messenger of Math.* 48(1918), 54-56. The first calculation of the Gauss sums (1) by application of the residue theorem to the integral $\int e^{2\pi i z^2/n}(e^{2\pi i z}-1)^{-1}dz$ was by L. KRONECKER, "Summirung der Gauss'schen Reihen $\sum_{h=0}^{h=n-1} e^{\frac{2h^2\pi i}{n}}$," *Jour. für Reine und Angew. Math.* 105(1889), 267-268 (and also in *Werke 4*, 295-300).

For the determination of the Gauss sums (1) we need a boundedness assertion about the function $e^{uz}(e^z-1)^{-1}$, $0 \le u \le 1$, which we will derive beforehand in subsection 1. This estimate also makes possible a simple determination (subsection 4) of the real Fourier series of the Bernoulli polynomials, which will contain the Euler formulas from 11.3.1 as a special case.

1. Estimation of $\dfrac{e^{uz}}{e^z-1}$ for $0 \le u \le 1$. The function $\varphi(z) := (e^z-1)^{-1}$ has a first-order pole at $2\pi i\nu$ for each $\nu \in \mathbb{Z}$ and no other singularities. About each of these poles we place a closed disc $\overline{B}_\nu$ of positive radius $r < 1$. Letting $Z := \mathbb{C} \setminus \bigcup_{\nu \in \mathbb{Z}} B_\nu$ denote the plane thus infinitely perforated, we have

Lemma. *The function $e^{uz}\varphi(z)$ is bounded in the set*

$$\{(u,z) \in \mathbb{R} \times \mathbb{C} : 0 \le u \le 1 \,,\, z \in Z\}.$$

Proof. Since u is real, the function $|e^{uz}\varphi(z)| = e^{u\Re z}|\varphi(z)|$ has period $2\pi i$ in z. It therefore suffices to establish its boundedness in $[0,1] \times S$, with S the perforated strip $\{z \in \mathbb{C} : |\Im z| \le \pi \,,\, |z| \ge r\}$ shown in the figure. In the compact subset $[0,1] \times \{z \in S : |\Re z| \le 1\}$ the function is continuous and consequently bounded. Let $z := x + iy$. In case $x \ge 1$, we have

$$|\varphi(z)| \le \frac{1}{|e^z|-1} = \frac{1}{e^x-1} \le 2e^{-x},$$

while if $x \le -1$, it follows quite trivially that $|\varphi(z)| \le \frac{1}{1-e^{-1}}$. We see therefore that

$$|e^{uz}\varphi(z)| \le \begin{cases} 2e^{(u-1)x} & \text{for} \quad x \ge 1 \\ (1-e^{-1})^{-1}e^{ux} & \text{for} \quad x \le -1. \end{cases}$$

Since $0 \le u \le 1$, $e^{(u-1)x} \le 1$ for $x \ge 1$ and $e^{ux} \le 1$ for $x \le -1$. The boundedness of $e^{uz}\varphi(z)$ in $[0,1] \times S$ is thus proven.

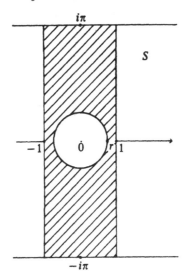

2. Calculation of the Gauss sums $G_n := \sum_0^{n-1} e^{\frac{2\pi i}{n} \nu^2}$, $n \geq 1$. One
verifies by direct calculation from the defining sum that

$$G_1 = 1 , \quad G_2 = 0 \quad \text{and} \quad G_3 = 1 + e^{\frac{2\pi i}{3}} + e^{\frac{2\pi i}{3} \cdot 4} = i\sqrt{3}.$$

In order to determine G_n generally, we introduce the entire function

$$G_n(z) := \sum_0^{n-1} \exp \frac{2\pi i}{n}(z+\nu)^2 , \qquad n \geq 1,$$

whose values at $z = 0$ are the Gauss sums. In order to be able to apply
the residue theorem we give up a little holomorphy and consider, as did
MORDELL, the functions $M_n(z) := \dfrac{G_n(z)}{e^{2\pi i z} - 1}$, which are meromorphic in
$\mathbb{C}$. We choose $r > 0$ large and discuss M_n in the parallelogram P having
vertices $-\frac{1}{2} - cr$, $\frac{1}{2} - cr$, $\frac{1}{2} + cr$ and $-\frac{1}{2} + cr$, where $c := e^{i\pi/4} = \frac{1}{\sqrt{2}}(1+i)$
(thus $c^2 = i$).

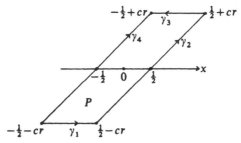

M_n has only one pole in P; it is at 0, is of the first order and the residue
there is $(2\pi i)^{-1}G_n(0)$. Therefore the residue theorem says that

(a) $\displaystyle\int_{\partial P} M_n(\zeta)d\zeta = G_n(0)$, with (cf. figure) $\partial P = \gamma_1 + \gamma_2 + \gamma_3 - \gamma_4$.

Let us set for short $I(r) := \int_{\gamma_2} M_n(\zeta)d\zeta - \int_{\gamma_4} M_n(\zeta)d\zeta$ and first show that

(b) $$\lim_{r \to \infty} I(r) = (1 + (-i)^n)c\sqrt{\tfrac{n}{2\pi}}\int_{-\infty}^{\infty} e^{-t^2} dt.$$

Proof. One checks easily that $G_n(z+1) - G_n(z) = e^{\frac{2\pi i}{n}z^2}(e^{4\pi iz} - 1)$, and as a result

$$M_n(z+1) - M_n(z) = e^{\frac{2\pi i}{n}z^2}(e^{2\pi iz} + 1).$$

Since $\int_{\gamma_2} M_n(\zeta)d\zeta = \int_{\gamma_4} M_n(\zeta + 1)d\zeta$, it follows that

$$I(r) = \int_{\gamma_4} e^{\frac{2\pi i}{n}\zeta^2}(e^{2\pi i\zeta} + 1)d\zeta.$$

Because of the identities $\frac{2\pi i}{n}\zeta^2 + 2\pi i\zeta = \frac{2\pi i}{n}(\zeta + \tfrac{1}{2}n)^2 - \tfrac{1}{2}\pi in$ and $e^{-\frac{1}{2}\pi in} = (-i)^n$, it further follows that

$$I(r) = \int_{\gamma_4} e^{\frac{2\pi i}{n}\zeta^2} d\zeta + (-i)^n \int_{\gamma_4} e^{\frac{2\pi i}{n}(\zeta + \tfrac{1}{2}n)^2} d\zeta.$$

Since γ_4 is parameterized as $\zeta(t) = -\tfrac{1}{2} + ct$, $t \in [-r, r]$, and $c^2 = i$, the last equality reads

$$I(r) = c\int_{-r}^{r} e^{-\frac{2\pi i}{n}\left(t - \frac{1}{2c}\right)^2} dt + (-i)^n c \int_{-r}^{r} e^{-\frac{2\pi i}{n}\left(t + \frac{1}{2c}(n-1)\right)^2} dt.$$

From this follows the equation (b), because the two integrals on the right side have the same limit (by translation-invariance of the error integral proved in 12.4.3(1)).   □

We next show that

(c) $$\lim_{r \to \infty} \int_{\gamma_1} M_n(\zeta)d\zeta = \lim_{r \to \infty} \int_{\gamma_3} M_n(\zeta)d\zeta = 0.$$

Proof. Since γ_1 and $-\gamma_3$ are given by $t \mapsto t - cr$ and $t \mapsto t + cr$, $t \in I := [-\tfrac{1}{2}, \tfrac{1}{2}]$, respectively, it suffices to show that $\lim_{r \to \infty} |M_n(t \pm cr)|_I = 0$. On the basis of lemma 1, $\varphi(2\pi iz) := (e^{2\pi iz} - 1)^{-1}$ is bounded on γ_1 and γ_3 independent of $r \geq 1$. Since $M_n(z)$ is comprised of a fixed number of summands $\exp(ai(z + \nu)^2) \cdot \varphi(2\pi iz)$, $0 \leq \nu < n$, (where $a := \frac{2\pi}{n}$) and $\Re[ai(t \pm cr + \nu)^2] = -ar^2 \mp \sqrt{2}(t + \nu)ar$, we only have to show that

$$\left| \exp ai(t \pm cr + \nu)^2 \right|_I = e^{-ar^2} \left(\max_{t \in I} e^{\mp \sqrt{2}(t+\nu)ar} \right)$$

tends to 0 as $r \to \infty$. However, since $a > 0$ and the exponent in the second factor on the right is only linear, this is clear.   □

From the equations (a), (b), (c) now follows directly that

$$G_n(0) = (1 + (-i)^n) \cdot \frac{1+i}{\sqrt{2\pi}} \sqrt{\frac{n}{2\pi}} \int_{-\infty}^{\infty} e^{-t^2} dt.$$

As we already know that $G_1(0) = 1$, the identity $\int_{-\infty}^{\infty} e^{-t^2} dt = \sqrt{\pi}$ is confirmed anew and therewith equation (1) of the introduction:

$$\boxed{\sum_{0}^{n-1} e^{2\pi i \nu^2/n} = \frac{1 + (-i)^n}{1 - i} \sqrt{n}}$$

a special case of which is

$$\sum_{0}^{n-1} e^{\frac{2\pi i}{n} \nu^2} = \sqrt{(-1)^{\frac{1}{2}(n-1)} n} \qquad \text{for odd integers } n.$$

Gauss' sum formula can be generalized, in the form of a "reciprocity formula" valid for all natural numbers $m, n \geq 1$:

$$\sum_{\nu=0}^{n-1} e^{\frac{m \pi i}{n} \nu^2} = e^{\frac{\pi i}{4}} \sqrt{\frac{n}{m}} \sum_{\nu=0}^{m-1} e^{\frac{-n\pi i}{m} \nu^2},$$

which reduces to (1) when $m = 2$. On this matter the reader should compare [Lin], p.75.

3. Direct residue-theoretic proof of the formula $\int_{-\infty}^{\infty} e^{-t^2} dt = \sqrt{\pi}$.

It is intriguing to determine the value of the error integral in as simple a way as possible using the residue theorem. Frontal attack using e^{-z^2} alone leads nowhere because e^{-z^2} has no non-zero residues. Instead of Mordell's auxiliary function M_1 we will consider the function

$$g(z) := e^{-z^2} / (1 + \exp(-2az)) \in \mathcal{M}(\mathbb{C}) , \text{ with } a := (1 + i)\sqrt{\pi/2}.$$

Since $a^2 = i\pi$, a is a period of $\exp(-2az)$; it follows from this that

$$(*) \qquad\qquad g(z) - g(z + a) = e^{-z^2}.$$

g has poles precisely at the points $-\frac{1}{2}a + na$, $n \in \mathbb{Z}$, and each is simple. Of these only the point $-\frac{1}{2}a$ lies in the strip determined by the real axis and the horizontal line through a (see the figure below). We have

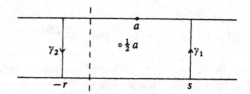

$$\operatorname{res}_{\frac{1}{2}a} g = \frac{\exp(-\frac{1}{4}a^2)}{-2a\exp(-a^2)} = -\frac{i}{2\sqrt{\pi}}.$$

It now follows from the residue theorem, on taking account of (∗), that

$$\int_{-r}^{s} e^{-x^2}\,dx + \int_{\gamma_1} g(\zeta)d\zeta + \int_{\gamma_2} g(\zeta)d\zeta = 2\pi i\,\operatorname{res}_{\frac{1}{2}a} g = \sqrt{\pi}.$$

The integrals along γ_1 and γ_2 converge to 0 with increasing s and r (proof!), so that the value of the desired improper integral follows.

The proof reproduced here is from H. KNESER [14], p. 121. In the older literature it was occasionally maintained that the error integral was not susceptible to evaluation via the residue calculus; see, for example, G. N. WATSON, *Complex Integration and Cauchy's Theorem*, Cambridge Tracts in Mathematics and Mathematical Physics 15, London 1914 (reprinted by Hafner Publishing Co., New York 1960), p.79; also E. T. COPSON *An Introduction to the Theory of Functions of a Complex Variable*, Oxford, At the Clarendon Press 1935 (reprinted 1944 and 1946), p.125.

An interesting presentation of this and related problems was given in 1945 by G. PÓLYA in "Remarks on computing the probability integral in one and two dimensions," pp. 63-78 of *Proceedings Berkeley Symposium on Mathematical Statistics and Probability*, Berkeley and Los Angeles 1949; and pp. 209-224, vol. 4 of his *Collected Papers*. PÓLYA integrates $\int e^{\pi i \zeta^2}\tan(\pi\zeta)d\zeta$ along the parallelogram with vertices $R + iR,\ -R - iR,\ -R + 1 - iR,\ R + 1 + iR$.

4. Fourier series of the Bernoulli polynomials. Let $w_0, w_1, w_2, \ldots$ be a sequence in $\mathbb{C}$ having *no accumulation point* in $\mathbb{C}$, let f be holomorphic in $\mathbb{C} \setminus \{w_0, w_1, w_2, \ldots\}$, γ_n a sequence of simply closed paths, $\{k_n\}$ a strictly monotone sequence in $\mathbb{N}$ with the property that of the w_ν precisely $w_0, w_1, \ldots, w_{k_n}$ lie in Int γ_n, for each $n \in \mathbb{N}$. If in fact each γ_n avoids all the w_ν, then it is immediate from the residue theorem that

$$(*) \qquad \lim_{n\to\infty} \sum_{0}^{k_n} \operatorname{res}_{w_\nu} f = \frac{1}{2\pi i}\lim_{n\to\infty}\int_{\gamma_n} f(\zeta)d\zeta,$$

whenever the limit on either side exists.   □

We will apply formula (∗) to the following family of functions which are all meromorphic in $\mathbb{C}$:

$$h_k(z) := z^{-k-1} F(w, z)\,,\ \text{with}\ F(w, z) := ze^{wz}(e^z - 1)^{-1}\,,\ k \geq 1.$$

For now $w \in \mathbb{C}$ is arbitrary. h_k is holomorphic in $\mathbb{C} \setminus 2\pi i\mathbb{Z}$ and every point $2\pi i\nu$, $\nu \neq 0$, is a simple pole, the residue of h_k there being $(2\pi i\nu)^{-k} e^{2\pi i\nu w}$. Since for z near 0 we have (cf. 7.5.4)

$$F(w, z) = \sum_0^\infty \frac{B_\mu(w)}{\mu!} z^\mu,$$

the point $0 \in \mathbb{C}$ is a pole of h_k of order $k+1$ and $\mathrm{res}_0 h_k = \dfrac{B_k(w)}{k!}$, $B_\mu(w)$ designating the μth Bernoulli polynomial. The poles which lie in the disc of radius $(2n+1)\pi$ centered at 0 are just $0, \pm 2\pi i, \dots, \pm 2\pi in$. Consequently, for $\gamma_n := \partial B_{(2n+1)\pi}(0)$, we have according to $(*)$

$$\frac{B_k(w)}{k!} + \sum_{\nu=1}^\infty [(2\pi i\nu)^{-k} e^{2\pi i\nu w} + (-2\pi i\nu)^{-k} e^{-2\pi i\nu w}] = \frac{1}{2\pi i} \lim_{n\to\infty} \int_{\gamma_n} h_k(\zeta) d\zeta,$$

provided the limit on the right exists. Now

$$\left| \int_{\gamma_n} h_k(\zeta) d\zeta \right| \leq |h_k|_{\gamma_n} L(\gamma_n)$$

$$\leq |z^{-k}|_{\gamma_n} |e^{wz} \varphi(z)|_{\gamma_n} L(\gamma_n) = \frac{2\pi}{((2n+1)\pi)^{k-1}} |e^{wz} \varphi(z)|_{\gamma_n}.$$

Since γ_n lies in the perforated plane Z, the sequence $|e^{wz}\varphi(z)|_{\gamma_n}$ is bounded, according to lemma 1, for every real number $w = u$ with $0 \leq u \leq 1$. For such u and each $k > 1$ the integrals therefore converge to 0 as $n \to \infty$. Upon writing x instead of w, we find we have proved

For all real x with $0 \leq x \leq 1$ and all $k \geq 2$

$$B_k(x) = \frac{-k!}{(2\pi i)^k} \sum_1^\infty \frac{1}{\nu^k} [e^{2\pi i\nu x} + (-1)^k e^{-2\pi i\nu x}].$$

Passing over to cos and sin, we get for even and odd indices the *real Fourier series for the Bernoulli polynomials*:

$$B_{2k}(x) = (-1)^{k-1} \frac{2(2k)!}{(2\pi)^{2k}} \sum_{\nu=1}^\infty \frac{\cos 2\pi\nu x}{\nu^{2k}} \qquad \text{for } 0 \leq x \leq 1 \,, \, k \geq 1$$

$$B_{2k+1}(x) = (-1)^{k-1} \frac{2(2k+1)!}{(2\pi)^{2k+1}} \sum_{\nu=1}^\infty \frac{\sin 2\pi\nu x}{\nu^{2k+1}} \qquad \text{for } 0 \leq x \leq 1 \,, \, k \geq 1.$$

One can show (e.g., via finer appraisals of $e^{uz}\varphi(z)$) that the latter formula remains valid for $k = 0$ as well, that is, gives the Fourier sine series of $B_1(x) = x - 1/2$, but only for $0 < x < 1$. On this point compare also [14], p.122.

Recall that $B_n(0)$ is the nth Bernoulli number. Since for odd n the above Fourier (sine) series vanishes at 0, we recover the known fact that the Bernoulli numbers with odd subscript are all 0. But for even n by contrast the Fourier (cosine) series yields anew the Euler formulas from 11.3.1.

Short Biographies of ABEL, CAUCHY, EISENSTEIN, EULER, RIEMANN, and WEIERSTRASS

Niels Henrik ABEL, Norwegian mathematician: born 1802 on the island of Finnöy near Stavanger; entered Christiania University as a completely self-taught student in 1822; in 1824 published as a pamphlet at his own expense a proof of the insolubility by radicals of algebraic equations of degree five or greater; 1825/26 acquaintanceship with CRELLE[1] in Berlin; 1826/27 disappointing sojourn in Paris; 1827 world famous, but without a position, return to Christiania as "studiosus Abel"; 1829 died in poverty of tuberculosis in Froland near Arendahl just two days before the arrival of a letter from CRELLE announcing a position for him at Berlin; first obituary 1829 by CRELLE in volume 4 of his journal; 1830 awarded posthumously (and shared with JACOBI) the great prize of the Paris Academy. In 1922 MITTAG-LEFFLER wrote: *Viele große Männer sind einmal Studenten gewesen. Keiner ist mehr als Abel schon als Student in die Unsterblichkeit eingegangen.* (Many great men were once students. But none more than Abel attained immortality as a student.) The book *Niels Henrik Abel: Mathematician Extraordinary*, Chelsea Publ. Company (1974), New York, by O. ORE is worth reading.

Baron Augustin-Louis CAUCHY, French mathematician born 1789 in Paris; 1810 at the age of 21 *Ingénieur des Ponts et Chaussées* in Cherbourg under Napoléon I; after 1813 again in Paris; 1816 at the age of 27, member of the Academy of Sciences and soon thereafter professor at

---

[1]August Leopold CRELLE, 1780–1855; highway engineer and amateur mathematician, important participant in the 1838 construction of the first Prussian rail line (Berlin-Potsdam), promoter of promising young mathematicians and in particular patron of ABEL'S; encouraged by ABEL and the geometer Jakob STEINER he founded the first German mathematics periodical which lasted, *Journal für die Reine und Angewandte Mathematik*; also founded the architecture periodical *Journal für Baukunst*.

the École Polytechnique, later also at the Sorbonne and at the Collège de France; Chevalier de la Légion d'honneur. As a Catholic and adherent of the Bourbons CAUCHY refused to take the oath to the new government in 1830; he emigrated first to Freiburg (Switzerland), was for a time professor of mathematical physics at Turin and from 1833-38 tutor to the son of Charles X in Prague; 1838 return to Paris with the title of Baron and once again active in the Academy; from 1848 after the abolition of the loyalty oath again professor (of astronomy) at the Sorbonne; 1849 Knight of the Order "Pour le mérite dans les sciences et dans les arts"; died 1857 in Sceaux. — As early as 1868 a (rather hagiographic) two-volume Cauchy biography by C.-A. VALSON appeared, entitled *La vie et les travaux du baron Cauchy* with a foreword by C. HERMITE (reprint by Albert Blanchard (1970), Paris); for a recent biography see [H₂].

Ferdinand Gotthold Max EISENSTEIN, German mathematician: born 1823 in Berlin; 1843 matriculated at Berlin University; 1844 publication of 25 works in volumes 27, 28 of Crelle's journal; 1845 as a 3rd semester student awarded an honorary doctorate by Breslau University at Kummer's suggestion, and nearly recommended by GAUSS for the non-military category of the order "Pour le mérite"; 1846 priority dispute with JACOBI; 1847 *privatdozent* at Berlin — RIEMANN heard his lectures on elliptic functions; 1848 imprisoned in Spandau; 1849 curtailment of his "allowance" from 500 to 300 Taler per year "in consequence of his calumniations as a republican"; 1850 labelled "very red", DIRICHLET, JACOBI and A. VON HUMBOLDT propose him for a university professorship (without success); 1851 simultaneously with KUMMER, EISENSTEIN becomes a corresponding member of the Göttingen Society; 1852 ordinary member of the Berlin Academy of Science; died 1852 of tuberculosis, the 83 year-old A. VON HUMBOLDT paying him the last honors.

As early as 1847 a collection of mathematical papers of EISENSTEIN's with a flattering foreword by GAUSS was published (reprinted 1967 by Georg Olms Verlagsbuchhandlung, Hildesheim). The complete *Mathematische Werke* of EISENSTEIN weren't however published until 1975, in two volumes by Chelsea Publ. Comp., New York (2nd ed., 1989). A. WEIL's review of this in *Bull. Amer. Math. Soc.* **82**(1976), 658-663 is very worth reading: He has good and bad fairies at the child's cradle prophesying the heights and depths of EISENSTEIN's life and mathematical accomplishments.

In 1895 there appeared *Eine Autobiographie von Gotthold Eisenstein* edited by F. RUDIO in the *Zeitschr. Math. Phys.* **40**, suppl. 143-168 (also in vol. 2 of the *Math. Werke*, 879-904). Also very informative is the article by Kurt-R. BIERMANN, "Gotthold Eisenstein. Die wichtigsten Daten seines Lebens und Wirkens," *Jour. für Reine und Angew. Math.* **214**(1964), 19-30 (reproduced in vol. **2** of the *Math. Werke*, 919-929).

Leonhard EULER, Swiss mathematician: born 1707 in Basel; 1720 student at Basel; 1727 emigration to St. Petersburg where Czar Peter I had founded an academy in 1724; 1730 Professor of Physics, from 1733 Professor of Mathematics at St. Petersburg as successor of Daniel BERNOULLI; 1735 loss of sight in his right eye; 1741 emigration to Berlin; 1744 Director of the mathematical section of the Prussian Academy of Science; 1766 return to St. Petersburg, among other reasons because of strained relations with the Prussian king, who had little understanding for EULER's mathematical activity; 1771 blindness; died 1783 in St. Petersburg.

Nikolaus FUSS, a student who was married to one of Euler's granddaughters published his *Lobrede auf Herrn Leonhard Euler* in 1786 (reproduced in Euler's *Opera Omnia* 1st set., vol. 1, p. XLIII); recommended reading is chapter 8, "Analysis Incarnate", in E. T. BELL, [H3]; also of interest is the short biography *Leonhard Euler* by R. FUETER in Supplement Nr. 3 of the periodical *Elemente der Mathematik*, Basel 1948. A detailed evaluation of Euler is offered by the memorial volume of the canton of Basel: *Leonhard Euler 1707-1783, Beiträge zu Leben und Werk*, Birkhäuser Verlag (1983), Basel. In the *Éloge de M. Euler par le Marquis de Condorcet* (in the *Opera Omnia* 3rd ser., vol. 12, 287-310) Euler's death is described thus (p.309): "la pipe qu'il tenoit à la main lui échappa, et il cessa de calculer et de vivre (the pipe that he held in his hand slipped from him, and he ceased calculating and living)."

Georg Friedrich Bernhard RIEMANN, German mathematician: born 1826 in Breselenz, in the Lüchow-Dannenberg district; 1846 student at Göttingen, at first in theology; 1847-1849 student at Berlin, auditor of DIRICHLET and JACOBI, acquaintanceship with EISENSTEIN; 1849 return to Göttingen; 1850 assistant to W. WEBER in physics; 1851 doctoral degree with epoch-making dissertation *Grundlagen für eine allgemeine Theorie der Functionen einer veränderlichen complexen Grösse*; 1853 Habilitationsschrift *Über die Darstellbarkeit einer Function durch eine trigonometrische Reihe*, where among other things the RIEMANN-integral is to be found; 1854 Habilitation's lecture *Über die Hypothesen, welche der Geometrie zu Grunde liegen* (English translation by H. S. WHITE in vol. II of D. E. SMITH'S *A Source Book in Mathematics*, Dover Publ., Inc. (1958), New York), in which modern differential geometry was born; *Privatdozent* in Göttingen without salary; 1855 annual remuneration of 200 Taler; 1857 Extraordinarius in Göttingen with 300 Taler annual salary; 1859 DIRICHLET's successor in the Gauss chair, member of the Göttingen Society of Science and corresponding member of the Berlin Academy, publication of the work *Ueber die Anzahl der Primzahlen unter einer gegebenen Grösse* containing the still unproven conjecture about the zeros of the Riemann ζ-function; died 1866 of tuberculosis in Selasca, Italy; his tombstone, which is still intact in the cemetery at Biganzolo (Lago Maggiore) bears the inscription "Denen, die Gott lieben, müssen alle Dinge zum Besten dienen (...

for those who love God all things must serve for the best)" – Romans 8,28. *Bernhard Riemann's Lebenslauf*, written by his friend Richard DEDEKIND and reproduced in his *Werke*, 539-558, is worth reading, as is Felix KLEIN's assessment in "Riemann and his significance for the development of modern mathematics," *Bull. Amer. Math. Soc.* 1(1895), 165-180.

Karl Theodor Wilhelm WEIERSTRASS, German mathematician: born 1815 in Ostenfelde in the Warendorf district, Westphalia; 1834-38 dutiful son reluctantly studying for the civil service (and fencing and drinking) in Bonn; 1839-40 study of mathematics at the Münster academy, state examination under GUDERMANN; 1842-1848 teacher at the Progymnasium in Deutsch-Krone, West Prussia, of mathematics, penmanship and gymnastics; 1848-1855 teacher at the Gymnasium in Braunsberg, East Prussia; 1854 publication of trail-blazing results (gotten already in 1849) in the work "Zur Theorie de Abelschen Functionen," in vol. 47 of *Jour. für Reine und Angew. Math.*, thereupon honorary doctorate from the University of Königsberg and promotion to assistant headmaster; 1856 at the instigation of A. VON HUMBOLDT and L. CRELLE appointment as professor at the Industrial Institute (later Technical University) in Berlin; 1857 adjunct professor at the University of Berlin; after 1860 lectures often with more than 200 auditors; 1861 breakdown from over-work; 1864 at the age of almost fifty appointment to an ordinary professorship, created for him, at the University of Berlin; 1873/74 *rector magnificus* there, member of numerous academies at home and abroad, 1875 Knight of the German Nation of the Order "Pour le mérite dans les sciences et dans les arts"; 1885 stamping of a Weierstrass medal (for his 70th birthday); 1890 teaching activity halted by serious illness, confinement to a wheelchair; 1895 festive unveilling of his image in the national gallery (80th birthday); 1897 died in Berlin. —

As yet there is no exhaustive biography of WEIERSTRASS. But thoroughly worth reading are the article by P. DUGAC, "Eléments d'analyse de Karl Weierstrass", *Archive for History of the Exact Sciences* 10(1973), 41-176 and the lecture "Karl Weierstrass: Ausgewählte Aspekte seiner Biographie" by K. BIERMANN which is reprinted in *Jour. für Reine und Angew. Math.* 223(1966), 191-220. The personal remarks which A. KNESER makes in his article "Leopold Kronecker" (*Jahresber. DMV* 33(1925), 210-288) are very revealing; on pp. 211, 212 he describes the mathematical life at Berlin in the 1880's thus: "Der unbestrittene Beherrscher des ganzen Betriebs war zweifellos Weierstraß, eine königliche, in jeder Weise imponierende Gestalt. Man kennt den prachtvollen, weiß umlockten Schädel, das leuchtend blaue, etwas schief verhängte Auge des reinrassigen westfälischen Landkindes. Seine Vorlesungen hatten sich damals zu hoher auch äußerer Vollendung entwickelt, und nur selten kamen jene aufregenden Minuten, wo der große Mann stockte, auch der Zuspruch des treuen Gehilfen an der Tafel, etwa meines Freundes Richard Müller, ihm nicht auf den Weg helfen konnte, und nun versank er für einige Minuten in ein majestätisches Schweigen; zweihundert junge Augenpaare ruhten auf dem prachtvollen Schädelrund mit der andächtigen Vorstellung, daß hinter dieser glänzenden Hülle die höchste Wissenschaft arbeitete. Zweihundert Jünglinge waren es in der Tat, die bei Weierstraß die elliptischen Funktionen hörten und durchhörten mit dem vollen Bewußtsein, daß diese Dinge damals in keinem Staatsexamen vorkamen, ein glänzendes Zeugnis für den wissenschaftlichen Geist jener Zeit. Ja auch von den Anwendungen dieser Dinge wußte man wenig, obwohl deren schon sehr schöne vorlagen; die Lehre vom Primat der angewandten Mathematik, von der höheren Würdigkeit der Anwendungen gegenüber der reinen Mathematik, war damals noch nicht entdeckt. Auch an diesem großen Manne übte sich der Humor der Jugend; er galt als guter Weinkenner, und die Berliner, die über die hart westfälische Sprechweise des Meisters lästerten, zitierten als Musterausspruch, den man gehört haben wollte: Ein chutes Chlas Burchunder trink ich chanz chern. (The undisputed master of the whole operation was without doubt Weierstrass, a regal and in every way imposing figure. All knew the magnificent white-locked head, the shining blue eyes slightly drooping at the corners which belonged to the country boy of pure Westphalian stock. By this time his lectures had evolved to a high level of perfection in presentation as well as content and only seldom were those tense minutes experienced where the great man faltered and even the promptings of his faithful assistant at the blackboard, perhaps my friend Richard Müller, couldn't get him back on track; then he would sink into majestic silence for a few minutes; two hundred pairs of young eyes were riveted on the splendid brow with the devout conviction that behind that shining facade the greatest intellect was at work. There were in fact two hundred youths who attended and listened intently to Weierstrass's lectures on elliptic functions, fully aware that at that time such things never came up on any state examination, a dazzling testimonial to the intellectual spirit of the times. People even knew very little about the applications of these things, although there were already available some very beautiful ones. The doctrine of the primacy of applied mathematics, of the greater worth of applications as against pure mathematics, had not yet been discovered. The humor of the young was unleashed even on this great man: He was considered a connoisseur of wine and the Berliners, who mocked his westphalian pronunciation, claimed to have actually heard from him the following quintessential example: I'd kladly kulp a kood klass of Burkundy.)" – the k's here should be read as g's.

WEIERSTRASS, by his lectures in Berlin, influenced mathematics in Germany like no one else. The assistant headmaster from East Prussia became the "praeceptor mathematicus Germaniae."

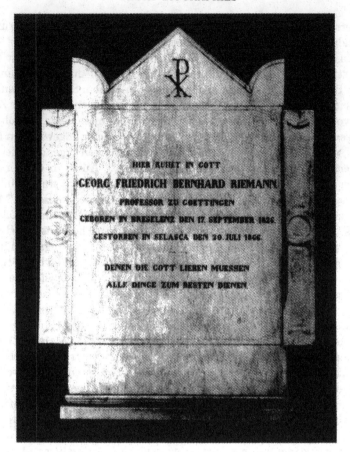

Photo: H. Götze.

DEDEKIND says in Bernhard RIEMANN's *Lebenslauf* that the headstone
was removed during a relocating of the cemetary. The panel carrying the
inscription was, however, not lost; visitors will find it at the entrance to
the Selasca cemetary as a remembrance of RIEMANN beyond the tomb and
through time.

Translation of the text on the gravestone: "Here rests in God, GEORG
FRIEDRICH BERNHARD RIEMANN, Professor at Göttingen. Born in
Breselenz on September 17, 1826. Died in Selasca on July 20, 1866. For
those who love God, all things must serve for the best."

Literature

Ecclesiastes XII, 12

Classical Literature on Function Theory

The textbook literature on function theory is inexhaustible and in the last few decades has become almost too vast even to survey. In Euler's time there was still no marked feeling for what later became known as "mathematical rigor". At that time the authors most read were BERNOULLI(S), DE L'HOSPITAL, MACLAURIN, LAGRANGE, *inter alii*. Today their books are only of historical interest; "these authors to a greater or lesser extent fall into the error of implicitly assuming unrestricted validity for their algebraic formulas and then drawing unwarranted conclusions from them."

In what follows, without any claim to completeness, some especially important classical treatises and textbooks are mentioned in alphabetical order; even though today many of them are forgotten. Particulars on quite a few other historically significant works will be found in the body of the text.

[A] ABEL, N.H., "Untersuchungen über die Reihe $1 + \frac{m}{1}x + \frac{m(m-1)}{1\cdot 2}x^2 + \frac{m(m-1)(m-2)}{1\cdot 2\cdot 3}x^3 + \cdots$ usw", *Jour. für Reine und Angew. Math.* **1** (1826), 311-339; also in his *Œuvres* **1**, 219-250 (in French) and in *Ostwald's Klassiker der Exakten Wissenschaften*, Nr. 71; English translation by K. MIWA in *Memoirs on Infinite Series*, Tokyo Mathematical and Physical Society (1891), Tokyo.

[BB] BRIOT, CH. and J.-C. BOUQUET, *Théorie des fonctions doublement périodiques et, en particulier, des fonctions elliptiques*, Mallet-Bachelier (1859), Paris; 2nd ed. 1875.

[Bu] BURKHARDT, H., *Einführung in die Theorie der analytischen Functionen einer complexen Veränderlichen*, Verlag von Veit & Comp.,

(1897), Leipzig; 3rd ed. 1908. English translation by S. E. RA-SOR, *Theory of Functions of a Complex Variable*, D. C. Heath & Co. (1913), Boston.

[Ca] CARATHÉODORY, C., "Untersuchungen über die konformen Abbildungen von festen und veränderlichen Gebieten," *Math. Annalen* **72** (1912), 107-144.

[C] CAUCHY, A. L., *Cours d'analyse de l'École Royale Polytechnique (Analyse algébrique)*. Paris, 1821. Reprinted by Wissenschaftliche Buchgesellschaft (1968), Darmstadt; also in his *Œuvres* (2) **3**, 1-331.

[C₁] CAUCHY, A. L., *Mémoire sur les intégrales définies*, 1814 (but not published until 1827); *Œuvres* (1) **1**, 319-506.

[C₂] CAUCHY, A. L., *Mémoire sur les intégrales définies, prises entre des limites imaginaires*, 1825; *Œuvres* (2) **15**, 41-89 (this volume wasn't published until 1974!); also reprinted in *Bull. Sci. Math.* (1) **7**(1874), 265-304 + **8**(1875), 43-55 & 148-159.

[E] EULER, L., *Introductio in Analysin Infinitorum* 1st volume, M. M. Bousquet (1748), Lausanne; also in his *Opera Omnia* (1) **8**. German translations by A. C. MICHELSEN 1788, Berlin and H. MASER 1885 published by Julius Springer under the title *Einleitung in die Analysis des Unendlichen* and reprinted in 1983 with a new introduction by Wolfgang WALTER. French translation by J. B. LABEY Chez Barrois (1796-97), Paris and reprinted in 1967 by Culture et Civilisation, Bruxelles. English translation (in two volumes) by John D. BLANTON, Springer-Verlag (1988 and 1989), Berlin and New York.

[Ei] EISENSTEIN, F. G. M., "Genaue Untersuchung der unendlichen Doppelproducte, aus welchen die elliptischen Functionen als Quotienten zusammengesetzt sind, und der mit ihnen zusammenhängenden Doppelreihen (als eine neue Begründungsweise der Theorie der elliptischen Functionen, mit besonderer Berücksichtigung ihrer Analogie zu den Kreisfunctionen)," *Jour. für Reine und Angew. Math.* **35**(1847), 153-274; also in his *Math. Werke* **1**, 357-478.

[G₁] GOURSAT, É., "Sur la definition générale des fonctions analytiques, d'après Cauchy," *Trans. Amer. Math. Soc.* **1**(1900), 14-16.

[G₂] GOURSAT, É., *Cours d'analyse mathématique*, Vol. 2, Gauthier-Villars (1905), Paris; 7th ed. 1949. English translation by E. R. HEDRICK and O. DUNKEL, *A Course in Mathematical Analysis*, Ginn & Co. (1916) Boston & New York; Reprinted by Dover Publ. Co. (1959), New York.

[Kr] KRONECKER, L., *Theorie der einfachen und der vielfachen Integrale*, E. NETTO, editor. B. G. Teubner (1894), Leipzig.

[Lan] LANDAU, E., *Darstellung und Begründung einiger neuerer Ergebnisse der Funktionentheorie*, Springer Verlag (1916), Berlin; 2nd ed. 1929; 3rd edition with supplements by D. GAIER, 1986.

[Lau] LAURENT, P. A., "Extension du théorème de M. Cauchy relatif à la convergence du développement d'une fonction suivant les puissances ascendantes de la variable," *Comptes Rendus Acad. Sci. Paris* **17**(1843), 348-349 (an announcement only). Cf. also pp. 938-940.

[Lin] LINDELÖF, E., *Le calcul des résidus et ses applications à la théorie des fonctions*, Gauthier-Villars (1905), Paris. Reprinted by Chelsea Publ. Co. (1947), New York.

[Liou] LIOUVILLE, J., "Leçons sur les fonctions doublement périodiques," 1847; published in *Jour. für Reine und Angew. Math.* **88**(1879), 277-310.

[M] MORERA, G., "Un teorema fondamentale nella teorica delle funzioni di una variabile complessa," *Rend. Reale Ist. Lomb. di scienze e lettere* (2) **19**(1886), 304-307.

[Os] OSGOOD, W. F., *Lehrbuch der Funktionentheorie I, II*, B. G. Teubner (1906), Leipzig; 5th ed. of vol. I, 1928. Reprinted by Chelsea Publ. Co. (1965), New York.

[P] PRINGSHEIM, A., *Vorlesungen über Funktionenlehre*. Part I: *Grundlagen der Theorie der analytischen Funktionen einer komplexen Veränderlichen*, 624 pp. (1925). Part II: *Eindeutige analytische Funktionen*, 600 pp. (1932). B. G. Teubner, Leipzig. Reprinted by Johnson Reprint Corp. (1968), New York.

[R] RIEMANN, B., "Grundlagen für eine allgemeine Theorie der Functionen einer veränderlichen complexen Grösse," Inaugural Dissertation (1851), Göttingen; *Werke*, 5-43.

[Sch] SCHOTTKY, F., "Über das Cauchysche Integral," *Jour. für Reine und Angew. Math.* **146**(1916), 234-244.

[W₁] WEIERSTRASS, K., "Darstellung einer analytischen Function einer complexen Veränderlichen, deren absoluter Betrag zwischen zwei gegebenen Grenzen liegt," Münster 1841; first published 1894 in the *Math. Werke* **1**, 51-66.

[W₂] WEIERSTRASS, K., "Zur Theorie der Potenzreihen," Münster 1841; first published 1894 in the *Math. Werke* **1**, 67-74.

[W₃] WEIERSTRASS, K., "Zur Theorie der eindeutigen analytischen Functionen," *Abhandlungen der Königl. Preuß. Akademie der Wissenschaften zu Berlin* (1876), 11-60; *Math. Werke* **2**, 77-124.

[W₄] WEIERSTRASS, K., "Zur Functionenlehre," *Monatsber. Königl. Preuß. Akademie der Wissenschaften zu Berlin* (1880), 719-743, Nachtrag (1881), 228-230; *Math. Werke* **2**, 201-233.

[W₅] WEIERSTRASS, K., *Einführung in die Theorie der analytischen Funktionen*, Schriftenreihe des Mathematischen Instituts Universität Münster, second series, vol. **38**(1986).

[W₆] WEIERSTRASS, K., *Einleitung in die Theorie der analytischen Funktionen*, Vorlesung Berlin 1878 in einer Mitschrift von Adolf HURWITZ. Bearb. von Peter ULLRICH (Dokumente zur Geschichte der Mathematik. Im Auftrag der Deutschen Mathematiker-Vereinigung, hrsg. von Winfried SCHARLAU; Band 4, Vieweg Verlag, Braunschweig Wiesbaden, 1988).

[We] WEIL, A., *Elliptic functions according to Eisenstein and Kronecker*, Ergebnisse der Mathematik **88**, Springer-Verlag (1976), Heidelberg.

[WW] WHITTAKER, E. T. and G. N. WATSON, *A Course of Modern Analysis*, Cambridge at the University Press, 1st ed. 1902; 4th ed. 1927.

We will close by commenting on some of these works in chronological order.

[E] **Euler** 1784: This is the first textbook on analysis and it can be read by students even today without great strain. The style of language and notations are nearly "modern", a considerable amount of our contemporary terminology having been first introduced here by EULER. Complex numbers are on the same footing as real numbers. Functions are analytic expressions (§4), hence holomorphic. In §28 the Fundamental Theorem of Algebra will be found (without proof). The binomial series is mentioned in §71 without further clarification as a "Theorema universale" and is extensively employed thereafter; oddly enough, EULER says nothing about a proof. The exponential function, logarithms and the circular functions are systematically treated for the first time in the *Introductio* and, via calculations with infinitely small numbers (§115 ff), they are developed into power series. The Euler formula $e^{ix} = \cos x + i \sin x$ occurs in §138; he derives his infinite product for the sine in §158 and the partial fraction representation for $\pi \cot \pi z$ in §178.

For EULER power series are just non-terminating polynomials. It is in the introduction to this book that he wrote the words with which our chapter 12 opens. EULER had such a mastery of the calculus of the infinitely small and the infinitely large that he is still envied for this skill today. "He is the great manipulator and pointed the way to thousands of results later established rigorously" (M. KLINE, [H₉], p.453).

The *Introductio* experienced many editions and was issued anew in 1922 as part of Euler's *Opera Omnia* by A. KRAZER and F. RUDIO. In their foreword these editors write: "... (ein) Werk, das auch heute noch verdient, nicht nur gelesen, sondern mit Andacht studiert zu werden. Kein Mathematiker wird es ohne reichen Gewinn aus der Hand legen. Dieses Werk ist nicht nur durch seinen Inhalt, sondern auch durch seine Sprache maßgebend geworden für die ganze Entwicklung der mathematischen Wissenschaft. (... a work which still deserves not only to be read but to be studied with assiduousness. No mathematician will put it down without having greatly benefitted. This work has become a landmark in the whole development of mathematical science, not only by virtue of its content but also by virtue of its language.)"

[C] **Cauchy** 1821: At the urging of LAPLACE and POISSON, CAUCHY wrote out his lecture course "pour la plus grande utilité des élèves (for the greater utility of the students)"; it may have been the first example of official lecture notes for students. The material is somewhat the same as in [E] but actual comparison reveals the new critical attitudes at work. Analysis is developed consistently *ab ovo* and, in principle, unimpeachably. This work exerted a guiding and lasting influence on the development of analysis and especially function theory throughout the 19th century. On account of its excellence it was very soon introduced into almost all the educational institutions of France and also became generally known in Germany; thus as early as 1828 the German translation by HUZLER (co-rector of the city high school in Königsberg) appeared. In his introduction (pp. i/ij) CAUCHY describes his program with these sentences: "Je traite successivement des diverses espèces de fonctions réelles ou imaginaires, des séries convergentes ou divergentes, de la résolution des équations, et de la décomposition des fractions rationnelles. (I treat successively various kinds of real or complex functions, convergent or divergent series, the resolution of equations and the decomposition of rational fractions.)" Thus the Cauchy convergence criterion for series is found in chapter VI; in chapters VII-X functions of a complex argument are introduced for the first time with, as a matter of principle, precise specification of their domains. To be sure, complex valued functions were not *consciously* introduced: the dominant role is always played by the *two* real functions u and v, not by the *single* complex function $u + \sqrt{-1}v$. The concept of continuity is carefully set out. The convergence of a series with complex terms is reduced to that of the series of corresponding absolute values (on p.240, e.g., it says that every complex power series has a circle of convergence whose radius is given by the familiar limit superior formula). In chapter X the Fundamental Theorem of Algebra is derived: the existence of zeros of a polynomial $p(z)$ is proved by considering the minima of the real function $|p(z)|^2$.

With the *Cours d'analyse* the age of rigor and the *arithmetization* of analysis begins. Only the important idea of (local) uniform convergence was missing, to give the work the finishing touch; in ignorance of this concept CAUCHY enunciates the incorrect theorem that convergent sequences of continuous functions always have continuous limit functions. Regarding the methods used in his book CAUCHY says (p.ij of the introduction): "Quant aux méthodes, j'ai cherche à leur donner toute la rigueur q'on exige en géométrie, de manière à ne jamais recourir aux raisons tirées de la généralité de l'algèbre. (I have sought to give them all the rigor that one demands in geometry, in such a way as to never have recourse to reasons drawn from the generality of algebra.)"

[A] **Abel** 1826: This work was written in Berlin, where in Crelle's library ABEL had become acquainted with Cauchy's *Cours d'analyse*. The latter work was his model; he writes: "Die vortreffliche Schrift von Cauchy, welche von jedem Analysten gelesen werden sollte, der die Strenge bei mathematischen Untersuchungen liebt, wird uns dabei zum Leitfaden dienen. (The splendid writing of Cauchy, which should be read by every analyst who loves rigor in mathematical investigations, will serve as our guide.)" Abel's work is itself a model of exact reasoning. It contains, among other things the Abel Lemma and the Abel Limit Theorem – and even some critical remarks about the *Cours d'analyse*.

[W_1–W_6] **Weierstrass** 1841 to 1880: His early publications did not become generally known to mathematicians until the appearance of his *Mathematische Werke*, beginning in 1894. From the decade of the 60's (last century) on WEIERSTRASS gave lecture courses in Berlin in the style that is prevalent today. He first gave his lectures on *Allgemeine Theorie der analytischen Functionen* during the 1863/64 winter – and indeed six hours per week (cf. *Math. Werke* **3**, pp. 355-60). Unfortunately, unlike CAUCHY, WEIERSTRASS never wrote his lectures out in book form, but there are transcriptions by his various pupils. Thus, for example, from H. A. SCHWARZ's hand we have an elaboration of his lectures on *Differentialrechnung* held at the Royal Industrial Institute in the 1861 summer semester. There is further a transcription by A. HURWITZ of his summer semester 1878 lectures *Einleitung in die Theorie der analytischen Funktionen* [W_6] and another by W. KILLING from a decade earlier [W_5].

Weierstrass's lectures soon became world-famous; when in 1873 – two years after the Franco-Prussian War – MITTAG-LEFFLER came to Paris to study, HERMITE said to him: "Vous avez fait erreur, Monsieur, vous auriez dû suivre les cours de Weierstrass à Berlin. C'est notre maître à tous. (You have made a mistake, sir; you should have attended Weierstrass' course in Berlin. He is the master of us all.)"

Weierstrass' name was misappropriated as a hallmark by several second-rate mathematicians. Thus in 1887 a book entitled *Theorie der analytischen Funktionen* by one Dr. O. BIERMANN appeared in the Teubner press; its foreword contained the passage "Der Plan dieses Werkes ist Herrn Weierstrass bekannt. (The plan of this work is known to Mr. Weierstrass.)" ITZIGSOHN (the German translator of Cauchy's *Cours d'analyse*), outraged at this, wrote to BURKHARDT in 1888: "Herr(n) Biermann (soll) in nächster Zeit gründlich heimgeleuchtet werden, weil er den Glauben erwecken wollte, er habe im Einverständnis mit Herrn Professor Weierstrass dessen Funktion-Theorie veröffentlicht. Herr Prof. (so!) Biermann hat *niemals* Funktionentheorie bei Herrn Prof. Weierstrass gehört. Das Werk ist alles, nur nicht die Weierstrass'sche Funktionentheorie. (Mr. Biermann should be roundly taken to task as soon as possible because he's trying to create the belief that he had an understanding with Weierstrass about publishing the latter's function theory. Herr Prof. (indeed!) Biermann *never* attended Prof. Weierstrass' lectures on function theory. This work is anything but Weierstrassian function theory.)" WEIERSTRASS expressed himself on the matter in 1888 in a letter to SCHWARZ as follows: "Dr. Biermann, Privatdocent in Prag, besuchte mich am Tage vor oder nach meinem 70sten Geburtstag. Er theilte mir mit, daß er die Absicht habe, eine 'allgemeine Funktionentheorie' auf der in meinen Vorlesungen gegebenen Grundlage zu schreiben und fragte mich, ob ich ihm die Benutzung meiner Vorlesungen für diesen Zweck gestatte. Ich antwortete ihm, daß er sich wohl eine zu schwierige Aufgabe gestellt habe, die ich selbst zur Zeit noch nicht zu lösen getraute. Da er aber ... die angegebene Frage wiederholte, sagte ich ihm zum Abschiede: 'Wenn Sie aus meinen Vorlesungen etwas gelernt haben, so kann ich Ihnen nicht verbieten, davon in angemessener Weise Gebrauch zu machen.' Er hatte sich mir als früherer Zuhörer vorgestellt und ich nahm selbstverständlich an, daß er meine Vorlesung über Funktionenlehre gehört habe. Dies ist aber nicht der Fall ... Er hat also sein Buch nach dem Hefte eines anderen gearbeitet. Eine derartige Buchmacherei kann nicht geduldet werden. (Dr. Biermann, lecturer in Prague, paid me a visit the day before, or maybe it was the day after, my 70th birthday. He let me know that he had in mind writing a 'general function theory' based on the foundations given in my lectures and asked me if I would permit him to use my lectures for this purpose. I answered him that he'd really set himself too difficult a task, one that I didn't even trust myself to resolve at that time. But because he repeated the proferred question, I told him in parting: 'If you've learned something from my lectures, I can't forbid you to make appropriate use of it.' He'd represented himself as having earlier been an auditor of my lectures and I naturally assumed he'd attended my lectures on function theory. But such is not the case ... He'd done his book in other words from the class-notes of others. Generating books this way can't be abided.)"

[R] **Riemann** 1851: The ideas in this trail-blazing but tightly written work had already been developed by RIEMANN during the 1847 Fall recess. Judging by reactions, it had no effect at first; its new ideas spread only slowly and quite gradually. The dissertation got a very knowledgeable evaluation from GAUSS, who, when RIEMANN visited him, even indicated that for some years he had been preparing a paper which treated the same material without in fact being quite as restricted. Characteristic of the reception which Riemann's work initially found is the following incident, which is related by Arnold SOMMERFELD in his *Vorlesungen über theoretische Physik* (vol. 2: Mechanik der deformierbaren Medien, reprinting of the 6th edition, 1978 Verlag Harri Deutsch, Thun & Frankfurt-am-Main; see §19.7, p.124): "Adolf WÜLLNER, der langjährige verdiente Vertreter der Experimentalphysik an der Technischen Hochschule in Aachen traf in den siebziger Jahren auf dem Rigi mit WEIERSTRASS and HELMHOLTZ zusammen. WEIERSTRASS hatte die RIEMANNsche Dissertation zum Ferienstudium mitgenommen und klagte, daß ihm, dem Funktiontheoretiker, die RIEMANNschen Methoden schwer verständlich seien. HELMHOLTZ bat sich die Schrift aus und sagte beim nächsten Zusammentreffen, ihm schienen die RIEMANNschen Gedankengänge völlig naturgemäß und selbstverständlich zu sein. (Adolf WÜLLNER, the venerable representative of experimental physics at the Technical University of Aachen, met WEIERSTRASS and HELMHOLTZ on Mt. Rigi sometime in the 1870's. WEIERSTRASS had brought along RIEMANN's dissertation for vacation study and complained that for him as a function theorist RIEMANN's methods were difficult to understand. HELMHOLTZ asked to be allowed to take the manuscript and at their next meeting he exclaimed that RIEMANN's thought processes seemed completely natural and self-evident to him.)"

[BB] **Briot** and **Bouquet** 1859: In the first 40 pages general function theory is developed, relying heavily on the works of CAUCHY. Holomorphic functions are still called "synectic", as they had been by CAUCHY. In the 2nd edition, which appeared in 1875 under the shorter title *Théorie des fonctions elliptiques*, the authors replace the word "synectic" with "holomorphic". This is the first textbook on function theory. HERMITE in 1885 considered it to be one of the most significant "publications analytiques de notre époque." This work of BRIOT and BOUQUET was, as the authors say in their foreword, strongly inspried by the classical lectures [Liou] of LIOUVILLE on elliptic functions, WEIERSTRASS was even of the opinion that everything essential was the work of LIOUVILLE.

[Os] **Osgood** 1906: This is the first textbook on function theory in German which experienced wide circulation (Burkhardt's book [Bu] having had little success); in spite of its over 600-page length, it enjoyed 5 editions. The foreword to the first edition begins with the ambitious statement: "Der erste Band dieses Werkes will eine systematische Entwicklung der Funktionentheorie auf Grundlage der Infinitesimalrechnung und in engster Fühlung mit der Geometrie und der mathematischen Physik geben. (The first volume of this work wants to give a systematic development of function theory based on the infinitesimal calculus in the closest possible contact with geometry and mathematical physics.)"

Incidentally, the second volume of this work (in two parts, 1929 and 1932) was the first textbook on functions of several complex variables.

[P] **Pringsheim** 1925/1932: As a convinced advocate of the Weierstrass power series calculus, PRINGSHEIM built the theory methodically on the Weierstrass definition of a holomorphic function as a system of overlapping power series. Complex contour integration is not developed until page 1108; as a substitute up to that point a method of *means* is used, which had its origin in the arithmetic averaging technique that WEIERSTRASS had used to prove the Cauchy inequalities (cf. 8.3.5) and that in fact CAUCHY himself had used. With the help of these means PRINGSHEIM proves (cf. pp. 386 ff) that complex-differentiable functions (with derivative hypothesized to be continuous!) can be developed into power series. He sees the advantage of his integration-free treatment as being "daß grundlegende Erkenntnisse, die dort als sensationelle Ergebnisse eines geheimnisvollen, gleichsam Wunder wirkenden Mechanismus erscheinen, hier ihre natürliche Erklärung durch Zurückführung auf die bescheidenere Wirksamkeit der vier Spezies finden (that basic insights which there seem like sensational results of a mysterious and, as it were, miraculous mechanism, here find their natural explanation by being brought within the more modest scope of the four species [of operations, viz., addition, subtraction, multiplication and division])" –from the foreword to volume 1. But PRINGSHEIM's program never managed to carry the day; on this point compare §12.1.5 of the present book.

Textbooks on Function Theory

[1] AHLFORS, L., *Complex Analysis*, McGraw-Hill (3rd ed. 1979), New York.

[2] BEHNKE, H. and F. SOMMER, *Theorie der analytischen Funktionen einer komplexen Veränderlichen*, Grundlehren **77**, Springer-Verlag (1955), Berlin. Paperback 3rd ed. 1976.

[3] BIEBERBACH, L., *Einführung in die konforme Abbildung*. Sammlung Göschen 768/768a, Walter de Gruyter (6th ed. 1967), Berlin. English trans. of 4th ed. by F. STEINHARDT, *Conformal Mapping*, Chelsea (1953), New York.

[4] BIEBERBACH, L., *Lehrbuch der Funktionentheorie*, 2 volumes. B. G. Teubner (1922/1930), Leipzig. Reprinted by Chelsea (1945), New York and by Johnson Reprint Corp. (1968), New York.

[5] CARATHÉODORY, C., *Theory of Functions of a Complex Variable* I (2nd ed. 1958), II (2nd ed. 1960); trans. from German by F. STEINHARDT. Chelsea, New York.

[6] CARTAN, H., *Elementary theory of analytic functions of one or several complex variables* (trans. from French by J. STANDRING and H. B. SHUTRICK). Éditions Scientifiques Hermann, Paris and Addison-Wesley, Reading (1963).

[7] CONWAY, J. B., *Functions of One Complex Variable*, Graduate Texts in Mathematics **11**, Springer-Verlag (2nd ed. 1978), New York.

[8] DINGHAS, A., *Vorlesungen über Funktionentheorie*, Grundlehren **110**. Springer-Verlag (1961), Berlin.

[9] DIEDERICH, K. and R. REMMERT, *Funktionentheorie I*, Heidelberger Taschenbücher Band **103**. Springer-Verlag (1973), Berlin.

[10] FISCHER, W. and I. LIEB, *Funktionentheorie*, Vieweg und Sohn (1980), Braunschweig.

[11] HEINS, M., *Complex Function Theory*, Pure and Applied Mathematics **28**. Academic Press (1968), New York.

[12] HURWITZ, A. and R. COURANT, *Allgemeine Funktionentheorie und elliptische Funktionen*, Grundlehren **3**. Springer-Verlag (4th ed. 1964), Berlin.

[13] JÄNICH, K., *Einführung in die Funktionentheorie*, Hochschultext, Springer-Verlag (2nd ed. 1980), Berlin.

[14] KNESER, H., *Funktionentheorie*, Studia Mathematica Bd. **13**, Vandenhoeck & Ruprecht (2nd ed. 1966), Göttingen.

[15] KNOPP, K., *Theory and Application of Infinite Series* (trans. from German by R. C. H. YOUNG) , Blackie & Son, Ltd. (2nd ed. 1951), Glasgow; reprinted by Dover Publications (1989), New York.

[16] KNOPP, K., *Theory of Functions*, Part I (1945), Part II (1947), (trans. from German by F. BAGEMIHL), Dover Publications, New York.

[17] LANG, S., *Complex Analysis*, Addison-Wesley (1977), Reading. 2nd ed., Graduate Texts in Mathematics **103**, Springer-Verlag (1985), New York.

[18] NEVANLINNA, R., *Analytic Functions* (trans. from German by P. EMIG), Grundlehren **162**. Springer-Verlag (1970), New York.

[19] *Numbers*, by H.-D. EBBINGHAUS, *et al.* (trans. from German by H. L. S. ORDE), Graduate Texts in Mathematics **123**, *Readings in Mathematics*, Springer-Verlag (1990), New York.

In the following books the reader will find many exercises of varying degrees of difficulty (with solutions):

JULIA, G. *Exercises d'analyse II.* Gauthier-Villars (2nd ed. 1958, reprinted 1969), Paris.

KNOPP, K., *Problem Book in the Theory of Functions* vol. I (trans. from German by L. BERS, 1948), vol. II (trans. from German by F. BAGEMIHL, 1953), Dover Publications, New York.

KRZYŻ, J. G., *Problems in Complex Variable Theory*, Elsevier (1971), New York, London, Amsterdam.

MITRINOVIĆ, D. S. and J. D. KEČKIĆ, *The Cauchy Method of Residues: Theory and Applications*, D. Reidel (1984), Dordrecht.

PÓLYA, G. and G. SZEGÖ, *Problems and Theorems in Analysis*, vol. I (trans. from German by D. AEPPLI) Grundlehren **193** (1972), vol. II (trans. from German by C. E. BILLIGHEIMER) Grundlehren **216** (1976). Springer-Verlag, New York.

Literature on the History of Function Theory and of Mathematics

[H₁] ARNOLD, W. and H. WUSSING (editors), *Biographien bedeutender Mathematiker*, Aulis Verlag Deubner & Co. KG (1978, 2nd ed., 1985), Cologne.

[H₂] BELHOSTE, B., *Augustin-Louis Cauchy: A Biography* (translated from French by Frank RAGLAND) Studies in the History of Mathematics and Physical Sciences, vol. **16**. Springer-Verlag (1991), New York.

[H₃] BELL, E. T., *Men of Mathematics*, Simon & Schuster (1937), New York.

[H₄] BOTTAZZINI, U., *The 'Higher Calculus': A History of Real and Complex Analysis from Euler to Weierstrass* (trans. from Italian by W. VAN EGMOND), Springer-Verlag (1986), New York.

[H₅] BOYER, C. B., *A History of Mathematics*, John Wiley & Sons (1968), New York; paperback reprint, Princeton University Press (1985), Princeton.

[H₆] DIEUDONNÉ, J. (editor), *Abrégé d'histoire des mathématiques 1700-1900*, Hermann (1978), Paris.

[H₇] GRABINER, J. V., *The Origins of Cauchy's Rigorous Calculus*, The MIT Press (1981), Cambridge, Massachusetts.

[H₈] KLEIN, F., *Development of Mathematics in the 19th Century* (trans. from German by M. ACKERMAN), Lie Groups: History, Frontiers and Applications, vol. **IX**. Math. Sci. Press (1979), Brookline, Massachusetts.

[H₉] KLINE, M., *Mathematical Thought from Ancient to Modern Times*, Oxford University Press (1972), New York.

[H₁₀] MARKUSCHEWITZ, A. I., *Skizzen zur Geschichte der analytischen Funktionen* (trans. from Russian by W. FICKER), Hochschulbücher für Mathematik Bd. **16**, Deutscher Verlag der Wissenschaften (1955), Berlin.

[H₁₁] NEUENSCHWANDER, E., "Über die Wechselwirkungen zwischen der französischen Schule, Riemann und Weierstrass. Eine Übersicht mit zwei Quellenstudien," *Arch. Hist. Exact Sciences* **24**(1981), 221-255 and a slightly abridged English version in *Bull. Amer. Math. Soc.* (2) **5**(1981), 87-105. This article contains 142 literature references.

[S₁₂] SMITHIES, F., *Cauchy and the Creation of Complex Function Theory*, Cambridge University Press (1997), Cambridge.

Symbol Index

Name Index

Subject Index

Graduate Texts in Mathematics

Printed in the United States
By Bookmasters